Dairy Chemistry and Biochemistry

P.F. Fox • T. Uniacke-Lowe
P.L.H. McSweeney • J.A. O'Mahony

Dairy Chemistry and Biochemistry

Second Edition

 Springer

P.F. Fox
School of Food and Nutritional Sciences
University College
Cork, Ireland

P.L.H. McSweeney
School of Food and Nutritional Sciences
University College
Cork, Ireland

T. Uniacke-Lowe
School of Food and Nutritional Sciences
University College
Cork, Ireland

J.A. O'Mahony
School of Food and Nutritional Sciences
University College
Cork, Ireland

ISBN 978-3-319-37436-9   ISBN 978-3-319-14892-2   (eBook)
DOI 10.1007/978-3-319-14892-2

Springer Cham Heidelberg New York Dordrecht London
© Springer International Publishing Switzerland 2015
Softcover reprint of the hardcover 2nd edition 2015

Printed on acid-free paper

Springer International Publishing AG Switzerland is part of Springer Science+Business Media (www.springer.com)

Preface to First Edition

Milk has been the subject of scientific study for about 150 years and, consequently, is probably the best characterized, in chemical terms, of our major foods. It is probably also the most complicated and serves as the raw material for a very large and diverse family of food products. Dairy science has existed as a university discipline for more than 100 years; it is the oldest sector of food science (and technology), with the exception of brewery science. Since dairy chemistry is a major facet of dairy science, it might be expected to have been the subject of numerous books. This is, in fact, not so. During the past 40 years, as far as we are aware, only six books or series on dairy chemistry have been published in English, i.e. *Principles of Dairy Chemistry* (Jenness and Patton 1959), *Dairy Chemistry and Physics* (Walstra and Jenness 1984), *Fundamentals of Dairy Chemistry* (Webb and Johnson 1964; Webb et al. 1974; Wong et al. 1988), *Developments in Dairy Chemistry* (Fox, four volumes, 1982, 1983, 1985, 1989), *Advanced Dairy Chemistry* (Fox, three volumes, 1992, 1995, 1997) and *Handbook of Milk Composition* (Jensen 1995). Of these, *Principles of Dairy Chemistry* and *Dairy Chemistry and Physics* were written essentially for senior undergraduate students. The other four books/series were focussed principally on lecturers, researchers, senior postgraduate students and senior production management. Thus, at present there is a lack of books written at senior undergraduate/junior postgraduate level specializing in dairy chemistry/science. This book is intended to fill that gap and should be useful to graduates working in the dairy industry as it is to those still studying.

This book assumes a knowledge of chemistry and biochemistry but not of dairy chemistry. As the title suggests, the book has a stronger biochemical orientation than either *Principles of Dairy Chemistry* or *Dairy Chemistry and Physics*. In addition to a fairly in-depth treatment of the chemistry of the principal constituents of milk, i.e. water, lactose, lipids, proteins (including enzymes), salts and vitamins, various more applied aspects are also covered, e.g. heat-induced changes, cheese, protein-rich products and the applications of enzymes in dairy technology. The principal physical properties are also described.

To facilitate the reader, the structure of various molecules mentioned frequently in the text is given in appendices but we emphasize that a good general knowledge of chemistry and biochemistry is assumed. The chemical composition of the principal dairy products is also included.

This book does not cover the technology of various dairy products, although brief manufacturing protocols for some products are included to facilitate discussion; however, a number of textbooks on various aspects of dairy technology are referenced. Neither are the chemical analyses, microbiology and nutritional aspects of dairy products covered, except in a very incidental manner. The effects of dairy husbandry on the composition and properties of milk are discussed briefly, as is the biosynthesis of milk constituents; in both cases, some major textbooks are referenced.

We hope that the book will answer some of your questions on the chemistry and biochemistry of milk and milk products and encourage you to undertake more extensive study of these topics.

The highly skilled and enthusiastic assistance of Ms Anne Cahalane and Ms Brid Considine in the preparation of the manuscript and of Professor D. M. Mulvihill and Dr. Nora O' Brien for critically and constructively reviewing the manuscript is gratefully acknowledged and very much appreciated.

Cork, Ireland   P.F. Fox
                                                                P.L.H. McSweeney

Preface to Second Edition

Since the publication of the first edition of this book by Chapman & Hall in 1998, there has been considerable progress on several aspects of the subject. The book was reprinted by Kluwer Academic/Plenum Publishers but not revised and is out of print. All topics covered in the first edition are retained, revised and expanded in the second edition. A new chapter "Bioactive Compounds in Milk" has been added and a full chapter has been devoted to "Fermented Milk Products", which were part of the chapter "Chemistry and Biochemistry of Cheese and Fermented Milk" in the first edition. The book is focussed on undergraduate and junior post-graduate students but should also be useful for teaching staff in dairy/food science/technology, researchers and industrial personnel, and for those changing direction. The book assumes a sound knowledge of general and physical chemistry and of biochemistry. The manufacture of the various dairy products mentioned is not described in detail, but the book provides appropriate references on dairy technology. The principles of the main analytical methods for lactose, lipids, proteins and milk salts are presented, but the methods are not described in detail. The nutritional and microbiological aspects of milk and dairy products are discussed only in so far as they are affected by, or affect, the chemistry and biochemistry of milk and dairy products. The effects of dairy husbandry on the composition and properties of milk are discussed briefly, as is the biosynthesis of milk constituents.

We expect that the book will answer some of your questions on the chemistry and biochemistry of milk and milk products and stimulate your interest in studying these subjects in greater detail.

Cork, Ireland

P.F. Fox
T. Uniacke-Lowe
P.L.H. McSweeney
J.A. O'Mahony

General References on Dairy Chemistry

Alais, C. (1974). *Science du Lait. Principes des Techniques Laitieres* (3rd ed.). Paris: SEP Editions.

Associates of Rogers (1928). *Fundamentals of dairy science.* American Chemical Society Monograph No. 41. New York: The Chemical Catalog. [1935, New York: Reinhold Publishing; 1955, 2nd ed., New York: Reinhold Publishing].

Cayot, P., & Lorient, D. (1998). *Structure et Technofonctions des Proteins du Lait.* Paris: Lavoisier Technique and Documentation.

Davis, J. G., & MacDonald, F. J. (1955). *Richmond's dairy chemistry* (5th ed.). London: Charles Griffin and Company.

Fleischmann, W. (1870). *Leherbuch der Milchwissenschaft.* Bremen: M. Heinsius. [7 editions up to 1932].

Fox, P. F. (Ed.). (1982–1989). *Developments in dairy chemistry* (Vols. 1–4). London: Elsevier Applied Science Publishers.

Fox, P. F. (Ed.). (1992–1997). *Advanced dairy chemistry* (Vols. 1–3). London: Elsevier Applied Science Publishers and Chapman and Hall.

Fox, P. F., & McSweeney, P. L. H. (1998). *Dairy chemistry and biochemistry.* London: Chapman and Hall.

Fox, P. F., & McSweeney, P. L. H. (2003). *Advanced dairy chemistry* (Vol. 1, *Proteins,* 3rd ed.). New York: Kluwer Academic/Plenum Publishers

Fox, P. F., & McSweeney, P. L. H. (2006). *Advanced dairy chemistry* (Vol. 1, *Lipids,* 3rd ed.). New York: Springer.

Jenness, R., & Patton, S. (1959). *Principles of dairy chemistry.* New York: John Wiley & Sons.

Jensen, R. G. (ed.) (1995). *Handbook of milk composition.* San Diego: Academic Press.

Ling, E. R. (1930). *A textbook of dairy chemistry.* London: Chapman and Hall. (3 further reprints/editions, 1944, 1949, 1956).

McKenzie, H. A. (1970, 1971). *Milk proteins: Chemistry and molecular biology* (Vols. 1 and 2). New York: Academic Press.

McSweeney, P. L. H., & Fox, P. F. (2009). *Advanced dairy chemistry* (Vol. 3, *Lactose, water, salts and minor constituents,* 3rd ed.). New York: Springer

McSweeney, P. L. H., & Fox, P. F. (2013). *Advanced dairy chemistry* (Vol. 1A, *Proteins, basic aspects,* 4th ed.). New York: Springer.

McSweeney, P. L. H., & O'Mahoney, J. A. (Eds.). (2015). *Advanced dairy chemistry* (Vol. 1, *Proteins, Part B*). New York: Springer.

Richmond, H. D. (1899). *Dairy chemistry: A practical handbook for dairy chemists and others having control of dairies.* London: C Griffin and Company. (published in 5 editions, the 5th, revised by Davis and MacDonald in 1955, see above).

Singh, H., Boland, B., & Thompson, A. (2014). *Milk proteins: From expression to food* (2nd ed.). San Diego: Academic Press.

Snyder, H. (1897). *The chemistry of dairying.* Easton, PA: Chemical Publishing Company.

Thompson, A., Boland, B., & Singh, H. (2009). *Milk proteins: From expression to food.* San Diego: Academic Press.

Walstra, P., Geurts, T. J., Noomen, A., Jellema, A., & van Boeckel, M. A. J. S. (1999). *Dairy technology: Principles of milk properties and processes.* Marcel Dekker, New York.

Walstra, P., & Jenness, R. (1984). *Dairy chemistry and physics.* New York: John Wiley & Sons.

Walstra, P., Wouters, J. T. H., & Geurts, T. J. (2005). *Dairy science and technology.* Oxford: CRC/Taylor and Francis.

Webb, B. H., & Johnson, A. H. (Eds.). (1964). *Fundamentals of dairy chemistry.* Westport, CT, USA: AVI.

Webb, B. H., Johnson, A. H., & Alford, J. A. (Eds.). (1974). *Fundamentals of dairy chemistry* (2nd ed.). Westport, CT, USA: AVI.

Wong, N. P., Jenness, R., Keeney, M., & Marth, E. H. (Eds.). (1988). *Fundamentals of dairy chemistry* (3rd ed.). New York: Van Nostrand Reinhold.

Contents

Chapter 1
Production and Utilization of Milk

1.1 Introduction

Milk is a fluid secreted by the female of all mammalian species, of which there are more than 4,000, for the primary function of meeting the complete nutritional requirements of the neonate of the species. In addition, milk serves several physiological functions for the neonate. Most of the non-nutritional functions of milk are served by proteins and peptides which include immunoglobulins, enzymes and enzyme inhibitors, binding or carrier proteins, growth factors and antibacterial agents. Because the nutritional and physiological requirements of each species are more or less unique, the composition of milk shows very marked inter-species differences. Of the more than 4,000 species of mammal, the milks of only ~180 have been analysed and of these, the data for only about 50 species are considered to be reliable (sufficient number of samples, representative sampling, adequate coverage of the lactation period). Not surprisingly, the milk of the principal dairying species, i.e., cow, goat, sheep and buffalo, and the human are among those that are well characterized. The gross composition of milks from selected species are summarized in Table 1.1; very extensive data on the composition of bovine and human milk are contained in Jensen (1995).

1.2 Composition and Variability of Milk

In addition to the principal constituents listed in Table 1.1, milk contains several hundred minor constituents, many of which, e.g., vitamins, metal ions and flavour compounds, have a major impact on the nutritional, technological and sensoric properties of milk and dairy products. Many of these effects will be discussed in subsequent chapters.

Milk is a very variable biological fluid. In addition to interspecies differences (Table 1.1), the milk of any particular species varies with the individuality of the

© Springer International Publishing Switzerland 2015
P.F. Fox et al., *Dairy Chemistry and Biochemistry*,
DOI 10.1007/978-3-319-14892-2_1

Table 1.1 Composition, %, of milks of some species

Species	Total solids	Fat	Protein	Lactose	Ash	Gross energy (kJ/kg)	Days to double birth weight
Human	12.2	3.8	1.0	7.0	0.2	2,763	120–180
Cow	12.7	3.7	3.4	4.8	0.7	2,763	30–47
Goat	12.3	4.5	2.9	4.1	0.8	2,719	12–19
Sheep	19.3	7.4	4.5	4.8	1.0	4,309	10–15
Pig	18.8	6.8	4.8	5.5	0.9	3,917	9
Horse	11.2	1.9	2.5	6.2	0.5	1,883	40–60
Donkey	11.7	1.4	2.0	7.4	0.5	1,966	30–50
Reindeer	33.1	16.9	11.5	2.8	1.5	6,900	22–25
Domestic rabbit	32.8	18.3	11.9	2.1	1.8	9,581	4–6
Bison	14.6	3.5	4.5	5.1	0.8	–	100–115
Indian elephant	31.9	11.6	4.9	4.7	0.7	3,975	100–260
Polar bear	47.6	33.1	10.9	0.3	1.4	16,900	2–4
Grey seal	67.7	53.1	11.2	0.7	0.8	20,836	5

animal, the breed (in the case of commercial dairying species), health (mastitis and other diseases), nutritional status, stage of lactation, age, interval between milkings, etc. The protein content of milk varies considerably between species and reflects the growth rate of the young. Bernhart (1961) found a linear correlation between the % calories derived from protein and the logarithm of the days to double birth weight for 12 mammalian species. For humans, one of the slowest growing and slowest maturing species, it takes 120–180 days to double birth weigh and only 7 % of calories come from protein. In contrast, carnivores can double their birth weigh in as little as 7 days and obtain >30 % of their energy from protein. Equid species take between 30 and 60 days to double their birth weight and, like humans, have an exceptionally low level of protein in their milk (Table 1.1). The high calorific value of some species e.g., polar bear and grey seal is due mainly to the lipid content.

In a bulked factory milk supply, variability due to many of these factors is evened out but some variability persists and may be quite large in situations where milk production is seasonal. Not only do the concentrations of the principal and minor constituents vary with the above factors, the actual chemistry of some of the constituents also varies, e.g., the fatty acid profile is strongly influenced by diet. Some of the variability in the composition and constituents of milk can be adjusted or counteracted by processing technology, e.g., standardization of fat content, but some differences may persist. The variability of milk and the consequent challenges will become apparent in subsequent chapters.

From a physicochemical viewpoint, milk is a very complex fluid. The constituents of milk occur in three phases. Quantitatively, most of the mass of milk is a true solution of lactose, organic and inorganic salts, vitamins and other small molecules in water. In this aqueous solution are dispersed proteins, some at the molecular level (whey proteins), others as large colloidal aggregates, ranging in diameter from 50 to 600 nm (the caseins), and lipids which exist in an emulsified state, as globules ranging in diameter from 0.1 to 20 μm. Thus, colloidal chemistry is very important in the study of milk, e.g., surface chemistry, light scattering and rheological properties and phase stability.

Milk is a dynamic system owing to: the instability of many of its structures, e.g., the milk fat globule membrane; changes in the conformation and solubility of many constituents with temperature and pH, especially of the inorganic salts but also of proteins; the presence of various enzymes which can modify constituents through lipolysis, proteolysis or oxidation/reduction; the growth of microorganisms, which can cause major changes either directly through their growth, e.g., changes in pH or redox potential (Eh) or through enzymes they excrete; and the interchange of gases with the atmosphere, e.g., CO_2. Milk was intended to be consumed directly from the mammary gland and to be expressed from the gland at frequent intervals. However, in dairying operations, milk is stored for various periods, ranging from a few hours to several days, during which it is cooled (and perhaps heated) and agitated. These treatments will cause at least some physical changes and permit some enzymatic and microbiological changes which may alter the properties of milk. Again, it may be possible to counteract some of these changes.

1.3 Classification of Mammals

The essential characteristic distinguishing mammals from other animal species is the ability of the female of the species to produce milk in specialized organs (mammary glands) for the nutrition of its newborn.

The class *Mammalia* is divided into three sub-classes:

1. Prototheria
   This sub-class contains only one order, Monotremes, the species of which are egg-laying mammals, e.g., duck-billed platypus and echidna, and are indigenous only to Australasia. They possess many (perhaps 200) mammary glands grouped in two areas of the abdomen; the glands do not terminate in a teat and the secretion (milk) is licked by the young from the surface of the gland.

2. Marsupials
   The young of marsupials are born live (viviparous) after a short gestation and are "premature" at birth to a greater or lesser degree, depending on the species. After birth, the young are transferred to a pouch where they reach maturity, e.g., kangaroo and wallaby.

   In marsupials, the mammary glands, which vary in number, are located within the pouch and terminate in a teat. The mother may nurse two off-spring, differing widely in age, simultaneously from different mammary glands that secrete milk of very different composition, designed to meet the different specific requirements of each off-spring.

3. Eutherians
   About 95 % of all mammals belong to this sub-class. The developing embryo *in utero* receives nourishment via the placental blood supply (they are referred to as placental mammals) and is born at a high, but variable, species-related state of maturity. All eutherians secrete milk, which, depending on the species, is more or less essential

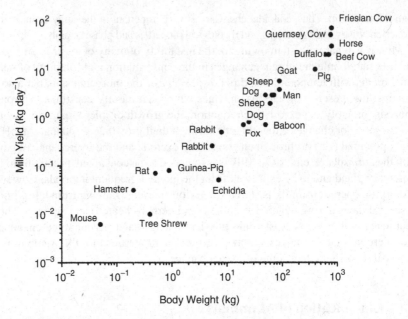

Fig. 1.1 Relation between daily milk yield and maternal body weight for some species (modified from Linzell 1972)

for the development of the young; the young of some species are born sufficiently mature to survive and develop without milk.

The number and location of mammary glands varies with species from 2, e.g., human, goat and sheep, to 14–16 for the pig. Each gland is anatomically and physiologically separate and is emptied via a teat.

The wide interspecies variation in the composition (Table 1.1) and the chemistry of the constituents of milk, as discussed elsewhere, renders milk species-specific, i.e., designed to meet the requirements of the young of that species. There is a surprisingly good relationship between milk yield and maternal body weight (Fig. 1.1); species bred for commercial milk production, e.g., dairy cow and goat, fall above the line.

1.4   Structure and Development of Mammary Tissue

The mammary glands of all species have the same basic structure and all are located external to the body cavity (which greatly facilitates research on milk biosynthesis). Milk constituents are synthesized in specialized epithelial cells (secretory cells or mammocytes, Fig. 1.2d) from molecules absorbed from the blood. The secretory cells are grouped as a single layer around a central space, the lumen, to form more or less spherical or pear-shaped bodies, known as alveoli (Fig. 1.2c). The milk is secreted from these cells into the lumen of the alveoli. When the lumen is full, the myoepithelial cells surrounding each alveolus contract under the influence of oxytocin and the

Milk-producing tissue of a cow, shown at progressively larger scale

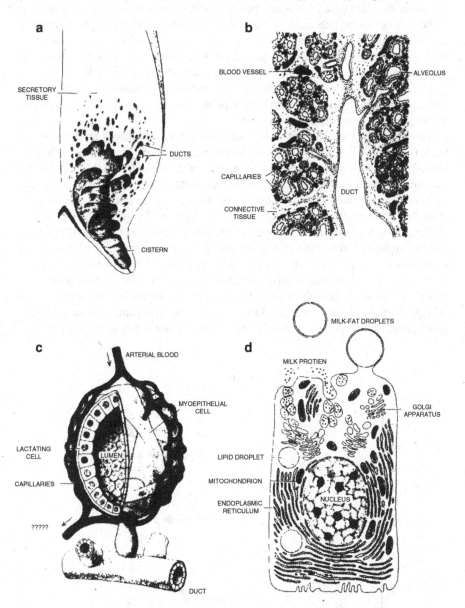

Fig. 1.2 (**a**) A longitudinal section of one of the four quarters of a mammary gland; (**b**) arrangement of the alveoli and the duct system that drains them; (**c**) single alveolus consisting of an elliptical arrangement of lactating cells surrounding the lumen, which is linked to the duct system of the mammary gland; (**d**) a lactating cell; part of the cell membrane becomes the membrane covering fat droplets; *dark circular bodies* in the vacuoles of Golgi apparatus are protein particles, which are discharged into the lumen (From Patton 1969)

milk is drained via a system of arborizing ducts towards sinuses or cisterns (Fig. 1.2a) which are the main collection points between suckling or milking. The cisterns lead to the outside *via* the teat canal. Groups of alveoli, which are drained by a common duct, constitute a lobule; neighbouring lobules are separated by connective tissue (Fig. 1.2b). The secretory elements are termed the "lobule-alveolar system" to distinguish them from the duct system. The whole gland is shown in Fig. 1.2a.

Milk constituents are synthesized from components obtained from the blood; consequently, the mammary gland has a plentiful blood supply and an elaborate nervous system to regulate excretion.

The substrates for milk synthesis enter the secretory cell across the basal membrane (outside), are utilized, converted and interchanged as they pass inwards through the cell and the finished milk constituents are excreted into the lumen across the apical membrane. Myoepithelial cells (spindle shaped) form a "basket" around each alveolus (Fig. 1.2c) and are capable of contracting on receiving an electrical, hormonally-mediated, stimulus, thereby causing ejection of milk from the lumen into the ducts.

Development of mammary tissue commences before birth, but at birth the gland is still rudimentary. It remains rudimentary until puberty when very significant growth occurs in some species; in all species the mammary gland is fully developed at puberty. In most species, the most rapid phase of mammary gland development occurs at pregnancy and continues through pregnancy and parturition, to reach a maximum at peak milk production. The data in Fig. 1.3 show the development pattern of the mammary gland in the rat, the species that has been studied most thoroughly in this regard.

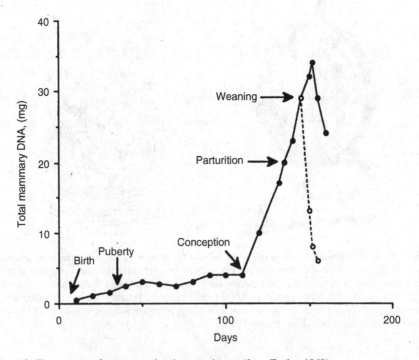

Fig. 1.3 Time-course of mammary development in rats (from Tucker 1969)

Fig. 1.4 The hormonal
control of mammary
development in rats.
Oest oestrogen, *Prog*
progesterone, *GH* growth
hormone, *PL* prolactin,
C corticosteroids

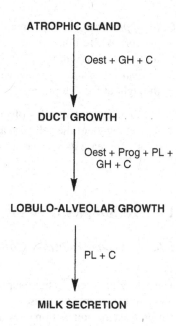

Mammary development is under the regulation of a complex set of hormones. Studies involving endocrinectomy (removal of different endocrine organs) show that the principal hormones are oestrogen, progesterone, growth hormone, prolactin and corticosteroids (Fig. 1.4).

The anatomy, growth, development, involution and the gene network controlling these are described in a series of articles in *Encyclopedia of Dairy Sciences*, volume 3, pp 328–351 (Fuquay et al. 2011)

1.5   Ultrastructure of the Secretory Cell

The structure of the secretory cell is essentially similar to that of other eukaryotic cells. In their normal state, the cells are roughly cubical, ~10 μm in cross section. It is estimated that there are ~5×10^{12} mammary cells in the udder of the lactating cow. A diagrammatic representation of the cell is shown in Fig. 1.2d. It contains a large nucleus towards the base of the cell and is surrounded by a cell membrane, the plasmalemma. The cytoplasm contains the usual range of organelles:

Mitochondria: principally involved in energy metabolism [tricarboxylic acid (Kreb's) cycle].

Endoplasmic reticulum: located towards the base of the cell and to which are attached *ribosomes*, giving it a rough appearance (hence the term, rough endoplasmic reticulum, RER). Many of the biosynthetic reactions of the cell occur in the RER.

Golgi apparatus: a smooth membrane system located toward the apical region of the cell, where much of the assembly and "packaging" of synthesised material for excretion occur.

Lysozyomes: capsules of enzymes (mostly hydrolytic) distributed fairly uniformly throughout the cytoplasm.

Fat droplets and vesicles of material for excretion are usually apparent toward the apical region of the cell. The apical membrane possesses microvilli which serve to greatly increase its surface area.

1.6   Techniques Used to Study Milk Synthesis

1.6.1   *Arterio-Venous Concentration Differences*

The artery and vein system supplying the mammary gland (Fig. 1.5) are readily accessible and may be easily cannulated to obtain blood samples for analysis. Differences in composition between arterial and venous blood give a measure of the

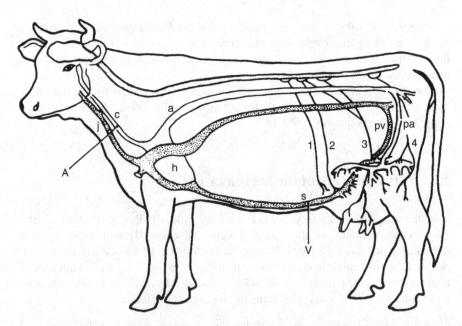

Fig. 1.5 The blood vessel and nerve supply in the mammary glands of a cow. Circulatory system (arteries, *white*; veins, *stippled*): *h* heart, *a* abdominal aorta, *pa* external pudic artery, *pv* external pudic vein, *s* subcutaneous abdominal vein, *c* carotid artery, *j* jugular vein. Nerves: *1* first lumbar nerve, *2* second lumbar nerve, *3* external spermatic nerve, *4* perineal nerve. *A* and *V* show blood sampling points for arteriovenous (AV) difference determinations (Mepham 1987)

constituents used in milk synthesis. The total amount of a constituent used may be determined if the blood flow rate is known, which may be easily done by infusing a known volume of cold saline solution into a vein and measuring the temperature of blood a little further down-stream. The extent to which the blood temperature is reduced is inversely proportional to blood flow rate.

1.6.2   Isotope Studies

Injection of radioactively labelled substrates, e.g., glucose, into the blood stream permits assessment of the milk constituents into which that substrate is incorporated. It may also be possible to study the intermediates through which biosynthesis proceeds.

1.6.3   Perfusion of Isolated Gland

In many species, the entire gland is located such that it may be readily excised intact and undamaged. An artificial blood supply may be connected to cannulated veins and arteries (Fig. 1.6); if desired, the blood supply may be passed through an artificial kidney. The entire mammary gland may be maintained active and secreting milk for several hours; substrates may be readily added to the blood supply for study.

1.6.4   Tissue Slices

The use of tissue slices is a standard technique in all aspects of metabolic biochemistry. The tissue is cut into slices, sufficiently thin to allow adequate rates of diffusion in and out of the tissue. The slices are submerged in physiological saline to which substrates or other compounds may be added.

Changes in the composition of the slices and/or incubation medium give some indication of metabolic activity but extensive damage may be caused to the cells on slicing; the system is so artificial that data obtained by the tissue slice technique may not pertain to the physiological situation. However, the technique is widely used at least for introductory, exploratory experiments.

1.6.5   Cell Homogenates

Cell homogenates are an extension of the tissue slice technique, in which the tissue is homogenised. As the tissue is completely disorganized, only individual biosynthetic reactions may be studied in such systems; useful preliminary work may be done with homogenates.

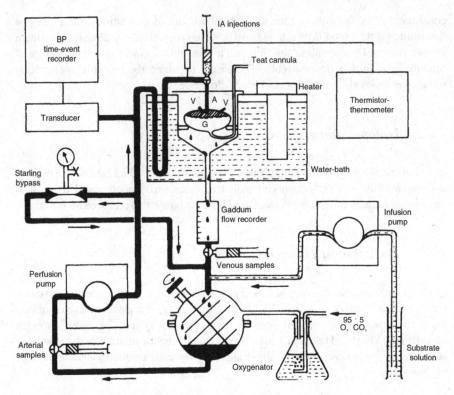

Fig. 1.6 Diagram of circuit for perfusion of an isolated mammary gland of a guinea-pig, *G* mammary gland, *A* artery, *V* veins (from Mepham 1987)

1.6.6 Tissue Culture

Tissue cultures are useful for preliminary or specific work but are incomplete.

In general, the specific constituents of milk are synthesized from small molecules absorbed from the blood. These precursors are absorbed across the basal membrane but very little is known about the mechanism by which they are transported across the membrane. Since the membrane is rich in lipids and precursors are mostly polar with poor solubility in lipid, it is unlikely that the precursors enter the cell by simple diffusion. It is likely, in common with other tissues, that there are specialized carrier systems to transport small molecules across the membrane; such carriers are probably proteins.

The mammary gland of the mature lactating female of many species is by far the most metabolically active organ of the body. For many small mammals, the energy input required for the milk secreted in a single day may exceed that required to develop a whole litter *in utero*. A cow at peak lactation yielding 45 kg milk/day secretes approximately 2 kg lactose and 1.5 kg each of fat and protein per day. This compares with the daily weight gain for a beef animal of 1–1.5 kg/day, 60–70 % of which is water. In large measure, a high yielding mammal is subservient to the needs of its mammary gland to which it must supply not only the precursors for the synthesis of milk constituents but also an adequate level of high-energy-yielding substrates (ATP, UTP, etc.) required to drive the necessary synthetic reactions. In addition, minor constituents (vitamins and minerals) must be supplied.

1.7   Biosynthesis of Milk Constituents

The constituents of milk can be grouped into four general classes according to their source:

– Organ (mammary gland) and species-specific (e.g., most proteins and lipids)
– Organ but not species-specific (lactose)
– Species but not organ-specific (some proteins)
– Neither organ- nor species-specific (water, salts, vitamins)

The principal constituents (lactose, lipids and most proteins) of milk are synthesised in the mammary gland from constituents absorbed from blood. However, considerable modification of constituents occurs in the mammary gland; the constituents are absorbed from blood through the basal membrane, modified (if necessary) and synthesised into the finished molecule (lactose, triglycerides, proteins) within the mammocyte (mainly in the endoplasmic reticulum) and excreted from the mammocyte through the apical membrane into the lumen of the alveolus. The biosynthesis of the principal constituents of milk is described in a series of articles in the *Encyclopedia of Dairy Sciences*, volume 3, pp 352–380 (Fuquay et al. 2011) and in the appropriate chapter.

1.8   Production and Utilization of Milk

Sheep and goats were domesticated early during the Agricultural Revolution, 8–10,000 years ago. Cattle were domesticated later but have become the principal dairying species, especially *Bos taurus* in the most intense dairying areas, dairy sheep and goats are widespread and very important in arid regions, especially around the Mediterranean. Buffalo are important in some regions, especially in India and Egypt. Dromedary camels are important dairy animals in North Africa and the Middle East. Mare's milk is used extensively in central Asia and is receiving attention in Europe for special dietary purposes since its composition is closer to that of human milk than is bovine milk. Donkeys are also used for milk production on a small scale in several European countries, China and Ethiopia. Yak are particularly important in Western China, Mongolia and Tibet where they are used for transport, milk, meat and hides. Reindeer are of major significance in sub-Arctic regions. Approximately 85, 14, 2 and 1.5 % of world milk production is obtained from cattle, buffalo, goats and sheep, respectively, with very small proportions obtained from camels, yak, horse, donkey and reindeer.

The animal species used for milk production are described in the *Encyclopedia of Dairy Sciences*, volume 1 pp 284–380 and the composition and properties of milk of selected species in volume 3, pp 478–631 (Fuquay et al. 2011).

Some milk and dairy products are consumed in most, or all, regions of the World but they are major dietary items in Europe, North and South America, Australia, New Zealand and some Middle Eastern countries. Total milk production in 2013 was estimated to be 780×10^6 tonnes, of which 156, 100, 100 and 29×10^6 tonnes were produced in the European Union, Eastern Europe, North America and the Pacific region, respectively (FAO 2013). The European Union and some other countries operate milk production quotas which are restricting growth in those areas; the quota system in the EU will cease in 2015 and milk production in the region is expected to increase.

Data on the consumption of milk and dairy products in countries that are members of the International Dairy Federation (IDF) are summarized in Table 1.2. Milk and dairy products are quite important in several countries that are not included in this table since they are not members of the IDF.

Because milk is perishable and its production was, traditionally, seasonal, milk surplus to immediate requirements was converted to more stable products, traditional examples being butter or ghee, fermented milk and cheese; smaller amounts of dried milk products were produced traditionally by sun-drying. These traditional products are still very important and many new variants thereof have been introduced. In addition, several new products have been developed during the past 150 years, e.g., sweetened condensed milk, sterilized concentrated milk, a range of milk powders, UHT-sterilized milk, ice creams, infant foods and milk protein products.

One of the important developments in dairy technology in recent years has been the fractionation of milk into its principal constituents, e.g., lactose, milk fat fractions and milk protein products (caseins, caseinates, milk protein concentrates, whey protein concentrates, whey protein isolates), mainly for use as functional proteins but recently some milk proteins, e.g., whey protein isolates, lactoferrin lactoperoxidase and immunoglobulins are marketed as "nutraceuticals", i.e., proteins for specific physiological and/or nutritional functions, As a raw material, milk has many attractive features:

1. Milk was designed for animal nutrition and hence contains the necessary nutrients in easily digestible forms (although the balance is designed for the young of a particular species) and free of toxins. No other single food, except the whole carcass of an animal, including the bones, contains the complete range of nutrients at adequate concentrations.
2. The principal constituents of milk, i.e., lipids, proteins and carbohydrates, can be readily fractionated and purified by relatively simple methods, for use as food ingredients.
3. Milk itself is readily converted into products with highly desirable organoleptic and physical characteristics and its constituents have many very desirable and some unique physicochemical (functional) properties.

Table 1.2 Consumption of liquid milk (L/caput/annum), cheese butter and fermented milks (kg/caput/annum)

Country	Milk	Cheese	Butter	Fermented milk
European Union				
Austria	75.2	19.2	5.0	21.8
Belgium	48.9	15.3	2.5	10.5
Croatia	71.2	9.6	1.0	16.9
Cyprus	97.9	18.1	1.9	12.4
Czech Republic	56.7	16.6	5.2	16.3
Denmark	87.2	16.1	1.8	48.2
Estonia	120.9	20.8	4.1	–
Finland	128.3	23.7	4.5	38.6
France	52.6	26.2	7.4	29.9
Germany	53.3	24.3	6.2	30.5
Greece	47.6	23.4	0.7	–
Hungary	49.0	11.5	1.0	13.9
Ireland	135.6	6.7	2.4	–
Italy	52.7	20.9	2.3	8.8
Latvia	91.9	16.0	2.8	–
Lithuania	29.5	16.3	2.8	–
Luxemburg	36.8	24.4	6.1	–
Netherlands	47.5	19.4	3.3	45.0
Poland	40.9	11.4	4.1	7.8
Portugal	78.5	9.6	1.8	26.6
Slovakia	53.2	10.1	2.9	13.8
Spain	80.6	9.3	0.6	29.1
Sweden	89.2	19.7	1.8	36.4
United Kingdom	102.9	11.2	3.4	–
Other European				
Iceland	96.0	25.2	4.9	37.9
Norway	83.9	17.7	3.2	25.5
Switzerland	64.9	21.1	5.2	31.4
Russia	70.0	6.6	2.8	–
Ukraine	19.3	4.2	2.1	11.7
Africa and Asia				
China	15.4	0.1	0.1	1.9
Egypt	23.7	9.4	0.7	–
India	40.0	2.4[a]	3.6	–
Iran	18.4	4.7	0.3	47.3
Israel	53.7	17.1	0.9	23.3
Japan	30.6	2.1	0.6	–
Kazakhstan	25.6	2.5	1.1	–
Mongolia	8.9	0.3	0.6	–
South Africa	23.1	1.5	0.3	1.8

(continued)

Table 1.2 (continued)

Country	Milk	Cheese	Butter	Fermented milk
South Korea	34.9	2.0	0.2	9.3
Turkey	16.0	7.2	0.7	–
Americas				
Argentina	41.1	11.2	1.4	12.8
Brazil	57.2	3.6	0.4	–
Canada	77.0	12.1	2.8	8.2
Chile	22.3	8.1	1.2	–
Colombia	59.4	0.9	0.1	–
Mexico	34.8	3.1	0.3	5.3
Uruguay	67.1	6.0	1.6	–
USA	74.0	15.2	2.5	–
Oceania				
Australia	105.9	11.8	4.0	6.7
New Zealand	6.0	6.7	4.7	–

Milk, cheese and butter data from IDF (2010, 2011) and Statistics Canada (2012)
Fermented milk data from IDF (2009)
[a]From Jayadevan (2013)

4. The modern dairy cow is a very efficient convertor of plant material; average national yields, e.g., in the USA and Israel, are about 8,000 kg per annum, with individual cows producing up to 20,000 kg per annum. In terms of kg protein that can be produced per hectare, milk production, especially by modern cows, is much more efficient than meat production (Fig. 1.7) but less efficient than some plants (e.g., cereals and soy beans). However, the functional and nutritional properties of milk proteins are superior to those of soy protein and since cattle, and especially sheep and goats, can thrive under farming conditions not suitable for growing cereals or soy beans, dairy animals need not be competitors with humans for use of land although high-yielding dairy cows are fed products that could be used for human foods. In any case, dairy products improve the "quality of life", which is a desirable objective *per se*.
5. One of the limitations of milk as a raw material is its perishability—it is an excellent source of nutrients for microorganisms as well as for humans. However, this perishability is readily overcome by a well-organized, efficient dairy industry.

Milk is probably the most adaptable and flexible of all food materials, as will be apparent from Table 1.3 which shows the principal families of milk-based foods—some of these families contain several hundred different products.

Many of the processes to which milk is subjected cause major changes in the composition (Table 1.4), physical state, stability, nutritional and sensoric attributes of the product; some of these changes will be discussed in later chapters.

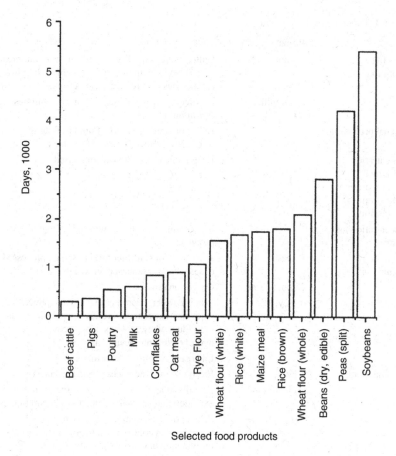

Fig. 1.7 Number of days of protein supply for a moderately active man produced per hectare yielding selected food products

1.9   Trade in Milk Products

Milk and dairy products have been traded for thousands of years and are now major items of trade. There is considerable international trade in dairy products, principally, whole and skim milk powders, cheese, butter, whey protein powders and infant formulae. Data for the production, exports and imports (million tonnes milk equivalent) of milk for 2012 are provided by FAO (2013) and are summarized in Table 1.5.

Traditionally, dairy products (cheese, fermented milks, butter) were produced on an artisanal level, as is still the case in underdeveloped regions and to some extent in highly developed dairying countries. Industrialization commenced during the nineteenth Century and dairy manufacturing is now a well organized industry. One of the features of the past few decades has been the amalgamation of smaller dairy companies both within countries, and, recently, internationally. The 20 largest dairy companies are listed in Table 1.6. An notable feature of these data is that the top 20

Table 1.3 Diversity of dairy products

Process	Primary product	Further products
Centrifugal Separation	Cream	Butter, butter oil, ghee, anhydrous milk fat; creams of various fat content: coffee creams, whipping creams, dessert creams; cream cheeses
	Skim milk	Powders, casein, cheese, protein concentrates and infant formulae
Thermal processing		HTST or super-pasteurization, UHT-sterilized or in-container sterilized
Concentration Thermal evaporation *or* Membrane filtration		Evaporated or sweetened condensed milk
Concentration and drying		Whole milk powders; infant formulae; dietary products
Enzymatic coagulation	Cheese	1,000 varieties; further products, e.g., processed cheese, cheese sauces, cheese dips
	Rennet casein	Cheese analogues
	Whey	Whey powders, demineralized whey powders, whey protein concentrates, whey protein isolates, individual whey proteins, whey protein hydrolyzates, nutraceuticals
		Lactose and lactose derivatives
Acid coagulation	Cheese	Fresh cheeses and cheese-based products
	Acid casein	Functional applications, e.g., coffee creamers, meat extenders; nutritional applications, cream liquers
	Whey	Whey powders, demineralized whey powders, whey protein concentrates, whey protein isolates, individual whey proteins, whey protein hydrolyzates, nutraceuticals
Fermentation		Various fermented milk products, e.g., yoghurt, buttermilk, acidophilus milk, bioyoghurt
Freezing		Ice cream (numerous types and formulations), frozen yoghurt
Miscellaneous		Chocolate products

companies process only 24.2 % of total milk production and the largest company only 3.0 %. Such developments have obvious advantages in terms of efficiency and standardization of product quality but pose the risk of over-standardization with the loss of variety. Greatest diversity occurs with cheeses and fortunately in this case, diversity is being preserved and even extended.

Table 1.4 Approximate composition (%) of some dairy products

Product	Moisture	Protein	Fat	Sugars[a]	Ash
Light whipping cream	63.5	2.2	30.9	3.0	0.5
Butter	15.9	0.85	81.1	0.06	2.1
Anhydrous butter oil	0.2	0.3	99.5	0.0	0.0
Ice cream[b]	60.8	3.6	10.8	23.8	1.0
Evaporated whole milk	74.0	6.8	7.6	10.0	1.5
Sweetened condensed milk	27.1	7.9	8.7	54.4	1.8
Whole milk powder	2.5	26.3	26.7	38.4	6.1
Skim milk powder	3.2	36.2	0.8	52.0	7.9
Whey powder[c]	3.2	12.9	1.1	74.5	8.3
Casein powder	7.0	88.5	0.2	0.0	3.8
Cottage cheese, creamed	79.0	12.5	4.5	2.7	1.4
Quarg	72.0	18.0	8.0	3.0	–
Camembert cheese	51.8	19.8	24.3	0.5	3.7
Blue cheese	42.4	21.4	28.7	2.3	5.1
Cheddar cheese	36.7	24.9	33.1	1.3	3.9
Emmental cheese	36.0	28.9	30.0	–	–
Parmesan cheese	29.2	35.7	24.8	3.2	6.0
Mozzarella cheese	54.1	19.4	31.2	2.2	2.6
Processed cheese[d]	39.2	22.1	31.2	1.6	5.8
Acid whey	93.9	0.6	0.2	4.2	–

[a]Total carbohydrate
[b]Hardened vanilla, 19 % fat
[c]Cheddar (sweet) whey
[d]American pasteurized processed cheese

Table 1.5 Production, exports and imports (millions of tonnes of milk equivalents) for 2012 (FAO 2013)

Country	Production	Imports	Exports
Asia	90.2	27.8	5.7
Africa	45.8	8.8	1.2
Central America	16.5	4.4	0.5
South America	68.2	3.8	3.8
North America	99.3	1.7	5.4
Europe[a]	216.3	5.9	16.2
Oceania	29.3	0.85	20.7
World	765.6	53.4	53.4

[a]Trade between EU countries is not included

Table 1.6 Ranking of global dairy processors, 2011 (from Jesse 2013)

Rank	Company name	Headquarter country	Turnover ($US Bil.)	Milk intake (Mil. MT)	Mkt share (% worldmilk prod.)
1	Nestlé	Switzerland	19.1	14.9	2.1
2	Parmalat	France	16.9	15.0	2.1
3	Fonterra Coop.	New Zealand	16.4	21.6	3.0
4	Danone	France	15.6	8.2	1.1
5	Friesland-Campina	Netherlands	13.4	10.1	1.4
6	DFA[a]	USA	13.0	17.1	2.4
7	Dean Foods	USA	13.0	12.1	1.7
8	Arla Group	Denmark	12.0	12.0	1.7
9	Kraft Foods	USA	7.5	7.8	1.1
10	Saputo	Canada	7.0	6.3	0.9
11	Müller	Germany	6.5	4.4	0.6
12	DMK[b]	Germany	6.4	6.9	1.0
13	Mengniu Dairy Co.	China	5.8	4.1	0.6
14	Yili	China	5.8	4.0	0.6
15	Groupe Sodiaal	France	5.7	4.1	0.6
16	Bongrain SA	France	5.5	3.6	0.5
17	Land O' Lakes, Inc.	USA	4.3	5.9	0.8
18	Glanbia Group	Ireland	3.9	6.0	0.8
19	California Dairies	USA	3.0	4.6	0.6
20	Amul (Coop.)	India	2.5	4.0	0.6

[a]Dairy Farmers of America
[b]Deutsches Milchkontor GmbH

References

Bernhart, F. W. (1961). Correlations between growth-rate of the suckling of various species and the percentage of total calories from protein in the milk. *Nature, 191*, 358–360.

FAO. (2013). *Food outlook*. Rome: FAO.

Fuquay, J., Fox, P. F., & McSweeney, P. L. H. (Eds.) (2011). *Encyclopedia of dairy sciences* (Vol. 3, pp. 328–351, 352–380, 478–631). Oxford: Elsevier Academic Press.

IDF. (2009). *The world dairy situation*. Bulletin 438/2009. International Dairy Federation, Brussels.

IDF. (2010). *The world dairy situation*. Bulletin 446/2010. International Dairy Federation, Brussels.

IDF. (2011). *The world dairy situation*. Bulletin 451/2011. International Dairy Federation, Brussels.

Jayadevan, G. R. (2013). A strategic analysis of cheese and cheese products market in India. *Indian Journal of Research, 2*, 247–250. http://theglobaljournals.com/paripex/file.php?val=March_20 13_1363940771_5f9c2_83.pdf.

Jensen, R. G. (Ed.). (1995a). *Handbook of milk composition*. San Diego, CA: Academic.

Jesse, E. V. (2013, February). *International dairy notes*. The Babcock Institute Newsletter. College of Agricultural and Life Sciences, University of Wisconsin, Madison, USA.

Linzell, J. L. (1972). Milk yield, energy loss, and mammary gland weight in different species. *Dairy Science Abstracts, 34*, 351–360.

Mepham, T. B. (1987a). *Physiology of lactation*. Milton Keynes, UK: Open University Press.

Patton, S. (1969). Milk. *Scientific American, 221*, 58–68.

Statistics Canada. (2012). *Dairy statistics*. Ottawa: Government of Canada.

Tucker, H. A. (1969). Factors affecting mammary gland cell numbers. *Journal of Dairy Science, 52*, 720–729.

Suggested Reading

Cowie, A. T., & Tindal, J. S. (1972). *The physiology of lactation*. London, UK: Edward Arnold.

Fuquay, J., Fox, P. F., & McSweeney, P. L. H. (Eds.). (2011b). *Encyclopedia of dairy sciences* (2nd ed., Vol. 1–4). Oxford, UK: Academic.

Jensen, R. G. (Ed.). (1995b). *Handbook of milk composition*. San Diego, CA: Academic.

Larson, B. L. & Smith, V. R. (1974–1979). *Lactation: A comprehensive treatise* (Vols. 1–4). New York: Academic Press.

Mepham, T. B. (1975). *The secretion of milk* (Studies in biology series, Vol. 60). London, UK: Edward Arnold.

Mepham, T. B. (Ed.). (1983). *Biochemistry of lactation*. Amsterdam: Elsevier.

Mepham, T. B. (1987b). *Physiology of lactation*. Milton Keynes, UK: Open University Press.

Park, Y. W., & Haenlein, G. F. W. (Eds.). (2013). *Milk and dairy products in human nutrition*. Chichester, UK: Wiley Blackwell.

Singh, H., Boland, M., & Thompson, A. (Eds.). (2014). *Milk proteins: From expression to food* (2nd ed.). Amsterdam: Academic.

Chapter 2
Lactose

2.1 Introduction

Lactose is the principal carbohydrate in the milk of most mammals, exceptions are the California sea lion and the hooded seal, which are the only significant sources. Milk contains only trace amounts of other sugars, including glucose (50 mg/l) and fructose and glucosamine, galactosamine and N-acetyl neuraminic acid as components of glycoproteins and glycolipids. The milk of all species that have been studied contain oligosaccharides which are major constituents of the milk of some species, including human. This chapter will concentrate on the chemistry and properties of lactose with a short section on oligosaccharides.

The concentration of lactose in milk varies widely between species (Table 2.1). The lactose content of cows' milk varies with the breed of cow, individual animals, udder infection (mastitis) and stage of lactation. The concentration of lactose decreases progressively and significantly during lactation (Fig. 2.1); this behaviour contrasts with the trends for lipids and proteins, which, after decreasing during early lactation, increase strongly during the second half of lactation. The concentration of lactose in milk is inversely related to the concentrations of lipids and proteins (Fig 2.2) (Jenness and Sloan 1970; Jenness and Holt 1987). The principal function of lactose and lipids is as sources of energy; since lipids are ~2.2 times more energy-dense than lactose, when a highly caloric milk is required, e.g., by animals in a cold environment (marine mammals and polar bears), this is achieved by increasing the fat content of the milk. The inverse relationship between the concentrations of lactose and lipids and protein reflects the fact that the synthesis of lactose draws water into the Golgi vesicles, thereby diluting the concentrations of proteins and lipids (Jenness and Holt 1987).

Mastitis causes an increased level of NaCl in milk and depresses the secretion of lactose. Lactose, along with sodium, potassium and chloride ions, plays a major role in maintaining the osmotic pressure in the mammary system. Thus, any increase or decrease in lactose content (a secreted constituent, i.e., formed within the mammary gland, which is isotonic with blood) is compensated for by an increase or decrease

© Springer International Publishing Switzerland 2015
P.F. Fox et al., *Dairy Chemistry and Biochemistry*,
DOI 10.1007/978-3-319-14892-2_2

Table 2.1 Concentration (%) of lactose in the milk of selected species

Species	Lactose	Species	Lactose	Species	Lactose
California sea lion	0.0	Mouse (house)	3.0	Cat (domestic)	4.8
Hooded seal	0.0	Guinea pig	3.0	Pig	5.5
Black bear	0.4	Dog (domestic)	3.1	Horse	6.2
Dolphin	0.6	Sika deer	3.4	Chimpanzee	7.0
Echidna	0.9	Goat	4.1	Rhesus monkey	7.0
Blue whale	1.3	Elephant (Indian)	4.7	Man	7.0
Rabbit	2.1	Cow	4.8	Donkey	7.4
Red deer	2.6	Sheep	4.8	Zebra	7.4
Grey seal	2.6	Water buffalo	4.8	Green monkey	10.2
Rat (Norwegian)	2.6				

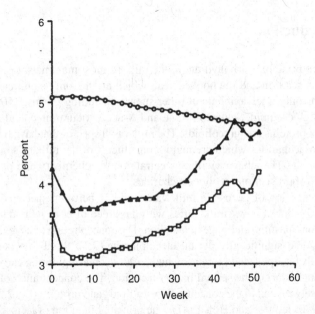

Fig. 2.1 Changes in the concentrations of fat (*closed triangle*), protein (*empty square*) and lactose (*open circle*) in milk during lactation

in the soluble salt constituents (excreted) (Fig. 2.3). This osmotic relationship partly explains why certain milks with a high lactose content have a low ash content and *vice versa* (Table 2.2).

Similarly, there is an inverse relationship between the concentrations of lactose and chloride, which is the basis of Koestler's chloride-lactose test for abnormal milk:

$$\text{Koestler Number} = \frac{\%\text{Chloride} \times 100}{\%\text{Lactose}}$$

A Koestler Number <2 indicates normal milk while a value >3 is considered abnormal.

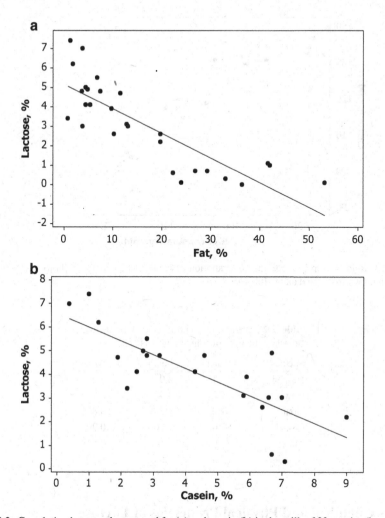

Fig. 2.2 Correlation between lactose and fat (**a**) and casein (**b**) in the milk of 23 species (based on the data of Jenness and Sloan 1970)

Lactose plays an important role in milk and milk products:

1. It is an essential constituent in the production of fermented dairy products.
2. It contributes to the nutritive value of milk and its products; however, many non-Europeans have limited or zero ability to digest lactose in adulthood, leading to **lactose intolerance**.
3. It affects the texture of certain concentrated and frozen products.
4. It is involved in heat-induced changes in the colour and flavour of highly heated milk products.
5. Its changes in state (amorphous vs. crystalline) have major implications for the production and stability of many dehydrated milk products.

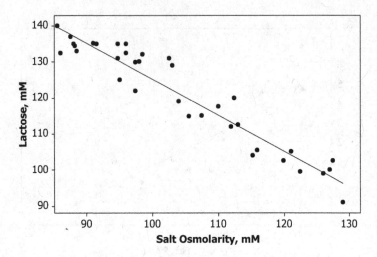

Fig. 2.3 Relationship between the concentration of lactose (mM) and osmolarity (mM) due to salts (redrawn from the data of Holt 1985)

Table 2.2 Average concentration (%) of lactose and ash in the milk of some mammals

Species	Water	Lactose	Ash
Human	87.4	6.9	0.21
Cow	87.2	4.9	0.70
Goat	87.0	4.2	0.86
Camel	87.6	3.26	0.70
Mare	89.0	6.14	0.51
Reindeer	63.3	2.5	1.40

2.2   Chemical and Physical Properties of Lactose

2.2.1   Structure of Lactose

Lactose is a disaccharide consisting of galactose and glucose, linked by a β1-4 glycosidic bond (Fig. 2.4). Its systematic name is 0-β-D-galactopyranosyl-(1-4)-α-D-glucopyranose (α-lactose) or 0-β-D-galactopyranosyl-(1-4)-β-D-glucopyranose (β-lactose). The hemiacetal group of the glucose moiety is potentially free (i.e., lactose is a **reducing** sugar) and may exist as an α- or β-anomer. In the structural formula of the α-form, the hydroxyl group on the C_1 of glucose is *cis* to the hydroxyl group at C_2 (oriented downward).

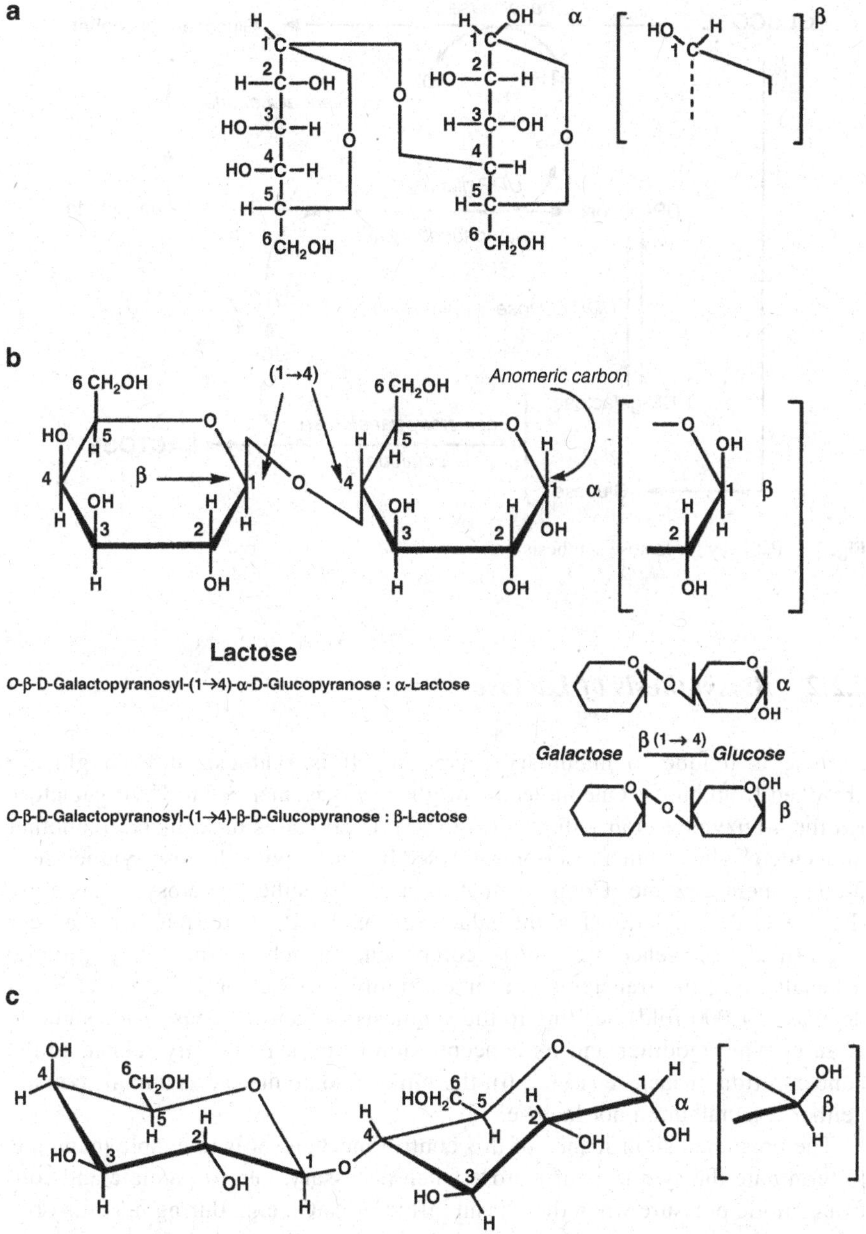

Fig. 2.4 Structural formulae of α- and β-lactose (**a**) Open chain, (**b**) Fischer projection, (**c**) Haworth projection and (**d**) conformational formula

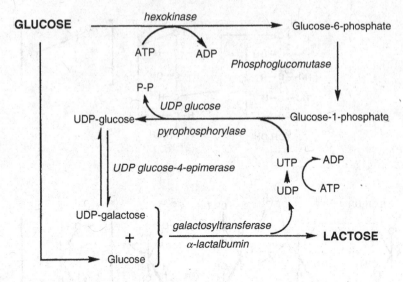

Fig. 2.5 Pathway for lactose synthesis

2.2.2   Biosynthesis of Lactose

Lactose is unique to mammary secretions. It is synthesized from glucose absorbed from blood. One molecule of glucose is isomerized to UDP-galactose *via* the 4-enzyme Leloir pathway (Fig. 2.5). UDP-Gal is then linked to another molecule of glucose in a reaction catalysed by the enzyme, lactose synthetase, a 2-component enzyme. Component A is a non-specific galactosyl transferase (EC 2.4.1.22) which transfers the galactose from UDP-gal to a number of acceptors. In the presence of the B component, which is the whey protein, α-lactalbumin, the transferase becomes highly specific for glucose (its K_M is decreased 1,000-fold), leading to the synthesis of lactose. Thus, α-lactalbumin is an enzyme modifier and its concentration in milk is directly related to the concentration of lactose (Fig. 2.6); the milk of some marine mammals contain neither α-lactalbumin nor lactose.

The presumed significance of this control mechanism is to enable mammals to terminate the synthesis of lactose when necessary, i.e., to regulate and control osmotic pressure when there is an influx of NaCl, e.g., during mastitis or in late lactation (lactose and NaCl are the major determinants of the osmotic pressure of milk, which is isotonic with blood, the osmotic pressure of which is essentially constant). The ability to control osmotic pressure is sufficiently important to justify an elaborate control mechanism and "wastage" of the enzyme modifier.

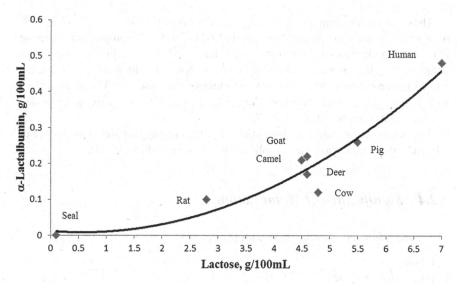

Fig. 2.6 Correlation between lactose and α-lactalbumin concentrations in the milk of eight species (adapted from Ley and Jenness 1970)

2.2.3 *Lactose Equilibrium in Solution*

The configuration around the C_1 of glucose (i.e., the anomeric C) is not stable and can readily change (**mutarotate**) from the α- to the β-form and *vice versa* when the sugar is in solution as a consequence of the fact that the hemiacetal form is in equilibrium with the open chain aldehyde form which can be converted into either of the two isomeric forms (Fig. 2.4).

When either isomer is dissolved in water, there is a gradual change from one form to the other until equilibrium is established, i.e., mutarotation occurs. These changes may be followed by measuring the change in optical rotation with time until, at equilibrium, the specific rotation is +55.4°.

The composition of the mixture at equilibrium may be calculated as follows:

Specific rotation: $[\alpha]_D^{20}$
α-form +89.4°
β-form +35.0°
Equilibrium mixture +55.4°
Let equilibrium mixture = 100
Let x% of the lactose be in the α-form
Then (100 - x)% is the β-form
At equilibrium:
$89.4x + 35(100 - x) = 55.4 \times 100$
x = 37.5
100 - x = 62.5

Thus, the equilibrium mixture at 20 °C is composed of 62.7 % β and 37.3 % α-lactose. The equilibrium constant, β/α, is 1.68 at 20 °C. The proportion of lactose in the α-form increases as the temperature is increased and the equilibrium constant consequently decreases. The equilibrium constant is not influenced by pH, but the rate of mutarotation is dependent on both temperature and pH. The change from α- to β- is 51.1, 17.5 and 3.4 % complete at 25, 15 and 0 °C, respectively, in 1 h and is almost instantaneous at about 75 °C.

The rate of mutarotation is slowest at pH 5.0, increasing rapidly at more acid or alkaline values; equilibrium is established in a few minutes at pH 9.0.

2.2.4 *Significance of Mutarotation*

The α and β forms of lactose differ with respect to:

Solubility
Crystal shape and size
Hydration of the crystalline form, which leads to **hygroscopicity**
Specific rotation
Sweetness

Many of these characteristics are discussed in the following sections.

2.2.5 *Solubility of Lactose*

The solubility characteristics of the α- and β-isomers are distinctly different. When α-lactose is added in excess to water at 20 °C, about 7 g per 100 g water dissolve immediately. Some α-lactose mutarotates to the β anomer to establish the equilibrium ratio 62.7β:37.3α; therefore, the solution becomes unsaturated with respect to α and more α-lactose dissolves and some mutarotetes to β-lactose. These two processes (mutarotation and solubilization of α-lactose) continue until two criteria are met: ~7 g α-lactose are in solution and the β/α ratio is 1.6:1.0. Since the β/α ratio at equilibrium is about 1.6 at 20 °C, the final solubility is 7 g + (1.6 × 7) g = 18.2 g per 100 g water.

When β-lactose is dissolved in water, the initial solubility is ~50 g per 100 g water at 20 °C. Some β-lactose mutarotates to α to establish a ratio of 1.6:1. At equilibrium, the solution would contain 30.8 g β and 19.2 g α/100 ml; therefore, the solution is supersaturated with α-lactose, some of which crystallizes, upsetting the equilibrium and leading to further mutarotation of β to α. These two events, i.e., crystallization of α-lactose and mutarotation of β, continue until the same two criteria are met, i.e., ~7 g of α-lactose in solution and a β/α ratio of 1.6:1. Again, the final solubility is ~18.2 g lactose per 100 g water. Since β-lactose is much more soluble than α and mutarotation is slow, it is possible to form more highly concentrated solutions by dissolving β- rather than α-lactose. In either case, the final solubility of lactose is the same (18.2 g/100 g of water).

The solubility of lactose as a function of temperature is summarized in Fig. 2.7. The solubility of α-lactose is more temperature dependent than that of β-lactose and the solubility curves intersect at 93.5 °C. A solution at 60 °C contains approximately 59 g lactose per 100 g water. Suppose that a 50 % solution of lactose (~30 g β- and 20 g α-) at 60 °C is cooled to 15 °C. At this temperature, the solution can contain only 7 g of α- or a total of 18.2 g of lactose per 100 g water at equilibrium. Therefore, lactose will crystallize very slowly out of solution as irregularly-sized crystals which may give rise to a sandy, gritty texture.

2.2.6   Crystallization of Lactose

As discussed in Sect. 2.2.5, the solubility of lactose is temperature dependent and solutions are capable of being highly supersaturated before spontaneous crystallization occurs and even then, crystallization may be slow. In general, supersolubility at any temperature equals the saturation (solubility) value at a temperature 30 °C higher. The insolubility of lactose, coupled with its capacity to form supersaturated solutions, is of considerable practical importance in the manufacture of concentrated milk products.

In the absence of nuclei and agitation, solutions of lactose are capable of being highly supersaturated before spontaneous crystallization occurs. Even in such solutions, crystallization occurs with difficulty. Solubility curves for lactose are shown in Fig. 2.8 and are divided into unsaturated, metastable and labile zones. Cooling a saturated solution or continued concentration beyond the saturation point, leads to supersaturation and produces a metastable area where crystallization does not occur readily. At higher levels of supersaturation, a labile area is observed where crystallization occurs readily. The pertinent points regarding supersaturation and crystallization are:

1. Neither nucleation nor crystal growth occurs in the unsaturated region.
2. Growth of crystals can occur in both the metastable and labile areas.
3. Nucleation occurs in the metastable area only if seeds (centres for crystal growth) are added.
4. Spontaneous crystallization can occur in the labile area without the addition of seeding material.

The rate of nucleation is slow at low levels of supersaturation and in highly supersaturated solutions owing to the high viscosity of the solution. The stability of a lactose "glass" (see Sect. 2.2.6.4) is due to the low probability of nuclei forming at very high concentrations.

Once a sufficient number of nuclei have formed, crystal growth occurs at a rate influenced by:

(a) Degree of supersaturation.
(b) Surface area available for deposition.
(c) Viscosity.
(d) Agitation.
(e) Temperature.
(f) Mutarotation, which is slow at low temperatures.

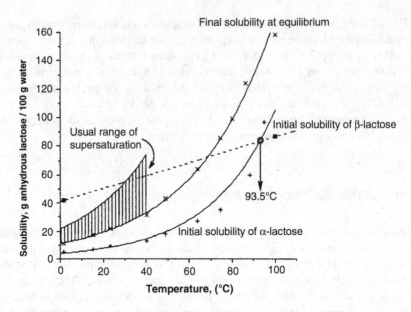

Fig. 2.7 Solubility of lactose in water (modified from Jenness and Patton 1959)

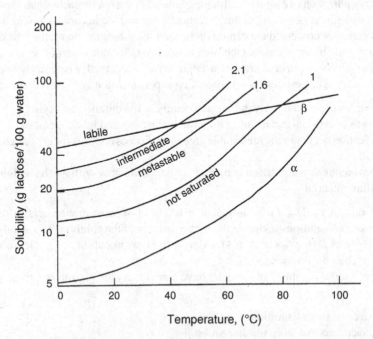

Fig. 2.8 Initial solubility of α-lactose and β-lactose, final solubility at equilibrium (*line 1*), and supersaturation by a factor 1.6 and 2.1 (α-lactose excluding water of crystallization) (modified from Walstra and Jenness 1984)

Fig. 2.9 The most common crystal form of α-lactose hydrate

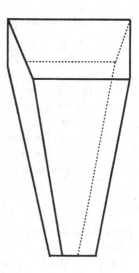

2.2.6.1   α-Hydrate

α-Lactose crystallises as a monohydrate containing 5 % water of crystallization and can be prepared by concentrating an aqueous lactose solutions to supersaturation and allowing crystallization to occur below 93.5 °C. The α-hydrate is the stable solid form at ambient temperatures and in the presence of small amounts of water below 93.5 °C, all other forms change to it. The α-monohydrate has a specific rotation in water at 20 °C of +89.4°. It is soluble only to the extent of 7 g per 100 g water at 20 °C. It forms a number of crystal shapes, depending on the conditions of crystallization; the most common type when fully developed is tomahawk-shaped (Fig. 2.9). Crystals of lactose are hard and dissolve slowly. In the mouth, crystals less than 10 μm are undetectable, but above 16 μm they feel gritty or "sandy" and at 30 μm, a definite gritty texture is perceptible. The term "sandy" or "sandiness" is used to describe the defect in condensed milk, ice cream or processed cheese spreads where, due to poor manufacturing techniques, large lactose crystals are formed.

2.2.6.2   α-Anhydrous

Anhydrous α-lactose may be prepared by dehydrating α-hydrate *in vacuo* at a temperature between 65 and 93.5 °C; it is stable only in the absence of moisture.

2.2.6.3   β-Anhydride

Since β-lactose is less soluble than the α-isomer >93.5 °C, the crystals formed from aqueous solutions at a temperature above 93.5 °C are β-lactose which are anhydrous and have a specific rotation of 35°. β-Lactose is sweeter than α-lactose, but is not

Table 2.3 Some physical properties of the two common forms of lactose

Property	α-Hydrate	β-Anhydride
Melting point[a], °C	202	252
Specific rotation, $[\alpha]_D^{20}$	+89.4°	+35°
Solubility in water (g/100 ml) at 20 °C	8	55
Specific gravity (20 °C)	1.54	1.59
Specific heat	0.299	0.285
Heat of combustion (kJ mol^{-1})	5,687	5,946

[a]Decomposes; values vary with rate of heating, α-hydrate loses H_2O at 120 °C

Values on anhydrous basis, both forms mutarotate to +55.4°

appreciably sweeter than the equilibrium mixture of α- and β- normally found in solution.

Some properties of α- and β-lactose are summarized in Table 2.3. Mixed α/β crystals, e.g., $\alpha_5\beta_3$, can be formed under certain conditions. The relationship between the different crystalline forms of lactose is shown in Fig. 2.10.

2.2.6.4 Lactose Glass

When a lactose solution is dried rapidly (e.g., spray drying lactose-containing concentrates), viscosity increases so quickly that there is insufficient time for crystallization to occur. A non-crystalline amorphous form is produced containing α- and β-forms in the ratio at which they exist in solution. Lactose in spray dried milk exists as a concentrated syrup or amorphous glass which is stable if protected from air, but is very hygroscopic and absorbs water rapidly from the atmosphere, becoming sticky.

2.2.7 Problems related to Lactose Crystallization

The tendency of lactose to form supersaturated solutions that do not crystallize readily causes problems in many dairy products unless adequate controls are exercised. The problems are due primarily to the formation of large crystals, which cause sandiness, or to the formation of a lactose glass, which leads to hygroscopicity and caking (Fig. 2.11).

2.2.7.1 Dried Milk and Whey

Lactose is the major component of dried milk products: whole milk powder, skim milk powder and whey powder contain ~30, 50 and ~70 % lactose, respectively. Protein, fat and air are dispersed in a continuous phase of amorphous solid lactose.

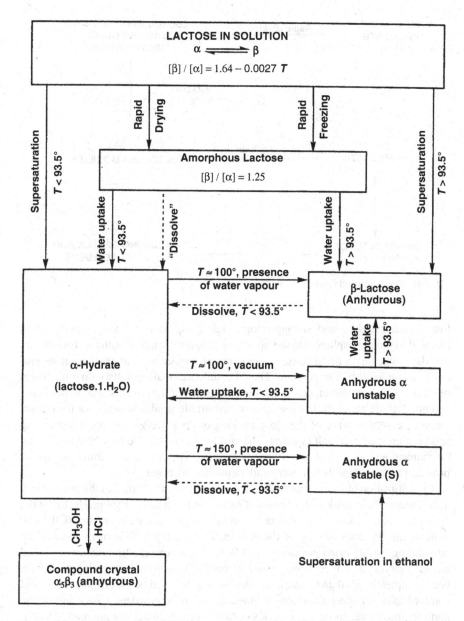

Fig. 2.10 Modifications of lactose (T, temperature in °C) from Walstra and Jenness 1984)

Consequently, the behaviour of lactose has a major impact on the properties of dried milk products (Schuck 2011).

In freshly-made powder, lactose is in an amorphous state with an α:β ratio of 1:1.6. This amorphous lactose glass is a highly concentrated syrup since there is not sufficient time during drying for crystallization to proceed normally. The glass has a

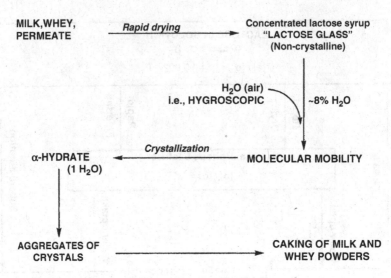

Fig. 2.11 Formation and crystallization of lactose glass

low vapour pressure and is hygroscopic, taking up moisture very rapidly when exposed to the atmosphere. On the uptake of moisture, dilution of the lactose occurs and the molecules acquire sufficient mobility and space to arrange themselves into crystals of α-lactose monohydrate. These crystals are small, usually with dimensions of <1 μm. Crevices and cracks exist along the edges of the crystals, into which other components are expelled. In these spaces, favourable conditions exist for the coagulation of casein because of the close packing of the micelles and the destabilizing action of concentrated salt systems. The fat globule membrane may be damaged by mechanical action, and Maillard browning, involving lactose and amino groups of protein, proceeds rapidly when crystallization has occurred.

Crystallization of lactose in dried milk particles causes "caking" of the powder into a hard mass. If a considerable portion of lactose in the freshly-dried product is in the crystalline state, caking of the powder on contact with moisture is prevented, thereby maintaining the dispersibility of the powder. Lactose crystallization is achieved by rehydrating freshly-dried powder to ~10 % H_2O, by exposure to moisture-saturated air, and redrying it or by removing powder from the main drying chamber before it has been completely dried and completing drying in a fluidized bed. This process is used commercially for the production of "instantized" milk powders. Clustering of the particles into loose, spongy aggregates occurs; these agglomerates are readily wettable and dispersible. They exhibit good capillary action and water readily penetrates the particles, allowing them to sink and disperse whereas the particles in non-instantized powder float due to their low density which contributes to their inability to overcome surface tension. Also, because of the small size of the particles in conventional spray-dried powders, close packing results in the formation of inadequate space for capillary action between the particles, thereby preventing uniform wetting. As a result, large masses of material are wetted on the outside, forming a barrier of highly concentrated product which prevents internal wetting and results in large undispersed

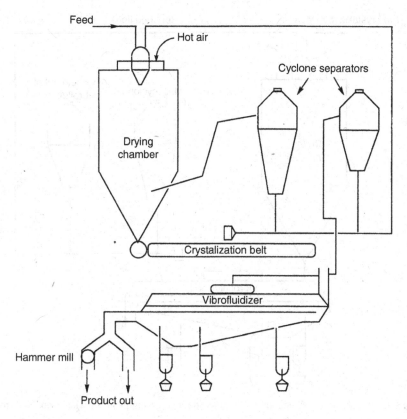

Fig. 2.12 Schematic representation of a low temperature drying plant for whey (modified from Hynd 1980)

clumps of powder. This problem is overcome by agglomeration and in this respect, lactose crystallization is important since it facilitates the formation of large sponge-like aggregates, with good capillary action and wettability.

The state of lactose has a major effect on the properties of spray dried whey powder manufactured by conventional methods, i.e., preheating, condensing to about 50 % total solids and drying to <4 % moisture. The powder is dusty and very hygroscopic and when exposed to ambient air, it has a pronounced tendency to cake owing to its very high lactose content (~70 %).

Problems arising from the crystallization of lactose in milk and whey powders may also be avoided or controlled by pre-crystallizing the lactose. Essentially, this involves adding finely-divided lactose powder which acts as nuclei on which the supersaturated lactose crystallises. Addition of 0.5 kg finely-ground lactose to the amount of concentrated product (whole milk, skim milk or whey) containing 1 tonne of lactose will induce the formation of ~10^6 crystals/ml, ~95 % of which will have dimensions <10 μm and 100 % <15 μm, i.e., too small to cause textural defects.

Diagrams of spray driers with instantizers attached are shown in Figs. 2.12 and 2.13.

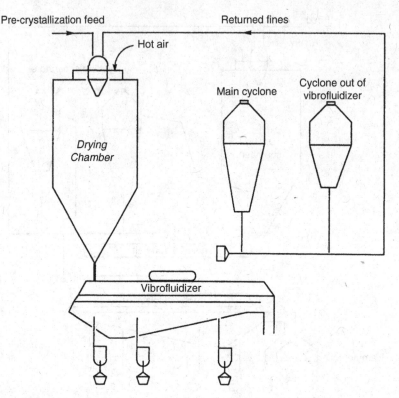

Fig. 2.13 Schematic representation of a straight through drying plant for whey (modified from Hynd 1980)

2.2.7.2 Thermoplasticity of Lactose

Unless certain precautions are taken during the drying of whey or other solutions containing a high concentration of lactose, the hot, semi-dry powder may adhere to the metal surfaces of the dryer, forming deposits, a phenomenon referred to as thermoplasticity. The principal factors which influence the temperature at which thermoplasticity occurs ("sticking temperature") are the concentrations of lactic acid, amorphous lactose and moisture in the whey powder.

Increasing the concentration of lactic acid from 0 to 16 % causes a linear decrease in sticking temperature (Fig. 2.14). The degree of pre-crystallization of lactose affects sticking temperature: a product containing 45 % pre-crystallized lactose has a sticking temperature of 60 °C while the same product with 80 % pre-crystallization sticks at 78 °C (Fig. 2.15). Pre-crystallization of the concentrate feed to the dryer thus permits considerably higher feed concentrations and drying temperatures. Pre-crystallization is routinely used in the drying of high-lactose products such as whey powder and demineralized whey powder.

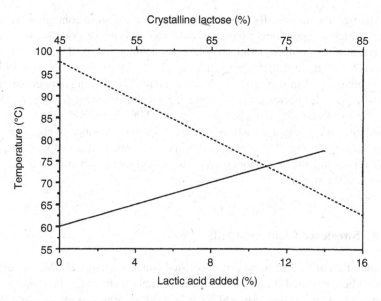

Fig. 2.14 Effect of added lactic acid (*dashed lines*) and degree of lactose crystallization (*dotted lines*) on the sticking temperature of whey powder (1.5–3.5 % moisture)

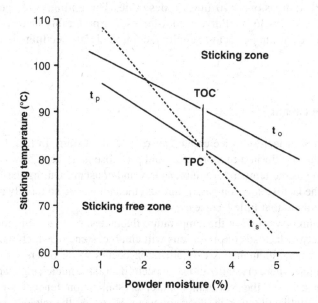

Fig. 2.15 Influence of moisture content on the temperature of powder in a spray dryer (t_p), dryer outlet temperature (t_o) and sticking temperature (t_s). The minimum product temperature required to avoid problems with sticking is at TPC with the corresponding dryer outlet temperature TOC (modified from Hynd 1980)

In practice, the most easily controlled factor is the moisture content of the whey powder, which is determined by the outlet temperature of the dryer (t_o, Fig. 2.15). However, as a result of evaporative cooling, the temperature of the particles in the dryer is lower than the outlet temperature (t_p, Fig. 2.15) and the difference between t_o and t_p increases with increasing moisture content. The sticking temperature for a given whey powder decreases with increasing moisture content (t_s, Fig. 2.15) and where the two curves (t_s and t_p) intersect (point TPC, Fig. 2.15) is the maximum product moisture content at which the dryer can be operated without product sticking during drying. The corresponding point on the outlet temperature curve (TOC) represents the maximum dryer outlet temperature which may be used without causing sticking.

2.2.7.3   Sweetened Condensed Milk

Crystallization of lactose occurs in sweetened condensed milk (SCM) and crystal size must be controlled if a product with a desirable texture is to be produced. As it comes from the evaporators, SCM is almost saturated with lactose. When cooled to 15–20 °C, 40–60 % of the lactose will eventually crystallize as α-lactose hydrate. There are 40–47 parts of lactose per 100 parts of water in SCM, consisting of about 40 % α- and 60 % β- (ex-evaporator). To obtain a smooth texture, crystals with dimensions <10 μm are desirable. The optimum temperature for crystallization is 26–36 °C. Pulverized α-lactose, or preferably lactose "glass", is used as seed. Continuous vacuum cooling, combined with seeding, gives the best product.

2.2.7.4   Ice Cream

Crystallization of lactose in ice cream causes a sandy texture. In freshly hardened ice cream, the equilibrium mixture of α- and β-lactose is in the "glass" state and is stable as long as the temperature remains low and constant. During the freezing of ice cream, the lactose solution passes through the labile zone so rapidly and at such a low temperature that little lactose crystallization occurs.

If ice cream is warmed or the temperature fluctuates, some ice will melt, and an infinite variety of lactose concentrations will emerge, some of which will be in the labile zone where spontaneous crystallization occurs while others will be in the metastable zone where crystallization can occur if suitable nuclei, e.g., lactose crystals, are present. At the low temperature, crystallization tendency is low and extensive crystallization usually does not occur. However, the nuclei formed act as seed for further crystallization when the opportunity arises and they tend to grow slowly with time, eventually causing a sandy texture. The defect is controlled by limiting the milk solids content or by using β-galactosidase to hydrolyse lactose.

2.2.7.5 Other Frozen Dairy Products

Although milk may become frozen inadvertently, freezing is not a common commercial practice. However, concentrated or unconcentrated milk is sometimes frozen commercially, e.g., to supply remote locations (as an alternative to dried or UHT milk), to store sheep's or goats' milk, production of which is seasonal, or human milk for infant feeding in emergencies (milk banks).

As will be discussed in Chap. 3, freezing damages the milk fat globule membrane, resulting in the release of "free fat". The casein system is also destabilized due to a decrease in pH and an increase in Ca^{2+} concentration, both caused by the precipitation of soluble CaH_2PO_4 and/or Ca_2HPO_4 as $Ca_3(PO_4)_2$, with the release of H^+ (see Chap. 5); precipitation of $Ca_3(PO_4)_2$ occurs on freezing because pure water crystallises, causing an increase in soluble calcium phosphate, with which milk is already saturated. Crystallization of lactose as α-hydrate during frozen storage aggravates the problem by reducing the amount of solvent water available.

In frozen milk products, lactose crystallization causes instability of the casein system. On freezing, supersaturated solutions of lactose are formed: e.g., in concentrated milk at −8 °C, 25 % of the water is unfrozen and it contains 80 g lactose per 100 g, whereas the solubility of lactose at −8 °C is only ~7 %. During storage at a low temperature, lactose crystallizes slowly as a monohydrate and consequently the amount of free water in the product is reduced.

The formation of supersaturated lactose solutions inhibits freezing, and consequently stabilizes the concentration of solutes in solution. However, when lactose crystallizes, water freezes and the concentration of other solutes increases markedly (Table 2.4).

The increase in calcium and phosphate leads to precipitation of calcium phosphate and a decrease in pH:

$$3\,Ca^{2+} + 2\,H_2PO_4^- \leftrightarrow Ca_3(PO_4)_2 + 4\,H^+$$

These changes in the concentration of Ca^{2+} and pH lead to destabilization of the casein micelles.

Table 2.4 Comparison of ultrafiltrate from liquid and frozen skim milk

Constituent	Ultrafiltrate of skim milk	Ultrafiltrate of liquid portion of frozen concentrated milk
pH	6.7	5.8
Chloride, mM	34.9	459
Citrate, mM	8.0	89
Phosphate, mM	10.5	84
Sodium, mM	19.7	218
Potassium, mM	38.5	393
Calcium, mM	9.1	59

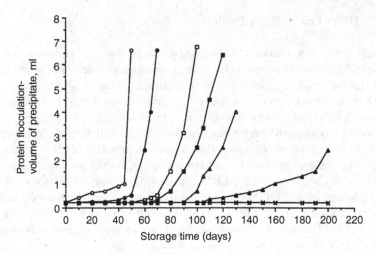

Fig. 2.16 Effect of lactose hydrolysis on the stability of milk to freezing (modified from Tumerman et al. 1954)

Any factor that accelerates the crystallization of lactose shortens the storage life of the product. At very low temperatures (<−23 °C), neither lactose crystallization nor casein flocculation occurs, even after long periods. Enzymatic hydrolysis of lactose by β-galactosidase before freezing retards or prevents lactose crystallization and casein precipitation in proportion to the extent of the hydrolysis (Fig. 2.16).

2.3 Production of Lactose

In comparison with sucrose (the annual production of which is 175×10^6 tonnes, US Department of Agriculture) and glucose or glucose-fructose syrups, only relatively small quantities of lactose are produced. However, it attracts commercial interest because it has some interesting properties and is readily available from whey, a by-product in the production of cheese or casein. World production of cheese is $\sim 19 \times 10^6$ tonnes, the whey from which contains $\sim 8 \times 10^6$ tonnes of lactose; $\sim 0.3 \times 10^6$ tonnes of lactose are contained in the whey produced during casein manufacture. According to Affertsholt-Allen (2007), only about 325,000 tonnes of lactose are used annually in the EU and 130,000 tonnes in the USA, i.e., only ~ 7 % of that potentially available. Much larger amounts are used in whey and demineralized whey powders.

Production of lactose essentially involves concentrating whey or UF permeate under vacuum, crystallization of lactose from the concentrate, recovery of the crystals by centrifugation and drying of the crystals (Fig. 2.17). The first-crop crystals are

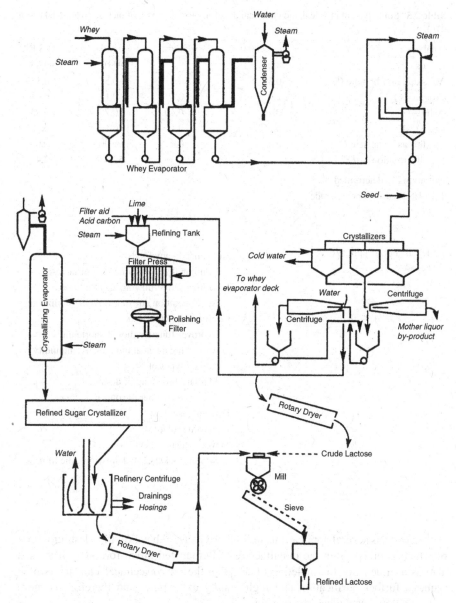

Fig. 2.17 Schematic representation of plant for the manufacture of crude and refined lactose from sweet whey

usually contaminated with riboflavin and are therefore yellowish; a higher grade, and hence more valuable, lactose is produced by redissolving and recrystallizing the crude lactose (Table 2.5). Lactose may also be recovered by precipitation with $Ca(OH)_2$, especially in the presence of ethanol, methanol or acetone (Paterson 2009, 2011).

Table 2.5 Some typical physical and chemical data for various grades of lactose (from Nickerson 1974)

Analysis	Fermentation	Crude	Edible	U.S.P.[b]
Lactose (%)	98.0	98.4	99.0	99.85
Moisture, non-hydrate (%)	0.35	0.3	0.5	0.1
Protein (%)	1.0	0.8	0.1	0.01
Ash (%)	0.45	0.40	0.2	0.03
Lipid (%)	0.2	0.1	0.1	0.001
Acidity, as lactic acid (%)	0.4	0.4	0.06	0.04
Specific rotation $[\alpha]D^{20}$	a	a	52.4°	52.4°

[a]Not normally determined
[b]*USP* US Pharmacopoeia grade

Table 2.6 Food applications of lactose

Humanized baby foods
Demineralized whey powder or lactose
Instantizing/free-flowing agent in foods
Agglomeration due to lactose crystallization
Confectionery products
Improves functionality of shortenings
Anticaking agent at high relative humidity
Certain types of icing
Maillard browning, if desired
Accentuates other flavours (chocolate)
Flavour adsorbant
Flavour volatiles
Flavour enhancement
Sauces, pickles, salad dressings, pie fillings

Lactose has several applications in food products (Table 2.6), the most important of which is probably in the manufacture of humanized infant formulae. It is used also as a diluent for the tableting of drugs in the pharmaceutical industry (which requires further purification and high quality extra pure, and therefore is more expensive) and as the base for plastics.

Among sugars, lactose has a low level of sweetness (Table 2.7), which is generally a disadvantage but is advantageous in certain applications. When properly crystallized, lactose has low hygroscopicity (Table 2.8), which makes it an attractive sugar for use in icings for confectionary products.

Table 2.7 Relative sweetness of sugars (concentration, %, required to give equivalent sweetness) (from Nickerson 1974)

Sucrose	Glucose	Fructose	Lactose
0.5	0.9	0.4	1.9
1.0	1.8	0.8	3.5
2.0	3.6	1.7	6.5
2.0	3.8	–	6.5
2.0	3.2	–	6.0
5.0	8.3	4.2	15.7
5.0	8.3	4.6	14.9
5.0	7.2	4.5	13.1
10.0	13.9	8.6	25.9
10.0	12.7	8.7	20.7
15.0	17.2	12.8	27.8
15.0	20.0	13.0	34.6
20.0	21.8	16.7	33.3

Table 2.8 Relative humectancy of sucrose, glucose and lactose (% moisture absorbed at 20 °C)

Sugar	Relative humidity		100 %
	60 %		
	1 h	9 days	25 days
Lactose	0.54	1.23	1.38
Glucose	0.29	9.00	47.14
Sucrose	0.04	0.03	18.35

2.4 Derivatives of Lactose

Although the demand for lactose has been strong in recent years, it is unlikely that a profitable market exists for all the lactose potentially available. Since the disposal of whey or UF permeate by dumping into waterways is no longer permitted, ways of utilizing lactose have been sought for several years. For many years, the most promising of these was considered to be hydrolysis to glucose and galactose, but other modifications are attracting increasing attention.

2.4.1 Enzymatic Modification of Lactose

Lactose may be hydrolysed to glucose and galactose by enzymes (β-galactosidases, commonly called lactase) or by acids. Commercial sources of β-galactosidase are moulds (especially *Aspergillus* spp.), the enzymes from which have acid pH optima, and yeasts (*Kluyveromyces*) which produce enzymes with neutral pH optima. When β-galactosidases became commercially available, they were considered to have considerable commercial potential as a solution to the "whey problem" and for the treatment of lactose intolerance (see Sect. 2.6.1), but for various reasons their commercialization has not been as great as expected. The very extensive literature on various aspects of β-galactosidases and on their application in free or immobilized form has been reviewed by Mahoney (1997) and Playne and Crittenden (2009). Technological challenges in the production of glucose-galactose syrups have been overcome but the process is not very successful commercially. Glucose-galactose syrups are not economically competitive with glucose or glucose-fructose syrups produced by hydrolysis of maize starch, unless the latter are heavily taxed. As discussed in Sect. 2.6.1, an estimated 70 % of the adult human population have inadequate intestinal β-galactosidase activity and are therefore lactose intolerant; the problem is particularly acute among Asians and Africans. Pre-hydrolysis of lactose was considered to offer the potential to develop new markets for dairy products in those countries. Various protocols are available: addition of β-galactosidase to milk in the home, pre-treatment of milk at the factory with free or immobilized enzyme or aseptic addition of sterilized free β-galactosidase to UHT milk, which appears to be particularly successful. However, the method is not used widely and it is now considered that the treatment of milk with β-galactosidase will be commercially successful only in niche markets.

Glucose-galactose syrups are about three times sweeter than lactose (70 % as sweet as sucrose) and hence lactose-hydrolysed milk could be used in the production of ice-cream, yoghurt or other sweetened dairy products, permitting the use of less sucrose and reducing caloric content. However, such applications have had limited commercial success.

The glucose moiety can be isomerized to fructose by the well-established glucose isomerization process to yield a galactose-glucose-fructose syrup with increased sweetness. Another possible variation would involve the isomerization of lactose to lactulose (galactose-fructose) which can be hydrolysed to galactose and fructose by some β-galactosidases.

β-Galactosidase has transferase as well as hydrolase activity and produces oligosaccharides (galactooligosaccharides, Fig. 2.18) which are later hydrolysed (Fig. 2.19). This property may be a disadvantage since the oligosaccharides are not digestible by humans and reach the large intestine where they are fermented by bacteria, leading to the same problem caused by lactose. However, they stimulate the growth of *Bifidobacterium* in the lower intestine; a product (oligonate, 6′galactosyl lactose) is produced commercially by the Yokult Company in Japan for addition to infant formulae. Other commercial preparations of galacto-oligosaccharides (GOS) include Vivinal® GOS, which is manufactured by Friesland Campina, the Netherlands, and when combined with fructo-oligosaccharides (FOS) has been clinically-proven to have health benefits such as aiding in the relief of eczema, allergies and gastrointestinal discomfort. Generally similar GOS-based products are available from Clasado Biosciences, UK. Some galactooligosaccharides have interesting functional properties and may find commercial applications (see Ganzle 2011b).

2.4.2 Chemical Modifications

Several interesting derivatives can be produced from lactose (see Ganzle 2011a).

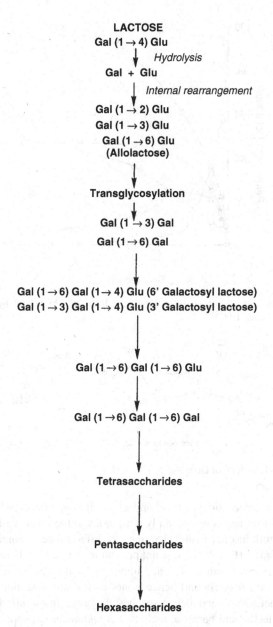

Fig. 2.18 Possible reaction products from the action of β-galactosidase on lactose (from Smart 1993)

2.4.2.1 Lactulose

Lactulose is an epimer of lactose in which the glucose moiety is isomerized to fructose (Fig. 2.20). The sugar does not occur naturally and was first synthesized by Montgomery and Hudson in 1930. It can be produced under mild alkaline conditions via the Lobry de Bruyn-Alberda van Ekenstein reaction and at a low yield as

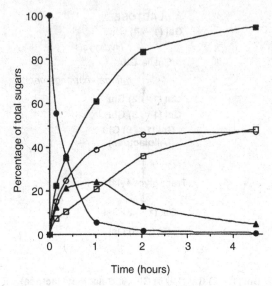

Fig. 2.19 Production of oligosaccharides during the hydrolysis of lactose by β-galactosidase (modified from Mahoney 1997)

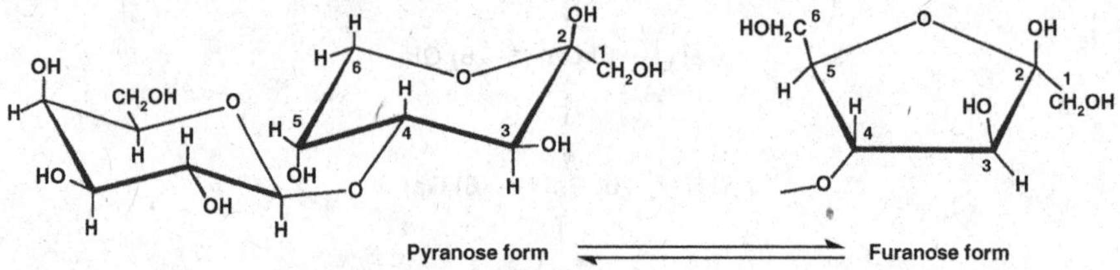

Fig. 2.20 Chemical structure of lactulose

a by-product of β-galactosidase action on lactose. It is produced on heating milk to sterilizing conditions and is a commonly used index of the severity of the heat treatment to which milk has been subjected, e.g., to differentiate in-container sterilized milk from UHT milk (Fig. 2.21); it is not present in raw or HTST pasteurized milk.

Lactulose is sweeter than lactose and about 60 % as sweet as sucrose. It is not metabolized by oral bacteria and hence is not cariogenic. It is not hydrolysed by intestinal β-galactosidase and hence reaches the large intestine where it can be metabolised by lactic acid bacteria, including *Bifidobacterium* spp. and serves as a bifidus factor. For this reason, lactulose has attracted considerable attention as a means of modifying the intestinal microflora, reducing intestinal pH and preventing the growth of undesirable putrefactive bacteria (Fig. 2.22). It is now commonly added to infant formulae to simulate the bifidogenic properties of human milk—apparently, 20,000 tonnes per annum are now used for this and similar applications. Lactulose is also reported to suppress the growth of certain tumour cells (Tamura et al. 1993).

Lactulose is usually used as a 50 % syrup but a crystalline trihydrate, which has very low hygroscopicity, is available.

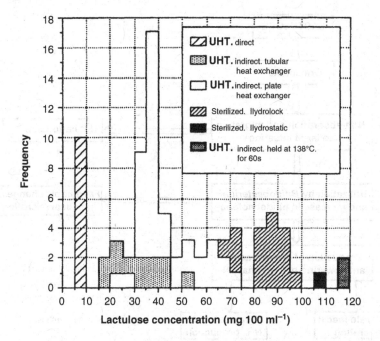

Fig. 2.21 Concentration of lactulose in heated milk products (modified from Andrews 1989)

2.4.2.2 Lactitol

Lactitol (4-O-β-D-galactopyranosyl-D-sorbitol), is a sugar alcohol produced on reduction of lactose (Fig. 2.23), usually using Raney nickel; it does not occur naturally. It can be crystallized as a mono- or di-hydrate. Lactitol is not metabolized by higher animals; it is relatively sweet and hence has potential as a non-nutritive sweetener. It is claimed that lactitol reduces the absorption of sucrose, reduces blood and liver cholesterol levels and is anti-cariogenic. It has applications in low-calorie foods (jams, marmalade, chocolate, baked goods); it is non-hygroscopic and can be used to coat moisture-sensitive foods, e.g., candies.

It can be esterified with 1 or more fatty acids (Fig. 2.23) to yield a family of food emulsifiers, analogous to the sorbitans produced from sorbitol.

2.4.2.3 Lactobionic Acid

This derivative is produced by oxidation of the free carbonyl group of lactose (Fig. 2.24), chemically (Pt, Pd or Bi), electrolytically, enzymatically or by fermentation. It has a sweet taste, which is very unusal for an acid. Its lactone crystallizes readily. Lactobionic acid has found only limited application; its lactone could be used as an acidogen but it is probably not cost-competitive with gluconic acid-δ-lactone. It is used in preservation solutions for organs (to prevent swelling) prior to transplantation, and in skin-care products.

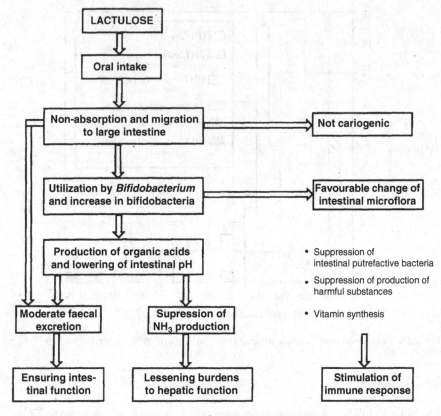

Fig. 2.22 Significance of lactulose in health (modified from Tamura et al. 1993)

2.4.2.4 Lactosyl Urea

Urea can serve as a cheap source of nitrogen for cattle but its use is limited because NH_3 is released too quickly, leading to a toxic level of NH_3 in the blood. Reaction of urea with lactose yields lactosyl urea (Fig. 2.25), from which NH_3 is released more slowly.

2.4.3 Fermentation Products

Lactose is readily fermented by lactic acid bacteria, especially *Lactococcus* spp. and *Lactobacillus* spp., to lactic acid, and by some species of yeast, e.g., *Kluyveromyces*, to ethanol (Fig. 2.26). Lactic acid may be used as a food acidulant, as a feed-stock

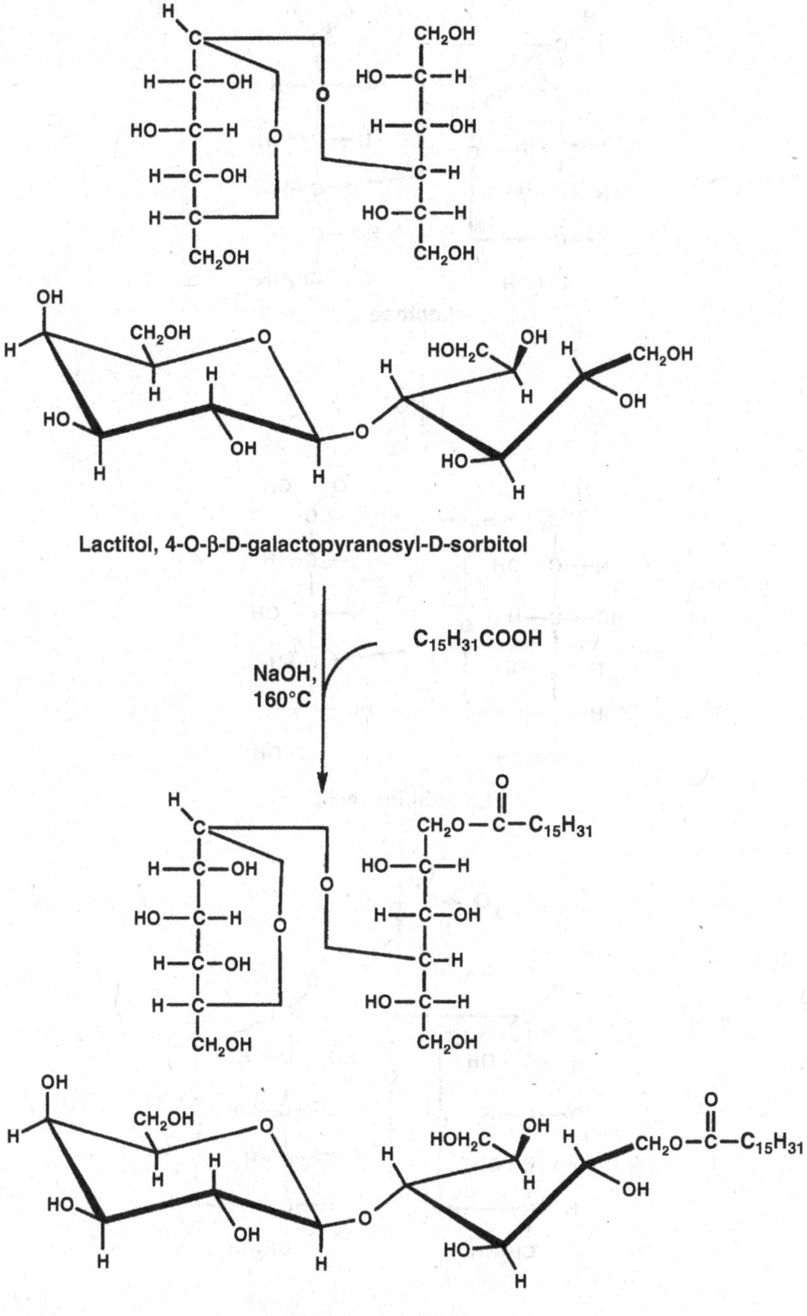
Lactitol, 4-O-β-D-galactopyranosyl-D-sorbitol
C15H31COOH
NaOH,
160°C

Lactitol monoester

Fig. 2.23 Structure of lactitol and its conversion to lactyl palmitate

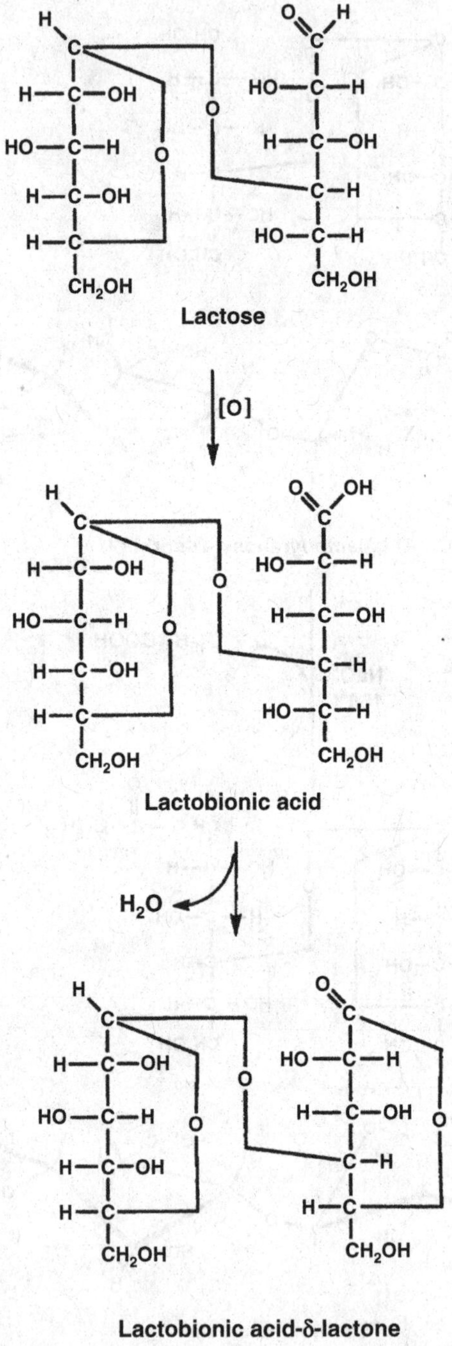

Fig. 2.24 Structure of lactobionic acid and its δ-lactone

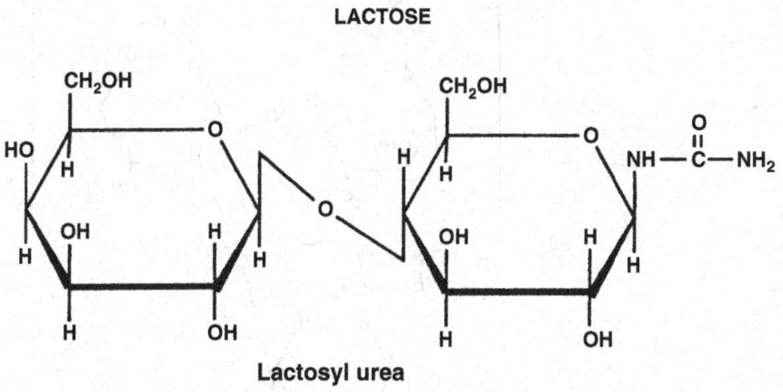

Fig. 2.25 Structure of lactosyl urea

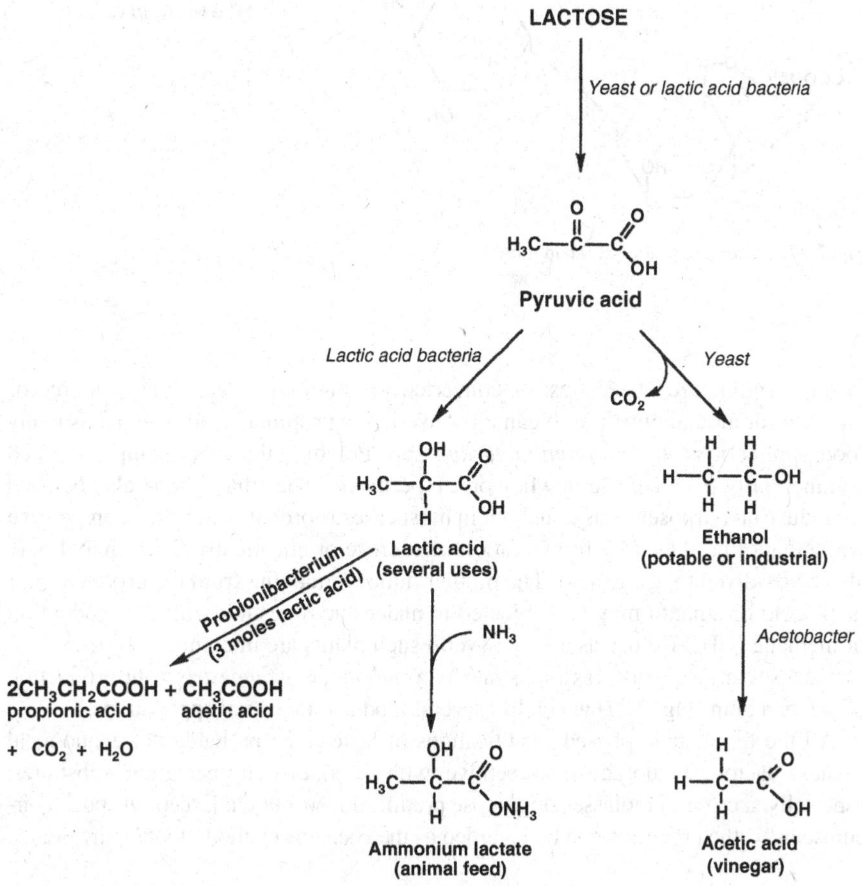

Fig. 2.26 Fermentation products from lactose

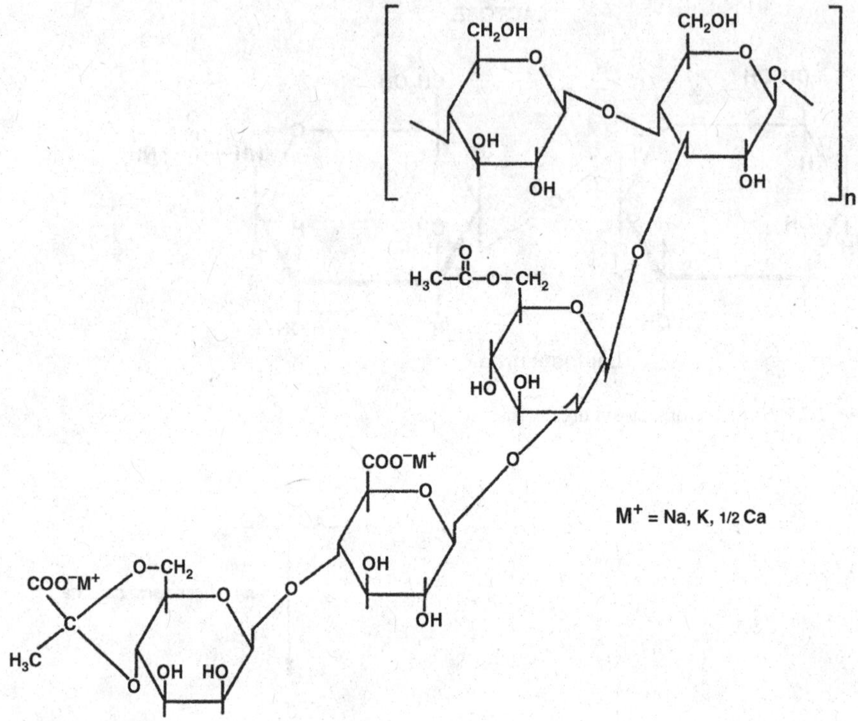

Fig. 2.27 Repeating unit of xanthan gum

in the manufacture of plastics, or converted to ammonium lactate as a source of nitrogen for animal nutrition. It can be converted to propionic acid, which has many food applications, by *Propionibacterium* spp. Potable ethanol is being produced commercially from lactose in whey or UF permeate. The ethanol may also be used for industrial purposes or as a fuel but in most cases is probably not cost-competitive with ethanol produced by fermentation of sucrose or chemically. The ethanol may also be oxidized to acetic acid. The mother liquor remaining from the production of lactic acid or ethanol may be subjected to anaerobic digestion with the production of methane (CH_4) for use as a fuel; several such plants are in commercial use.

Lactose can also be used as a substrate for *Xanthomonas campestris* in the production of xanthan gum (Fig. 2.27) which has several food and industrial applications.

All the fermentation-based modifications of lactose are probably not economical because lactose is not cost-competitive with alternative fermentation substrates, especially sucrose in molasses or glucose produced from starch. Except in special circumstances, the processes can be regarded as the cheapest method of whey disposal.

2.5 Lactose and the Maillard Reaction

As a reducing sugar, lactose can participate in the Maillard reaction, leading to non-enzymatic browning (see O'Brien 1997, 2009; Nursten 2011). The Maillard reaction involves interaction between a carbonyl (in this case, lactose) and an amino group (in foods, principally the ε-NH$_2$ group of lysine in proteins) to form a glycosamine (lactosamine) (Fig. 2.28). The glycosamine may undergo an Amadori rearrangement to form a 1-amino-2-keto sugar (Amadori compound) (Fig. 2.29).

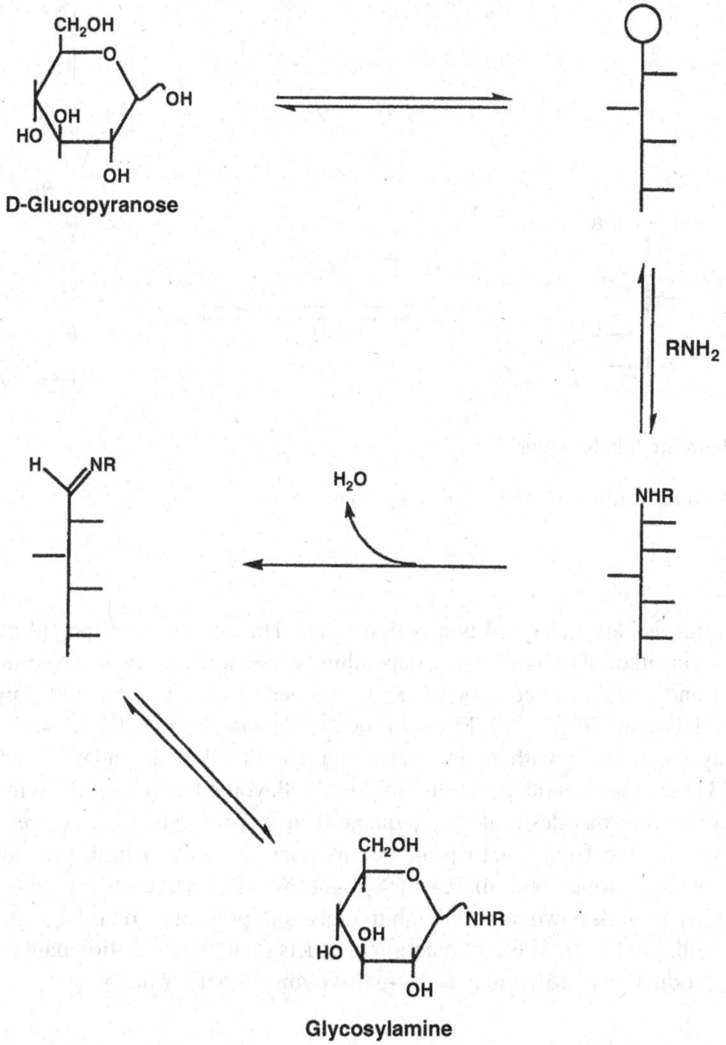

Fig. 2.28 Formation of glycosylamine, the initial step in Maillard browning

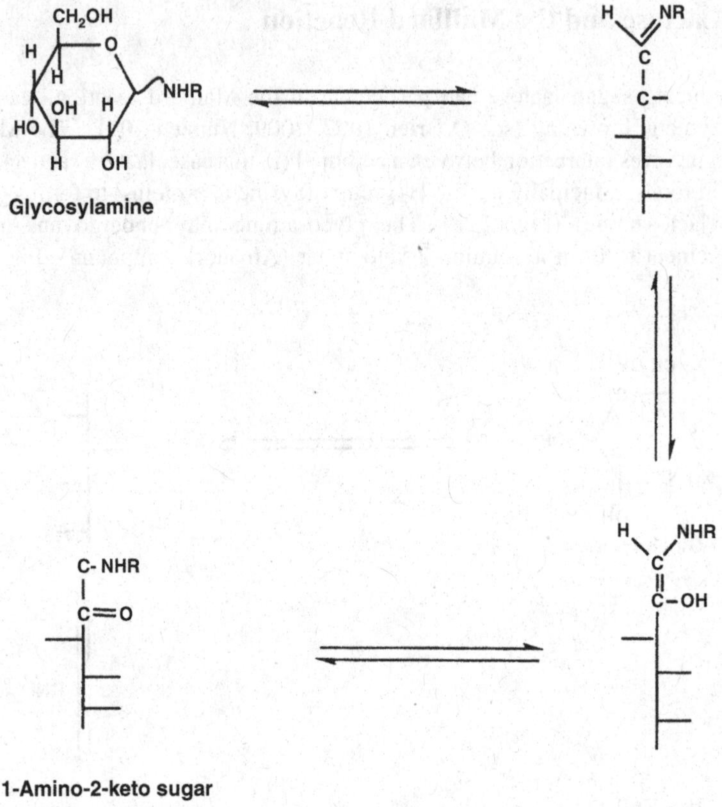

Fig. 2.29 Amadori rearrangement of a glycosylamine

The reaction is base-catalysed and is first order. The Amadori compound may be degraded via either of two pathways, depending on pH, to a variety of active alcohol, carbonyl and dicarbonyl compounds and ultimately to brown-coloured polymers called melanoidins (Fig. 2.30). Many of the intermediates are (off-) flavoured. The dicarbonyls can react with amino acids via the Strecker degradation pathway (Fig. 2.31) to yield another family of highly flavoured compounds While the Maillard reaction has desirable consequences in many foods, e.g., coffee, bread crust, toast, french fried potato products, its consequences in milk products are negative, e.g., brown colour, off-flavours, slight loss of nutritive value (lysine), loss of solubility in milk powders (although it appears to prevent or retard age-gelation in UHT milk products). Maillard reaction products (MRP) have antioxidant properties; the production of MRP may be a small-volume outlet for lactose.

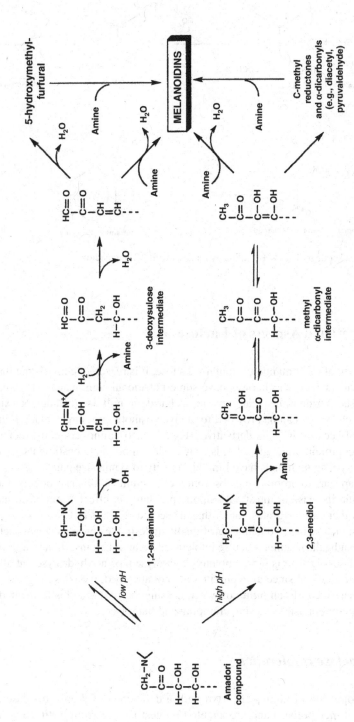

Fig. 2.30 Pathways for the Maillard browning reaction

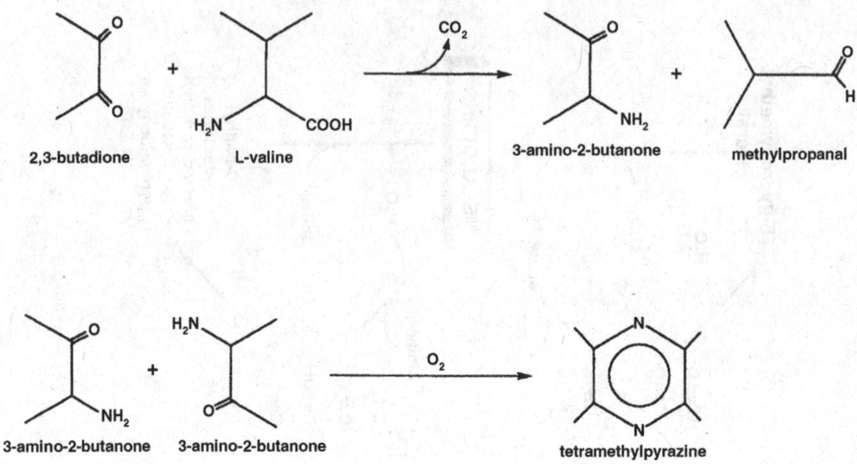

Fig. 2.31 Strecker degradation of L-valine by reaction with 2,3-butanedione

2.6 Nutritional Aspects of Lactose

Since the milk of most mammals contains lactose, it is reasonable to assume that it or its constituent monosaccharides have some nutritional significance. The secretion of a disaccharide rather than a monosaccharide in milk is advantageous since twice as much energy can be provided for a given osmotic pressure. Galactose may be important because it or its derivatives, e.g., galactosamine, are constituents of several glycoproteins and glycolipids, which are important constituents of cell membranes; young mammals have limited capacity to synthesize galactose.

Lactose appears to promote the absorption of calcium but this is probably due to a non-specific increase in intestinal osmotic pressure, an effect common to many sugars and other carbohydrates, rather than a specific effect of lactose.

However, lactose has two major nutritionally undesirable consequences—lactose intolerance and galactosemia. Lactose intolerance is caused by insufficient intestinal β-galactosidase—lactose is not completely hydrolysed, or not hydrolysed at all, in the small intestine and since disaccharides are not absorbed, it passes into the large intestine where it causes an influx of water, causing diarrhoea, and is fermented by intestinal microorganisms, causing cramping and flatulence.

2.6.1 Lactose Intolerance

A small proportion of babies are born with a deficiency of β-galactosidase (inborn error of metabolism) and are unable to digest lactose from birth. In normal infants (and other neonatal mammals), the specific activity of intestinal

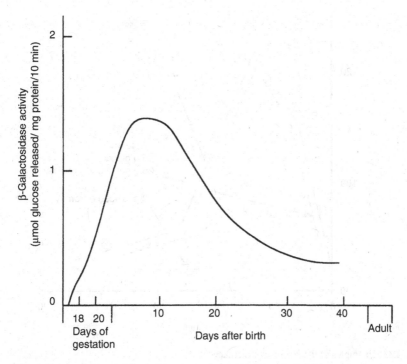

Fig. 2.32 β-Galactosidase activity in homogenates from the intestine of the developing rat

β-galactosidase increases to a maximum at parturition (Fig. 2.32), although total activity continues to increase for some time post-partum due to increasing intestinal area. However, in late childhood, total activity decreases and in an estimated 70 % of the world's population, decreases to a level which causes lactose intolerance among adults. Only northern Europeans and a few African tribes, e.g., Fulami, can consume milk with impunity; the inability to consume lactose appears to be the normal pattern in humans and other species and the ability of northern Europeans to do so presumably reflects positive selective pressure for the ability to consume milk as a source of calcium (better bone development) (see Ingram and Swallow 2009; Swallow 2011).

Lactose intolerance can be diagnosed by (1) jujunal biopsy, with assay for β-galactosidase or (2) administration of an oral dose of lactose followed by monitoring blood glucose level or pulmonary hydrogen level. A test dose of 50 g lactose in water (equivalent to 1 l of milk) is normally administered to a fasting patient; the dose is rather excessive and gastric emptying is faster for a fasted than a fed subject—the presence of other constituents in the meal will delay gastric emptying. Blood glucose level will increase in a lactose-tolerant subject shortly after consuming lactose or a lactose-containing product but not if the subject has a deficiency of

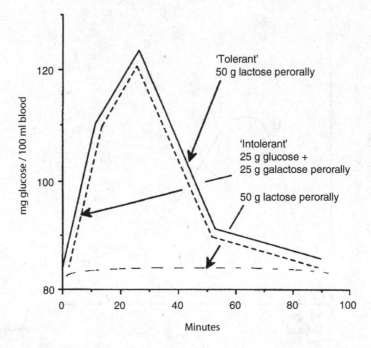

Fig. 2.33 Examples of the "lactose intolerance" test

β-galactosidase (Fig. 2.33). Pulmonary H_2 increases in lactose-intolerant subjects because lactose is metabolised by bacteria in the large intestine, with the production of H_2, which is absorbed and exhaled through the lungs.

Milk can be suitably modified for lactose-intolerant subjects by:

1. Ultrafiltration, which also removes valuable minerals and vitamins, and therefore the milk must be supplemented with these.
2. Fermentation to yoghurt or other fermented product in which ~25 % of the lactose is metabolised by lactic acid bacteria, and which contains bacterial β-galactosidase and is also discharged more slowly from the stomach due to its texture.
3. Conversion to cheese, which is essentially free of lactose.
4. Treatment with exogenous β-galactosidase, either domestically by the consumer or the dairy factory, using free or immobilized enzyme; several protocols for treatment have been developed (Fig. 2.34). Lactose-hydrolysed milks are technologically successful and commercially available but have not led to large increases in the consumption of milk in countries where lactose intolerance is widespread, presumably due to cultural and economic factors. However, there are niche markets for such products.

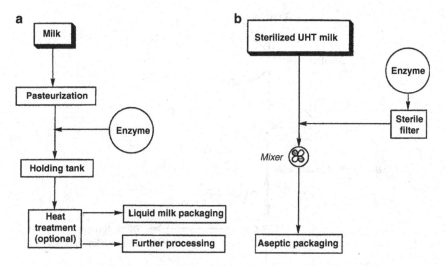

Fig. 2.34 (**a**) Scheme for manufacture of low-lactose milk using a "high" level soluble β-galactosidase. (**b**) Scheme for the manufacture of low-lactose milk by addition of a low level of soluble β-galactosidase to UHT-sterilized milk (redrawn from Mahoney 1997)

2.6.2 Galactosemia

Glactosemia is caused by the inability to metabolise galactose due to a hereditary deficiency of galactokinase or galactose-1-phosphate (Gal-1-P): uridyl transferase (Fig. 2.35). Lack of the former enzyme leads to the accumulation of galactose which is metabolised via other pathways, leading, among other products, to galactitol which accumulates in the lens of the eye, causing cataract in 10–20 years (in humans) if consumption of galactose-containing foods (milk, legumes) is continued. The incidence is about 1:40,000. The lack of Gal-1-P: uridyl transferase leads to the accumulation of Gal and Gal-1-P. The latter interferes with the synthesis of glycoproteins and glycolipids (important for membranes, e.g., in the brain) and results in irreversible mental retardation within 2–3 months if the consumption of galactose-containing foods is continued. The incidence of this disease, often called "classical galactosemia", is about 1 in 60,000.

The ability to metabolise galactose decreases on aging (after 70 years), leading to cataract; perhaps this, together with the fact that mammals normally encounter lactose only while suckling, explains why many people lose the ability to utilise lactose at the end of childhood.

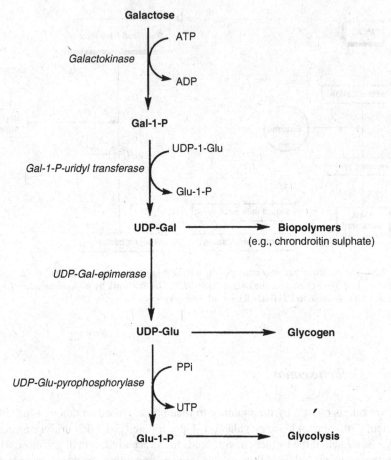

Fig. 2.35 Pathways for the metabolism of galactose

2.7   Determination of Lactose Concentration

Lactose may be quantified by methods based on one of five principles:

1. Polarimetry
2. Oxidation-reduction titration
3. Colorimetry
4. Chromatography
5. Enzymatically

2.7.1 Polarimetry

The specific rotation, $[\alpha]_D^{20}$, of lactose in solution at equilibrium is 55.4° expressed on an anhydrous basis (52.6° on a monohydrate basis). The specific rotation is defined as the optical rotation of a solution containing 1 g/ml in a 1 dm polarimeter tube; it is affected by temperature (20 °C is usually used; indicated by superscript) and wavelength [usually the sodium D line (589.3 nm) is used; indicated by a subscript].

$$[\alpha]_D^{20} = a/lc$$

where: a is the measured optical rotation, l is the light path in dm and c is the concentration as g/ml

It is usually expressed as:

$$[\alpha]_D^{20} = 100\,a/lc$$
$$\text{where} : c = g/100\,ml$$

The milk sample must first be defatted and de-proteinated, usually by treatment with mercuric nitrate [$Hg(NO_3)_2$]. In calculating the concentration of lactose, a correction should be used for the concentration of fat and protein in the precipitate, i.e., 0.92 for whole milk and 0.96 for skimmed milk.

2.7.2 Oxidation and Reduction Titration

Lactose is a reducing sugar, i.e., it is capable of reducing appropriate oxidising agents, two of which are usually used, i.e., alkaline copper sulphate ($CuSO_4$ in sodium potassium tartrate; Fehling's solution) or Chloroamine-T (2.1).

$$\begin{array}{c} HNCL \\ | \\ O=S=O \\ | \end{array}$$

$$CH_3$$
Chloroamine-T (2.1)

For analysis by titration with Fehling's solution, the sample is treated with lead acetate to precipitate protein and fat, filtered and the filtrate titrated with alkaline $CuSO_4$, while heating. The reactions involved are summarized in Fig. 2.36.

Cu_2O precipitates and may be recovered by filtration and weighed; the concentration of lactose can then be calculated since the oxidation of one mole of lactose (360 g) yields one mole of Cu_2O (143 g). However, it is more convenient to add an excess of a standard solution of $CuSO_4$ to the lactose-containing solution; the

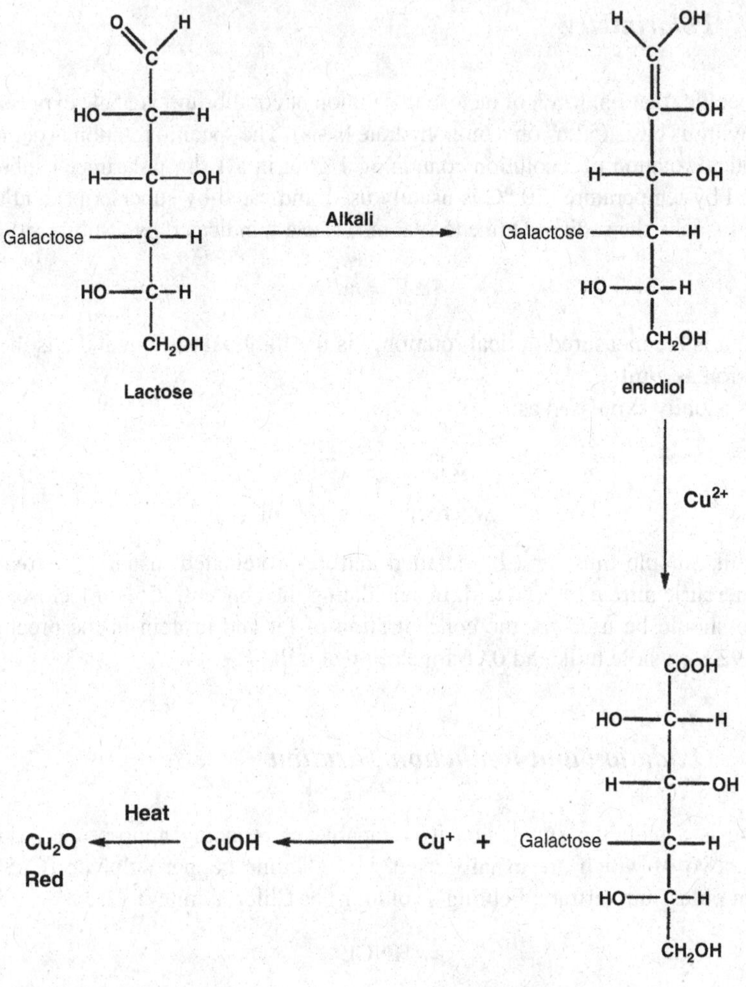

Fig. 2.36 Oxidation of lactose by alkaline copper sulfate (Fehling's reagent)

solution is cooled and the excess $CuSO_4$ determined by reaction with KI and titrating the liberated I_2 with standard sodium thiosulphate ($Na_2S_2O_3$) using starch as an indicator.

$$2CuSO_4 + 4KI \rightarrow CuI_2 + 2K_2SO_4 + I_2$$
$$I_2 + 2Na_2S_2O_3 \rightarrow 2NaI + Na_2S_2O_6$$

The end point in the Fehling's is not sharp and the redox determination of lactose is now usually performed using Chloramine-T rather than $CuSO_4$ as oxidising agent. The reactions involved are as follows:

$$CH_3C_6H_4SO_2NClH + H_2O + KI\,(\text{excess})$$
$$\leftrightarrow CH_3C_6H_4SO_2NH_2 + HCl + KIO\,(\text{K hypoiodate})$$

$$KIO + \text{lactose}\ (-CHO) \rightarrow KI + \text{lactobionic acid}\,(-COOH)$$

$$KI + KIO \rightarrow 2KOH + I_2$$

The I_2 titrated with standard $Na_2S_2O_3$

$$I_2 + 2Na_2S_2O_3 \rightarrow 2NaI + Na_2S_4O_6$$
$$\text{(thiosulphate)}$$

One ml of 0.04 N thiosulphate is equivalent to 0.0072 g lactose monohydrate or 0.0064 g anhydrous lactose.

The sample is deproteinized and defatted using phosphotungstic acid.

2.7.3 Infrared (IR) Spectroscopy

Stretching of the –O–H bond of lactose (and other sugars) by IR radiation of 9.5 μm, permits the quantitative determination of lactose. As discussed in Chaps. 3 and 4, respectively, the ester bond of triglycerides absorbs IR radiation at 5.7 μm and the peptide bond of proteins absorbs IR radiation at 6.46 μm. Thus, in a single scan, the concentrations of fat, protein and lactose in milk can be determined by IR spectroscopy using an Infra Red Milk Analyzer (IRMA).

Such instruments are now widely used in the dairy industry.

2.7.4 Colorimetric Methods

Reducing sugars, including lactose, react on boiling with phenol (2.2) or anthrone (2.3) in strongly acidic solution (70 %, v/v, H_2SO_4) to give a coloured solution (2.1 and (2.3).

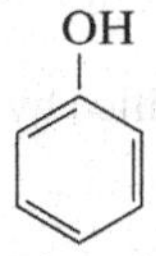

OH

Phenol   (2.2)

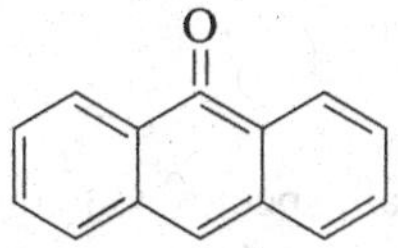

O

Anthrone   (2.3)

The complex with anthrone absorbs maximally at 625 nm. The concentration of lactose is determined from a standard curve prepared using a range of lactose concentrations.

The method is very sensitive but must be performed under precisely controlled conditions.

2.7.5   Chromatographic Methods

While lactose may be determined by gas liquid chromatography, high performance liquid chromatography (HPLC), using an ion-exchange column and a refractive index detector, is now usually used.

2.7.6   Enzymatic Methods

Enzymatic methods are very sensitive but are rather expensive, especially for a small number of samples.

Lactose is first hydrolysed by β-galactosidase to glucose and galactose. The glucose may be quantified by reaction with:

1. Glucose oxidase using a platinum electrode or the H_2O_2 generated may be quantified by using a peroxidase and a suitable dye acceptor
   or
2. Glucose-6-phosphate dehydrogenase (G-6-P-DH):

$$\text{D-Glucose} + \text{ATP} \xrightarrow{\text{Hexokinase}} \text{Gluconate-6-P} \xrightarrow{\text{G-6-DH, NADP}^+} \text{Gluconate-6-P} + \text{NADPH} + \text{H}^+$$

The concentration of NADPH produced may be quantified by measuring the increase in absorbance at 334, 340 or 365 nm.

Alternatively, the galactose produced may be quantified using galactose dehydrogenase (Gal-DH).

$$\text{D-galactose} + \text{NAD}^+ \xrightarrow{\text{Gal-DH}} \text{Galactonic acid} + \text{NADH} + \text{H}^+$$

The NADH produced may be quantified by measuring the increase in absorbance at 334, 340 or 365 nm.

2.8   Oligosaccharides

The milk of most, and probably all, species contains other free saccharides, mainly oligosaccharides (OSs), the concentration, proportions and types of which show large interspecies differences. The concentration of OSs is higher in colostrum than in milk. General reviews on the OSs in milk include Newburg and Newbauer (1995), Mehra and Kelly (2006), and Urashima et al. (2001, 2009, 2011).

Almost all of the OSs have lactose at the non-reducing end, they contain three to eight monosaccharides, they may be linear or branched, and contain either or both of two unusual monosaccharides, fucose (a 6-deoxyhexose) and N-acetylneuraminic acid. Fucose occurs quite widely in tissues of mammals and other animals where it serves a wide array of functions (Becker and Lowe 2003). Its significance in the OSs in milk is not clear; perhaps it is to supply the neonate with preformed fucose.

The OSs are synthesized in the mammary gland, catalyzed by special transferases that transfer galactosyl, sialyl, N-acetylglucosaminyl, or fucosyl residues from nucleotide sugars to the core structures. These transferases are not affected by α-La and are probably similar to the transferases that catalyze the glycosylation of lipids and proteins.

The milk of all species examined contains OSs, but the concentration varies markedly. The highest levels are in the milk of monotremes, marsupials, marine mammals, humans, elephants, and bears. With the exception of humans and elephants, the milk of these species contains little or no lactose, and OSs are the principal carbohydrates.

The milk of the echidna contains mainly the trisaccharide, fucosyllactose, while that of the platypus contains mainly the tetrasaccharide, difucosyllactose. Among marsupials, the best studied is the Tammar wallaby; presumably, its lactation pattern and milk composition are typical of marsupials. A low level of lactose is produced at the start of lactation, but about 7 days after birth, a second galactosyltransferase appears and tri- to penta-saccharides are produced, which by ~180 days are the principal saccharides. During this period the content is high, ~50 % of total solids, and the level of lipids is low (~15 % of total solids). At about 180 days, the carbohydrate decreases to a very low level and consists mainly of monosaccharides, while the level of lipids increases to ~60 % of total solids (Sharp et al. 2011).

Human milk contains ~130 OSs, at a total concentration of ~15 g/L; these are considered to be important for neonatal brain development. Bear milk contains little lactose but a high level of total sugars (mainly OSs) −1.7 and 28.6 g/kg, respectively (Oftedal 2013). Elephant milk contains ~50 and 12 g/kg of lactose and OSs, respectively, a few days post- partum, but as lactation progresses, the concentration of lactose decreases while that of OSs increases (e.g., 12 and 18 g/kg, respectively), at 47 days (Osthoff et al. 2005). The milk of seals contains both lactose and OSs, but the milk of the Californian sea lion, Northern fur seal, and Australian fur seal contain neither, probably because they contain no α-La (Urashima et al. 2001).

Bovine, ovine, caprine, and equine milk contain relatively low levels of OSs, which have been characterized (see Urashima et al. 2001, 2009, 2011). Caprine milk contains about ten times as much OSs as bovine and ovine milk, and a process for their isolation by nanofiltration has been reported (Martinez-Ferez et al. 2006). Possible methods for producing OSs similar to those found in human milk, by fermentation or by transgenic animals or by recovering OSs from cow's milk whey or UF permeate were discussed by Mehra and Kelly (2006) and O'Mahony and Touhy (2013).

OSs with bactericidal properties were probably the saccharides present in the mammary secretions of early mammals; the high level of OSs in the milk of monotremes and marsupials conforms with their secretion early in evolution. It is proposed that the primitive mammary glands of the first common ancestor of mammals produced lysozyme (a predecessor of α-La), and a number of glycosyltransferases but little or no α-La. This resulted in the production of a low level of lactose that was utilized in the synthesis of OSs and did not accumulate (Urashima et al. 2009). Initially, the OSs served mainly as bactericidal agents but later became a source of energy for the neonate. Both of these functions persist for monotremes, marsupials, and some eutherians such as bears, elephants, and marine mammals. However, most eutherians evolved to secrete predominantly lactose as an energy source, due to the synthesis of an increased level of α-La, while OSs continued to play a bactericidal role. Human and elephant milk, both of which contain high levels of lactose and OSs, seem to be anomalous. Work on the OSs of a wider range of species is needed to explain this situation.

The significance of OSs is not clear, but the following aspects may be significant: For any particular level of energy, they have a smaller impact on osmotic pressure than smaller saccharides, they are not hydrolyzed by β-galactosidase, and fucosidase or neuraminidase is not secreted in the intestine. Hence the OSs are not hydrolyzed and absorbed in the gastro-intestinal tract, and they function as soluble fiber and prebiotics that affect the microflora of the large intestine. It is claimed that they prevent the adhesion of pathogenic bacteria in the intestine; galactose, and especially N-acetylneuraminic acid, are important for the synthesis of glycolipids and glycoproteins, which are vital for brain development. It has therefore been suggested that the OSs are important for brain development (see Kunz and Rudloff 2006).

There is considerable interest in the development of OS-enriched ingredients from bovine milk (O'Mahony and Touhy 2013), primarily for infant formula applications. This interest has been spurred by the demonstrated bioactive functionality of these compounds in humans (Kunz and Rudloff 2006).

In addition to lactose and free OSs, the milk of all species examined contains small amounts of monosaccharides and some milk proteins, especially κ-casein, are glycosylated, and there are low levels of highly glycosylated glycoproteins, especially mucins, and glycolipids in the milk fat globule membrane.

References

Affertsholt-Allen, T. (2007). Market developments and industry challenges for lactose and lactose derivatives. *IDF lactose symposium*, 14–16 May. Moscow, Russia.

Andrews, G. (1989). Lactulose in heated milk. In P. F. Fox (Ed.), *Heat-induced changes in milk, bulletin 238* (pp. 45–52). Brussels: International Dairy Federation.

Becker, C. J., & Lowe, J. R. (2003). Fucose: Biosynthesis and biological function in mammals. *Glycobiology, 13*, 41R–53R.

Ganzle, M. G. (2011a). Lactose derivatives. In J. W. Fuquay, P. F. Fox, & P. L. H. McSweeney (Eds.), *Encyclopedia of dairy sciences* (2nd ed., Vol. 3, pp. 202–208). Oxford: Academic Press.

Ganzle, M. G. (2011b). Galactooligosaccharides. In J. W. Fuquay, P. F. Fox, & P. L. H. McSweeney (Eds.), *Encyclopedia of dairy sciences* (2nd ed., Vol. 3, pp. 209–216). Oxford: Academic Press.

Holt, C. (1985). The milk salts. Their secretion, concentrations and physical chemistry. In P. F. Fox (Ed.), *Developments in dairy chemistry* (Lactose and minor constituents, Vol. 3, pp. 143–181). London: Elsevier Applied Science.

Hynd, J. (1980). Drying of whey. *Journal of the Society of Dairy Technology, 33*, 52–54.

Ingram, C. J. E., & Swallow, D. M. (2009). Lactose intolerance. In P. L. H. McSweeney & P. F. Fox (Eds.), *Advanced dairy chemistry* (Lactose, water, salts and minor constituents 3rd ed., Vol. 3, pp. 203–229). New York: Springer.

Jenness, R., & Holt, C. (1987). Casein and lactose concentrations in milk of 31 species are negatively correlated. *Experimentia, 43*, 1015–1018.

Jenness, R., & Patton, S. (1959). Lactose. In *Principles of dairy chemistry* (pp. 73-100). New York: Wiley

Jenness, R., & Sloan, R. E. (1970). The composition of milk of various species: A review. *Dairy Science Abstracts, 32*, 599–612.

Kunz, C., & Rudloff, S. (2006). Health promoting aspects of milk oligosaccharides. A review. *International Dairy Journal, 16*, 1341–1346.

Ley, J. M., & Jenness, R. (1970). Lactose synthetase activity of α-lactalbumins from several species. *Archives of Biochemistry and Biophysics, 138*, 464–469.

Mahoney, R. R. (1997). Lactose: Enzymatic modification. In P. F. Fox (Ed.), *Advanced dairy chemistry – 3 – lactose, water, salts and vitamins* (2nd ed., pp. 77–125). London: Chapman & Hall.

Martinez-Ferez, A., Rudloff, S., Gaudix, A., Henkel, C. A., Pohlentz, G., Boza, J. J., Gaudix, E. M., & Kunz, C. (2006). Goats' milk as a natural source of lactose-derived oligosaccharides: Isolation by membrane technology. *International Dairy Journal, 16*, 173–181.

Mehra, R., & Kelly, P. (2006). Milk oligosaccharides: Structural and technological aspects. *International Dairy Journal, 16*, 1334–1340.

Newberg, D. S., & Newbauer, S. H. (1995). Carbohydrates in milk: Analysis, quantities and significance. In R. G. Jensen (Ed.), *Handbook of milk composition* (pp. 273–349). San Diego: Academic Press.

Nursten, H. (2011). Maillard reaction. In J. W. Fuquay, P. F. Fox, & P. L. H. McSweeney (Eds.), *Encyclopedia of dairy sciences* (2nd ed., Vol. 3, pp. 217–235). Oxford: Academic Press.

O'Brien, J. (1997). Reaction chemistry of lactose: Non-enzymatic degradation pathways and their significance in dairy products. In P. F. Fox (Ed.), *Advanced dairy chemistry* (Lactose, water, salts and vitamins 2nd ed., Vol. 3, pp. 155–231). London: Chapman & Hall.

O'Brien, J. (2009). Non-enzymatic degradation pathways of lactose and their significance in dairy products. In P. L. H. McSweeney & P. F. Fox (Eds.), *Advanced dairy chemistry* (Lactose, water, salts and minor constituents 3rd ed., Vol. 3, pp. 231–294). New York: Springer.

Oftedal, O. T. (2013). Origin and evolution of the major constituents in milk. In P. L. H. McSweeney & P. F. Fox (Eds.), *Advanced dairy chemistry* (4th edn, Vol. 1A, pp. 1–42). New York: Springer.

O'Mahony, J. A., & Touhy, J. J. (2013). Further applications of membrane filtration in dairy processing. In A. Tamime (Ed.), *Membrane processing: Dairy and beverage applications* (pp. 225–261). West Sussex: Blackwell.

Osthoff, G., de Waal, H. O., Hugo, A., de Wit, M., & Botes, P. (2005). Milk composition of a free-ranging African elephant (*Loxodonta Africana*) cow in early lactation. *Comparative Biochemistry and Physiology, 141*, 223–229.

Paterson, A. H. J. (2009). Lactose: Production and applications. In J. W. Fuquay, P. F. Fox, & P. L. H. McSweeney (Eds.), *Encyclopedia of dairy sciences* (2nd ed., Vol. 3, pp. 196–201). Oxford: Academic Press.

Paterson, A. H. J. (2011). Production and uses of lactose. In P. L. H. McSweeney & P. F. Fox (Eds.), *Advanced dairy chemistry* (Lactose, water, salts and minor constituents 3rd ed., Vol. 3, pp. 105–120). New York: Springer.

Playne, M. J., & Crittenden, R. G. (2009). Galactosaccharides and other products derived from lactose. In P. L. H. McSweeney & P. F. Fox (Eds.), *Advanced dairy chemistry* (Lactose, water, salts and minor constituents 3rd ed., Vol. 3, pp. 121–201). New York: Springer.

Schuck, P. (2011). Lactose crystallization. In J. W. Fuquay, P. F. Fox, & P. L. H. McSweeney (Eds.), *Encyclopedia of dairy sciences* (2nd ed., Vol. 3, pp. 182–195). Oxford: Academic Press.

Sharp, J. A., Menzies, K., Lefevre, C., & Nicholas, K. R. (2011). Milk of monotremes and marsupial. In J. W. Fuquay, P. F. Fox, & P. L. H. McSweeney (Eds.), *Encyclopedia of dairy sciences* (2nd ed., Vol. 1, pp. 553–562). Oxford: Academic Press.

Smart, J. B. (1993). Transferase reactions of β-galactosidases – new product opportunities. In *Lactose hydrolysis, bulletin 239* (pp. 16–22). Brussels: International Dairy Federation.

Swallow, D. M. (2011). Lactose intolerance. In J. W. Fuquay, P. F. Fox, & P. L. H. McSweeney (Eds.), *Encyclopedia of dairy sciences* (2nd ed., Vol. 3, pp. 236–240). Oxford: Academic Press.

Tamura., Y., Mizota, T., Shimamura, S., & Tomita, M. (1993). Lactulose and its application to food and pharmaceutical industries. In *Lactose hydrolysis, bulletin 239* (pp. 43–53) Brussels: International Dairy Federation.

Tumerman, L., Fram, H., & Cornely, K. W. (1954). The effect of lactose crystallization on protein stability in frozen concentrated milk. *Journal of Dairy Science, 37*, 830–839.

Urashima, T., Saito, T., Nakarmura, T., & Messer, M. (2001). Oligosaccharides of milk and colostrums in non-human mammals. *Glycoconjugate Journal, 18*, 357–371.

Urashima, T., Kitaoka, M., Asakuma, S., & Messer, M. (2009). Indigenous oligosaccharides in milk. In P. L. H. McSweeney & P. F. Fox (Eds.), *Advanced dairy chemistry* (Lactose, water, salts and minor constituents 3rd ed., Vol. 3, pp. 295–349). New York: Springer.

Urashima, T., Asakuma, S., Kitaoka, M., & Messer, M. (2011). Indigenous oligosaccharides in milk. In J. W. Fuquay, P. F. Fox, & P. L. H. McSweeney (Eds.), *Encyclopedia of dairy sciences* (2nd ed., Vol. 3, pp. 241–273). Oxford: Academic Press.

Walstra, P., & Jenness, R. (1984a). *Dairy chemistry and physics*. New York: Wiley.

Suggested Reading

Fox, P. F. (Ed.). (1985). *Developments in dairy chemistry – 3 – lactose and minor constituents*. London: Elsevier Applied Science Publishers.

Fox, P. F. (Ed.). (1997). *Advanced dairy chemistry – 3 – lactose, water, salts and vitamins* (2nd ed.). London: Chapman & Hall.

Fuquay, J. W., Fox, P. F., & McSweeney, P. L. H. (Eds.). (2011). *Encyclopedia of dairy sciences* (2nd ed., Vol. 3, pp. 173–273). Oxford: Academic Press.

Holsinger, V. H. (1988). Lactose. In N. P. Wong (Ed.), *Fundamentals of dairy chemistry* (pp. 279–342). New York: Van Nostrand Reinhold Co.

IDF. (1993). *Proceedings of the IDF workshop on lactose hydrolysis, bulletin 289*. Brussels: International Dairy Federation.

Jenness, R., & Patton, S. (1959). Lactose. In *Principles of dairy chemistry* (pp. 73-100). New York: Wiley.

Labuza, T. P., Reineccius, G. A., Monnier, V. M., O'Brien, J., & Baynes, J. W. (Eds.). (1994). *Maillard reactions in chemistry, food and health*. Cambridge: Royal Society of Chemistry.

McSweeney, P. L. H., & Fox, P. F. (Eds.). (2009). *Advanced dairy chemistry* (Lactose, water, salts and minor constituents 3rd ed., Vol. 3). New York: Springer.

Nickerson, T. A. (1965). Lactose. In B. H. Webb & A. H. Johnson (Eds.), *Fundamentals of dairy chemistry* (pp. 224–260). Westport, CT: AVI Publishing Co. Inc.

Nickerson, T. A. (1974). Lactose. In B. H. Webb, A. H. Johnson, & J. A. Alford (Eds.), *Fundamentals of dairy chemistry* (pp. 273–324). Westport, CT: AVI Publishing Co. Inc.

Walstra, P., & Jenness, R. (1984b). *Dairy chemistry and physics*. New York: Wiley.

Walstra, P., Geurts, T. J., Noomen, A., Jellema, A., & van Boekel, M. A. J. S. (1999). *Dairy technology: Principles of milk processing and processes*. New York: Marcel Dekker, Inc.

Walstra, P., Wouters, J. F., & Geurts, T. J. (2006). *Dairy science and technology*. Boca Raton, FL: CRC Press.

Yang, S. T., & Silva, E. M. (1995). Novel products and new technologies for use of a familiar carbohydrate, milk lactose. *Journal of Dairy Science, 78*, 2541–2562.

Chapter 3
Milk Lipids

3.1 Introduction

The milk of all mammals contains lipids but the concentration varies widely between species from ~2 % to >50 % (Table 3.1). The principal function of dietary lipids is to serve as a source of energy for the neonate and the fat content in milk largely reflects the energy requirements of the species, e.g., land animals indigenous to cold environments and marine mammals secrete high levels of lipids in their milk.

Milk lipids are also important (1) as a source of essential fatty acids (i.e., fatty acids which cannot be synthesised by higher animals, especially linoleic acid, $C_{18:2}$) and fat-soluble vitamins (A, D, E, K), (2) for the flavour and rheological properties of dairy products and foods in which they are used. Because of its wide range of fatty acids, the flavour of milk fat is superior to that of other fats. In certain products and after certain processes, fatty acids serve as precursors of very flavourful compounds such as methyl ketones and lactones. Unfortunately, lipids also serve as precursors of compounds that cause off-flavour defects (hydrolytic and oxidative rancidity) and as solvent for compounds in the environment which may cause off-flavours.

Until recently, the economic value of milk was based mainly or totally on its fat content, which is still true in some cases. This practice was satisfactory when milk was used mainly or solely for butter production. Possibly, the reason for paying for milk on the basis of its fat content, apart from its value for butter production, is that relatively simple quantitative analytical methods were developed for fat earlier than for protein or lactose. Because of its economic value, there has long been commercial pressure to increase the yield of milk fat per cow by nutrition or genetic means.

To facilitate the reader, the nomenclature, structure and properties of the principal fatty acids and of the principal lipid classes are summarized in Appendices A, B and C. The structure, properties and functions of the fat-soluble vitamins, A, D, E and K, are discussed in Chap. 6.

© Springer International Publishing Switzerland 2015
P.F. Fox et al., *Dairy Chemistry and Biochemistry*,
DOI 10.1007/978-3-319-14892-2_3

Table 3.1 The fat content of milks from various species (gL^{-1}) (from Christie 1995)

Species	Fat content	Species	Fat content
Cow	33–47	Marmoset	77
Buffalo	47	Rabbit	183
Sheep	40–99	Guinea-pig	39
Goat	41–45	Snowshoe hare	71
Musk-ox	109	Muskrat	110
Dall-sheep	32–206	Mink	134
Moose	39–105	Chinchilla	117
Antelope	93	Rat	103
Elephant	85–190	Red kangaroo	9–119
Human	38	Dolphin	62–330
Horse	19	Manatee	55–215
Monkeys	10–51	Pygmy sperm whale	153
Lemurs	8–33	Harp seal	502–532
Pig	68	Bear (four species)	108–331

3.2 Factors that Affect the Fat Content of Bovine Milk

Bovine milk typically contains ~3.5 % fat but the level varies widely, depending on several factors, including: breed, individuality of the animal, stage of lactation, season, nutritional status, type of feed, health and age of the animal, interval between milkings and the point during milking when the sample is taken.

Of the common European breeds, milk from Jersey cows contains the highest level of fat and that from Holstein/Friesians the lowest (Fig. 3.1). The data in Fig. 3.1 also show the very wide range of fat content in individual-cow samples.

The fat content of milk decreases during the first 4–6 weeks after parturition and then increases steadily throughout the remainder of lactation, especially toward the end (Fig. 3.2). For any particular population, fat content is highest in winter and lowest in summer, due partly to the effect of environmental temperature. Production of creamery (manufacturing) milk in Ireland, New Zealand and parts of Australia is very seasonal; lactational, seasonal and possibly nutritional effects coincide, leading to large seasonal changes in the fat content of milk (Fig. 3.3), and also in the levels of protein and lactose.

For any individual animal, fat content decreases slightly during successive lactations, by ~0.2 % over a typical productive lifetime (~5 lactations). In practice, this factor usually has no overall effect on the fat content of a bulk milk supply because herds normally include cows of various ages. The concentration of fat (and of all other milk-specific constituents) decreases markedly on mastitic infection due to impaired synthesizing ability of the mammary tissue; the effect is clear-cut in the case of clinical mastitis but is less so for sub-clinical infection.

Milk yield is reduced by underfeeding but the concentration of fat usually increases, with little effect on the amount of fat produced. Diets low in roughage have a marked depressing effect on the fat content of milk, with little effect on milk yield.

Fig. 3.1 Range of fat content
in the milk of individual cows
of four breeds (from Jenness
and Patton 1959)

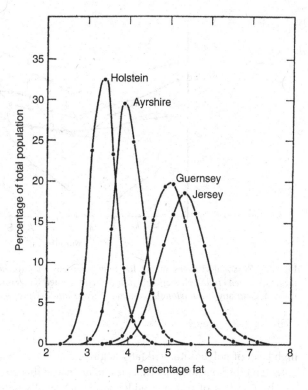

Fig. 3.2 Typical changes
in the concentrations of fat
(*filled circle*), protein
(*filled square*) and lactose
(*open circle*) in bovine milk
during lactation

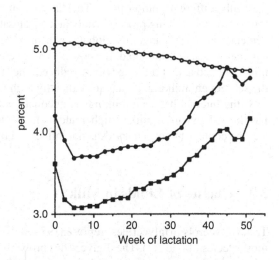

Ruminants synthesize milk fat mainly from carbohydrate-derived precursors; addition of fat to the diet usually causes slight increases in the yield of both milk and fat, with little effect on the fat content of milk. Feeding of some fish oils (e.g., cod liver oil, in an effort to increase the concentrations of vitamins A and D in milk) has a very marked (~25 %) depressing effect on the fat content of milk, apparently due to the

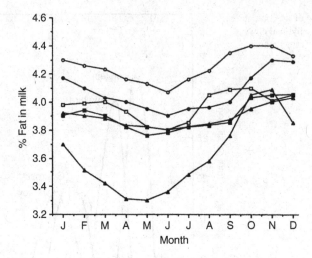

Fig. 3.3 Seasonal changes in the fat content of bovine milk in some European countries—Denmark (*open circle*), Netherlands (*filled circle*), United Kingdom (*empty square*), France (*filled square*), Germany (*open triangle*), Ireland (*filled triangle*) (from An Foras Taluntais 1981)

high level of polyunsaturated fatty acids (the effect is eliminated by hydrogenation), although oils from some fish species do not cause this effect.

The quarters of a cow's udder are anatomically separate and secrete milk of markedly different composition. The fat content of milk increases continuously throughout the milking process while the concentrations of the various non-fat constituents show no change; fat globules appear to be partially trapped in the alveoli and their passage is hindered. If a cow is incompletely milked, the fat content of the milk obtained at that milking will be reduced; the "trapped" fat will be expressed at the subsequent milking, giving an artificially high value for fat content.

If the interval between milkings is unequal (as they usually are in commercial farming), the yield of milk is higher and its fat content lower after the longer interval; the content of non-fat solids is not influenced by milking interval.

3.3   Classes of Lipids in Milk

Triacylglycerols (triglycerides) represent 97–98 % of the total lipids in the milks of most species (Table 3.2). The diglycerides probably represent incompletely synthesized lipids in most cases although the value for the rat probably also includes partially hydrolyzed triglycerides, as indicated by the high concentration of free fatty acids, suggesting damage to the milk fat globule membrane (MFGM) during milking and storage. The very high level of phospholipids in mink milk probably indicates the presence of mammary cell membranes.

Table 3.2 Composition of individual simple lipids and total phospholipids in milks of some species (wt% of the total lipids)

Lipid class	Cow	Buffalo	Human	Pig	Rat	Mink
Triacylglycerols	97.5	98.6	98.2	96.8	87.5	81.3
Diacylglycerols	0.36	0.7	0.7	2.9		1.7
Monoacylglycerols	0.027	T	0.1	0.4		T
Cholesteryl esters	T	0.1	T	0.06	–	T
Cholesterol	0.31	0.3	0.25	0.6	1.6	T
Free fatty acids	0.027	0.5	0.4	0.2	3.1	1.3
Phospholipids	0.6	0.5	0.26	1.6	0.7	15.3

From Christie (1995); *T* trace

Table 3.3 Composition of the phospholipids in milk from various species (expressed as mol% of total lipid phosphorus)

Species	Phosphatidyl-choline	Phosphatidyl ethanolamine	Phosphatidyl-serine	Phosphatidyl-inositol	Sphingo-myelin	Lysophospho-lipids[a]
Cow	34.5	31.8	3.1	4.7	25.2	0.8
Sheep	29.2	36.0	3.1	3.4	28.3	
Buffalo	27.8	29.6	3.9	4.2	32.1	2.4
Goat	25.7	33.2	6.9	5.6[b]	27.9	0.5
Camel	24.0	35.9	4.9	5.9	28.3	1.0
Ass	26.3	32.1	3.7	3.8	34.1	
Pig	21.6	36.8	3.4	3.3	34.9	
Human	27.9	25.9	5.8	4.2	31.1	5.1
Cat	25.8	22.0	2.7	7.8[b]	37.9	3.4
Rat	38.0	31.6	3.2	4.9	19.2	3.1
Guinea-pig	35.7	38.0	3.2	7.1[b]	11.0	2.0
Rabbit	32.6	30.0	5.2	5.8[b]	24.9	0.4
Mouse[c]	32.8	39.8	10.8	3.6	12.5	
Mink	52.8	10.0	3.6	6.6	15.3	8.3

From Christie (1995)
[a]Mainly lysophosphatidylcholine but also lysophosphatidylethanolamine
[b]Also contains lysophosphatidylethanolamine
[c]Analysis of milk fat globule membrane phospholipids

Although phospholipids represent <1 % of total lipid, they play a particularly important role, being present mainly in the MFGM and other membraneous material in milk. The principal phospholipids are phosphatidyl choline, phosphatidyl ethanolamine and sphingomyelin (Table 3.3). Trace amounts of other polar lipids, including ceramides, cerebrosides and gangliosides, are also present. Phospholipids represent a considerable proportion of the total lipid of buttermilk and skimmilk (Table 3.4), reflecting the presence of proportionately larger amounts of membrane material in these products.

Table 3.4 Total fat and phospholipid content of some milk products

Product	Total lipid, %, w/v	Phospholipids, %, w/v	Phospholipid as %, w/w, of total lipids
Whole milk	3–5	0.02–0.04	0.6–1.0
Cream	10–50	0.07–0.18	0.3–0.4
Butter	81–82	0.14–0.25	0.16–0.29
Butter oil	~100	0.02–0.08	0.02–0.08
Skim milk	0.03–0.1	0.01–0.06	17–30
Buttermilk	2	0.03–0.18	10

Table 3.5 Vitamin A activity and β-carotene in milk of different breeds of cows

	Channel Island breeds		Non-Channel Island breeds	
	Summer	Winter	Summer	Winter
Retinol (μl l^{-1})	649	265	619	412
β-Carotene (μl l^{-1})	1,143	266	315	105
Retinol/β-carotene ratio	0.6	11.0	2.0	4.0
Contribution (%) of β-carotene to vitamin A activity	46.8	33.4	20.3	11.4

Modified from (Cremin and Power 1985)

Cholesterol (Appendix C) is the principal sterol in milk (>95 % of total sterols); the level (~0.3 %, w/w, of total lipids) is low compared with many other foods. Most of the cholesterol is in the free form, with <10 % as cholesteryl esters. Several other sterols, including steroid hormones, occur at trace levels.

Several hydrocarbons occur in milk in trace amounts. Of these, carotenoids are the most significant. In quantitative terms, carotenes occur at only trace levels in milk (typically ~200 μg/l) but they contribute 10–50 % of the Vitamin A activity in milk (Table 3.5) and are responsible for the yellow colour of milk fat. The carotenoid content of milk varies with breed (milk from Channel Island breeds contains two to three times as much β-carotene as milk from other breeds) and very markedly with season (Fig. 3.4). The latter reflects differences in the carotenoid content of the diet (since they are derived totally from the diet); fresh pasture, especially if it is rich in clover and alfalfa, is much richer in carotenoids than hay or silage (due to oxidation on conservation) or cereal-based concentrates. The higher the carotenoid content of the diet, the more yellow will be the colour of milk and milk fat, e.g., butter from cows on pasture is yellower than that from cows on winter feed, especially if the pasture is rich in clover (New Zealand butter is more yellow than Irish butter which in turn is more yellow than mainland European or US butter). Sheep, goats and buffaloes do not transfer carotenoids to their milks which are, consequently, much whiter than bovine milk. This may reduce the acceptability of dairy products (e.g., cheeses, butter, cream, ice cream) made from bovine milk in regions where goat sheep or buffalo milk is traditional (the carotenoids may be bleached by using peroxides, e.g., H_2O_2 or benzoyl peroxide, or masked, e.g., with chlorophyll or titanium oxide).

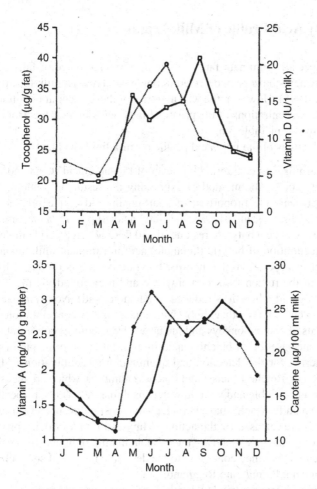

Fig. 3.4 Seasonal variations in the concentration of β-carotene (*open diamond*) and of Vitamins A (*closed triangle*), D (*open circle*) and E (*empty square*) in milk and milk products (from Cremin and Power 1985)

Milk contains significant concentrations of fat-soluble vitamins (Table 3.5; Fig. 3.4) and milk and dairy products make a significant contribution to the dietary requirements for these vitamins in western countries. The actual form of the fat-soluble vitamins in milk appears to be uncertain and their concentration varies widely with breed of animal, feed and stage of lactation, e.g., the Vitamin A activity of colostrum is ~30 times higher than that of mature milk.

Several prostaglandins occur in milk but it is not known whether they play a physiological role; they may not survive storage and processing in a biologically active form. Human milk contains prostaglandins E and F at concentrations 100-fold higher than human plasma and these may have a physiological function, e.g., gut motility.

3.4   Fatty Acid Profile of Milk Lipids

Milk fat, especially ruminant fats, contain a very wide range of fatty acids: >400 and 184 fatty acids have been detected in bovine and human milk fats, respectively (see Christie 1995). However, the vast majority of these occur at only trace concentrations. The concentrations of the principal fatty acids in milk fats from a range of species are shown in Table 3.6.

Notable features of the fatty acid profile of milk lipids include:

1. Ruminant milk fats contain a high level of butanoic acid ($C_{4:0}$) and other short chain fatty acids. The method of expressing the results in Table 3.6 (%, w/w) under-represents the proportion of short-chain acids—if expressed as mol%, butanoic acid represents ~10 % of all fatty acids (up to 15 % in some samples), i.e., there could be a butyrate residue in ~30 % of all triglyceride molecules. The high concentration of butyric (butanoic) acid in ruminant milk fats arises from the direct incorporation of β-hydroxybutyrate (which is produced by microorganisms in the rumen from carbohydrate and transported via the blood to the mammary gland where it is reduced to butanoic acid). Non-ruminant milk fats contain no butanoic or other short-chain acids; the low concentration of butyrate in milk fats of some monkeys and the brown bear requires confirmation.

   The concentration of butanoic acid in milk fat is the principle of the widely used criterion for the detection and quantitation of adulteration of butter with other fats, i.e., Reichert Meissl and Polenski numbers, which are measures of the volatile water-soluble and volatile water-insoluble fatty acids, respectively.

   Short chain fatty acids have strong, characteristic flavours and aromas. When these acids are released by the action of lipases in milk or dairy products, they impart strong flavours which are undesirable in milk or butter (they cause hydrolytic rancidity) but they contribute to the desirable flavour of some cheeses, e.g., Blue, Romano, Parmigiano Reggiano.

2. Ruminant milk fats contain low levels of polyunsaturated fatty acids (PUFAs) in comparison with monogastric milk fats. This is because a high proportion of the fatty acids in monogastric milk fats is derived from dietary lipids (following digestion and absorption) via blood. Unsaturated fatty acids in the diet of ruminants (grass contains considerable levels of PUFAs) are hydrogenated by rumen microorganisms unless protected by encapsulation (see Sect. 3.16.1). The low levels of PUFAs in bovine milk fat is considered to be nutritionally undesirable.

3. The milk fat from marine mammals contain high levels of long chain, highly unsaturated fatty acids, presumably reflecting the requirement that the lipids of these species remain liquid at the low temperature of their environments.

4. Ruminant milk fats are also rich in medium chain fatty acids. These are synthesized in the mammary gland via the usual malonyl CoA pathway (see Sect. 3.5) and are released from the synthesizing enzyme complex by thioacylases; presumably, the higher levels of medium chain acids in ruminant milk fats compared with those of monogastric animals reflect higher thioacylase activity in the mammary tissue of the former.

5. The fatty acid profile of bovine milk fat shows a marked seasonal pattern, especially when cows are fed on pasture in summer. Data for Irish milk fat are shown in Fig. 3.5;

Table 3.6 Principal fatty acids (wt% of total) in milk triacylglycerols or total lipids from various species (from Christie 1995)

Species	4:0	6:0	8:0	10:0	12:0	14:0	16:0	16:1	18:0	18:1	18:2	18:3	C$_{20}$-C22
Cow	3.3	1.6	1.3	3.0	3.1	9.5	26.3	2.3	14.6	29.8	2.4	0.8	T
Buffalo	3.6	1.6	1.1	1.9	2.0	8.7	30.4	3.4	10.1	28.7	2.5	2.5	T
Sheep	4.0	2.8	2.7	9.0	5.4	11.8	25.4	3.4	9.0	20.0	2.1	1.4	–
Goat	2.6	2.9	2.7	8.4	3.3	10.3	24.6	2.2	12.5	28.5	2.2	–	–
Musk-ox	T	0.9	1.9	4.7	2.3	6.2	19.5	1.7	23.0	27.2	2.7	3.0	0.4
Dall-sheep	0.6	0.3	0.2	4.9	1.8	10.6	23.0	2.4	15.5	23.1	4.0	4.1	2.6
Moose	0.4	T	8.4	5.5	0.6	2.0	28.4	4.3	4.5	21.2	20.2	3.7	–
Blackbuck antelope	6.7	6.0	2.7	6.5	3.5	11.5	39.3	5.7	5.5	19.2	3.3		
Elephant	7.4	–	0.3	29.4	18.3	5.3	12.6	3.0	0.5	17.3	3.0	0.7	–
Human	–	T	T	1.3	3.1	5.1	20.2	5.7	5.9	46.4	13.0	1.4	T
Monkey (mean of 6 species)	0.4	0.6	5.9	11.0	4.4	2.8	21.4	6.7	4.9	26.0	14.5	1.3	–
Baboon	–	0.4	5.1	7.9	2.3	1.3	16.5	1.2	4.2	22.7	37.6	0.6	–
Lemur macaco	–	–	0.2	1.9	10.5	15.0	27.1	9.6	1.0	25.7	6.6	0.5	–
Horse	–	T	1.8	5.1	6.2	5.7	23.8	7.8	2.3	20.9	14.9	12.6	–
Pig	–	–	–	0.7	0.5	4.0	32.9	11.3	3.5	35.2	11.9	0.7	–
Rat	1.1	7.0	7.5	8.2	22.6	1.9	6.5	26.7	16.3	0.8	1.1	1.1	7.0
Guinea-pig	–	T	–	–	–	2.6	31.3	2.4	2.9	33.6	18.4	5.7	T
Marmoset	–	–	–	8.0	8.5	7.7	18.1	5.5	3.4	29.6	10.9	0.9	7.0
Rabbit	–	T	22.4	20.1	2.9	1.7	14.2	2.0	3.8	13.6	14.0	4.4	T
Cottontail rabbit	–	–	9.6	14.3	3.8	2.0	18.7	1.0	3.0	12.7	24.7	9.8	0.4
European hare	–	T	10.9	17.7	5.5	5.3	24.8	5.0	2.9	14.4	10.6	1.7	T
Mink	–	–	–	–	0.5	3.3	26.1	5.2	10.9	36.1	14.9	1.5	–
Chinchilla	–	–	–	–	T	3.0	30.0	–	–	35.2	26.8	2.9	–
Red kangaroo	–	–	–	–	0.1	2.7	31.2	6.8	6.3	37.2	10.4	2.1	0.1

(continued)

Table 3.6 (continued)

Species	4:0	6:0	8:0	10:0	12:0	14:0	16:0	16:1	18:0	18:1	18:2	18:3	C_{20}-C22
Platypus	–	–	–	–	–	1.6	19.8	13.9	3.9	22.7	5.4	7.6	12.2
Numbat	–	–	–	–	0.1	0.9	14.1	3.4	7.0	57.7	7.9	0.1	0.2
Bottle-nosed dolphin	–	–	–	–	0.3	3.2	21.1	13.3	3.3	23.1	1.2	0.2	17.3
Manatee	–	–	0.6	3.5	4.0	6.3	20.2	11.6	0.5	47.0	1.8	2.2	0.4
Pygmy sperm whale	–	–	–	–	–	3.6	27.6	9.1	7.4	46.6	0.6	0.6	4.5
Harp seal	–	–	–	–	–	5.3	13.6	17.4	4.9	21.5	1.2	0.9	31.2
Northern elephant seal	–	–	–	–	–	2.6	14.2	5.7	3.6	41.6	1.9	–	29.3
Polar bear	–	T	–	T	0.5	3.9	18.5	16.8	13.9	30.1	1.2	0.4	11.3
Grizzly bear	–	T	–	–	0.1	2.7	16.4	3.2	20.4	30.2	5.6	2.3	9.5

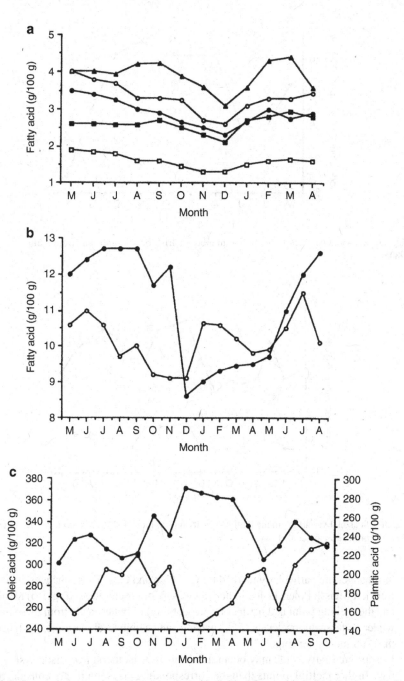

Fig. 3.5 Seasonal changes in the concentration of individual fatty acids in Irish bovine milk fat: (a) $C_{4:0}$ (*filled triangle*), $C_{6:0}$ (*filled square*), $C_{8:0}$ (*empty square*), $C_{10:0}$ (*filled circle*), $C_{12:0}$ (*open circle*) (b) $C_{14:0}$ (*open circle*), $C_{18:0}$ (*filled circle*) (c) $C_{16:0}$ (*filled circle*), $C_{18:1}$ (*open circle*) (from Cullinane et al. 1984a)

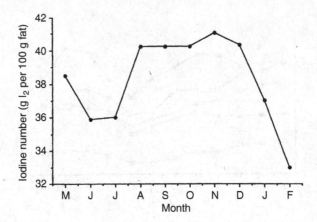

Fig. 3.6 Seasonal changes in the iodine number of Irish bovine milk fat (from Cullinane et al. · 1984a)

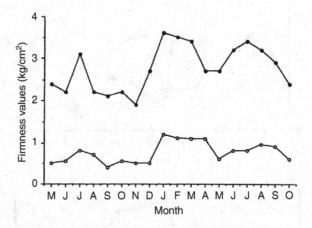

Fig. 3.7 Seasonal variations in the hardness of Irish milk fat at 4 °C (*filled circle*) or 15 °C (*open circle*) (from Cullinane et al. 1984b)

the changes are particularly marked for $C_{4:0}$, $C_{16:0}$ and $C_{18:1}$. These changes affect the Reichert Meissl, Polenski and iodine (a measure of unsaturation) (Fig. 3.6) numbers and the melting point and hardness (spreadability) of butter made from these milks: winter butter, with low levels of $C_{4:0}$ and $C_{18:1}$ and a high level of $C_{16:0}$ is much harder than summer butter (Fig. 3.7).

6. Unsaturated fatty acids may occur as *cis* or *trans* isomers; *trans* isomers, which have higher melting points than the corresponding *cis* isomers, are considered to be nutritionally undesirable. Bovine milk fat contains a low level (5 %) of *trans* fatty acids in comparison with chemically hydrogenated (hardened) vegetable oils, in which the value may be 50 % due to non-stereospecific hydrogenation.

7. Bovine milk fat contains low concentrations of keto and hydroxy acids (each at ~0.3 % of total fatty acids). The keto acids may have the carbonyl group ($C=O$) at various positions. The 3-keto acids give rise to methyl ketones on heating (high concentrations of methyl ketones are produced in blue cheeses through the oxidative activity of *Penicillium roqueforti*). The position of the hydroxy group on the hydroxy acids also varies; some can form lactones, e.g., the 4 and 5 hydroxy acids can form γ and δ lactones, respectively.

Lactones have strong flavours; traces of δ-lactones are found in fresh milk and contribute to the flavour of milk fat but higher concentrations may occur in dried milk or butter oil as a result of heating or prolonged storage and may cause atypical flavours.

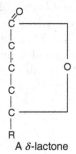

A δ-lactone

The fatty acids in the various polar lipids and cholesteryl esters are long-chain, saturated or unsaturated acids, with little or no acids $< C_{12:0}$ (Table 3.7; for further details see Christie 1995).

Table 3.7 The fatty acid composition of the cholesteryl ester, phosphatidylcholine and phosphatidylethanolamine in the milks of some species

Fatty acid composition (wt% of the total)

Fatty acid	Cow			Human			Pig		Mink			Mouse	
	CE	PC	PE	CE	PC	PE	PC	PE	CE	PC	PE	PC	PE
12:0	0.2	0.3	0.1	3.2	–	–			0.3	–	–	–	–
14:0	2.3	7.1	1.0	4.8	4.5	1.1	1.8	0.4	1.1	1.3	0.8	–	4.5
16:0	23.1	32.2	11.4	23.8	33.7	8.5	39.9	12.4	25.4	26.4	20.6	20.3	8.9
16:1	8.8	3.4	2.7	1.5	1.7	2.4	6.3	7.3	4.4	1.1	1.2	–	2.7
18:0	10.6	7.5	10.3	8.0	23.1	29.1	10.3	12.3	14.7	20.8	29.3	30.0	18.0
18:1	17.1	30.1	47.0	45.7	14.0	15.8	21.8	36.2	35.7	31.7	27.8	13.9	19.8
18:2	27.1	8.9	13.5	12.4	15.6	17.7	15.9	17.8	13.5	17.4	19.1	22.8	17.2
18:3	4.2	1.4	2.3	T	1.3	4.1	1.5	1.9	2.6	2.2	0.5	–	–
20:3	0.7	1.0	1.7	–	2.1	3.4	0.3	0.7		–	–	–	–
20:4	1.4	1.2	2.7	T	3.3	12.5	1.3	6.6		–	–	8.9	20.0
22:6	–	–	0.1	–	0.4	2.6	0.2	1.6		–	–	1.8	6.3

CE cholesteryl esters, *PC* phosphatidylcholine, *PE* phosphatidylethanolamine, *T* trace amount (from Christie 1995)

3.5   Synthesis of Fatty Acids in Milk Fat

In non-ruminants, blood glucose is the principal precursor of fatty acids in milk fat; the glucose is converted to acetyl CoA in the mammary gland. In ruminants, acetate and β-hydroxybutyrate, produced by microorganisms in the rumen and transported *via* the blood, are the principal precursors; in fact, ruminant mammary tissue has low "ATP citrate lyase" activity which is required for fat synthesis from glucose. Blood glucose is low in ruminants and is conserved for lactose synthesis. The differences in fatty acid precursors are reflected in marked interspecies differences in milk fatty acid profiles. Restriction of roughage in the diet of ruminants leads to suppression of milk fat synthesis, possibly through a reduction in the available concentrations of acetate and β-hydroxybutyrate.

In all species, the principal precursor for fatty acid synthesis is acetyl CoA, derived in non-ruminants from glucose and in ruminants from acetate or oxidation of β-hydroxybutyrate. Acetyl CoA is first converted, in the cytoplasm, to malonyl CoA:

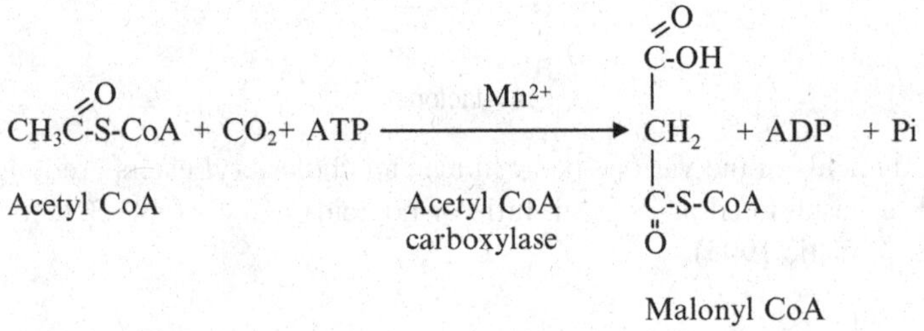

$$
\underset{\text{Acetyl CoA}}{CH_3\overset{\overset{\displaystyle O}{\|}}{C}\text{-S-CoA}} + CO_2 + ATP \quad \xrightarrow[\underset{\text{carboxylase}}{\text{Acetyl CoA}}]{\quad Mn^{2+} \quad} \quad \underset{\text{Malonyl CoA}}{\overset{\displaystyle \overset{O}{\|}}{\underset{\displaystyle \overset{|}{\underset{O}{C\text{-S-CoA}}}}{\underset{|}{\overset{\overset{\displaystyle O}{\|}}{C}\text{-OH}}{CH_2}}}} + ADP + Pi
$$

A reduced supply of bicarbonate (source of CO_2) depresses fatty acid synthesis.

Some β-hydroxybutyrate is reduced to butyrate and incorporated directly into milk fat; hence, the high level of this acid in ruminant milk fat.

In non-ruminants, the malonyl CoA is combined with an "acyl carrier protein" (ACP) which is part of a six-enzyme complex (molecular weight ~500 kDa) located in the cytoplasm. All subsequent steps in fatty acid synthesis occur attached to this complex; through a series of steps and repeated cycles, the fatty acid is elongated by two carbon units per cycle (Fig. 3.8, see Lehninger et al. 1993; Palmquist 2006).

The net equation for the synthesis of a fatty acid is:

$$n \text{ acetyl CoA} + 2(n-1)\text{ NADPH} + 2(n-1)H^+ + (n-1)\text{ ATP} + (n-1)\text{ CO}_2 \rightarrow$$

$$CH_3CH_2(CH_2CH_2)_{n-1}CH_2\overset{\overset{\displaystyle O}{\|}}{C}\text{-oA} + (n-1)\text{ CoA} + (n-1)\text{ ADP} + (n-1)\text{ P}_i$$

$$+ 2(n-1)\text{ NADP} + (n-1)\text{ CO}_2$$

The large supply of NADPH required for the above reactions is obtained through the metabolism of glucose-6-P *via* the pentose pathway.

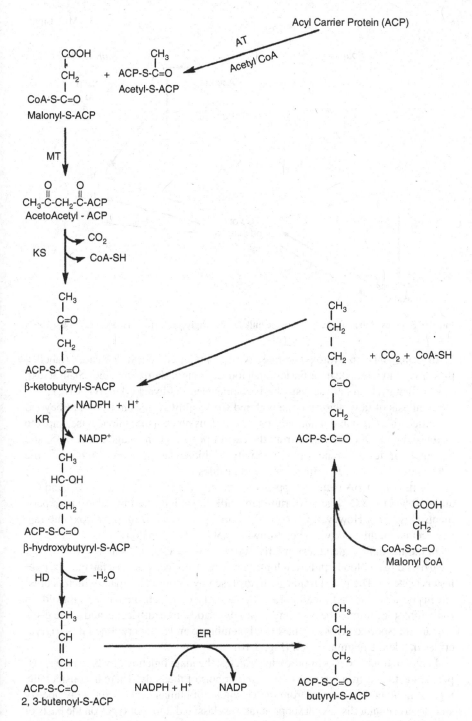

Fig. 3.8 One complete cycle and the first step in the next cycle of the events during the synthesis of fatty acids. *ACP* acyl carrier protein, a complex of 6 enzymes: i.e., *AT* acetyl CoA-ACP trans-acetylase, *MT* malonyl-CoA-ACP transferase, *KS* β-keto-ACP synthase, *KR* β-ketoacyl-ACP reductase, *HD* β-hydroxyacyl-ACP-dehydrase, *ER* enoyl-ACP reductase

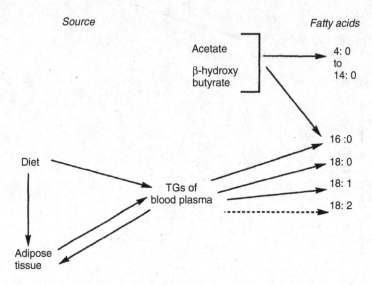

Fig. 3.9 Sources of the fatty acids in bovine milk fat. *TG* triglyceride (from Hawke and Taylor 1995)

In ruminants, β-hydroxybutyrate is the preferred chain initiator (labelled β-hydroxybutyrate appears as the terminal four carbons of short to medium chain acids), i.e., the first cycle in fatty acid synthesis commences at β-hydroxybutyryl-S-ACP.

Synthesis of fatty acids *via* the malonyl CoA pathway does not proceed beyond palmitic acid ($C_{16:0}$) and mammary tissue contains an enzyme, thioacylase, capable of releasing the acyl fatty acid from the carrier protein at any stage between C_4 and C_{16}. Interspecies differences in the activity of thioacylase account for some of the interspecies differences in milk fatty acid profiles.

The malonyl CoA pathway appears to account for 100 % of the C_{10}, C_{12} and C_{14} and ~50 % of the $C_{16:0}$ acids in ruminant milk fat as indicated by labelling experiments (Fig. 3.9). However, C_4, C_6 and C_8 are synthesized from β-hydroxybutyrate and acetate mainly *via* two other pathways not involving malonyl CoA.

In the mammary gland, essentially 100 % of $C_{18:0}$, $C_{18:1}$, $C_{18:2}$ and ≈50 % of C_{16} are derived from blood lipids (chylomicrons, free triglycerides, free fatty acids, cholesteryl esters). The blood lipids are hydrolysed by lipoprotein lipase which is present in the alveolar blood capillaries, the activity of which increases eightfold on initiation of lactation. The resulting monoglycerides, free fatty acids and some glycerol are transported across the basal cell membrane and re-incorporated into triglycerides inside the mammary cell (Fig. 3.10).

In blood, lipids exist as lipoprotein particles, the main function of which is to transport lipids to and from various tissues and organs of the body. There is considerable interest in blood lipoproteins from the viewpoint of human health, especially obesity and cardiovascular diseases. Lipoproteins are classified into four types on the basis of density, which is essentially a function of their triglyceride content, i.e., chylomicrons, very low density lipoprotein particles (VLDL), low density lipoprotein (LDL) particles and high density lipoprotein (HDL) particles, containing ~98, 90, 77 and 45 % total lipid, respectively (Fig. 3.11).

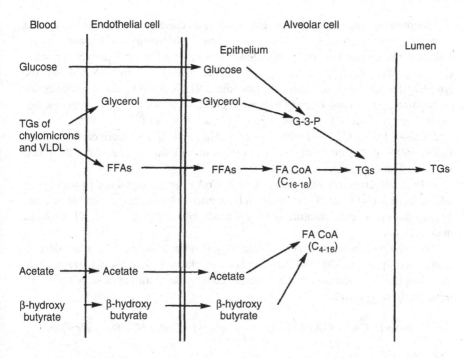

Fig. 3.10 Uptake of blood constituents by the mammary gland. *CoA* coenzyme A, *G-3-P* glycerol-3-phosphate, *FFA* free fatty acid, *FA* fatty acid, *TG* triglyeride, *VLDL* very low density lipoprotein (from Hawke and Taylor 1995)

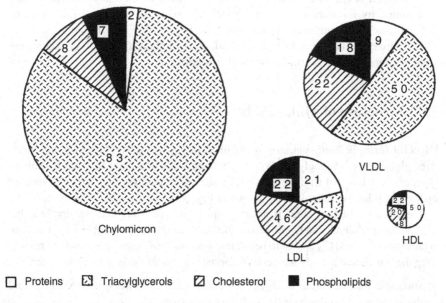

Fig. 3.11 Composition (%) of human serum lipoproteins. *VLDL* very low density lipoproteins, *LDL* low density lipoproteins, *HDL* high density lipoproteins

Lipoproteins, especially chylomicrons, are at an elevated level in the blood after eating, especially after a high-fat meal and give blood serum a milky appearance. They are also elevated during or after tension (so-called Racing Driver Syndrome). Chylomicrons, which are formed in the intestinal mucosa, are secreted into the lymph and enter the blood via the thoracic duct. VLDL lipoproteins are synthesised in intestinal mucosa and liver. LDL lipoproteins are formed at various sites, including mammary gland, by removing of triglycerides from VLDL.

Since ~50 % of $C_{16:0}$ and 100 % of $C_{18:0}$, $C_{18:1}$ and $C_{18:2}$ are derived from blood lipids, ~50 % of the total fatty acids in ruminant milk fat originate from the blood via diet or other organs.

In liver mitochondria, palmitic acid, as its CoA ester, is lengthened by successive additions of acetyl CoA. There is also a liver microsomal enzyme capable of elongating saturated and unsaturated fatty acids by addition of acetyl CoA or malonyl CoA.

The principal monoenoic acids, oleic ($C_{18:1}$) and palmitoleic ($C_{16:1}$), are derived from blood lipids but ~30 % of these acids are produced by microsomal enzymes (in the endoplasmic reticulum) in the secretory cells by desaturation of stearic and palmitic acids, respectively:

$$stearyl-CoA + NADPH + O_2 \xrightarrow{desaturase} oleoyl-CoA + NADP^+ + 2H_2O$$

Shorter chain unsaturated acids ($C_{10:1}$ to $C_{14:1}$) are probably also produced by the same enzyme.

Linoleic ($C_{18:2}$) and linolenic ($C_{18:3}$) acids cannot be synthesised by mammals and must be supplied in the diet, i.e., they are essential fatty acids (linoleic is the only true essential acid). These two polyenoic acids may be elongated and/or further desaturated by mechanisms similar to stearic → oleic, to provide a full range of polyenoic acids. A summary of these reactions is given in Fig. 3.12a, b.

δ-Hydroxy acids are produced by δ-oxidation of fatty acids and β-keto acids may arise from incomplete syntheses or via β-oxidation.

3.6 Structure of Milk Lipids

Glycerol for milk lipid synthesis is obtained in part from hydrolyzed blood lipids (free glycerol and monoglycerides), partly from glucose and a little from free blood glycerol. Synthesis of triglycerides within the cell is catalysed by enzymes located on the endoplasmic reticulum, as shown in Fig. 3.13.

Esterification of fatty acids is not random: C_{12}-C_{16} are esterified principally at the sn-2 position while C_4 and C_6 are principally at the sn-3 position (Table 3.8). The concentrations of C_4 and C_{18} appear to be the rate-limiting fatty acids because of the need to keep the lipid liquid at body temperature. Some features of the structures are notable:

1. Butanoic and hexanoic acids are esterified almost entirely, and octanoic and decanoic acids predominantly, at the sn-3 position.

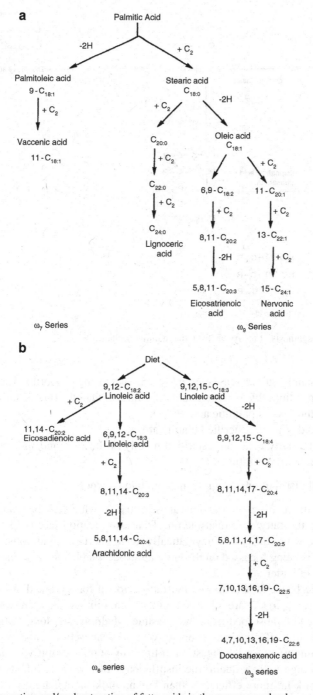

Fig. 3.12 Elongation and/or desaturation of fatty acids in the mammary gland

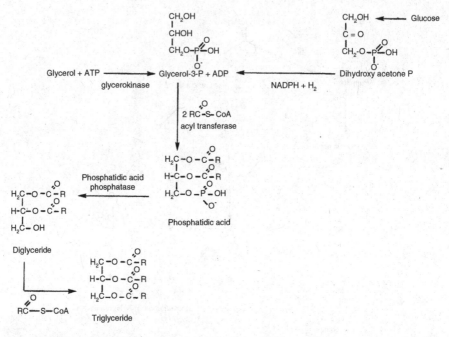

Fig. 3.13 Biosynthesis of triglycerides in the mammary gland

2. As the chain length increases up to $C_{16:0}$, an increasing proportion is esterified at the *sn*-2 position; this is more marked for human than for bovine milk fat, especially in the case of palmitic acid ($C_{16:0}$).
3. Stearic acid ($C_{18:0}$) is esterified mainly at *sn*-1.
4. Unsaturated fatty acids are esterified mainly at the *sn*-1 and *sn*-3 positions, in roughly equal proportions.

Fatty acid distribution is significant from two viewpoints:

1. It affects the melting point and hardness of the fat which can be reduced by randomizing the fatty acid distribution. Transesterification can be performed by treatment with $SnCl_2$ or enzymatically under certain conditions; increasing attention is being focussed on the latter as an acceptable means of modifying the hardness of butter.
2. Pancreatic lipase is specific for the fatty acids at the *sn*-1 and *sn*-3 positions. Therefore, $C_{4:0}$ to $C_{8:0}$ are released rapidly from milk fat; these are water-soluble and are readily absorbed from the intestine. Medium- and long-chain acids are absorbed more effectively as 2-monoglycerides than as free acids; this appears to be quite important for the digestion of lipids by human infants who have limited ability to digest lipids due to the insufficiency of bile salts. Infants metabolize human milk fat more efficiently than bovine milk fat, apparently owing to the very high proportion of $C_{16:0}$ esterified at *sn*-2 in the former. The effect of transesterification on the digestibility of milk fat by infants merits investigation.

Table 3.8 Composition of fatty acids (mol% of the total) esterified to each position of the triacyl-sn-glycerols in the milks of various species

Fatty acid	Cow			Human			Rat			Pig			Rabbit			Seal			Echidna		
	sn-1	sn-2	sn-3	sn-1	sn-2	sn-3	sn-1	sn-2	sn-3	sn-1	sn-2	sn-3	sn-1	sn-2	sn-3	sn-1	sn-2	sn-3	sn-1	sn-2	sn-3
4:0	–	–	35.4	–	–	–	–	–	–	–	–	–	–	–	–	–	–	–	–	–	–
6:0	–	0.9	12.9	–	–	–	–	–	–	–	–	–	–	–	–	–	–	–	–	–	–
8:0	1.4	0.7	3.6	–	–	–	3.7	5.7	10.0	–	–	–	–	19.2	33.7	38.9	–	–	–	–	–
10:0	1.9	3.0	6.2	0.2	0.2	1.1	10.1	20.0	26.0	–	–	–	–	22.5	22.5	26.1	–	–	–	–	–
12:0	4.9	6.2	0.6	1.3	2.1	5.6	10.4	15.9	15.1	–	–	–	–	3.5	2.8	1.8	0.3	0.2	–	–	–
14:0	9.7	17.5	6.4	3.2	7.3	6.9	9.6	17.8	8.9	2.4	6.8	3.7	2.7	2.1	2.6	0.7	23.6	3.8	1.7	0.9	0.4
16:0	34.0	32.3	5.4	16.1	58.2	5.5	20.2	28.7	12.6	21.8	57.6	15.4	24.1	12.7	23.8	6.1	31.0	1.0	31.5	9.0	27.9
16:1	2.8	3.6	1.4	3.6	4.7	7.6	1.8	2.1	1.8	6.6	11.2	10.4	4.1	1.3	1.5	1.1	16.8	14.1	–	–	–
18:0	10.3	9.5	1.2	15.0	3.3	1.8	4.9	0.8	1.5	6.9	1.1	5.5	6.9	3.5	0.9	1.9	0.7	1.0	16.8	2.1	14.3
18:1	30.0	18.9	23.1	46.1	12.7	50.4	24.2	3.3	11.8	49.6	13.9	51.7	40.8	16.6	3.8	11.4	19.4	45.4	33.1	57.6	39.8
18:2	1.7	3.5	2.3	11.0	7.3	15.0	14.1	5.2	11.6	11.3	8.4	11.5	15.6	15.1	6.4	9.7	2.3	2.8	4.1	18.3	4.9
18:3	–	–	–	0.4	0.6	1.7	1.2	0.5	0.7	1.4	1.0	1.8	3.4	3.5	2.0	2.3	0.5	0.7	1.0	2.9	2.0
C_{20}-C_{22}	–	–	–	–	–	–	–	–	–	–	–	–	–	–		–	0.8	28.7	–	–	–

From Christie (1995)

3.7   Milk Fat as an Emulsion

In 1674, Van Leeuwenhoek reported that the fat in milk exists as microscopic globules. Milk is an oil-in-water emulsion, the properties of which have a marked influence on many properties of milk, e.g., colour, mouthfeel and viscosity. The globules in bovine milk range in diameter from ~0.1 to ~20 μm, with a mean of ~3.5 μm (the range and mean vary with breed and health of the cow, stage of lactation, etc). The size and size distribution of fat globules in milk may be determined by light microscopy, light scattering, e.g., using a Malvern Mastersizer or electronic counting devices, e.g., the Coulter counter. The frequency distribution of globule number and volume as a function of diameter for bovine milk are summarized in Fig. 3.14. Although small globules are very numerous (~75 % of all globules have a diameter <1 μm), they represent only a small proportion of total fat volume or mass. The number average diameter of the globules in milk is only ~0.8 μm. The mean fat globule size in milk from Channel Island breeds (Jersey and Guernsey) is larger than that in milk from other breeds (the fat content of the former milks is also higher) and the mean globule diameter decreases throughout lactation (Fig. 3.15).

Milk contains ~15×10^9 globules per ml, with a total interfacial area of 1.2–2.5 m^2 per g fat.

Example Assume a fat content of 4.0 %, w/v, with a mean globule diameter of 3 μm.

$$\text{Volume of typical globule} \ = \frac{4}{3} \, \Pi r^3$$

$$= \frac{4}{3} \times \frac{22}{7} \times \frac{(3)^3}{2} \, \mu m^3$$

$$\sim 14 \ \mu m^3$$

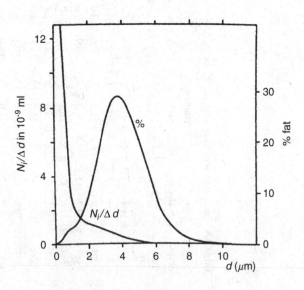

Fig. 3.14 Size distribution of the fat globules in bovine milk. N number per ml $\times 10^{-9}$, V volume (% of fat) (modified from Walstra and Jenness 1984)

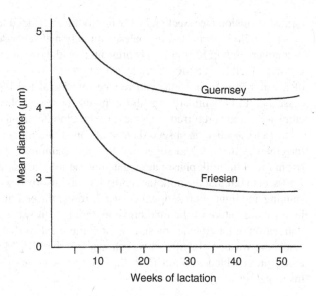

Fig. 3.15 Average diameter of the fat globules in milk of Guernsey or Friesian cows throughout lactation (from Walstra and Jenness 1984)

$$1 \text{ ml milk contains}: \quad 0.04 \text{ g fat}$$
$$= 4.4 \times 10^{10} \, \mu m^3$$

$$1 \text{ ml milk contains}: \quad \frac{4.4 \times 10^{10}}{14} \sim 3.14 \times 10^9 \text{ globules}$$

$$\text{Surface area of a typical globule} = 4 \Pi r^2$$
$$= 4 \times \frac{22}{7} \times \frac{9}{4} \, \mu m^2$$
$$= 28.3 \, \mu m^2$$

$$\text{Interfacial area per ml milk} = 28.3 \times \left(3.14 \times 10^9\right) \, \mu m^2$$
$$= 88.9 \times 10^9 \, \mu m^2$$
$$= 889 \text{ cm}^2 \approx 0.09 \text{ m}^2$$

$$\text{Interfacial area per g fat} = 88.9 \times 10^{-3} \times \frac{1}{0.04} \, m^2$$
$$= 2.22 \text{ m}^2$$

3.8   Milk Fat Globule Membrane

Lipids are insoluble in water and an interfacial tension therefore exists between the phases when lipids are dispersed (emulsified) in an aqueous medium (or *vice versa*). This tension *in toto* is very large, considering the very large interfacial area in a typical emulsion (see Sect. 3.7). Owing to the interfacial tension, the oil and water phases would be expected to coalesce quickly and separate. However, coalescence (but not creaming; Sect. 3.9.2) is prevented by the use of emulsifiers (surface active agents) which form a film around each fat globule (or each water droplet in the case of a water-in-oil emulsion) and reduce interfacial tension. In the case of unprocessed milk, the emulsifying film is much more complex than that in "artificial" emulsions, and is referred to as the milk fat globule membrane (MFGM).

In 1840, Ascherson observed an emulsion-stabilizing membrane surrounding the fat globules in milk and suggested that the membrane was "condensed" albumin (from the skim milk phase) aggregated at the fat/plasma interface. Babcock, in the 1880s, also felt that the milk fat emulsifier was adsorbed serum protein. Histological staining and light microscopy were used around the end of the nineteenth century to identify the nature of the membrane material but it was early recognised that contamination of fat globules by skim milk components presented a major problem. By analysing washed globules, it was shown that the MFGM contained phospholipids and protein which differed from the skim milk proteins [see Brunner (1974) for historical review)].

3.8.1 Isolation of the Fat Globule Membrane

The definition of what precisely constitutes the membrane leads to considerable difficulty and uncertainty. The outer boundary is assumed to constitute everything that travels with the fat globule when it moves slowly through milk; however, the outer regions of the membrane are loosely attached and some or all may be lost, depending on the extent of mechanical damage the globule suffers. The inner boundary is ill-defined and depends on the method of preparation; there is considerable discussion as to whether a layer of high melting point triglyceride, immediately inside the membrane, is part of the membrane or not. Some hydrophobic constituents of the membrane probably diffuse into the core of the globules while components of the plasma may adsorb at the outer surface. Since the membrane contains numerous enzymes, enzymatic changes may occur.

Several methods are available for isolating all or part of the membrane. The usual initial step involves separating a cream from milk by mechanical centrifugation (which may cause some damage) or by gravity. The cream is washed repeatedly (3–6 times) with water or dilute buffer by dilution and gravity separation; soluble salts and other small molecules are probably lost into the serum. Mechanical damage may remove the loosely-bound outer layers and may even cause some homogenisation and adsorption of serum constituents; small globules are lost during each washing cycle.

The washed cream is destabilized by churning or freezing or by using a detergent; then, the fat (mainly triglycerides) is melted and separated from the membrane material by centrifugation. Cross-contamination of membrane with core material may be considerable and methods must be carefully standardised. An elaborate scheme for the isolation and fractionation of the MFGM was developed by Brunner and co-workers (see Brunner 1974).

Table 3.9 Gross composition of the milk fat globule membrane

Component	mg/100 g fat globule	mg/m² fat globule surface	% (w/w) of total membrane
Protein	900	4.5	41
Phospholipid	600	3.0	27
Cerebrosides	80	0.4	3
Cholesterol	40	0.2	2
Neutral glycerides	300	1.5	14
Water	280	1.4	13
Total	2,200	11.0	100

From Mulder and Walstra (1974)

Treatment of washed cream with surfactants, usually sodium deoxycholate, releases part of the membrane, assumed to represent only the outer layer. Unless the treatment is carefully controlled, some inner material will be released also.

3.8.2   Gross Chemical Composition of FGM

Yields of 0.5–1.5 g MFGM/100 g fat have been reported; the range reflects variations in temperature history, washing technique, age, agitation, etc. The gross chemical composition of the membrane is reasonably well established and the relatively small differences reported are normally attributed to different methods used to isolate and fractionate the membrane material. The data in Table 3.9, from Mulder and Walstra (1974) and based on the investigations of many workers, give a reasonable estimate of the gross composition of the MFGM. A more detailed compositional analysis is provided by Keenan et al. (1983) (Table 3.10). Brunner (1965, 1974), Mulder and Walstra (1974a, b), Patton and Keenan (1975), Keenan et al. (1983), Keenan and Dylewski (1995) or Keenan and Mathur (2006) should be consulted for more detailed information on compositional.

3.8.3   The Protein Fraction

Depending on the preparative method used, the membrane may or may not contain skim milk proteins (i.e., caseins and whey proteins); if the membrane has been damaged prior to isolation, it may contain considerable amounts of these proteins. The membrane contains unique proteins which do not occur in the skim milk phase. Many of the proteins are glycoproteins and contain a considerable amount of carbohydrate (hexose, 2.8–4.15 %; hexosamine, 2.5–4.2 % and sialic acid, 1.3–1.8 %).

Sodium dodecylsulphate-polyacrylamide gel electrophoresis (SDS-PAGE), with silver staining of the gels, resolves MFGM proteins into as many as 60 discrete bands, ranging in molecular mass from 11 to 250 kDa (see Keenan and Dylewski 1995; Mather 2000; Keenan and Mathur 2006). Most of these proteins are present

Table 3.10 Composition of bovine milk fat globule membranes

Constituent class	Amount
Protein	25–60 % of dry weight
Total lipid	0.5–1.2 mg per mg protein
Phospholipid	0.13–0.34 mg per mg protein
Phosphatidyl choline	34 % of total lipid phosphorus
Phosphatidyl ethanolamine	28 % of total lipid phosphorus
Sphingomyelin	22 % of total lipid phosphorus
Phosphatidyl inositol	10 % of total lipid phosphorus
Phosphatidyl serine	6 % of total lipid phosphorus
Neutral lipid	56–80 % of total lipid
Hydrocarbons	1.2 % of total lipid
Sterols	0.2–5.2 % of total lipid
Sterol esters	0.1–0.8 % of total lipid
Glycerides	53–74 % of total lipid
Free fatty acids	0.6–6.3 % of total lipid
Cerebrosides	3.5 nmol per mg protein
Gangliosides	6–7.4 nmol sialic acid per mg protein
Total sialic acids	63 nmol per mg protein
Hexoses	0.6 μmol per mg protein
Hexosamines	0.3 μmol per mg protein
Cytochrome b_5 + P-420	30 pmol per mg protein
Uronic acids	99 ng per mg protein
RNA	20 μg per mg protein

From Keenan et al. (1983)

at very low concentrations (many are detectable only when gels are stained with silver but not with Coomassie blue). Some of these proteins may be genetic variants and since the MFGM contains a plasmin-like proteinase, some of the smaller polypeptides may be fragments of larger proteins. The three principal proteins, with MWs (by SDS-PAGE) of 155, 67 and 48 kDa, are xanthine oxidoreductase, butyrophilin and glycoprotein B, respectively; 5 or 6 glycoproteins have been detected by staining with Schiff's reagent. Xanthine oxidoreductase, which requires Fe, Mo and FAD as co-factors, is capable of oxidizing lipids via the production of superoxide radicals (see Chap. 10). It represents ~20 % of the MFGM protein and part is readily lost from the membrane, e.g., on cooling; isoelectrofocusing indicates at least 4 variants with isoelectric points (pI) in the range 7.0–7.5.

Butyrophilin, the principal MFGM protein and so-named because of its high affinity for milk lipids, is a very hydrophobic, difficult-to-solubilize (insoluble or only sparingly soluble in most protein solvents, including detergents) glycoprotein. Isoelectric focussing indicates at least 4 variants (pI 5.2–5.3). The amino acid sequence of butyrophilin has been determined and its gene has been cloned, which indicates that butyrophilin is synthesised with a leader sequence; it consists of 526 amino acids and has a molecular mass, without carbohydrate, of 56,460 Da. It tenaciously binds phospholipids and perhaps even contains covalently bound fatty acids.

It is located only at the apical cell surface of the mammary epithelial cells, suggesting a role in membrane envelopment of fat globules.

Several of the minor proteins of the MFGM have been isolated and partially characterized (see Keenan and Dylewski 1995; Keenan and Mathur 2006). A systematic nomenclature for the MFGM proteins based on their relative electrophoretic mobility on SDS-PAGE was proposed by Mather (2000). The proteins of the MFGM represent approximately 1 % of the total proteins in milk.

3.8.4   The Lipid Fraction

The membrane contains 0.5–1.0 % of the total lipid in milk and is composed principally of phospholipids and neutral lipids in the approximate ratio 2:1, with lesser amounts of other lipids (Tables 3.9 and 3.10); contamination with core lipid is a major problem. The phospholipids are principally phosphatidyl choline, phosphatidyl ethanolamine and sphingomyelin in the approximate ratio 2:2:1. The principal fatty acids in the phospholipids are $C_{14:0}$ (~5 %), $C_{16:0}$ (~25 %), $C_{18:0}$ (~14 %), $C_{18:1}$ (~25 %), $C_{18:2}$ (~9 %), $C_{22:0}$ (~3 %) and $C_{24:0}$ (~3 %). Thus, the membrane contains a significantly higher level of polyunsaturated fatty acids than milk fat generally and is, therefore, more susceptible to oxidation. The cerebrosides are rich in very long-chain fatty acids which possibly contribute to membrane stability. The membrane contains several glycolipids (Table 3.11).

The amount and nature of the neutral lipids present in the MFGM are uncertain because of the difficulty in defining precisely the inner limits of the membrane. It is generally considered to consist of 83–88 % triglyceride, 5–14 % diglyceride and 1–5 % free fatty acids. The level of diglyceride is considerably higher than in milk fat as a whole; diglycerides are relatively polar and are, therefore, surface-active. The fatty acids of the neutral lipid fraction are longer-chained than milk fat as a whole and in order of the proportion present are palmitic, stearic, myristic, oleic and lauric.

Table 3.11 Structures of glycosphingolipids of bovine milk fat globule membrane (from Keenan et al. 1983)

Glycosphingolipid	Structure
Glucosyl ceramide	β-Glucosyl-$(1 \rightarrow 1)$-ceramide
Lactosyl ceramide	β-Glucosyl-$(1 \rightarrow 4)$-β-glucosyl-$(1 \rightarrow 1)$-ceramide
GM_3 (Hematoside)	Neuraminosyl-$(2 \rightarrow 3)$-galactosyl-glucosyl-ceramide
GM_2	N-acetylgalactosaminyl-(neuraminosyl)-galactosyl-glucosyl-ceramide
GM_1	Galactosyl-N-acetylgalactosaminyl-(neuraminosyl)-galactosyl-glucosyl-ceramide
GD_3 (Disialohematoside)	Neuraminosyl-$(2 \rightarrow 8)$-neuraminosyl-$(2 \rightarrow 3)$-galactosyl-glucosyl-ceramide
GD_2	N-acetylgalactosaminyl-(neuraminosyl-neuraminosyl)-galactosyl-glucosyl-ceramide
GD_{1b}	Galactosyl-N-acetylgalactosaminyl-(neuraminosyl-neuraminosyl)-galactosyl-glucosyl-ceramide

Table 3.12 Enzymatic activities detected in bovine milk lipid globule membrane preparations from Keenan and Dylewski (1995)

Enzyme	EC number
Lipoamide dehydrogenase	1.6.4.3
Xanthine oxidase	1.2.3.2
Thiol oxidase	1.8.3.2
NADH oxidase	1.6.99.3
NADPH oxidase	1.6.99.1
Catalase	1.11.1.6
γ-Glutamyl transpeptidase	2.3.2.1
Galactosyl transferase	2.4.1
Alkaline phosphatase	3.1.3.1
Acid phosphatase	3.1.3.2
N^1-Nucleotidase	3.1.3.5
Phosphodiesterase I	3.1.4.1
Inorganic pyrophosphatase	3.6.1.1
Nucleotide pyrophosphatase	3.6.1.9
Phosphatidic acid phosphatase	3.1.3.4
Adenosine triphosphatase	3.6.1.15
Cholinesterase	3.1.1.8
UDP-glycosyl hydrolase	3.2.1
Glucose-6-phosphatase	3.1.3.9
Plasmin	3.4.21.7
β-Glucosidase	3.2.1.21
β-Galactosidase	3.2.1.23
Ribonuclease I	3.1.4.22
Aldolase	4.1.2.13
Acetyl-CoA carboxylase	6.4.1.2

Most of the sterols and sterol esters, vitamin A, carotenoids and squalene in milk are dissolved in the core of the fat globules but some are probably present in the membrane.

3.8.5 Other Membrane Components

Trace Metals: The membrane contains 5–25 % of the indigenous Cu and 30–60 % of the indigenous Fe of milk as well as several other elements, e.g., Co, Ca, Na, K, Mg, Mn, Mo, Zn, Se at trace levels; many of these metals are constituents of enzymes, e.g., Zn and Mg in alkaline phosphatase, Fe and Mo in xanthine oxidoreductase, Fe in catalase and lactoperoxidase.

Enzymes: The MFGM contains many enzymes (Table 3.12). These enzymes originate from the cytoplasm and membranes of the secretory cell and are present in the MFGM due to the mechanism of globule excretion from the cells (see Sect. 3.8.7).

3.8.6   Membrane Structure

Several early attempts to describe the structure of the MFGM included King (1955), Hayashi and Smith (1965), Peereboom (1969), Prentice (1969) and Wooding (1971). Although the structures proposed by these workers were inaccurate, they stimulated thinking on the subject. Keenan and Dylewski (1995), Keenan and Patton (1995), Keenan and Mathur (2006) and Mather (2011) should be consulted for recent reviews.

Understanding of the structure of the MFGM requires understanding three processes: the formation of lipid droplets from triglycerides synthesized in or on the sacroplasmic reticulum at the base of the cell, movement of the droplets (globules) through the cell and excretion of the globules from the cell into the lumen of the alveolus.

The MFGM originates from regions of the apical plasma membrane, and also from endoplasmic reticulum (ER) and perhaps other intracellular compartments. That portion of the MFGM derived from the apical plasma membrane, termed the primary membrane, has a typical bilayer membrane appearance, with electron-dense material on the inner membrane face. The components derived from ER appear to be a monolayer of proteins and polar lipids which covers the triacylglycerol-rich core lipids of the globule before its secretion. This monolayer or coat material compartmentalizes the core lipid within the cell and participates in intracellular fusions through which droplets grow in volume. Constituents of this coat also may be involved in interaction of droplets with the plasma membrane.

Milk lipid globules originate as small lipid droplets in the ER. Lipids, presumed to be primarily triacylglycerols, appear to accumulate at focal points on or in the ER membrane. This accumulation of lipids may be due to localized synthesis at these focal points, or to accretion from dispersed or uniformly distributed biosynthetic sites. It has been suggested that triacylglycerols accumulate between the layers of the bilayer membrane and are released from the ER into the cytoplasm as droplets coated with the outer (cytoplasmic) half of the ER membrane. A cell-free system has been developed in which ER isolated from lactating mammary gland can be induced to release lipid droplets which resemble closely droplets formed *in situ* in both morphology and composition. In this cell-free system, lipid droplets were formed only when a fraction of cytosol with a MW greater than 10 kDa was included in the incubation mixture, suggesting that cytosolic factors are involved in droplet formation or release from ER.

By whatever mechanism they are formed, on or in, and released from the ER, milk lipid globule precursors first appear in the cytoplasm as droplets with diameters <0.5 µm, with a triglyceride-rich core surrounded by a granular coat material that lacks bilayer membrane structure, but which appears to be thickened, with tripartite-like structure, in some regions. These small droplets, named microlipid droplets, appear to grow in volume by fusing with each other. Fusions give rise to larger droplets, called cytoplasmic lipid droplets, with a diameter >1 µm.

Droplets of different density and a lipid to protein ratios ranging from about 1.5:1 to 40:1 have been isolated from bovine mammary gland. Triglycerides are the major lipid class in droplets of all sizes and represent an increasingly greater proportion of total droplet mass in increasingly less dense droplet preparations. Surface coat material of droplets contains cholesterol and the major phospholipid classes found in milk, i.e., sphingomyelin, phosphatidylcholine, phosphatidylethanolamine, phosphatidylinositol and phosphatidylserine.

SDS-PAGE shows that micro and cytoplasmic lipid droplets have complex and similar polypeptide patterns. Many polypeptides with electrophoretic mobilities in common with those of intracellular lipid droplets are present also in milk lipid globules. Some polypeptides of the MFGM and intracellular lipid droplets share antigenic reactivity. Taken together, current information suggests that lipid droplet precursors of milk lipid globules originate in the ER and retain at least part of the surface material of droplets during their secretion as milk fat globules. The protein and polar lipid coat on the surface of lipid droplets stabilizes the triglyceride-rich droplet core, preventing coalescence in the cytoplasm. Beyond a stabilization role, constituents of the coat material may participate also in droplet fusions and in droplet-plasma membrane interactions. If elements of the cytoskeleton function in guiding lipid droplets from their sites of origin to their sites of secretion from the cell, coat constituents may participate in interaction with filamentous or tubular cytoskeletal elements.

Within mammary epithelial cells, one mechanism by which lipid droplets can grow is by fusion of microlipid droplets. Microlipid droplets can also fuse with cytoplasmic lipid droplets, providing triacylglycerols for continued growth of larger droplets. The size range of lipid globules in milk can be accounted for, at least in part, by a droplet fusion-based growth process. Small milk fat globules probably arise from secretion of microlipid droplets which have undergone no or a few fusions while larger droplets can be formed by continued fusions with microlipid droplets.

While evidence favours the view that lipid droplets grow by fusion, there is no evidence as to how this process is regulated to control the ultimate size distribution of milk lipid globules. The possibility that fusion is purely a random event, regulated only by the probability of droplet–droplet contact before secretion, cannot be ruled out. Insufficient evidence is available to conclude that fusion of droplets is the sole or major mechanism by which droplets grow. Other possible mechanisms for growth, e.g., lipid transfer proteins which convey triglycerides from their site of synthesis to growing lipid droplets, cannot be excluded.

Available evidence indicates that lipid droplets migrate from their sites of origin, primarily in basal region of the cell, through the cytoplasm to apical cell regions. This process appears to be unique to the mammary gland and in distinct contrast to lipid transit in other cell types, where triacylglycerols are sequestered within ER and the Golgi apparatus and are secreted as lipoproteins or chylomicrons that are conveyed to the cell surface *via* secretory vesicles.

Mechanisms which guide the unidirectional transport of lipid droplets are not yet understood. Evidence for possible involvement of microtubules and microfilaments, elements of the cytoskeletal system, in guiding this transit has been obtained, but this evidence is weak and is contradictory in some cases. Cytoplasmic microtubules

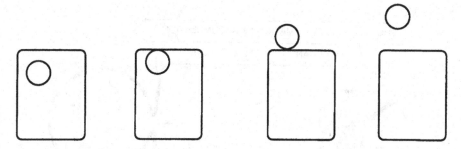

Fig. 3.16 Schematic representation of the excretion of a fat globule through the apical membrane of the mammary cell

are numerous in milk-secreting cells and the tubulin content of the mammary gland increases substantially prior to milk secretion. A general role for microtubules in the cytoplasm, and the association of proteins with force-producing properties with microtubules, provide a plausible basis for assuming the microtubules may be involved in lipid droplet translocation. Microfilaments, which are abundant in milk secreting cells, appear to be concentrated in apical regions.

3.8.7   Secretion of Milk Lipid Globules

The mechanism by which lipid droplets are secreted from the mammocyte was first described in 1959 by Bargmann and Knoop and has been confirmed by several investigators since (see Keenan and Dylewski 1995; Keenan and Mathur 2006). The lipid droplets are pushed through and become enveloped progressively by the apical membrane up to the point at which they are released from the cell, surrounded entirely by apical membrane (Fig. 3.16). Current concepts of the pathway by which lipid droplets originate, grow and are secreted are summarized diagrammatically in Fig. 3.17.

Lipid droplets associate with regions of the plasma membrane that are characterized by the appearance of electron-dense material on the cytoplasmic face of the membrane. Droplet surfaces do not contact the plasma membrane directly but rather the electron-dense cytoplasmic face material; which constituents of the latter recognize and interact with constituents on the droplet surface are not known. Immunological and biochemical studies have shown that butyrophilin and xanthine oxidoreductase, two of the principal proteins in the MFGM, are major constituents of the electron-dense material on the cytoplasmic face of apical plasma membrane. Butyrophilin, a hydrophobic, transmembrane glycoprotein that is characteristic of milk-secreting cells, is concentrated at the apical surface of these cells; it binds phospholipids tightly and is believed to be involved in mediating interaction between lipid droplets and apical plasma membrane. Xanthine oxidoreductase is distributed throughout the cytoplasm, but appears to be enriched at the apical cell surface.

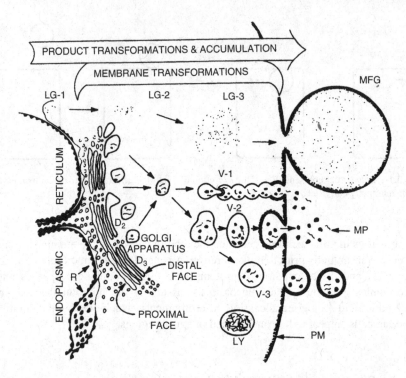

Fig. 3.17 Diagram summarizing the roles of components of the endo-membrane system of mammary epithelial cells in the synthesis and secretion of the constituents of milk. Intracellular lipid globules (LG-1, LG-2, LG-3). Lipid globules are discharged from the cell by progressive envelopment in regions of apical plasma membrane. MFG denotes a lipid globule being enveloped in plasma membrane. Milk proteins (MP) are synthesized on polysomes of endoplasmic reticulum and are transported, perhaps in small vesicles which bleb from endoplasmic reticulum, to dictyosomes (D_1, D_2, D_3) of the Golgi apparatus. These small vesicles may fuse to form the proximal cisterna of Golgi apparatus dictyosomes. Milk proteins are incorporated into secretory vesicles formed from cisternal membranes on the distal face of dictyosomes. Lactose is synthesized within cisternal luminae of the Golgi apparatus and is incorporated into secretory vesicles. Certain ions of milk are also present in secretory vesicles. Three different mechanisms for exocytotic interaction of secretory vesicle with apical plasma membrane have been described: (1) through the formation of a chain of fused vesicles (V-1); (2) by fusion of individual vesicles with apical plasma membrane (V-2), with integration of vesicle membrane into plasma membrane; (3) by direct envelopment of secretory vesicles in apical plasma membrane (V-3). Lysosomes (LY) may function in the degradation of excess secretory vesicle membrane (from Keenan et al. 1988)

In the secretion process, milk fat globules usually are enveloped compactly by apical plasma membrane but closure of the membrane behind the projecting fat droplet occasionally entrains some cytoplasm as a so-called crescent or signet between the membrane and the droplet surface. These crescents can vary from thin slivers of cellular material to situations in which the crescent represents a greater volume than does the globule core lipid. Except for nuclei, cytoplasmic crescents contain nearly all membranes and organelles of the milk-secreting cell.

Globule populations with a high proportion of crescents exhibit a more complex pattern of proteins (as indicated by SDS-PAGE) than low-crescent populations. Presumably, the many additional minor bands arise from cytoplasmic components in crescents. Crescents have been identified in association with the milk fat globules of all species examined to date but the proportion of globules with crescents varies between and within species; about 1 % of globules in bovine milk contain crescents.

Thus, the fat globules are surrounded, at least initially, by a membrane typical of eukaryotic cells. Membranes are a conspicuous feature of all cells and may represent 80 % of the dry weight of some cells. They serve as barriers separating aqueous compartments with different solute composition and as the structural base on which many enzymes and transport systems are located. Although there is considerable variation, the typical composition of membranes is ~40 % lipid and ~60 % protein. The lipids are mostly polar (nearly all the polar lipids in cells are located in the membranes), principally phospholipids and cholesterol in varying proportions. Membranes contain several proteins, perhaps up to 100 in complex membranes. Some of the proteins, referred to as extrinsic or peripheral, are loosely attached to the membrane surface and are easily removed by mild extraction procedures. The intrinsic or integral proteins, ~70 % of the total protein, are tightly bound to the lipid portion and are removed only by severe treatment, e.g., by SDS or urea.

Electron microscopy shows that membranes are 7–9 nm thick, with a trilaminar structure (a light, electron-sparse layer, sandwiched between two dark electron-dense layers, Fig. 3.18). The phospholipid molecules are arranged in a bi-layer structure (Fig. 3.18); the non-polar hydrocarbon chains are orientated inward where they "wriggle" freely and form a continuous hydrocarbon base; the hydrophilic

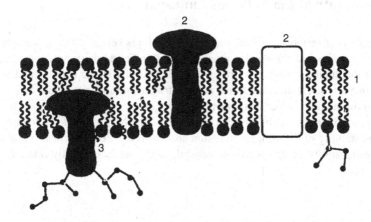

Fig. 3.18 Simple diagrammatic representation of the MFGM and schematic representation of a trilaminar cell membrane which is derived from the apical membrane of the mammary cell and forms the outer layer of the milk fat globule membrane following expression from the mammary cell, but which is more or less extensively lost on ageing. (1) Phospholipid/glycolipid, (2) protein, (3) glycoprotein. Modified from Mather (2011)

regions are orientated outward and are relatively rigid. In this bi-layer, individual lipid molecules can move laterally, endowing the bilayer with fluidity, flexibility, high electrical resistance and low permeability to polar molecules. Some of the globular membrane proteins are partially embedded in the membrane, penetrating into the lipid phase from either side, others are completely buried within it, while others transverse the membrane. The extent to which a protein penetrates into the lipid phase is determined by its amino acid composition, sequence, secondary and tertiary structure. Thus, membrane proteins form a mosaic-like structure in an otherwise fluid phospholipid bi-layer, i.e., the *fluid-mosaic* model (Fig. 3.18).

Thus, the milk fat globules are surrounded and stabilized by a structure which includes the trilaminar apical membrane (which is replaced by Golgi membranes on secretion of proteins and lactose). The inner face of the membrane has a dense proteinaceous layer, probably acquired within the secretory cell during movement of the globule from the rough endoplasmic reticulum at the base of the cell, where the triglycerides are synthesised, to the apex of the cell. A layer of high melting triglycerides may be present inside this proteinaceous layer. Much of the trilaminar membrane is lost on ageing of the milk, especially if it is agitated; the membrane thus shed is present in the skim milk as vesicles (or microsomes), which explains the high proportion of phospholipids in skim milk. A succession of structural models of the MFGM have been published, including, McPherson and Kitchen (1983), Keenan et al. (1983), Keenan and Dylewski (1995), Keenan and Patton (1995), Keenan and Mathur (2006) and Mather (2011) (Fig 3.19). Since the MFGM is a dynamic unstable structure, it is probably not possible to describe a structure which is applicable in all situations and conditions.

3.9 Stability of the Milk Fat Emulsion

The stability, or instability, of the milk fat emulsion is very significant with respect to many physical and chemical characteristics of milk and dairy products. The stability of emulsion depends strongly on the integrity of the MFGM and as discussed in Sect. 3.8.7, this membrane is quite fragile and is more or less extensively changed during dairy processing operations.

In the following, some of the principal aspects and problems related to or arising from the stability of the milk fat emulsion are discussed. Some of these relate to the inherent instability of emulsions in general, others are specifically related to the milk system.

3.9.1 Emulsion Stability in General

Lipid emulsions are inherently unstable systems due to:

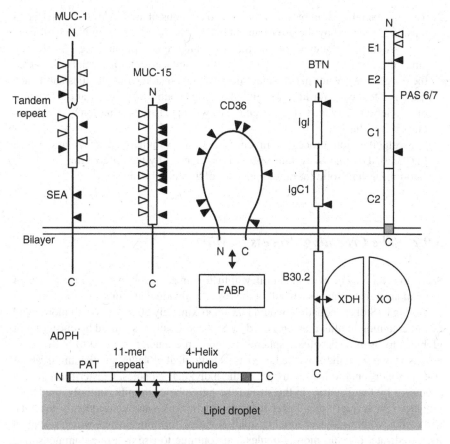

Fig 3.19 The major proteins in the bovine milk fat globule membrane. Shown are MUC-1, MUC-15, CD36, BTN, and PAS 6/7 in the bilayer and the locations of XDH/XO, FABP, and ADPH between the bilayer and the lipid droplet surface. Modified from Mather (2011)

1. The difference in density between the lipid and aqueous phases (~0.9 and 1.036 g cm⁻³, respectively, for milk), which causes the fat globules to float or cream according to the Stokes' equation:

$$V = \frac{2r^2 \left(\rho_1 - \rho_2 \right) g}{9\eta}$$

Where V = rate of creaming
  r = radius of fat globules
  ρ_1, ρ_2 = densities of the continuous and dispersed phases, respectively
  g = acceleration due to gravity
  η = viscosity of the system
  If creaming is not accompanied by other changes, it is readily reversible by gentle agitation.

2. The interfacial tension between the oil and aqueous phases. Although interfacial tension is reduced by the use of an emulsifier, the interfacial film may be imperfect. When two globules collide, they may adhere (flocculate), e.g., by sharing emulsifier, or they may coalesce due to the Laplace principle which states that the pressure is greater inside small globules than inside large globules and hence there is a tendency for large fat globules (or gas bubbles) to grow at the expense of smaller ones. Taken to the extreme, this will lead to the formation of a continuous mass of fat.

Destabilization processes in emulsions are summarized schematically in Fig. 3.20. The rate of destabilization is influenced by the fat content, shear rate (motion), liquid:solid fat ratio, inclusion of air and globule size.

3.9.2   The Creaming Process in Milk

A cream layer may be evident in milk within 20 min after milking. The appearance of a cream layer, if formed as a result of the rise of individual globules of 4 μm diameter according to Stokes' equation, would take approximately 50 h. The much more rapid rate of creaming in milk than predicted by Stoke's equation is caused by clustering of globules to form approximate spheres, ranging in diameter from 10 to 800 μm. As milk is drawn from the cow, the fat exists as individual globules and the initial rate of rise is proportional to the radius (r^2) of the individual globules. Cluster formation is promoted by the disparity in the size of the fat globules in milk. Initially, the larger globules rise several times faster than the smaller ones and consequently overtake and collide with the slower-moving small globules, forming clusters which rise at an increased rate, pick up more globules and continue to rise at a rate commensurate with the increased radius. The creaming of clusters only approximates to Stokes' equation since they are irregular in geometry and contain a considerable amount of occluded serum and therefore $\Delta \rho$ is variable and smaller than for a single globule.

In 1889, Babcock postulated that creaming of cows' milk resulted from an agglutination-type reaction, similar to the agglutination of red blood cells; this hypothesis has been confirmed. Creaming is enhanced by adding blood serum or colostrum to milk; the responsible agents are immunoglobulins (Ig, which are present at high levels in colostrum), especially IgM. Because these Igs aggregate and precipitate at low temperature (<37 °C) and redisperse on warming, they are often referred to as cryoglobulins. Aggregation is also dependent on ionic strength and pH. When aggregation of the cryoglobulins occurs in the cold they may precipitate onto the surfaces of large particles, e.g., fat globules, causing them to agglutinate, probably through a reduction in surface (electrokinetic) potential. The cryoprecipitated globulins may also form a network in which the fat globules are entrapped. The clusters can be dispersed by gentle stirring and are completely dispersed on warming to 37 °C or higher. Creaming is strongly dependent on temperature and does not occur above 37 °C (Fig. 3.21). The milk of buffalo, sheep and goat do not exhibit flocculation and the milk of some cows exhibit little or none, apparently a genetic trait.

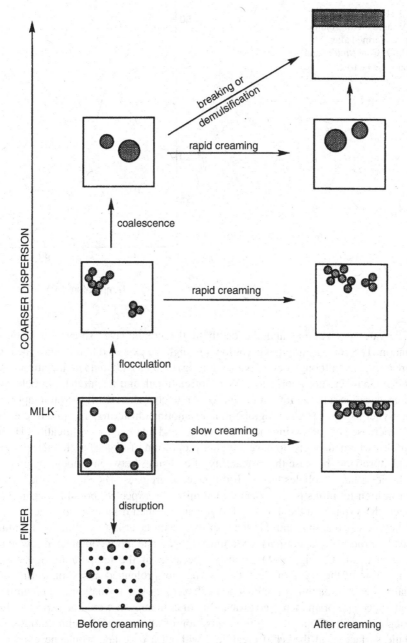

Fig. 3.20 Schematic representation of different forms of emulsion destabilization (modified from Mulder and Walstra 1974)

Fig. 3.21 Effect of temperature on the volume of cream formed after 2 h (modified from Mulder and Walstra 1974a, b)

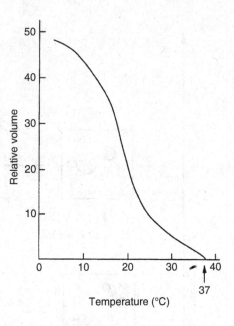

The rate of creaming and the depth of the cream layer show considerable variation. The concentration of cryoglobulin might be expected to influence the rate of creaming and although colostrum (rich in Ig) creams well and late lactation milk (deficient in Ig) creams poorly, there is no correlation in mid-lactation milks between Ig concentration and the rate of creaming. An uncharacterized lipoprotein appears to act synergistically with cryoglobulin in promoting clustering. The rate of creaming is increased by increasing the ionic strength and retarded by acidification. High-fat milks, which also tend to have a higher proportion of larger fat globules, cream quickly, probably because the probability of collisions between globules is greater and because large globules tend to form larger aggregates. The depth of the cream layer in high-fat milks is also greater than might be expected, possibly because of greater 'dead space' in the intrices of aggregates formed from large globules.

The rate of creaming and the depth of the cream layer are very markedly influenced by processing operations. Creaming is faster and more complete at low temperatures (<20 °C; Fig. 3.21), probably because of the temperature-dependent precipitation of the cryoglobulins. Gentle (but not prolonged) agitation during the initial stages of creaming promotes and enhances cluster formation and creaming, possibly because of an increased probability of collisions. It would be expected that stirring cold milk would lead to the deposition of all the cryoglobulin onto the fat globule surfaces and the rapid creaming, without a time lag, would be expected when stirring ceased. However, milk so treated does not cream at all or only slightly after a prolonged lag period. If cold, creamed milk is agitated gently, the clusters are dispersed and do not reform unless the milk is rewarmed to ~40 °C and then recooled, i.e., the whole cycle repeated. Violent agitation is detrimental to creaming, possibly due to denaturation of the cryoglobulins and/or alteration to the fat globule surface.

If milk is separated $\geq 40\ ^\circ$C, the cryoglobulins are present predominantly in the serum while they are in the cream produced at lower temperatures. Agglutination and creaming are impaired or prevented by heating (e.g., 70 $^\circ$C$\times$30 min or 77 $^\circ$C$\times$20 s) owing to denaturation of the cryoglobulins; addition of Igs to heated milk restores creaming (except after very severe heat treatment, e.g., 2 min at 95 $^\circ$C or equivalent). Homogenization prevents creaming, not only due to the reduction of fat globule size but also to some other factor since a blend of raw cream and homogenized skim-milk does not cream well. In fact two types of euglobulin appear to be involved in agglutination, one of which is denatured by heating, the other by homogenization. Thus, a variety of factors which involve temperature changes, agitation or homogenization influence the rate and extent of creaming.

3.10 Influence of Processing Operations on the Fat Globule Membrane

As discussed in Sect. 3.8.7, the milk fat globule membrane (MFGM) is relatively fragile and susceptible to damage during a range of processing operations; consequently, emulsion stability is reduced by dislodging interfacial material by agitation, homogenization, heat treatment, concentration, drying or freezing. Rearrangement of the membrane increases the susceptibility of the fat to hydrolytic rancidity, light-activated off-flavours and "oiling-off" of the fat but reduces susceptibility to metal-catalysed oxidation. The influence of the principal dairy processing operations on MFGM and concomitant defects are discussed below.

3.10.1 Milk Supply: Hydrolytic Rancidity

The production of milk on the farm and transportation to the processing plant are potentially major causes of damage to the MFGM. Damage to the membrane may occur at several stages of the milking operation: foaming due to air sucked in at teat-cups, agitation due to vertical sections (risers) in milk pipe lines, constrictions and/or expansion in pipelines, pumps, especially if not operating at full capacity, surface coolers, agitators in bulk tanks and freezing of milk on the walls of bulk tanks. While some oiling-off and perhaps other physical damage to the milk fat emulsion may accrue from such damage, by far the most serious consequence is the development of hydrolytic rancidity. The extent of lipolysis is commonly expressed as 'acid degree value' (ADV) of the fat as millimoles of free fatty acids per 100 g fat; an ADV >1 is undesirable and is perceptible by taste to most people.

The principal lipase in bovine milk is a lipoprotein lipase (LPL; see Chap. 8) which is associated predominantly with the casein micelles and is isolated from its substrate, milk fat, by the MFGM, i.e., the enzyme and its substrate are compartmentalized. However, even slight damage to the membrane permits contact between enzyme and substrate, resulting in hydrolytic rancidity. The enzyme is optimally active at ~37 °C and ~pH 8.5 and is stimulated by divalent cations, e.g., Ca^{2+}. (Ca^{2+} complex free fatty acids, which are strongly inhibitory). The initial turnover of milk LPL is ~3,000 s^{-1}, i.e., 3,000 fatty acid molecules are liberated per second per mole of enzyme (milk usually contains 1–2 mg lipase/l, i.e., 10–20 nM) which, if fully active, is sufficient to induce rancidity in ~10 s. This never happens in milk due to a variety of factors, e.g., the pH, ionic strength and, usually, the temperature are not optimal; the lipase is bound to the casein micelles; the substrate is not readily available; milk probably contains lipase inhibitors, including caseins. The activity of lipase in milk is not correlated with its concentration due to the various inhibitory and adverse factors.

Machine milking, especially pipe-line milking systems, markedly increases the incidence of hydrolytic rancidity unless adequate precautions are taken. The effectors are the clawpiece and the tube taking the milk from the clawpiece to the pipeline; damage at the clawpiece may be minimised by proper regulation of air intake and low-line milking installations cause less damage than high-line systems but the former are more expensive and less convenient for operators. Larger diameter pipelines (e.g., 5 cm) reduce the incidence of rancidity but may cause cleaning problems and high milk losses. The receiving jar, pump (diaphragm or centrifugal, provided they are operated properly) and type of bulk tank, including agitator, transportation in bulk tankers or preliminary processing operations (e.g., pumping and refrigerated storage) at the factory make little if any contribution to hydrolytic rancidity.

The frequency and severity of lipolysis increases in late lactation, possibly owing to a weak MFGM and the low level of milk produced (which may aggravate agitation); this problem is particularly acute when milk production is seasonal, e.g., as in Ireland or New Zealand.

The lipase system can also be activated by cooling freshly drawn milk to 5 °C, rewarming to 30 °C and recooling to 5 °C. Such a temperature cycle may occur under farm conditions, e.g., addition of a large quantity of warm milks to a small volume of cold milk. It is important that bulk tanks are emptied completely at each collection (this practice is also essential for the maintenance of good hygiene). No satisfactory explanation for temperature activation is available but changes in the physical state of fat (liquid/solid ratio) have been suggested; damage/alteration of the globule surface and binding of lipoprotein cofactor may also be involved.

Some cows produce milk which is susceptible to a defect known as "spontaneous rancidity"—no activation treatment, other than cooling of the milk, is required; the frequency of such milks may be as high as 30 % of the population. Suggested causes of spontaneous rancidity include:

1. A second lipase located in the membrane rather than on the casein micelles; there is no evidence for this and it is unlikely.
2. A weak membrane which does not adequately protect the fat from the normal LPL.
3. A high level of lipoprotein co-factor or proteose peptone 3 which facilitate attachment of the LPL to the fat surface; this appears to be the most probable cause.

Mixing of normal milk with susceptible milk in a ratio of 4:1 prevents spontaneous rancidity and the problem is, therefore, not serious except in small or abnormal herds. The incidence of spontaneous rancidity increases with advancing lactation and with dry feeding.

3.10.2   Mechanical Separation of Milk

Gravity creaming is relatively efficient, especially in the cold (a fat content of 0.1 % in the skim phase may be obtained). However, it is slow and inconvenient for industrial-scale operations. The benefits of centrifugal separation of fat from milk were recognised in the 1860s and several attempts to develop a separator were made during the 1860s and 1870s. The first successful separator was produced in 1878 by the Swede, Gustav de Laval, whose company still prospers. The milk separator has changed markedly since its development; schematic representations of a modern separator are shown in Figs. 3.22 and 3.23.

In centrifugal separation, g in Stokes' equation is replaced by centrifugal force, $\omega^2 R$,

where

$$\omega \ = \ \text{centrifugal speed in radians sec}^{-1} \left(2\Pi \ \text{radians} = 360° \right)$$

$$R \ = \ \text{distance (cm) of the particle from the axis of rotation}$$

or

$$\frac{(2\Pi S)^2 R}{(60)^2}$$

$$\text{where } S \ = \ \text{bowl speed in r.p.m.}$$

Inserting this value for g into Stokes' equation and simplifying gives:

$$V = \frac{0.00244 \ (\rho_1 - \rho_2) r^2 S^2 R}{\eta}$$

Thus, the rate of separation is influenced by the radius of the fat globules, the radius and speed of the separator bowel, the difference between the density of the continuous and dispersed phases and the viscosity of the milk; temperature influences r, $(\rho_1 - \rho_2)$ and η.

Fat globules <2 μm in diameter are incompletely removed by cream separators and since the average size of fat globules decreases with advancing lactation (Fig. 3.15), the efficiency of separation decreases concomitantly. The % fat in cream is regulated by manipulating the ratio of cream to skim-milk streams from the separator, which in effect regulates back-pressure. With any particular separator operating

Fig. 3.22 Flow of cream and skim milk in the space between a pair of discs in centrifugal separator (**a**); a stack of discs (**b**) (from Towler 1994)

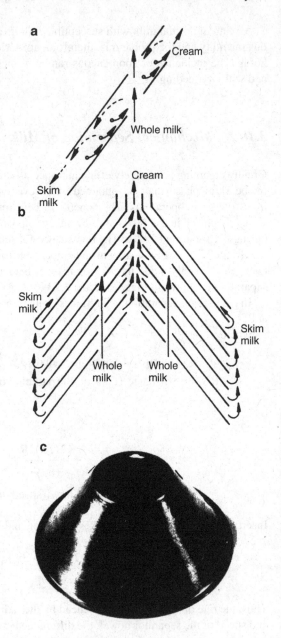

under more or less fixed conditions, temperature is the most important variable affecting the efficiency of separation via its effects on r, η and $(\rho_1 - \rho_2)$. The efficiency of separation increases with temperature, especially in the range 20–40 °C. In the past, separation was usually performed ≥40 °C but modern separators are very efficient even at a low temperature.

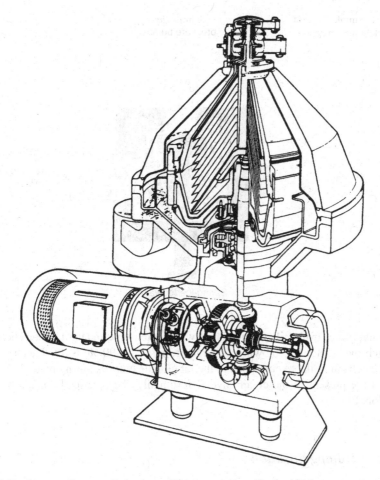

Fig. 3.23 Cutaway diagram of a modern milk separator (from Towler 1994)

As discussed in Sect. 3.9.2, the cryoglobulin are entirely in the serum phase at temperatures above ~37 °C, as a result of which creams prepared at these temperatures have poor natural creaming properties and the skim milk foams copiously due to the presence of cryoglobuins. Following separation at a low temperature (<10–15 °C), most of the cryoglobulins remain in the cream phase. Considerable incorporation of air and foaming may occur during separation, especially with older machines, causing damage to the MFGM. The viscosity of cream produced by low-temperature separation is much higher than that produced at higher temperatures, due to the presence of cryoglobulins and clustering of fat globules in the former.

Centrifugal force is also applied in the clarification and bactofugation of milk. Clarification is used principally to remove somatic cells and physical dirt, while bactofugation, in addition to removing these, also removes 95–99 % of the bacterial cells present. One of the principal applications of bactofugation is the removal of

Fig. 3.24 Simple diagram
of a milk homogenizer

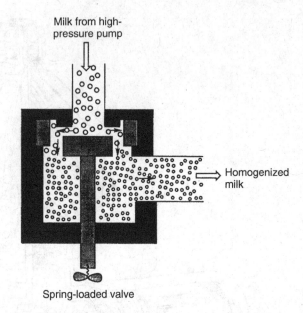

Milk from high-
pressure pump

Homogenized
milk

Spring-loaded valve

clostridial spores from milk intended for Swiss, Dutch-type and hard Italian cheeses, in which they cause late blowing. A large proportion (~90 %) of the bacteria and somatic cells in milk are entrapped in the fat globule clusters during natural creaming and are present in the cream layer; presumably, they are agglutinated by the cryoglobulins.

3.10.3 Homogenization

Homogenization is widely practised in the manufacture of liquid milk and milk products. The process essentially involves forcing milk through a small orifice (Fig. 3.24) at a high pressure (13–20 M Nm^{-2}), usually at about 40 °C (at this temperature, the fat is liquid; homogenization is less effective at lower temperatures when the fat is partially solid). The principal effect of homogenization is to reduce the average diameter of the fat globules to <1 μm (the vast majority of the globules in homogenized milk are <2 μm) (Fig. 3.25). Reduction is achieved through the combined action of shearing, impingement, distention and cavitation. Following a single passage of milk through a homogenizer, the fat globules occur in clumps, causing an increase in viscosity; a second stage homogenization at a lower pressure (e.g., 3.5 M Nm^{-2}) disperses the clumps and reduces the viscosity. Clumping arises from incomplete coverage of the greatly increased emulsion interfacial area during the short passage time through the homogenizer valve, resulting in the sharing of casein micelles by neighbouring globules.

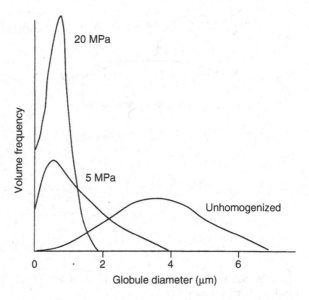

Fig. 3.25 Effect of homogenization on the size (volume distribution) of fat globules in milk (modified from Mulder and Walstra 1974)

In the dairy industry, the valve homogenizer is the principal type of homogenizer, but there are several other types of homogenizer which may be used for certain products, including (see Huppertz 2011):

- High-speed mixers, e.g., Silverston type, in which the fat globules are reduced through the shearing action of fast-moving blades.
- Colloid mills, which are particularly useful for very viscous material.
- Microfluidizers, in which two liquid streams, e.g., oil and an aqueous phase, are forced to collide in a reaction chamber at an angle of 180°, at a pressure up to 300 MPa.
- Ultrasonic homogenizers operating at a frequency of 20–100 kHz.
- Membrane homogenizers.

Reducing the average diameter of the fat globules to 1 μm results in a four- to sixfold increase in the fat/plasma interface. There is insufficient natural membrane to completely coat the newly formed surface or insufficient time for complete coverage to occur and consequently the globules in homogenized milk are coated by a membrane which consists mostly of casein (93 % of dry mass, with some whey proteins, which are adsorbed less efficiently than the caseins) (Fig. 3.26). The membrane of homogenized milk contains 2.3 g protein/100 g fat (~10 mg protein m^{-2}), which is very considerably higher than the level of protein in the natural membrane (0.5–0.8 g/100 g fat), and is estimated to be ~15 nm thick. The casein content in the serum phase of homogenized milk is reduced by about 6–8 %.

Homogenization causes several major changes in the properties of milk:

1. Homogenized milk does not cream naturally and the fat is recovered only poorly by mechanical separation. This is due in part to the smaller average size of the fat

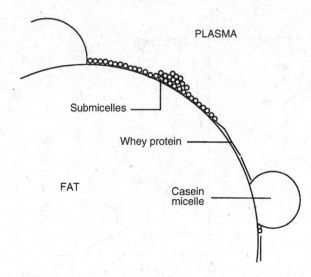

PLASMA

Submicelles

Whey protein

FAT

Casein micelle

Fig. 3.26 Schematic representation of the membrane of fat globules in homogenized milk (modified from Walstra 1983)

globules but failure of the globules in homogenized milk to form aggregates, due mostly to the agitation-induced denaturation of some immunoglobulins, is mainly responsible for the failure to cream.

2. As discussed in Sect. 3.10.1, homogenized milk is very susceptible to hydrolytic rancidity because the artificial membrane does not isolate the fat from the lipase; consequently, homogenized milk must be pasteurized before or immediately after homogenization. Homogenized milk is also more susceptible to sunlight oxidized flavour, which is due to the production of methional from methionine, but is less susceptible to metal-catalysed lipid oxidation; the latter is presumably because the phospholipids, which are very susceptible to oxidation (highly unsaturated) and are located largely in the natural membrane (which contains prooxidants, e.g., xanthine oxidoreductase and metals) are more uniformly distributed after homogenization and, therefore, are less likely to propagate lipid oxidation.

3. Homogenized milk is whiter due to finer dispersion of the fat (and thus greater light scattering) and its flavour is more bland.

4. The heat stability of whole milk is reduced by homogenization, as is the strength (curd tension) of rennet-induced gels; these changes will be discussed in more detail in Chaps. 9 and 12. Viscosity is increased by homogenization. Homogenized milk has improved foaming characteristics, a feature which may be due to the release of foam-promoting proteins from the natural membrane or to reduction in fat globule size—small globules are less likely to damage foam lamellae. Homogenization reduces surface tension, possibly due to inclusion of very surface-active proteins in the artificial membrane and to changes in the fat globule

surface. Homogenized milk drains cleanly from the sides of a glass bottle or drinking glass. Milk for homogenization should be clarified to avoid sedimentation of leucocytes during storage.

The efficiency of homogenization may be assessed by microscopic examination or more effectively by a particle sizer, e.g., Malvern Mastersizer.

3.10.4   Heating

Normal HTST pasteurization causes very little change in the fat globule membrane or in the characteristics of milk fat dependent on the membrane. However, excessively high pasteurization temperatures denature the cryoglobulins and aggregation of the fat globules and creaming are impaired or prevented. Severe treatments, e.g., $80\,°C \times 15$ min, remove lipid and protein material from the membrane, the fat globules are partially denuded and may coalesce, forming large clumps of fat and resulting in defects such as cream plug in milk or cream (see Sect. 3.11).

Processes such as thermal evaporation also cause membrane damage, especially since many of these treatments also involve vigorous agitation in high velocity heating systems. Since milk for concentrated and dehydrated milk products is normally homogenized, damage to the natural membrane is of little significance.

3.11   Physical Defects in Milk and Cream

In addition to the flavour defects initiated or influenced by damage to the fat globule membrane, such damage also results in a variety of physical defects in milk and especially in cream. The more important of these are "oiling off", "cream plug" and "age thickening".

"Oiling off", characterized by the appearance of globules of oil or fat on the surface of coffee or tea when milk and especially cream is added, is due to membrane damage during processing, resulting in "free fat"; low pressure homogenization re-emulsifies the free fat and eliminates the defect.

"Cream plug" is characterized by the formation of a layer of solid fat on the surface of cream or milk in bottles; the defect is due to a high level of "free fat" which forms interlocking crystals on cooling and is most common in high-fat creams. Cream plug is common in unhomogenized, pasteurized, late lactation milk, presumably due to a weak MFGM.

"Age thickening" is due essentially to a high level of free fat, especially in high-fat creams; the product becomes very viscous due to interlocking of crystals of free fat.

Two somewhat related instability problems are "feathering" and "bitty" cream. "Feathering" is characterized by the appearance of white flecks when milk or cream is poured on hot coffee and is a form of heat-induced coagulation; the white flecks are mainly destabilized protein. The heat stability of cream and its resistance to feathering are reduced by:

1. Single stage homogenization.
2. High homogenization pressure at a low temperature.
3. High concentrations of Ca^{2+} in the cream or water.
4. A high ratio of fat to serum solids, i.e., high-fat creams.
5. High temperature and low pH of the coffee.

Protein–lipid interaction is enhanced by homogenization while high temperatures, low pH and high divalent cation concentration induce aggregation of the casein-coated fat globules into large visible particles. Stability may be improved by:

1. Using fresh milk.
2. Adding disodium phosphate or sodium citrate, which sequester Ca^{2+}, increase protein charge and dissociate casein micelles.
3. Standardizing the cream with buttermilk, which is a good emulsifier owing to its high content of phospholipids.

"Bitty cream" is caused by the hydrolysis of phospholipids of the fat globule membrane by phospholipases secreted by bacteria, especially *Bacillus cereus*, but also by psychrotrophs; the partially denuded globules coalesce when closely packed, as in cream or in the cream layer of milk, forming aggregates rather than a solid mass of fat.

3.11.1   Free Fat

"Free fat" may be defined as non-globular fat, i.e., fat globules from which the membrane has been totally or partially removed. Damage to fat globules may be determined by measuring the level of free fat present. The fat in undamaged globules is not extractable by apolar solvents because it is protected by the membrane, damage to which permits extraction, i.e., the amount of fat extractable by apolar solvents is termed "free fat".

Free fat may be determined by a modified Rose-Gottlieb method or by extraction with carbon tetrachloride (CCl_4). In the standard Rose-Gottlieb method, the emulsion is destabilized by the action of ammonia and ethanol and the fat is then extracted with ethyl/petroleum ether. The free fat in a sample may be determined by omitting the destabilization step, i.e., by extracting the product directly with fat solvent, and expressed as the percentage of free fat in the sample or as a percentage of total fat. Alternatively, the sample may be extracted with CCl_4. In both methods, the sample is shaken with the fat solvent; the duration and severity of shaking must be carefully standardized if reproducible results are to be obtained.

Other methods used to quantify free fat include: centrifugation in Babcock or Gerber butyrometers at 40–60 °C (the free fat is read off directly on the graduated scale); release of membrane-bound enzymes, especially xanthine oxidoreductase or alkaline phosphatase or the susceptibility of milk fat to hydrolysis by added lipase (e.g., from *Geotrichum candidum*).

3.12 Churning

It has been known since prehistoric times that if milk, and especially cream, is agitated, the fat aggregates to form granules (grains) which are converted to butter by kneading (Fig. 3.27). Buttermaking has been a traditional method for a very long time in temperate zones for conserving milk fat; in tropical regions, butter grains or cream are heated to remove all the water; the resulting product is called "ghee", a crude form of butter oil. Global production of butter is about 9.3×10^6 tonnes p.a. (USDA 2014).

The cream used for butter may be fresh (~pH 6.6) or ripened (fermented; ~pH 4.6), yielding "sweet cream" and "ripened cream (lactic)" butter, respectively. Sweet-cream butter is most common in English-speaking countries but ripened cream butter is more popular elsewhere. Traditionally, the cream for ripened cream butter was fermented by the indigenous microflora, which was variable. Product quality and consistency were improved by the introduction in the 1880s of cultures (starters) of selected lactic acid bacteria, which produce lactic acid from lactose and diacetyl (the principal flavour component in ripened cream butter) from citric acid. A flavour concentrate, containing lactic acid and diacetyl, is now frequently used in the manufacture of ripened cream butter, to facilitate production schedules and improve consistency.

Butter manufacture or churning essentially involves phase inversion, i.e., the conversion of the oil-in-water emulsion of cream to a water-in-oil emulsion. Inversion is achieved by some form of mechanical agitation which denudes some of the globules of their stabilizing membrane; the denuded globules coalesce to form butter grains, entrapping some globular fat. The butter grains are then kneaded ("worked") which releases fat liquid at room temperature. Depending on temperature and on the method and extent of working, liquid fat may represent 50–95 % of total fat. The liquid fat forms the continuous phase in which fat globules, fat crystals, membrane material, water droplets and small air bubbles are dispersed (Fig. 3.28, Table 3.13). NaCl may be added (to ~2 %) to modify flavour but more

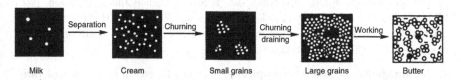

Fig. 3.27 Schematic representation of the stages of butter production. *Black* indicates continuous aqueous phase and white indicates continuous fat phase (modified from Mulder and Walstra 1974)

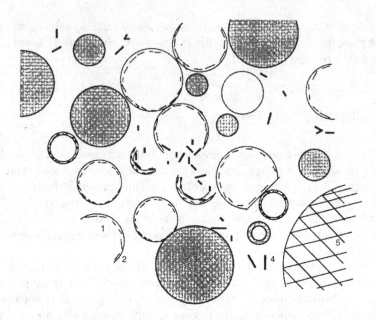

Fig. 3.28 Schematic representation of the structure of butter. (1) Fat globule, (2) membrane, (3) aqueous droplet, (4) fat crystals, (5) air cell (modified from Mulder and Walstra 1974a, b)

Table 3.13 Structural elements of conventional butter (modified from Mulder and Walstra 1974)

Element	Approximate number (ml^{-1})	Proportion of butter (%, v/v)	Dimensions	Remarks (μm)
Fat globules	10^{10}	10–50	2–8	Differ in composition; with complete or partial membrane
Fat crystal	10^{13}	10–40	0.01–2	Amount depends on temperature occur mainly in globules; at low temperature, form solid networks
Moisture	10^{10}	16	1–25	Differ in composition droplets
Air cells	10^{7}		5	>20

importantly as a preservative; added salt dissolves in the water droplets (to give ~12 % salt in moisture) which also contain contaminating bacteria. Usually, ripened cream butter is not salted.

The process of phase inversion has received considerable attention [see McDowall (1953) and Wilbey (1994) and Mortensen (2011a, b, 2014) for a detailed discussion]. Briefly, churning methods can be divided into: (1) traditional batch methods, (2) continuous methods.

1. The traditional method involves placing 30–40 % fat cream in a churn (of various shapes and design, Fig. 3.29) which is rotated gently. During rotation, air is incorporated and numerous small air bubbles are formed; fat globules are trapped

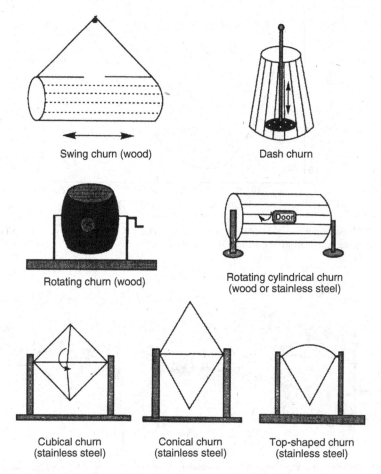

Fig. 3.29 Examples of butter churns

between the lamellae of the bubbles. As the bubbles grow, the lamellae become thinner and exert a shearing effect on the fat globules. Some globules become denuded of membrane and coalesce; the aggregated globules are cemented by liquid fat expressed from the globules. A portion of the liquid fat spreads over the surface of the air bubbles, causing them to collapse, releasing butter grains and **buttermilk** (representing the serum phase of cream plus the fat globule membrane).

When a certain degree of globular destabilization has occurred, the foam collapses rather abruptly and when the grains have grown to the requisite size, the buttermilk is drained off and the grains worked to a continuous mass. Proper working of the butter is essential for good quality—a fine dispersion of water droplets reduces the risk of microbial growth and other spoilage reactions (most water droplets are <5 µm). Working is also necessary to reduce the water content

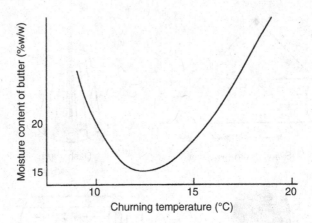

Fig. 3.30 Moisture content of traditional butter as a function of churning temperature, all other conditions being equal (from Mulder and Walstra 1974)

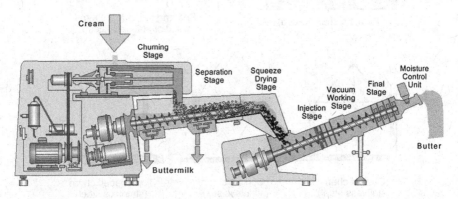

Fig. 3.31 Diagram of a Westfalia continuous buttermaker

to the legal limit, i.e., ≤16 %. The length of time required to churn cream, fat loss in the buttermilk and the moisture content of the butter are influenced by various factors, including temperature, pH, fat content of the cream, globule size and turning rate of the churn. Ripened cream churns faster than sweet cream. The temperature of the cream has a large effect on the moisture content of butter (Fig. 3.30).

2. Modern 'churns' operate continuously by either of two principles:

   (a) Processes using ~40 % fat cream (i.e., the Fritz process, e.g., Westfalia Separator AG) in which air is whipped into a thin film of cream in a Votator (Fig. 3.31). The process of phase inversion is essentially similar to traditional churning methods.

   (b) Processes using high-fat cream (80 % fat); although the fat in 80 % fat cream is still in an oil-in-water emulsion, it is a very unstable emulsion and is destabilized easily by chilling and agitation.

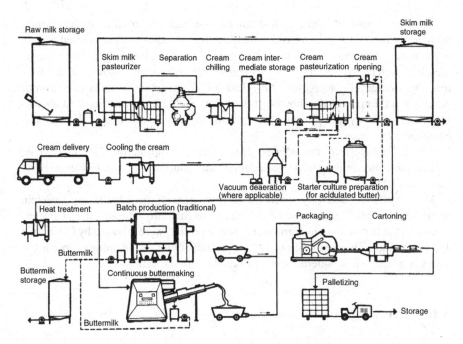

Fig. 3.32 Line diagram of a modern buttermaking plant (from Alfa-Laval Dairy Handbook)

The line diagram for a modern buttermaking plant is shown in Fig. 3.32.

All the methods of butter manufacture involve complete or partial removal of the fat globule membrane, most of which is lost in the buttermilk, which is, consequently, a good source of phospholipids and other emulsifiers.

3.12.1   Buttermilk

Assuming that butter is made from cream containing 40 % fat, an equal amount of buttermilk, i.e., 9.3×10^6 tonnes, is produced per annum. The typical composition of traditional buttermilk (to be distinguished from fermented skimmed milk) is approximately the same as that of skimmed milk but with a little more fat, i.e., about 4.9 % lactose, 3.4 % protein, 0.8 % ash and 0.6 % fat. The protein consists mainly of casein and whey proteins with lesser amounts of proteins of the MFGM. The lipids are rich in phospholipids (7 times more than skimmed milk), derived from the MFGM, making it a very good emulsifier and bestows it with desirable nutritional properties, i.e., it is a valuable dairy ingredient. The composition and properties of buttermilk obtained from sweet cream, sour (acid) cream or whey cream differ.

Some buttermilk is mixed with skimmed milk and converted to skimmed milk powder but buttermilk powder is a valuable dairy ingredient, which is used mainly in the bakery and dairy industries. In the bakery industry, it increases loaf volume, increases water sorption, improves softening and ameliorates staling. In the dairy industry, buttermilk increases the yield of Pizza cheese, enhances flavour development, improves the mouthfeel, body and meltability of reduced-fat cheese. It reduces viscosity and prevents fat crystallization in chocolate and is a valuable emulsifier for sauces.

Several lipids from the MFGM have desirable physiological effects, e.g., anticarcinogenic, antibacterial or antidepressant (Ward et al. 2006; Eyzaguirre and Corredig 2011).

Many colloidal and physicochemical properties of buttermilk, e.g., protein profile, heat stability, ethanol stability, rennet coagulability and micellar characteristics (size, zeta potential, hydration and protein profile) were reported by O'Connell and Fox (2000). The composition, viscosity, emulsifying and foaming properties of sweet, acid and whey buttermilk were studied by Sodini et al. 2006).

3.13 Freezing

Freezing and dehydration tend to destabilize all lipoprotein complexes, both natural and artificial. Thus, freezing of milk, and especially cream, results in damage to the MFGM which causes destabilization when the product is thawed. Most of the destabilizing effect is due to physico-chemical changes induced by dehydration of the lipoprotein complexes but some physical damage is also caused by ice crystals. The damage is manifest as oiling off and free fat formation. The extent of damage is proportional to fat concentration and moderately high-fat creams (50 %) are completely destabilized by freezing.

Frozen cream is produced commercially and is used mainly for the production of soups, butter-oil, butter, etc., where emulsion stability is not important. Damage may be reduced by:

1. Rapid freezing as thin blocks or continuously on refrigerated drums.
2. Homogenization and pasteurization before freezing.
3. Storage at a very low temperature (~−30 °C) and avoiding temperature fluctuations during storage.

3.14 Dehydration

The physico-chemical state of fat in milk powder particles, which markedly influences the wettability and dispersibility of the powder on reconstitution, depends on the manufacturing process. The fat occurs either in a finely emulsified or in a partly coalesced, de-emulsified state. In the latter case, the MFGM has been ruptured or completely removed, causing the globules to run together to form pools of free fat.

The amount of de-emulsified, "free fat" depends on the manufacturing method and storage conditions. Typical values for "free fat" (as % of total fat) in milk powders are: spray dried powders: 3.3–20 %; roller dried powders: 91.6–95.8 %; freeze dried powders: 43–75 %; foam dried powders: less than 10 %.

The high level of "free fat" in roller-dried powder is due to the effects of the high temperature to which milk is exposed on the roller surfaces and to the mechanical effect of the scraping knives. The free fat in roller-dried whole milk powder improves the texture of milk chocolate due to co-crystallization with the cocoa fat (Liang and Hartel 2004). In properly made and stored spray-dried powder, the fat globules are distributed throughout the powder particles. The amount of free fat depends on the total fat content, and may be about 25 % of total fat. Homogenization pre-drying reduces the level of free fat formed.

Further liberation of "free fat" may occur under adverse storage conditions. If powder absorbs water it becomes "clammy" and lactose crystallizes, resulting in the expulsion of other milk components from the lactose crystals into the spaces between the crystals. De-emulsification of the fat may occur due to the mechanical action of sharp edges of lactose crystals on the MFGM. If the fat is liquid at the time of membrane rupture or if it becomes liquid during storage, it will adsorb onto the powder particles, forming a water-repellant film around the particles.

The state of fat in powder has a major influence on wettability, i.e., the ease with which the powder particles make contact with water. Adequate wettability is a pre-requisite for good dispersibility. Free fat has a water-repelling effect on the particles during dissolution, making the powder difficult to reconstitute. Clumps of fat and oily patches appear on the surface of the reconstituted powder, as well as greasy films on the walls of containers. The presence of "free fat" on the surface of the particles tends to increase the susceptibility of fat to oxidation. A scum of fat-protein complexes may appear on the surface of reconstituted milk; the propensity to scum formation is increased by high storage temperatures.

3.15 Lipid Oxidation

Lipid oxidation, leading to oxidative rancidity, is a major cause of deterioration in milk and dairy products. The subject has been reviewed by Richardson and Korycka-Dahl (1983) and O'Connor and O'Brien (1995, 2006).

3.15.1 Autocatalytic Mechanism

Lipid oxidation is an autocatalysed free radical chain reaction which is normally divided into three phases: initiation, propagation and termination (Fig. 3.33).

The initial step involves abstracting a hydrogen atom from a fatty acid, forming a fatty acid (FA) free radical, e.g.,

$$CH_3 ----CH_2 - CH = CH - \overset{\cdot}{CH} - CH = CH - CH_2 ----COOH$$

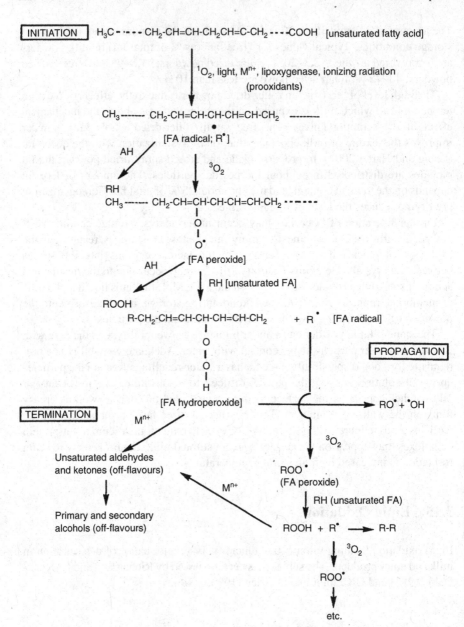

Fig. 3.33 Autooxidation of fatty acids, FA (fatty acid); AH (antioxidant); M^{n+} (metal ion)

Although saturated fatty acids may lose a H• and undergo oxidation, the reaction principally involves unsaturated fatty acids, especially polyunsaturated fatty acids (PUFA), the methylene, $-CH_2-$, group between double bonds being particularly sensitive:

$$C_{18:3} \gg C_{18:2} \gg C_{18:1} > C_{18:0}$$

The polar lipids in milk fat are richer in PUFA than neutral lipids and are concentrated in the fat globule membrane in juxta position with several pro-oxidants and are, therefore, particularly sensitive to oxidation.

The initiation reaction is catalyzed by singlet oxygen ($^1O_2$, produced by ionizing radiation and other factors), polyvalent metal ions that can undergo a monovalent oxidation/reduction reaction ($M^{n+1} \rightarrow M^n$), especially iron and copper (the metal may be free or organically bound, for example, xanthine oxidoreductase, peroxidase, catalase or cytochromes) or light, especially in the presence of a photosensitizer, e.g., riboflavin [in the case of vegetable products, lipoxygenase is a major pro-oxidant but this enzyme is not present in milk or dairy products].

The FA free radical may abstract a H from a hydrogen donor, e.g., an antioxidant (AH), terminating the reaction, or may react with molecular triplet oxygen, $^3O_2$, forming an unstable peroxy radical. In turn, the peroxy radical may obtain a H from an antioxidant, terminating the reaction, or from another fatty acid, forming a hydroperoxide and another FA free radical, which continues the reaction.

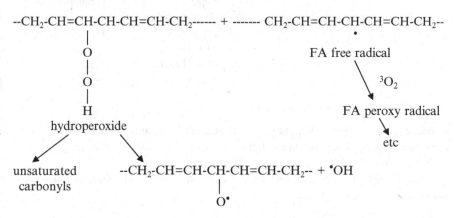

The intermediate products of lipid oxidation are themselves free radicals and more than one may be formed during each cycle; hence the reaction is autolcatalytic, i.e., the rate of oxidation increases with time, as shown schematically in Fig. 3.34. Thus, the formation of only very few (theoretically only one) free radicals by an exogenous agent is necessary to initiate the reaction. The reaction shows an induction period, the length of which depends on the presence of prooxidants and antioxidants.

The hydroperoxides are unstable and may break down to various products, including unsaturated carbonyls, which are mainly responsible for the off-flavours of oxidized lipids (the FA free radicals, peroxy radicals and hydroperoxides are

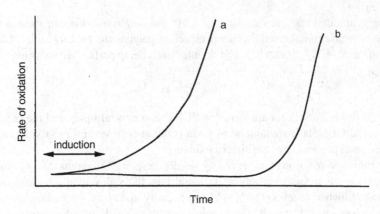

Fig. 3.34 Rate of oxidation in the absence (**a**) or presence (**b**) of an antioxidant

Table 3.14 Compounds contributing to typical oxidized flavour (from Richardson and Korycka-Dahl 1983)

Compounds	Flavours
Alkanals C_6-C_{11}	Green tallowy
2-Alkenals C_6-C_{10}	Green fatty
2,4-Alkadienals C_7-C_{10}	Oily deep-fried
3-*cis*-Hexenal	Green
4-*cis*-Heptenal	Cream/putty
2,6- and 3,6-Nonadienal	Cucumber
2,4,7-Decatrienal	Fishy, sliced beans
1-Octen-3-one	Metallic
1,5-*cis*-Octadien-3-one	Metallic
1-Octen-3-ol	Mushroom

flavourless). Different carbonyls vary with respect to flavour impact and since the carbonyls produced depend on the fatty acid being oxidized, the flavour characteristics of oxidized dairy product vary (Table 3.14).

The principal factors affecting lipid oxidation in milk and milk products are summarized in Table 3.15.

3.15.2   Pro-oxidants in Milk and Milk Products

Probably the principal pro-oxidants in milk and dairy products are metals, Cu and to a lesser extent Fe, and light. The metals may be indigenous, e.g., as part of xanthine oxidoreductase, lactoperoxidase, catalase or cytochromes, or may arise through contamination from equipment, water, soil, etc. Contamination with such metals can be reduced through the use of stainless steel equipment.

Metal-containing enzymes, e.g., lactoperoxidase and catalase, and cytochromes can act as pro-oxidants owing to the metals they contain rather than enzymatically;

Table 3.15 Major factors affecting the oxidation of lipids in milk and dairy products[a]

A. Potential pro-oxidants
1. Oxygen and activated oxygen species
Active oxygen system of somatic cells?
2. Riboflavin and light
3. Metals (e.g., copper and iron) associated with various ligands
Metallo-proteins
Salts of fatty acids
4. Metallo-enzymes (denatured?)
Xanthine oxidase
Lactoperoxidase, catalase (denatured)
Cytochrome P-420
Cytochrome b_5
Sulphydryl oxidase?
5. Ascorbate (?) and thiols (?) (via reductive activation of metals?)
B. Potential antioxidants
1. Tocopherols
2. Milk proteins
3. Carotenoids (β-carotene; bixin in anatto)
4. Certain ligands for metal pro-oxidants
5. Ascorbate and thiols
6. Maillard browning reaction products
7. Antioxidant enzymes (superoxide dismutase, sulphydryl oxidase)
C. Environmental and physical factors
1. Inert gas or vacuum packing
2. Gas permeability and opacity of packaging materials
3. Light
4. Temperature
5. pH
6. Water activity
7. Reduction potential
8. Surface area
D. Processing and storage
1. Homogenization
2. Thermal treatments
3. Fermentation
4. Proteolysis

[a]Many of these factors are interrelated and may even present paradoxical effects (e.g., ascorbate and thiols) on lipid oxidation (modified from Richardson and Korycka-Dahl, 1983)

the pro-oxidant effect of these enzymes is increased by heating (although there are conflicting reports). Xanthine oxidoreductase, which contains Fe and Mo, can act both enzymatically and as a source of prooxidant metals.

Riboflavin is a potent photosensitizer and catalyses a number of oxidative reactions in milk, e.g., fatty acids, proteins (with the formation of 3-methyl thiopropanal from methionine which is responsible for light-induced off-flavour) and ascorbic acid. Milk and dairy products should be protected from light by opaque packaging (cardboard or foil) and exposure to UV light should be minimized.

Ascorbic acid is a very effective anti-oxidant but combinations of ascorbate and copper can be pro-oxidant depending on their relative concentrations. Apparently, ascorbate reduces Cu^{2+} to Cu^+.

3.15.3   Antioxidants in Milk

Antioxidants are molecules with an easily detachable H atom which they donate to fatty acid free radicals or fatty acid peroxy radicals, which would otherwise abstract a H form another fatty acid, forming another free radical. The residual antioxidant molecule (less its donatable H) is stable and antioxidants thus break the autocatalytic chain reaction.

Milk and dairy products contain several antioxidants, of which the following are probably the most important:

Tocopherols (Vitamin E), which are discussed in Chap. 6. The principal function of tocopherols *in vivo* is probably to serve as antioxidants. The concentration of tocopherols in milk and meat products can be increased by supplementing the animal's diet.

Ascorbic acid (Vitamin C): at low concentrations, as in milk, ascorbic acid is an effective antioxidant, but acts as a prooxidant at higher concentrations.

Superoxidase dimutase (SOD): This enzyme, which occurs in various body tissues and fluids, including milk, scavenges superoxide radicals (O_2^-) which are powerful prooxidants. SOD is discussed in Chap. 10.

Carotenoids can act as scavengers of free radicals but whether or not they act as antioxidants in milk is controversial.

The *thiol groups* of β-lactoglobulin and proteins of the fat globule membrane are activated by heating. Most evidence indicates that thiol groups have antioxidant properties but they may also produce active oxygen species which could act as prooxidants under certain circumstances. The caseins are also effective antioxidants, possibly *via* chelation of Cu.

Some *products of the Maillard reaction* are effective antioxidants.

The addition of synthetic antioxidants, e.g., β-hydroxyanisole or butylated hydroxytoluene, to dairy products is prohibited in most countries.

3.15.4   Spontaneous Oxidation

Between 10 and 20 % of raw individual-cow milk samples undergo oxidation rapidly while others are more stable. Milks have been classified into three categories, based on their propensity to lipid oxidation:

Spontaneous: milk which is labile to oxidation without added Cu or Fe.

Susceptible: milk which is susceptible to oxidation on addition of Cu or Fe but not without.

Non-susceptible: milk that does not become oxidized even in the presence of added Cu or Fe.

It has been proposed that spontaneous milks have a high content (10 times normal) of xanthine oxidoreductase (XOR). Although addition of exogenous XOR to non-susceptible milk induces oxidative rancidity, no correlation has been found between the level of indigenous XOR and susceptibility to oxidative rancidity. The Cu-ascorbate system appears to be the principal pro-oxidant in susceptible milk. A balance between the principal anti-oxidant in milk, α-tocopherol (see Chap. 6), and XOR may determine the oxidative stability of milk. The level of superoxide dismutase (SOD) in milk might also be a factor but there is no correlation between the level of SOD and the propensity to oxidative rancidity.

3.15.5 Other Factors that Affect Lipid Oxidation in Milk and Dairy Products

Like many other reactions, lipid oxidation is influenced by the water activity (a_w) of the system. Minimal oxidation occurs at $a_w \sim 0.3$. Low values of a_w (<0.3) are considered to promote oxidation because low amounts of water are unable to "mask" pro-oxidants as happens at monolayer a_w values ($a_w \sim 0.3$). Higher values of a_w facilitate the mobility of pro-oxidants while very high values of a_w, water may have a diluent effect.

Oxygen is essential for lipid oxidation. At oxygen pressures <10 kPa (≈ 0.1 atm; oxygen content ~10 mg kg^{-1} fat), lipid oxidation is proportional to O_2 content. Low concentrations of oxygen can be achieved by flushing with inert gas, e.g., N_2, the use of glucose oxidase (see Chap. 10) or by fermentation.

Lipid oxidation is increased by decreasing pH (optimum ~ pH 3.8), perhaps due to competition between H^+ and metal ions (M^{n+}) for ligands, causing the release of M^{n+}. The principal cause may be a shift of the Cu distribution, e.g., at pH 4.6, 30–40 % of the Cu accompanies the fat globules.

Homogenization markedly reduces the propensity to oxidative rancidity, perhaps due to redistribution of the susceptible lipids and pro-oxidants of the MFGM; however, the propensity to hydrolytic rancidity and sunlight oxidized flavour (due to the production of methional from methionine in protein) is increased.

NaCl reduces the rate of auto-oxidation in sweet-cream butter but increases it in ripened cream butter (pH ~ 5); the mechanism is unknown.

In addition to influencing the rate of lipid oxidation *via* activation of thiol groups and metallo-enzymes, heating milk may also affect oxidation *via* redistribution of Cu [which migrates to the FGM on heating] and possibly by the formation of Maillard browning products, some of which have metal chelating and antioxidant properties.

The rate of auto-oxidation increases with increasing temperature ($Q_{10} \sim 2$) but oxidation in raw and HTST-pasteurized milk is promoted by low temperatures whereas the reverse is true for UHT-sterilized products (i.e., the effect of temperature is normal). The reason(s) for this anomalous behaviour is unknown.

3.15.6 Measurement of Lipid Oxidation

In addition to organoleptic assessment, several chemical/physical methods have been developed to measure lipid oxidation. These include: peroxide value, thiobarbituric acid (TBA) value, ultraviolet absorption (at 233 nm), ferric thiocyanate, Kreis test, chemiluminescence, oxygen uptake and analysis of carbonyls by HPLC (see Rossell 1986; O'Connor and O'Brien 2006).

3.16 Rheology of Milk Fat

The rheological properties of many dairy products are strongly influenced by the amount and melting point of the fat present. The sensory properties of cheese are strongly influenced by fat content but the effect is even greater in butter in which hardness/spreadability is of major concern. The hardness of fats is determined by the ratio of solid to liquid fat which is influenced by: fatty acid profile, fatty acid distribution and processing treatments.

3.16.1 Fatty Acid Profile and Distribution

The fatty acid profile of ruminant fats (milk and adipose tissue) is relatively constant due to the "buffering" action of the rumen microflora that modify ingested lipids. However, the proportions of various fatty acids in milk lipids show seasonal/nutritional/lactational variations (Fig. 3.5) which are reflected in seasonal variations in the hardness of milk fat (Fig. 3.7).

The fatty acid profile can be modified substantially by feeding encapsulated (protected) polyunsaturated oils to cows. The oil is encapsulated in a film of formaldehyde-treated protein or in crushed oil-rich seeds. The encapsulating protein is digested in the abomasum, resulting in the release of the unsaturated lipid, a high proportion of the fatty acids of which are then incorporated into the milk (and adipose tissue) lipids. The technical feasibility of this approach has been demonstrated but it is not used widely.

The melting point of triglycerides is determined by the fatty acid profile and the position of the fatty acids in the triglyceride. The melting point of fatty acids increases with increasing length of the acyl chain (Fig. 3.35) and the number, position and isomeric form of double bonds. The melting point decreases as the number

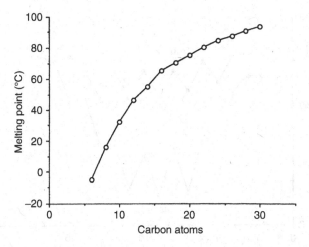

Fig. 3.35 Relationship between the melting point of fatty acids and their chain length

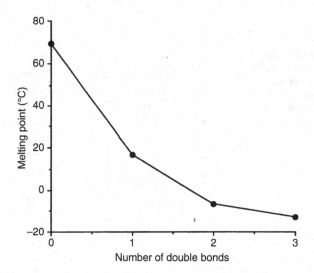

Fig. 3.36 Effect of introducing one or more double bonds on the melting point of octadecanoic acid

of double bonds in the molecule increases (Fig. 3.36) and *cis* isomers have lower melting points than the corresponding *trans* isomers (Fig. 3.37). The melting point of both *cis* and *trans* isomers increases as the double bond moves from the carboxyl group towards the ω-carbon.

Symmetrical triglycerides have a higher melting point than an asymmetrical molecule containing the same fatty acids (Table 3.16).

As discussed in Sect. 3.6, the fatty acids in milk fat are not distributed randomly and the melting point may be modified by randomizing the fatty acid distribution by transesterification using a lipase or chemical catalysts.

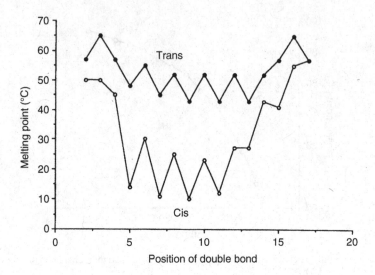

Fig. 3.37 Effect of the position of the double bond on the melting point of octadecenoic acid

Table 3.16 Effect on the	Symmetrical		Asymmetrical	
melting point of shortening a single fatty acid chain of triglyceride from 18 to 0 carbon atoms and of esterification position (symmetrical orasymmetrical)	Glyceride	MP (°C)	Glyceride	MP (°C)
	18-18-18	73.1	18-18-18	73.1
	18-16-18	68	18-18-16	65
	18-14-18	62.5	18-18-14	62
	18-12-18	60.5	18-18-12	54
	18-10-18	57	18-18-10	49
	18-8-18	51.8	18-18-8	47.6
	18-6-18	47.2	18-18-6	44
	18-4-18	51	18-18-4	–
	18-2-18	62	18-18-2	55.2
	18-0-18	78	18-18-0	68

3.16.2   Process Parameters

3.16.2.1   Temperature Treatment of Cream

The melting point of lipids is strongly influenced by the crystalline form, α, β, β^1, which is influenced by the structure of the triglycerides and by the thermal history of the product. The hardness of butter can be reduced by subjecting the cream to one of a variety of temperature programmes, that may be automated. The classical example of this is the Alnarp process, a typical example of which involves cooling pasteurized cream to ~8 °C, holding for ~2 h, warming to 20 °C, holding for ~2 h and then cooling to ~10 °C for churning. More complicated schedules may be justified in certain cases.

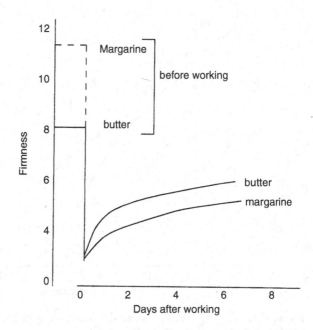

Fig. 3.38 Effect of microfixing on the hardness of butter and conventional margarine (from Mulder and Walstra 1974)

All these treatments exert their effect by controlled crystal growth, e.g., larger, fewer crystals adsorb less liquid fat and there is less formation of mixed (liquid-solid) crystals due to reduced supercooling.

3.16.2.2 Work Softening (Microfixing)

The liquid fat in butter crystallizes during cold storage after manufacture, forming an interlocking crystal network and resulting in increased hardness. Firmness can be reduced by 50–55 % by disrupting this network, e.g., by passing the product through a small orifice (Fig. 3.38), a process known as "microfixing"(the hardness of margarine can be reduced by 70–75 % by a similar process; the greater impact of disrupting the crystal network on the hardness of margarine, makes margarine appear to be more spreadable than butter even when both contain the same proportion of solid fat). Microfixing is relatively more effective when a strong crystal network has formed, i.e., when setting is at an advanced stage, e.g., after storage at 5 °C for 7 days. The effect of microfixing is reversed on storage or by warming/cooling, i.e., is essentially a reversible phenomenon (Fig. 3.38).

3.16.2.3 Fractionation

The melting and spreading characteristics of butter can be altered by fractional crystallization, i.e., controlled crystallization of molten fat or crystallization from a solution of fat in an organic solvent (e.g., ethanol or acetone). Cleaner, sharper

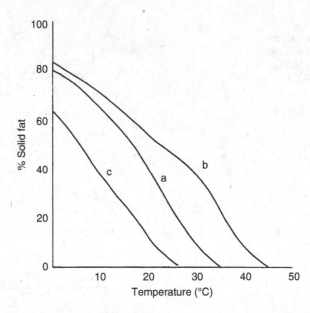

Fig. 3.39 Melting point curves of unfractionated milk fat (**a**), fraction solid at 25 °C (**b**), fraction liquid at 25 °C (**c**) (from Mulder and Walstra 1974)

fractionation is obtained in the latter but solvents may not be acceptable for use with foods. The crystals formed may be removed by centrifugation (special centrifuges have been developed) or filtration. Early studies on fractional crystallization involved removing the high-melting point fraction for use in other applications, the mother liquor being used as a modified butter spread. This approach shifts the melting point—temperature curve to lower temperatures without significantly changing its shape (Fig. 3.39). While the resulting butter has acceptable spreadability at low temperatures, its "stand-up" properties are unsatisfactory, i.e., it becomes totally liquid at too low a temperature. A better approach is to blend low and high melting point fractions by which an ideal melting curve can be approached. The problem of finding economic uses for the middle melting point fraction remains.

3.16.2.4 Blending

Blends of vegetable oils and milk fat offer an obvious solution to the problem of butter hardness—any desired hardness values can be obtained. Such products were introduced in the 1960s and are now used widely in many countries. These products may be produced by blending an emulsion of the oil with dairy cream for the manufacture of butter or by blending the oil directly with butter.

In addition to modifying the rheological properties of butter, blends of milk fat and vegetable oils can be produced at a reduced cost (depending on the price paid for milk fat) and have an increased content of polyunsaturated fatty acids, which probably has a nutritional advantage. Oils rich in $\omega - 3$ fatty acids, which are considered to have desirable nutritional properties, may be included in the blend although these oils may be susceptible to oxidative rancidity.

3.16.2.5   Low-Fat Spreads

Spreads containing 40 % fat (milk fat or blends of milk fat and vegetable oils), ~3–5 % protein and selected emulsifiers are now commonly available in many countries. These products have good spreadability and reduced calorie density (see Keogh 1995).

3.16.2.6   High Melting Point Products

Butter may be too soft for use as a shortening in certain applications; a more suitable product may be produced by blending butter and lard or tallow.

3.17   Analytical Methods for the Quantitative Determination of Milk Fat

When milk was processed (into butter or cheese) on the producing farm, determination of its fat content was not important, but with the development of creameries after the mid-nineteenth century, the farmer was paid for milk on the basis of its butter-making potential, i.e., on its fat content. Initially, butter-making potential was estimated by churning a sample of the milk and determining the amount of resulting butter; this was a very cumbersome approach. The first analytical method for determining the fat content of foods, and similar materials, was developed by Franz Soxhlet in 1879, and is still the standard reference method. A weighed sample of food is placed in a heavy filter paper thimble and extracted continuously with ethyl ether until fat extraction is complete, up to 24 h. The ether is then evaporated off and the residue of fat in the flask weighed. A diagram of the Soxhlet apparatus is shown in Fig. 3.40a.

The Soxhlet method is not suitable for liquids, including milk, and several ether extraction methods were developed for determining the fat content of milk and dairy products. The first of these was developed by B. Röse in 1884 and modified by E. Gottlieb in 1892; this method, known as the Röse-Gottlieb method, is the standard reference method for determining the fat content of milk and dairy products. A diagram of the Röse-Gottlieb apparatus in shown in Fig. 3.40b. In the case of milk, the globular fat is demulsified by treatment with NH_4OH and ethanol and the "free" fat extracted using a mixture of ethyl and petroleum ether.

The Röse-Gottlieb method is slow and tedious (it requires about 8 h to complete an enalysis) and a special apparatus was developed by Timothy Mojonnier in 1922

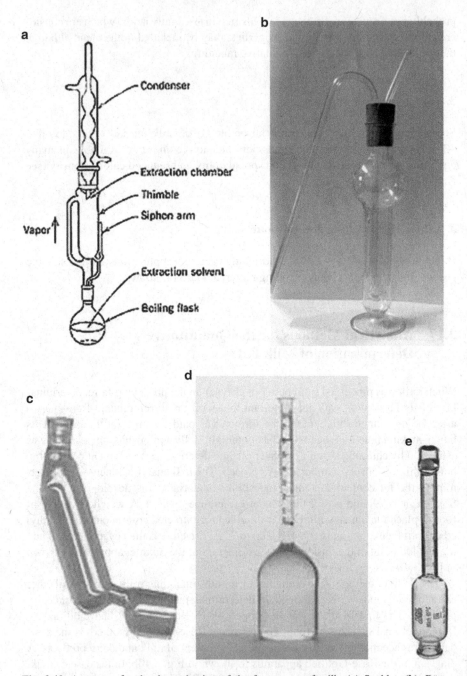

Fig. 3.40 Apparatus for the determination of the fat content of milk. (**a**) Soxhlet, (**b**). Röse-Gottlieb, (**c**) Mojonnier, (**d**) Babcock, (**e**) Gerber

to facilitate and speed-up the analysis. With Hugh Troy, Mojonnier established the Mojonnier Company which produced special glasswear (Fig. 3.40c), centrifuges to facilitate phase separation and evaporation and drying equipment. Mojonnier and Troy published a book *"The Technical Control of Dairy Products"*, in Chicago in 1925. The Mojonier version of the Röse-Gottlieb method is that usually used in the dairy industry.

The Röse-Gottlieb method, and the Mojonnier modification are not amenable for the analysis of large numbers of milk samples, such as required at a creamery, to facilitate which, two rapid volumetric methods were developed. In 1890, Dr. S.M. Babcock, developed a method from the determination of the fat content of milk and dairy products, involving dissolving the protein with concentrated H_2SO_4 and measuring the volume of the fat in a special calibrated glass tube, a butyrometer (Fig. 3.40d). Until recently, the Babcock method was the usual method used for the determination of fat in milk in the USA and many other countries.

In 1891, Dr. N. Gerber developed a method which is similar in principle to that of Babcock but he included n-butanol to clarify the fat column and used different butyrometers (Fig. 3.40e). The Gerber method became the usual method for determination of the fat content of milk in Europe.

Concentrated H_2SO_4 is very corrosive and various alternative, especially detergent were used instead but these methods were short lived. Around 1960, light scattering (turbidometric) were developed, e.g., the "Milkotester" for determination of the fat content of milk. These were used widely for a period but were replaced by infrared spectroscopy from about 1970. The ester bond of triglycerides absorbs IR radiation at 5.7 μm, the peptide bond of proteins absorbs IR radiation at 6.46 μm (the amide II band) and the –O–H of lactose absorbs at 9.5 μm. Thus, the three principal constituents of milk, lipids, proteins and lactose, can be determined quantitatively in a single IR scan and has become a widely used method for milk analysis.

3.18   Appendix A: Principal Fatty Acids in Milk Fat

Abbreviated designation	Structure	Systematic name	Common name	Melting point (°C)	Odour threshold value mg/kg
Saturated					
$C_{4:0}$	$CH_3(CH_2)_2COOH$	Butanoic acid	Butyric acid	−7.9	0.5–10
$C_{6:0}$	$CH_3(CH_2)_4COOH$	Hexanoic acid	Caproic acid	−3.9	3
$C_{8:0}$	$CH_3(CH_2)_6COOH$	Octanoic acid	Caprylic acid	16.3	3
$C_{10:0}$	$CH_3(CH_2)_8COOH$	Decanoic acid	Capric acid	31.3	10
$C_{12:0}$	$CH_3(CH_2)_{10}COOH$	Dodecanoic acid	Lauric acid	44.0	10
$C_{14:0}$	$CH_3(CH_2)_{12}COOH$	Tetradecanoic acid	Myristic acid	54.0	
$C_{16:0}$	$CH_3(CH_2)_{14}COOH$	Hexadecanoic acid	Palmitic acid	62.9	
$C_{18:0}$	$CH_3(CH_2)_{16}COOH$	Octadecanoic acid	Stearic acid	69.6	
Unsaturated					
	ω9-Family				
16:1	$CH_3(CH_2)_5CH=CH-CH_2-(CH_2)_6-COOH$	Δ9-Hexadecenoic acid	Palmitoleic acid	0.5	
18:1	$CH_3(CH_2)_7CH=CH-CH_2-(CH_2)_6-COOH$	Δ9-Octadecenoic acid	Oleic acid	13.4	
	ω6-Family				
18:2	$CH_3(CH_2)_4-(CH=CH-CH_2)_2-(CH_2)_6-COOH$	Δ9,12-Octadecdienoic acid	Linoleic acid	−5.0	
18:3	$CH_3(CH_2)_4-(CH=CH-CH_2)_3-(CH_2)_3-COOH$	Δ6,9,12-Octadectrienoic acid	γ-Linoleic acid		
20:4	$CH_3(CH_2)_4-(CH=CH-CH_2)_4-(CH_2)_2-COOH$	Δ5,8,11,14-Ecosatetraenoic acid	Arachidonic acid	−49.5	
	ω3-Family				
18:3	$CH_3-CH_2-(CH=CH-CH_2)_3-(CH_2)_6-COOH$	Δ9,12,15-Octadectrienoic acid	α-Linolenic acid	−11.0	
	Δ9-Family				

3.19 Appendix B

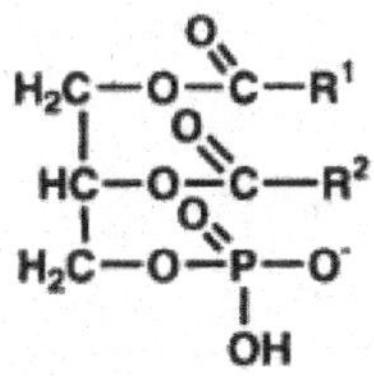

A Phosphatidic acid

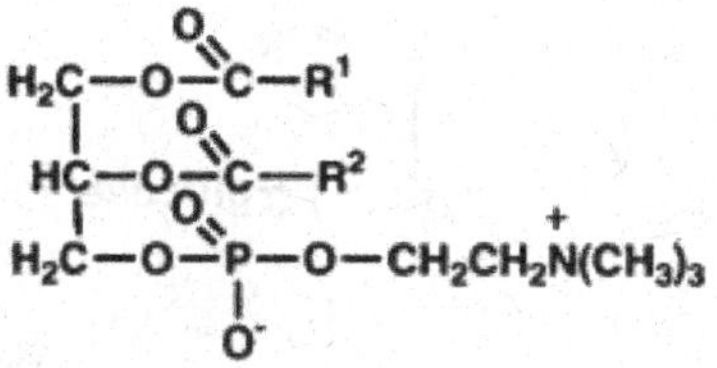

A phosphatidylcholine (lecithin)

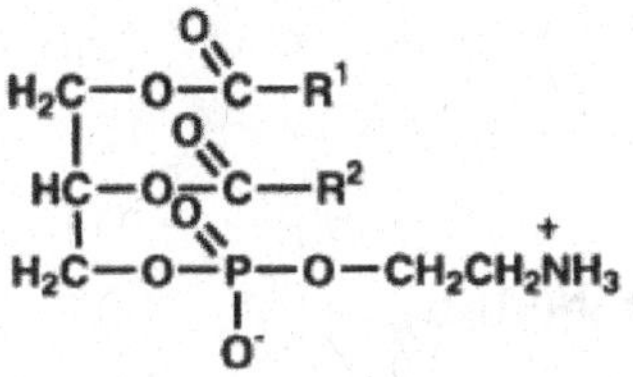

A Phosphatidylethanolamine

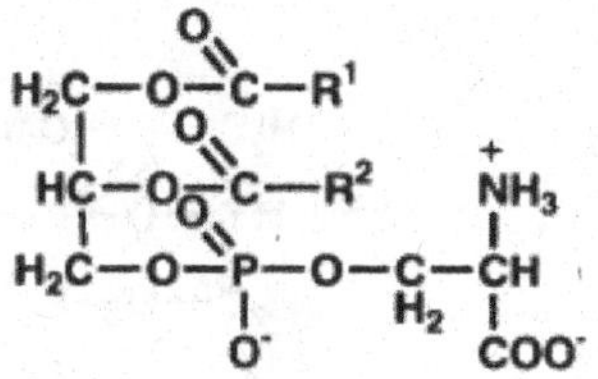

A phosphatidylserine

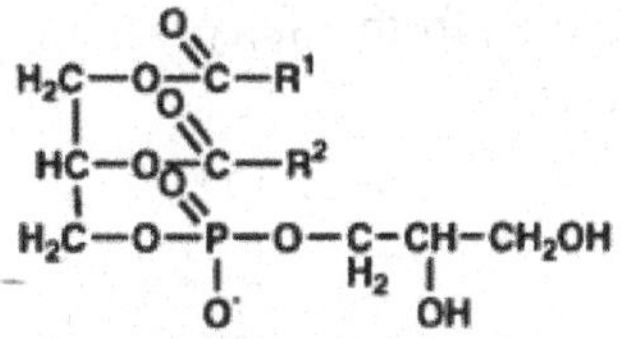

A phosphatidylglycerol

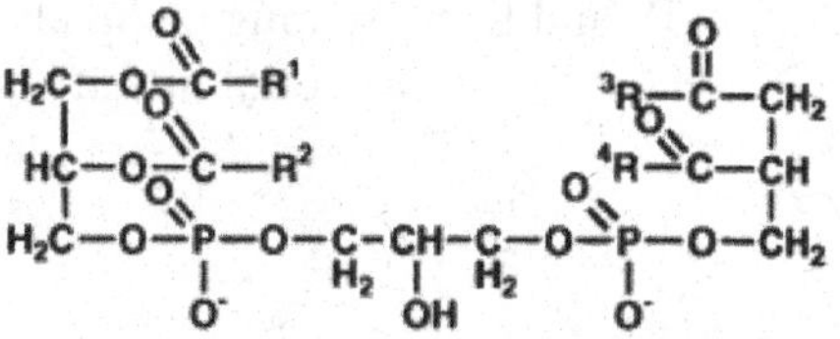

A disphosphatidylglycerol (cardiolipin)

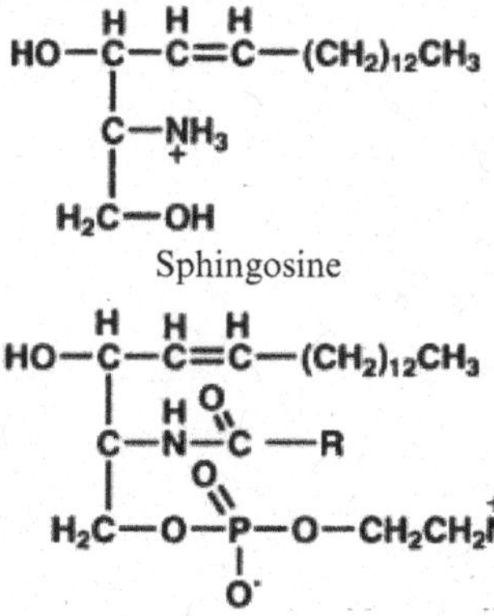

Sphingosine

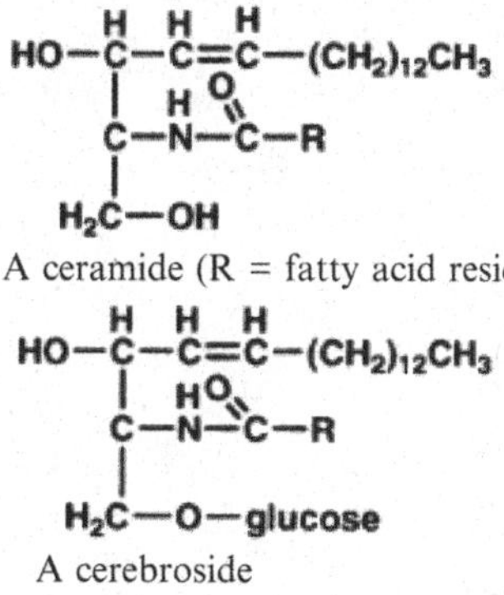

A ceramide (R = fatty acid residue)

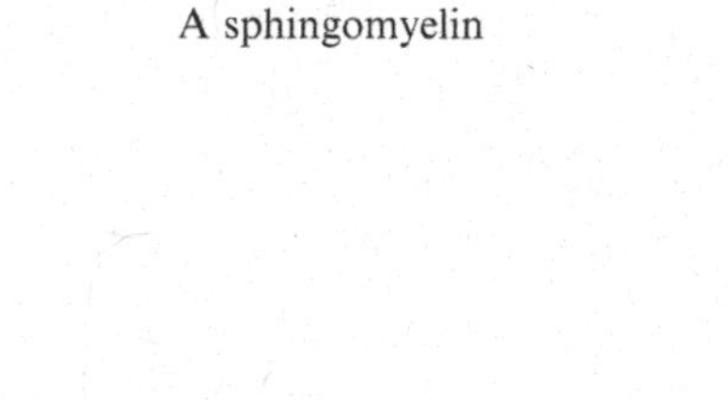

A sphingomyelin

A cerebroside

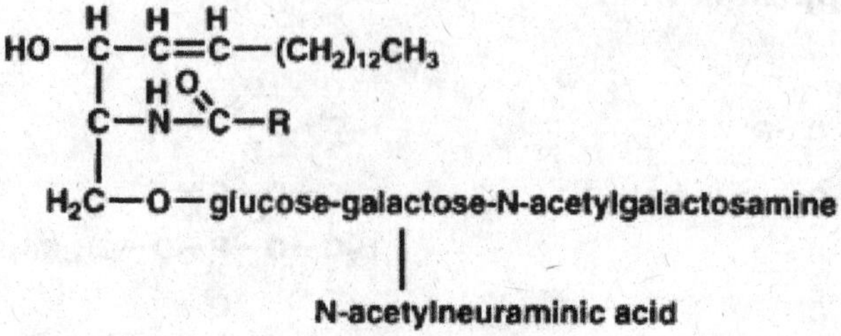

$$HO-\underset{\underset{\underset{\underset{H_2C-O-}{|}}{\overset{|}{C}-\overset{O}{\overset{\|}{N}}-\overset{}{C}-R}}{|}}{\overset{H}{\overset{|}{C}}}-\overset{H}{\overset{}{C}}=\overset{H}{\overset{}{C}}-(CH_2)_{12}CH_3$$

$$H_2C-O-\text{glucose-galactose-N-acetylgalactosamine}$$

$$|$$

N-acetylneuraminic acid

A ganglioside

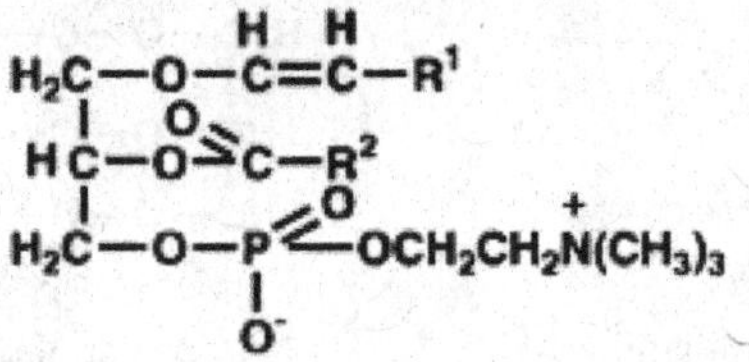

$$H_2C-O-\overset{H}{\overset{|}{C}}=\overset{H}{\overset{|}{C}}-R^1$$
$$H\overset{}{C}-O\overset{O}{\gtrless}C-R^2$$
$$H_2C-O-P\overset{O}{\overset{\angle}{\underset{|}{\underset{O^-}{}}}}OCH_2CH_2\overset{+}{N}(CH_3)_3$$

R^1 and R^2 = are long chain alkyl groups derived from a fatty alcohol or fatty acid, respectively.

A plasmalogen

3.20 Appendix C

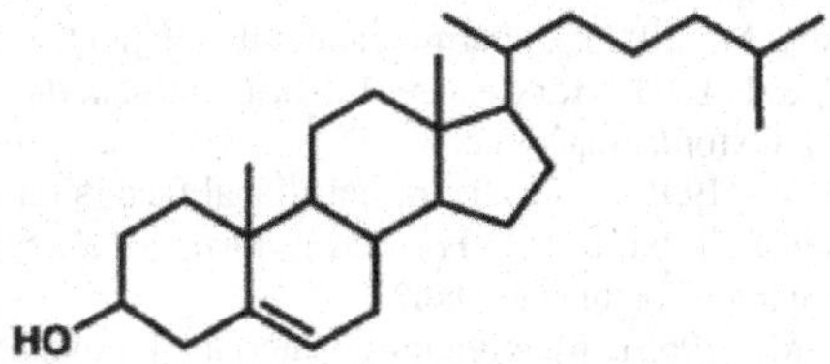

Cholesterol

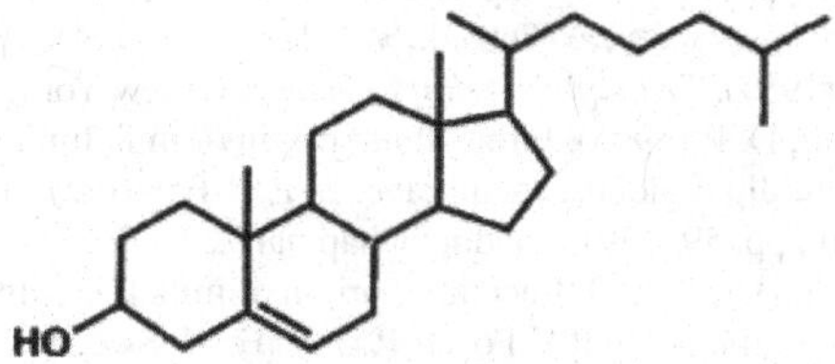

7-Dehydrocholesterol

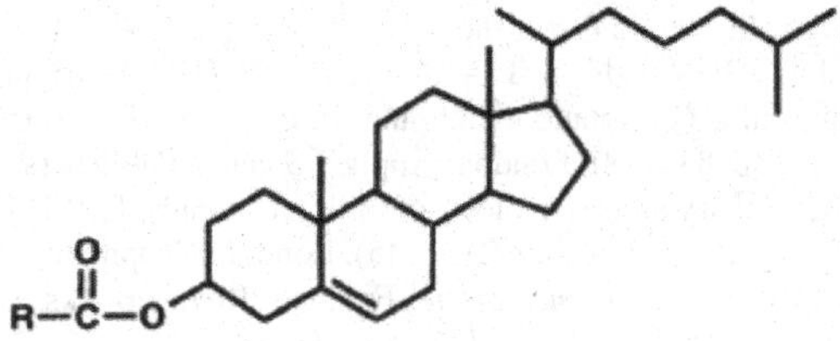

Cholesteryl ester

References

An Foras Taluntais. (1981). *Chemical composition of milk in Ireland.* Dublin: An Foras Taluntais.

Brunner, J. R. (1965). Physical equilibria in milk: The lipid phase. In B. H. Webb & A. H. Johnson (Eds.), *Fundamentals of dairy chemistry* (pp. 403–505). Westport, CT: AVI Publishing Co., Inc.

Brunner, J. R. (1974). Physical equilibria in milk: The lipid phase. In B. H. Webb, A. H. Johnson, & J. A. Alford (Eds.), *Fundamentals of dairy chemistry* (2nd ed., pp. 474–602). Westport, CT: AVI Publishing Co., Inc.

Christie, W. W. (1995). Composition and structure of milk lipids. In P. F. Fox (Ed.), *Advanced dairy chemistry – 2 – lipids* (2nd ed., pp. 1–36). London: Chapman & Hall.

Cremin, F. H., & Power, P. (1985). Vitamins in bovine and human milks. In P. F. Fox (Ed.), *Developments in dairy chemistry – 3 – lactose and minor constituents* (pp. 337–398). London: Elsevier Applied Science Publishers.

Cullinane, N., Aherne, S., Connolly, J. F., & Phelan, J. A. (1984a). Seasonal variation in the triglyceride and fatty acid composition of Irish butter. *Irish Journal of Food Science and Technology, 8,* 1–12.

Cullinane, N., Condon, D., Eason, D., Phelan, J. A., & Connolly, J. F. (1984b). Influence of season and processing parameters on the physical properties of Irish butter. *Irish Journal of Food Science and Technology, 8*, 13–25.

Eyzaguirre, R. Z., & Corredig, M. (2011). Buttermilk and milk fat globule membrane fractions. In J. W. Fuquay, P. F. Fox, & P. L. H. McSweeney (Eds.), *Encycliopedia of dairy sciences* (2nd ed., Vol. 3, pp. 691–697). Oxford: Academic.

Hawke, J. C., & Taylor, M. W. (1995). Influence of nutritional factors on the yield, composition and physical properties of milk fat. In P. F. Fox (Ed.), *Advanced dairy chemistry – 2 – lipids* (2nd ed., pp. 37–88). London: Chapman & Hall.

Hayashi, S., & Smith, L. M. (1965). Membranous material of bovine milk fat globules. 1. Comparison of membranous fractions released by deoxycholate and by churning. *Biochemistry, 4*, 2550–2557.

Huppertz, T. (2011). Other types of homogenizer. In J. W. Fuquay, P. F. Fox, & P. L. H. McSweeney (Eds.), *Encyclopedia of dairy sciences* (2nd ed., Vol. 2, pp. 761–764). Oxford: Academic Press.

Jenness, R., & Patton, S. (1959). *Principles of dairy chemistry*. New York: Wiley.

Keenan, T. W., & Dylewski, D. P. (1995). Intracellular origin of milk lipid globules and the nature and structure of the milk lipid globule membrane. In P. F. Fox (Ed.), *Advanced dairy chemistry – 2 – lipids* (2nd ed., pp. 89–130). London: Chapman & Hall.

Keenan, T. W., & Mathur, I. H. (2006). Intracellular origin of milk lipid globules and the nature of the milk lipid globule membrane. In P. F. Fox & P. L. H. H. McSweeney (Eds.), *Advanced dairy chemistry – 2 – lipids* (3rd ed., pp. 137–171). New York: Springer.

Keenan, T. W., & Patton, S. (1995). The structure of milk: Implications for sampling and storage. A. The milk lipid globule membrane. In R. G. Jensen (Ed.), *Handbook of milk composition* (pp. 5–50). San Diego: Academic Press, Inc.

Keenan, T. W., Dylewski, D. P., Woodford, T. A., & Ford, R. H. (1983). Origin of milk fat globules and the nature of the milk fat globule membrane. In P. F. Fox (Ed.), *Developments in dairy chemistry – 2 – lipids* (pp. 83–118). London: Applied Science Publishers.

Keogh, M. K. (1995). Chemistry and technology of milk fat spreads. In P. F. Fox (Ed.), *Advanced dairy chemistry – 2 – lipids* (2nd ed., pp. 213–245). London: Chapman & Hall.

King, N. (1955). *The milk fat globule membrane*. Farnham Royal, Bucks, UK: Commonwealth Agricultural Bureau.

Lehninger, A. L., Nelson, D. L., & Cox, M. M. (1993). *Principles of biochemistry* (2nd ed.). New York: Worth Publishers.

Liang, B., & Hartel, R. W. (2004). Effects of milk powders on milk chocolate. *Journal of Dairy Science, 87*, 20–31.

Mather, I. H. (2000). A review and proposed nomenclature of the major milk proteins of the milk fat globule membrane. *Journal of Dairy Science, 83*, 203–247.

Mather, I. H. (2011). Milk fat globule membrane. In J. W. Fuquay, P. F. Fox, & P. L. H. McSweeney (Eds.), *Encyclopedia of dairy sciences* (2nd ed., pp. 680–690). Oxford: Academic.

McDowall, F. H. (1953). *The buttermakers manual* (Vol. I and II). Wellington: New Zealand University Press.

McPherson, A. V., & Kitchen, B. J. (1983). Reviews of the progress of dairy science: The bovine milk fat globule membrane – its formation, composition, structure and behaviour in milk and dairy products. *Journal of Dairy Research, 50*, 107–133.

Mortensen, B. K. (2011a). Butter and other milk fat products. In J. W. Fuquay, P. F. Fox, & P. L. H. McSweeney (Eds.), *Encyclopedia of dairy sciences* (2nd ed., Vol. 1, pp. 492–499). Oxford: Academic.

Mortensen, B. K. (2011b). Modified butters. In J. W. Fuquay, P. F. Fox, & P. L. H. McSweeney (Eds.), *Encyclopedia of dairy sciences* (2nd ed., Vol. 1, pp. 500–505). Oxford: Academic.

Mulder, H., & Walstra, P. (1974). *The milk fat globule: Emulsion science as applied to milk products and comparable foods*. Wageningen: Podoc.

O'Connell, J. E., & Fox, P. F. (2000). Heat stability of buttermilk. *Journal of Dairy Science, 83*, 1728–1732.

O'Connor, T. P., & O'Brien, N. M. (1995). Lipid oxidation. In P. F. Fox (Ed.), *Advanced dairy chemistry – 2 – lipids* (2nd ed., pp. 309–347). London: Chapman & Hall.

O'Connor, T. P., & O'Brien, N. M. (2006). Lipid oxidation. In P. F. Fox & P. L. H. McSweeney (Eds.), *Advanced dairy chemistry – 2 – lipids* (3rd ed., pp. 557–600). New York: Springer.

Palmquist, D. L. (2006). Milk fat: Origin of fatty acids and influence of nutritional factors thereon. In P. F. Fox & P. L. H. McSweeney (Eds.), *Advanced dairy chemistry – 2 – lipids* (3rd ed., pp. 43–92). New York: Springer.

Patton, S., & Keenan, T. W. (1975). The milk fat globule membrane. *Biochimica et Biophysica Acta, 415,* 273–309.

Peereboom, J. W. C. (1969). Theory on the renaturation of alkaline milk phosphates from pasteurized cream. *Milchwissenschaft, 24,* 266–269.

Prentice, J. H. (1969). The milk fat globule membrane 1955–1968. *Dairy Science Abstracts, 31,* 353–356.

Richardson, T., & Korycka-Dahl, M. (1983). Lipid oxidation. In P. F. Fox (Ed.), *Developments in dairy chemistry – 2 – lipids* (pp. 241–363). London: Applied Science Publishers.

Rossell, J. B. (1986). Classical analysis of oils and fats. In R. J. Hamilton & J. B. Rossell (Eds.), *Analysis of oils and fats* (pp. 1–90). London: Elsevier Applied Science.

Sodini, I., Morin, P., Olabi, A., & Jimenez-Flores, R. (2006). Compositional and functional properties of buttermilk: A comparison between sweet, sour and whey buttermilk. *Journal of Dairy Science, 89,* 525–536.

Towler, C. (1994). Developments in cream separation and processing. In R. K. Robinson (Ed.), *Modern dairy technology* (2nd ed., Vol. 1, pp. 61–105). London: Chapman & Hall.

USDA. (2014). *Dairy: World markets and trade.* United States Department of Agriculture, Foreign Agricultural Service, Washington, DC.

Walstra, P. (1983). Physical chemistry of milk fat globules. In P. F. Fox (Ed.), *Developments in dairy chemistry – 2 – lipids* (pp. 119–158). London: Applied Science Publishers.

Walstra, P., & Jenness, R. (1984a). *Dairy chemistry and physics.* New York: Wiley.

Ward, R. E., Greman, J. B., & Corredig, M. (2006). Composition, applications, fractionation, technological and nutritional significance of milk fat globule material. In P. F. Fox & P. L. H. McSweeney (Eds.), *Advanced dairy chemistry – 2 – lipids* (3rd ed., pp. 213–244). New York: Springer.

Wilbey, R. A. (1994). Production of butter and dairy based spreads. In R. K. Robinson (Ed.), *Modern dairy technology* (2nd ed., Vol. 1, pp. 107–158). London: Chapman & Hall.

Wooding, F. B. P. (1971). The structure of the milk fat globule membrane. *Journal of Ultrastructure Research, 37,* 388–400.

Suggested Reading

Bauman, D. E., & Luck, A. J. (2006). Conjugated linoleic acid: Biosynthesis and nutritional significance. In P. F. Fox & P. L. H. McSweeney (Eds.), *Advanced dairy chemistry – 2 – lipids* (3rd ed., pp. 93–136). New York: Springer.

Deeth, H. C., & Fitz-Gerald, C. H. (2006). Lipolytic enzymes and hydrolytic rancidity. In P. F. Fox & P. L. H. McSweeney (Eds.), *Advanced dairy chemistry – 2 – lipids* (3rd ed., pp. 481–556). New York: Springer.

Fox, P. F. (Ed.). (1983). *Developments in dairy chemistry – 2 – lipids.* London: Applied Science Publishers.

Fox, P. F. (Ed.). (1995). *Advanced dairy chemistry – 2 – lipids* (2nd ed.). London: Chapman & Hall.

Fox, P. F., & McSweeney, P. L. H. (Eds.). (2006). *Advanced dairy chemistry – 2 – lipids* (3rd ed.). New York: Springer.

Freda, E. (2011). Butter: Properties and analysis. In J. W. Fuquay, P. F. Fox, & P. L. H. McSweeney (Eds.), *Encyclopedia of dairy sciences* (2nd ed., pp. 506–514). Oxford: Academic.

Fuquay, J. W., Fox, P. F., & McSweeney, P. L. H. (2011). Milk lipids. In *Encyclopedia of dairy sciences* (2nd ed., Vol. 3, pp. 649–740). Oxford: Academic Press.

Huppertz, T., & Kelly, A. L. (2006). Physical chemistry of milk fat globules. In P. F. Fox & P. L. H. McSweeney (Eds.), *Advanced dairy chemistry – 2 – lipids* (3rd ed., pp. 173–212). New York: Springer.

Keenan, T. W., Mather, I. H., & Dylewski, D. P. (1988). Physical equilibria: Lipid phase. In N. P. Wong (Ed.), *Fundamentals of dairy chemistry* (3rd ed., pp. 511–582). New York: van Nostrand Reinhold.

MacGibbon, A. K. M., & Reynolds, M. A. (2011). Milk lipids: Analytical methods. In J. W. Fuquay, P. F. Fox, & P. L. H. McSweeney (Eds.), *Encyclopedia of dairy sciences* (2nd ed., Vol. 3, pp. 698–703). Oxford: Academic.

MacGibbon, A. K. H., & Taylor, M. W. (2006). Composition and structure of bovine milk lipids. In P. F. Fox & P. L. H. McSweeney (Eds.), *Advanced dairy chemistry – 2 – lipids* (3rd ed., pp. 1–42). New York: Springer.

Mortensen, B. K. (2011c). Butter and other milk fat products: The product and its manufacture. In J. W. Fuquay, P. F. Fox, & P. L. H. McSweeney (Eds.), *Encyclopedia of dairy sciences* (2nd ed., Vol. 1, pp. 492–499). Oxford: Academic.

Mortensen, B. K. (2011d). Modified butters. In J. W. Fuquay, P. F. Fox, & P. L. H. McSweeney (Eds.), *Encyclopedia of dairy sciences* (2nd ed., Vol. 1, pp. 500–505). Oxford: Academic.

Mortensen, B. K. (2011e). Butter and other milk fat products. In J. W. Fuquay, P. F. Fox, & P. L. H. McSweeney (Eds.), *Encyclopedia of dairy sciences* (2nd ed., Vol. 1, pp. 515–521). Oxford: Academic.

Mortensen, B. K. (2011f). Milk fat-based spreads. In J. W. Fuquay, P. F. Fox, & P. L. H. McSweeney (Eds.), *Encyclopedia of dairy sciences* (2nd ed., Vol. 1, pp. 522–527). Oxford: Academic.

Mortensen, B. K. (2014). *Butter and related products.* Odense, Denmark: International Dairy Books.

Mulder, H., & Walstra, P. (1974). *The milk fat globule.* Wageningen: Podoc.

Walstra, P., & Jenness, R. (1984b). *Dairy chemistry and physics.* New York: Wiley Interscience.

Webb, B. H., & Johnson, A. H. (1965). *Fundamentals of dairy chemistry.* Westport, CT: AVI Publishing Co. Inc.

Webb, B. H., Johnson, A. H., & Alford, J. A. (Eds.). (1974). *Fundamentals of dairy chemistry* (2nd ed.). Westport, CT: AVI Publishing Co. Inc.

Wong, N. P. (Ed.). (1980). *Fundamentals of dairy chemistry – 1* (3rd ed.). Westport, CT: AVI Publishing Co. Inc.

Wright, A. J., & Marangoni, A. G. (2006). Crystallization and rheological properties of milk fat. In P. F. Fox & P. L. H. McSweeney (Eds.), *Advanced dairy chemistry – 2 – lipids* (3rd ed., pp. 245–291). New York: Springer.

Wright, A. J., Marangoni, A. G., & Hartel, R. W. (2011). Milk lipids: Rheological properties and their measurement. In J. W. Fuquay, P. F. Fox, & P. L. H. McSweeney (Eds.), *Encyclopedia of dairy sciences* (2nd ed., Vol. 3, pp. 704–710). Oxford: Academic.

Chapter 4
Milk Proteins

4.1 Introduction

Normal bovine milk contains about 3.5 % protein. The concentration changes significantly during lactation, especially during the first few days post-partum (Fig. 4.1); the greatest change occurs in the whey protein fraction (Fig. 4.2). The natural function of milk proteins is to supply young mammals with the essential amino acids required for the development of muscular and other protein-containing tissues, and with a number of biologically active proteins, e.g. immunoglobulins, vitamin-binding and metal-binding proteins and various protein hormones. The young of different species are born at very different states of maturity, and, consequently, have different nutritional and physiological requirements. These differences are reflected in the protein content of the milk of the species, which ranges from ~1 to ~20 % (Table 4.1). The protein content of milk is directly related to the growth rate of the young of that species (Fig. 4.3), reflecting the requirements of protein for growth.

The properties of many dairy products, in fact their very existence, depend on the properties of milk proteins, although the fat, lactose and especially the salts, exert very significant modifying influences. Casein products are almost exclusively milk protein while the production of most cheese varieties is initiated through the specific modification of proteins by proteolytic enzymes or isoelectric precipitation. The high heat treatment to which many milk products is subjected are possible only because of the exceptionally high heat stability of the principal milk proteins, the caseins. Traditionally, milk was paid for mainly on the basis of its fat content but milk payments are now usually based on the content of fat plus protein. Specifications for many dairy products include a value for protein content. Changes in protein characteristics, e.g., insolubility as a result of heat denaturation in milk powders or the increasing solubility of cheese proteins during ripening, are industrially important features of these products.

© Springer International Publishing Switzerland 2015
P.F. Fox et al., *Dairy Chemistry and Biochemistry*,
DOI 10.1007/978-3-319-14892-2_4

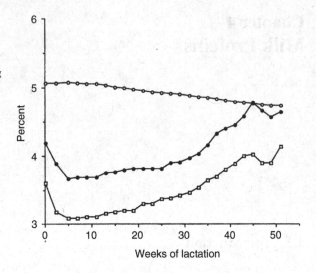

Fig. 4.1 Changes in the concentrations of lactose (*open circle*), fat (*filled circle*) and protein (*open square*) in bovine milk during lactation

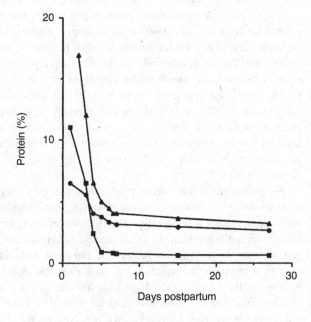

Fig. 4.2 Changes in the concentration of total protein (*filled triangle*) and of casein (*filled circle*) and whey proteins (*filled square*) in bovine milk during the early stage of lactation

It is assumed that the reader is familiar with the structure of proteins; for convenience, the structures of the amino acids found in milk are given in Appendix 4A. Throughout this chapter, the term cystine is used to indicate two disulphide-linked cysteines.

Table 4.1 Protein content (%) in the milk of some species

Species	Casein	Whey proteins	Total
Bison	3.7	0.8	4.5
Black bear	8.8	5.7	14.5
Buffalo	3.5–4.2	0.92	4.42–5.12
Camel (bactrian)	2.9	1.0	3.9
Cat			11.1
Cow	2.8	0.6	3.4
Domestic rabbit	9.3	4.6	13.9
Donkey	1.0	1.0	2.0
Echidna	7.3	5.2	12.5
Goat	2.5	0.4	2.9
Grey seal			11.2
Guinea-pig	6.6	1.5	8.1
Hare			19.5
Horse	1.3	1.2	2.5
House mouse	7.0	2.0	9.0
Human	0.4	0.6	1.0
Indian elephant	1.9	3.0	4.9
Pig	2.8	2.0	4.8
Polar bear	7.1	3.8	10.9
Red kangaroo	2.3	2.3	4.6
Reindeer	8.6	1.5	10.1
Rhesus monkey	1.1	0.5	1.6
Sheep	4.6	0.9	5.5

Fig. 4.3 Relationship between the growth rate (days to double birth weight) of the young of some species of mammal and the protein content (expressed as % of total calories derived from protein) of the milk of that species (from Bernhart 1961)

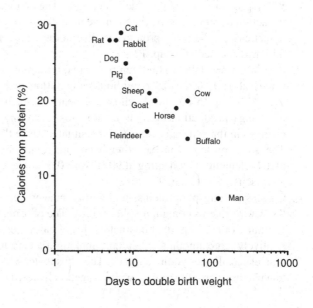

4.2 Heterogeneity of Milk Proteins

Initially, it was believed that milk contained only one type of protein but about 1880 it was shown by the Swedish scientist, Olav Hammarsten, that the proteins in milk can be fractionated into two well-defined groups. On acidification to pH 4.6 (the isoelectric pH) at around 30 °C, about 80 % of the total protein in bovine milk precipitates out of solution; this fraction is now called isoelectric (acid) casein and sometimes as casein *nach* Hammarsten. The proteins which remain soluble under these conditions are referred to as whey or serum protein or non-casein nitrogen. The ratio of casein to whey proteins shows large interspecies differences; in human milk, the ratio is ~40:60, in equine (mare's) milk it is 50:50 while in the milk of the cow, goat, sheep and buffalo it is ~80:20. Presumably, these differences reflect the nutritional and physiological requirements of the young of these species. There are several major differences between the caseins and whey proteins, of which the following are probably the most significant, especially from an industrial or technological viewpoint:

1. In contrast to the caseins, the whey proteins do not precipitate from solution when the pH of milk is adjusted to 4.6. This characteristic is used as the usual operational definition of casein. This difference in the properties of the two milk protein groups is exploited in the preparation of industrial casein and certain varieties of cheese (e.g., Cottage, Quarg and Cream cheese). Only the casein fraction of milk protein is normally incorporated into these products, the whey proteins being lost in the whey.
2. Chymosin and some other proteinases (known as rennets) cause a very slight, specific change in casein, resulting in its coagulation in the presence of Ca^{2+}. Whey proteins undergo no such alteration. The coagulability of casein through the action of rennets is exploited in the manufacture of most cheese varieties and rennet casein; the whey proteins are lost in the whey. The rennet coagulation of milk is discussed in Chap. 12.
3. Casein is very stable to high temperatures; milk may be heated at its natural pH (~6.7) at 100 °C for 24 h without coagulation and it withstands heating at 140 °C for up to 20 min. Such severe heat treatments cause many changes in milk, e.g., production of acids from lactose resulting in a decrease in pH and changes in the salt balance, which eventually cause the precipitation of casein. The whey proteins, on the other hand, are relatively heat labile, being completely denatured by heating at 90 °C for 10 min. Heat-induced changes in milk are discussed in Chap. 9.
4. Caseins are phosphoproteins, containing, on average, 0.85 % phosphorus, while the whey proteins contain no phosphorus. The phosphate groups are responsible for many of the important characteristics of casein, especially its ability to bind relatively large amounts of calcium, making it a very nutritionally valuable protein, especially for young animals. The phosphate, which is esterified to the protein *via* the hydroxyl group of serine, is generally referred to as organic phosphate.

Part of the inorganic phosphorus in milk is also associated with the casein in the form of colloidal calcium phosphate (~57 % of the inorganic phosphorus) (Chap. 5). The phosphate of casein is an important contributor to its remarkably high heat stability and to the calcium-induced coagulation of rennet-altered casein (although many other factors are involved in both cases).

5. Casein is low in sulphur (0.8 %) while the whey proteins are relatively rich (1.7 %) in sulphur. Differences in sulphur content become more apparent if one considers the levels of individual sulphur-containing amino acids. The sulphur of casein is present mainly in methionine, with a very low concentration of cysteine; in fact, the principal caseins contain *only* methionine. The whey proteins contain significant amounts of both cysteine and cystine in addition to methionine and these amino acids are responsible, in part, for many of the changes which occur in milk on heating, e.g., cooked flavour, increased rennet coagulation time (due to interaction between β-lactoglobulin and κ-casein) and the improved heat stability of milk pre-heated prior to sterilization.

6. Casein is synthesized in the mammary gland and is found nowhere else in nature. Some of the whey proteins (β-lactoglobulin and α-lactalbumin) are also synthesized in the mammary gland, while others (e.g., bovine serum albumin and some immunoglobulins) are derived from the blood.

7. The whey proteins are molecularly dispersed in solution or have simple quaternary structures, whereas the caseins have a complicated quaternary structure and exist in milk as large colloidal aggregates, referred to as micelles, with a particle mass of 10^6–10^9 Da.

8. Both the casein and whey protein groups are heterogeneous, each containing several different proteins.

4.2.1 Other Protein Fractions

In addition to the caseins and whey proteins, milk contains two other groups of proteins or protein-like material, i.e., the proteose-peptone fraction and the non-protein nitrogen (NPN) fraction. These fractions were recognized as early as 1938 by S.J. Rowland but until recently very little was known about them. Rowland observed that when milk was heated to 95 °C for 10 min, 80 % of the nitrogenous compounds in whey were denatured and co-precipitated with the casein when the pH of the heated milk was adjusted subsequently to 4.6. He considered that the heat-denaturable whey proteins represented the lactoglobulin and lactalbumin fractions and designated the remaining 20 % 'proteose-peptone'. The proteose peptone fraction, which is quite heterogenous (see Sect. 4.4.2) is precipitated by 12 % trichloroacetic acid (TCA) but some nitrogenous compounds remain soluble in 12 % TCA and are designated as non-protein nitrogen. A scheme for the fractionation of the principal groups of milk proteins, based on that of Rowland, is shown in Fig. 4.4.

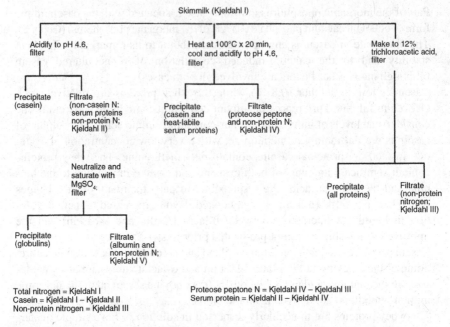

Total nitrogen = Kjeldahl I   Proteose peptone N = Kjeldahl IV – Kjeldahl III
Casein = Kjeldahl I – Kjeldahl II   Serum protein = Kjeldahl II – Kjeldahl IV
Non-protein nitrogen = Kjeldahl III

Fig. 4.4 Scheme for quantifying the principal protein fractions in milk

4.3   Preparation of Casein and Whey Proteins

Skim milk prepared by centrifugal separation (see Chap. 3) is used as the starting material for the preparation of casein and whey proteins.

4.3.1   Acid (Isoelectric) Precipitation

Acidification of milk to about pH 4.6 induces coagulation of the casein. Aggregation occurs at all temperatures, but below about 6 °C the aggregates are very fine and remain in suspension, although they can be sedimented by low-speed centrifugation. At a higher temperature (30–40 °C), the aggregates are quite coarse and precipitate readily from solution. At temperatures above about 50 °C, the precipitate tends to be stringy and difficult to handle.

For laboratory-scale production of casein, HCl is usually used for acidification; acetic or lactic acids are used less frequently. Industrially, HCl is also usually used; H_2SO_4 is used occasionally but the resulting whey is not suitable for animal feeding ($MgSO_4$ is a laxative). Lactic acid produced *in situ* by a culture of lactic acid bacteria may be used.

The inorganic colloidal calcium phosphate associated with casein in normal milk dissolves on acidification of milk to pH 4.6 so that if sufficient time is allowed for solution, isoelectric casein is essentially free of calcium phosphate. In the laboratory, best results are obtained by acidifying skim milk to pH 4.6 at 2 °C, holding for about 30 min and then warming to 30–35 °C. The fine aggregate formed at 2 °C allows time for the colloidal calcium phosphate to dissolve (Chap. 5). A moderately dilute acid (1 M) is preferred, since concentrated acid may cause localized coagulation. Acid production by a bacterial culture occurs slowly and allows time for colloidal calcium phosphate to dissolve. The casein is recovered by filtration or centrifugation and washed repeatedly with water to free the casein of lactose and salts. Thorough removal of lactose is essential since even traces of lactose will interact with casein on heating *via* the Maillard browning reaction (see Chap. 2), with undesirable consequences.

The procedure used for the industrial production of acid (isoelectric) casein is essentially the same as that used on a laboratory scale, except for many technological differences (Sect. 4.18.1). The whey proteins may be recovered from the whey by salting out, dialysis or ultrafiltration.

4.3.2 Centrifugation

Because casein occurs in milk as large aggregates, micelles, most (90–95 %) of the casein in milk is sedimented by centrifugation at $100,000 \times g$ for 1 h. Sedimentation is more complete at higher (30–37 °C) than at low (2 °C) temperature, at which some of the caseins dissociate from the micelles and are non-sedimentable. Casein prepared by centrifugation contains its original level of colloidal calcium phosphate and can be redispersed (by grinding the pellet with a mortar and pestle and stirring overnight in the cold) as micelles with properties essentially similar to the original micelles.

4.3.3 Centrifugation of Calcium-Supplemented Milk

Addition of $CaCl_2$ to about 0.2 M causes aggregation of the casein such that it can be recovered by low-speed centrifugation. If calcium is added at 90 °C, the casein forms coarse aggregates which precipitate readily. This principle is used in the commercial production of some 'casein co-precipitates' in which the whey proteins, denatured on heating milk at 90 °C for 10 min, co-precipitate with the casein. Such products have a very high ash content.

4.3.4 Salting-Out Methods

Casein can be precipitated from solution by any of several salts. Addition of $(NH_4)_2SO_4$ to milk to a concentration of 260 g L^{-1} causes complete precipitation of the casein and some whey proteins (immunoglobulins, Ig). Saturation of milk with

$MgSO_4$ or NaCl at 37 °C precipitates the casein and Igs while the major whey proteins remain soluble, provided they are not denatured. This characteristic is the basis of a commercial test used for the heat classification of milk powders which contain variable levels of denatured whey proteins.

4.3.5 Ultrafiltration

Ultrafiltration (UF) membranes with molecular weight cut off of between 10 and 20 kDa retain both the caseins and whey proteins while lactose and soluble salts are permeable. Milk protein concentrate (MPC) ingredients are manufactured using this approach, whereby the total protein fraction (i.e., caseins and whey proteins) of pasteurised skim milk is concentrated by UF prior to evaporation and spray drying to produce MPC powders. The volume concentration factor (VCF) and extent (if any) of diafiltration used during the manufacturing process determine the final protein concentration of the MPC powders. The protein content of MPC powders can range from 35 to 85 %, while the ingredients are normally classified as milk protein isolates (MPI's) once a protein concentration of ≥90 % is achieved. The MPC ingredients traded in the greatest quantities have a protein concentration ranging from 80 to 90 % and are used in the formulation of processed cheese, infant nutritional, clinical nutritional and elderly nutritional products. MPC ingredients are attracting increasing interest as they offer food formulators a soluble, nutritionally-attractive (i.e., reduced non-protein nitrogen and a good source of natural milk minerals) source of total milk protein with broad-ranging functionality. The main factors affecting the uptake of MPC ingredients is their poor rehydration and heat stability properties, although these are the basis of much on-going research.

4.3.6 Microfiltration

The casein micelles may be recovered from skim milk using microfiltration (MF) membranes with pore sizes ranging from 0.1 to 0.8 μm. Using this approach, the native casein micelles are retained by the membrane while the soluble constituents of milk, lactose, minerals and the whey proteins, are removed in the permeate. Bacteria and fat are also retained by membranes of such porosity and for this reason, such protein fractionation processing is normally conducted on skim milk. As with UF of skim milk, during MF, the VCF and extent (if any) of diafiltration used during the manufacturing process determine the final protein concentration and ratio of casein to whey protein in the retentate. With extensive concentration and diafiltration (DF), it is possible to generate a retentate which has a high protein content and is considerably enriched with respect to casein proteins. This retentate is normally dried to produce powders enriched in native casein micelles (>90 % of total protein) and is referred to as phosphocaseinate, native micellar casein, micellar casein concentrate or micellar casein isolate. Such ingredients have a range of applications including, fortification of cheese milk and clinical nutritional products. The permeate stream produced from this MF process is an excellent starting material for the enrichment of whey proteins, having essentially circumvented traditional cheese or casein manufacture. The MF permeate from skim milk is free of starter culture, rennet enzyme, changes in pH caused by fermentation, added colour (e.g., annatto) and has improved protein quality compared with traditional sweet whey. For these reasons, this whey protein stream is often referred to as native, virgin, unadulterated or ideal whey and is of growing interest for the generation of highly functional native whey protein ingredients (e.g., whey protein concentrate/isolate) for use in applications such as nutritional beverages and infant formula. This MF technology (normally using MF membranes with a porosity in the range 0.8–2.0 μm) is also used in the dairy processing industry for the removal of bacteria from milk in the production of extended shelf-life (ESL) milk. Membranes are available for protein fractionation on a laboratory, pilot or industrial scale. Industrially, whey protein-based products are prepared by UF (and possible DF) of whey (to remove lactose and salts), followed by spray drying; these products, referred to as whey protein concentrates/isolates contain 30–85 % protein.

4.3.7 Gel Filtration (Gel Permeation Chromatography)

Filtration through cross-linked dextrans (e.g., Sephadex, Pharmacia, Uppsala, Sweden) permits fractionation of molecules, including proteins, on a commercial scale. It is possible to separate the casein and whey proteins by gel filtration but the process is uneconomical on an industrial scale.

4.3.8 Precipitation with Ethanol

The caseins may be precipitated from milk by ~40 % ethanol while the whey proteins remain soluble; a lower concentration of ethanol may be used at a lower pH.

4.3.9 Cryoprecipitation

Casein, in a mainly micellar form, is destabilized and precipitated by freezing milk or, preferably, concentrated milk, at about −10 °C; casein prepared by this method has some interesting properties.

4.3.10 Rennet Coagulation

Casein may be coagulated and recovered as rennet casein by treatment of milk with a selected proteinase (rennet). However, one of the caseins, κ-casein, is hydrolysed during renneting and therefore the properties of rennet casein differ fundamentally from those of acid casein. Rennet casein, which contains the colloidal calcium phosphate of milk, is insoluble in water at pH 7 but can be dissolved by adding a calcium sequestering agent, usually citrates or polyphosphates. It has desirable functional properties for certain food applications, e.g., in the production of cheese analogues.

4.3.11 Other Methods for the Preparation of Whey Proteins

Highly purified whey protein preparations, referred to as whey protein isolates (containing 90–95 % protein), are prepared industrially from whey by ion exchange chromatography. Denatured (insoluble) whey proteins, referred to as lactalbumin, may be prepared by heating whey to 95 °C for 10–20 min at about pH 6.0; the coagulated whey proteins are recovered by centrifugation. The whey proteins may also be precipitated using $FeCl_3$ or polyphosphates (Sect. 4.18.6).

4.4 Heterogeneity and Fractionation of Casein

Initially, casein was considered to be a homogeneous protein. Heterogeneity was first demonstrated in the 1920s by Linderstrøm-Lang and co-workers, using fractionation with ethanol-HCl, and confirmed in 1936 by K.O. Pedersen, using analytical ultracentrifugation, and in 1939 by O. Mellander, using free boundary electrophoresis. Three components were demonstrated and named α-, β- and γ-casein in order of decreasing electrophoretic mobility and represented 75, 22 and 3 %, respectively, of whole casein. These caseins were successfully fractionated in 1952 by N.J. Hipp and collaborators based on differential solubilities in urea at ~pH 4.6 or in ethanol/water mixtures; the former was widely used although the possibility of forming artefacts through interaction of casein with cyanate produced from urea was of concern.

In 1956, D.F. Waugh and P. H. von Hippel showed that the α-casein fraction of Hipp et al. contains two proteins, one of which is precipitated by a low concentration of Ca^{2+} and was called α_s-casein (s = sensitive) while the other, which was insensitive to Ca^{2+}, was called κ-casein. α_s-Casein was later shown to contain two proteins which are now called α_{s1}- and α_{s2}-caseins. Thus, bovine casein contains four distinct gene products, designated α_{s1}-, α_{s2}-, β- and κ-caseins which represent approximately 37, 10, 35 and 12 % of whole casein, respectively.

Various chemical methods were developed to fractionate the caseins but none gives homogeneous preparations. Fractionation is now usually achieved by ion-exchange chromatography on, for example, DEAE-cellulose, using a urea-containing buffer; quite large (e.g., 10 g) amounts of caseinate can be fractionated by this method, with excellent results (Fig. 4.5a, b). Good results are also obtained by ion-exchange chromatography using a urea-free buffer at 2–4 °C. High performance ion-exchange chromatography (e.g., Pharmacia FPLC™ on Mono Q or Mono S) gives excellent results for small amounts of sample (Fig. 4.5c, d). Reversed-phase HPLC or hydrophobic interaction chromatography may also be used but are less effective than ion-exchange chromatography.

The caseins may be quantified by densitometrically scanning polyacrylamide gel electrophoretograms (Sect. 4.4.1) but better quantitative results are obtained by ion-exchange chromatography using urea-containing buffers. However, it should be realized that the specific absorbance of the individual caseins differs greatly (Table 4.2).

4.4.1   Resolution of Caseins by Electrophoresis

Zonal electrophoresis in starch gels containing 7 M urea was used by R. G. Wake and R.L. Baldwin in 1961 to resolve casein into about 20 bands (zones); the two principal bands were α_{s1}- and β-caseins. Incorporation of urea was necessary to dissociate extensive intermolecular hydrophobic bonding. Electrophoresis in polyacrylamide gels (PAGE), containing urea or sodium dodecyl sulphate (SDS), was introduced by R. F. Peterson in 1963; resolution was similar to starch gels (SGE) but since it is easier to use, PAGE has become the standard technique for analysis of caseins; a schematic representation of a urea-PAGE electrophoretogram of whole bovine casein is shown in Fig. 4.6 and the urea-PAGE of milk from a selection of species in Fig. 4.7.

Urea-PAGE, which resolves proteins mainly on the basis of charge is the preferred medium for resolving caseins. SDS-PAGE, which resolves proteins mainly on the basis of size, is not very effective for resolving caseins because the molecular mass of the four caseins are quite close and, owing to its high hydrophobicity, β-casein binds more SDS molecules than α_{s1}-casein and therefore has a higher mobility although it is the larger molecule. SDS-PAGE of whey proteins gives better results than urea-PAGE.

Owing to the presence of intermolecular disulphide bonds, κ-casein resolves poorly on SGE or PAGE unless it is reduced, usually by 2-mercaptoethanol ($HSCH_2CH_2OH$), or alkylated. Inclusion of a stacking gel improves the resolution of both urea-PAGE and SDS-PAGE. Electrophoretic techniques for the analysis of casein were reviewed by Swaisgood (1975), Strange et al. (1992), O'Donnell et al. (2004), Chevalier (2011a, b).

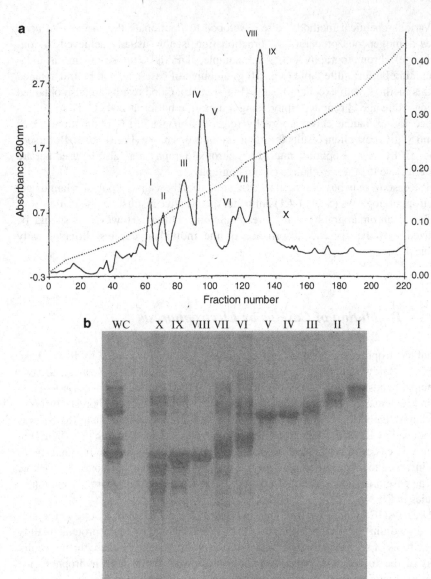

Fig. 4.5 (**a**) Chromatogram of sodium caseinate on an open column of DEAE cellulose anion exchanger. Buffer: 5 M urea in imidazole–HCl buffer, pH 7.0; gradient: 0–0.5 M NaCl. (**b**) Urea polyacrylamide gel electrophoretograms of the fractions from (**a**). (**c**) Chromatogram of sodium caseinate on a Pharmacia Mono Q HR5/5 anion exchange column. Buffer: 6 M urea in 5 mM bis-tris-propane/7 mM HCl, pH 7; gradient: 0–0.5 M NaCl. (**d**) Chromatogram of sodium caseinate on a Pharmacia Mono S HR5/5 cation exchange column. Buffer: 8 M urea in 20 mM acetate buffer, pH 5; gradient: 0–1.0 M NaCl

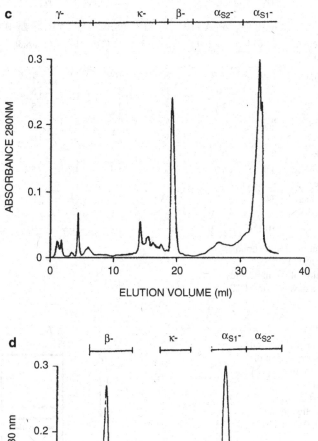

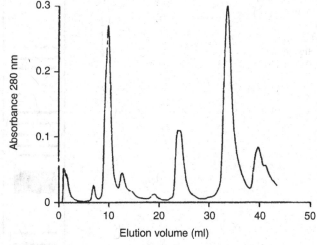

Fig. 4.5 (continued)

Table 4.2 Properties of some milk proteins (modified from Walstra and Jenness 1984)

Property	Caseins				Whey proteins		
	α_{s1}-B 8P	α_{s1}-A 11P	β-A² 5P	κ-B 1P	α-La-B	β-Lg-B	Serum albumin
Molecular weight	23,614	25,230	23,983	19,023	14,176	18,363	66,267
Residues/molecule							
Amino acids	199	207	209	169	123	162	582
Proline	17	10	35	20	2	8	34
Cysteine	0	2	0	2	8	5	35
Disulphides[a]	0	0	0	0	4	2	17
Phosphate	8	11	5	1	0	0	0
Carbohydrate	0	0	0	[b]	[c]	[d]	0
Hydrophobicity (kJ/residue)	4.9	4.7	5.6	5.1	4.7	5.1	4.3
Charged residues/mol	34	36	23	21	28	30	34
A_{280}	10.1	14.0[e]	4.5	10.5	20.9	9.5	6.6

[a]Intramolecular disulphide bonds
[b]Variable, see text
[c]A small fraction of the molecules
[d]O except for a dave variant (DV)
[e]A_{290}

Fig. 4.6 Schematic diagram of an electrophoretogram of sodium caseinate in a polyacrylamide gel containing 5 M urea in tris-hydroxymethylamine buffer, pH 8.9. 0 indicates origin

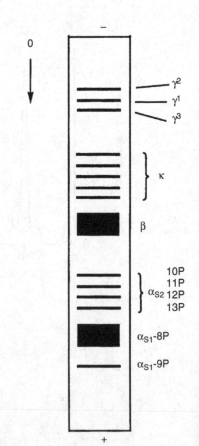

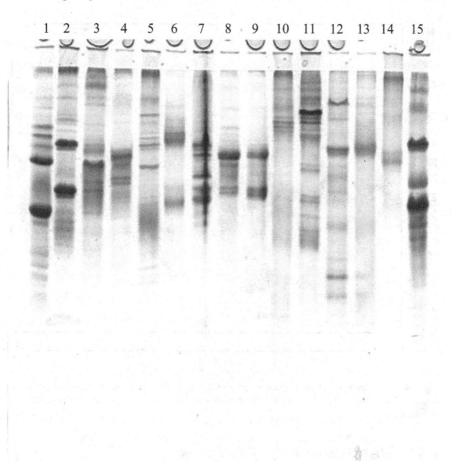

Fig. 4.7 Polyacrylamide gel electrophoretograms of the milk of a selection of species: *1*, Bovine. *2*, Camel. *3*, Equine. *4*, Asinine. *5*, Human. *6*, Rhinoceros. *7*, Porcine. *8*, Caprine. *9*, Ovine. *10*, Asian Elephant. *11*, African Elephant. *12*, Vervet monkey. *13*, Macaque monkey. *14*, Rat (casein). *15*, Canine

The protein bands in electrophoretograms may be visualized/stained with Amido Black at an acidic pH but Coomassie Blue G 250 gives better results and is easier to use (Shalabi and Fox 1987). Two-dimensional electrophoresis (SDS-PAGE in the first direction and isoelectric focussing in the second) gives excellent results (Fig. 4.8). 2-D electrophoresis coupled with mass spectrometry, commonly referred to as a proteomic approach, gives excellent results for the identification and quantification of proteins in mixtures (Chevalier 2011b).

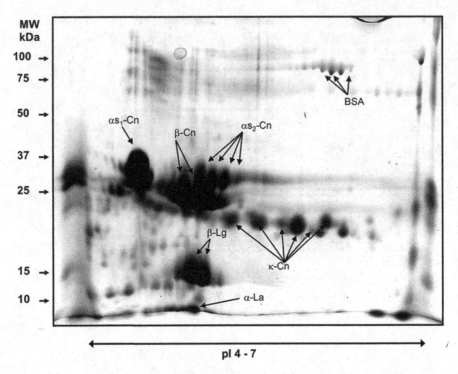

Fig. 4.8 Two-dimensional electrophoretogram of bovine milk under reducing conditions using isoelectric focusing in the range pH 4–7 for the first dimension and a 12 % acrylamide gel for the second dimension

4.4.2   Microheterogeneity of the Caseins

Each of the four caseins, α_{s1}- α_{s2}-, β- and κ-, exhibits variability, which we will refer to as microheterogeneity, arising from five causes:

Variability in the degree of phosphorylation. Each of the four caseins is phosphorylated (see Sect. 4.5.1) to a characteristic but variable level:

Casein	Number of phosphate residues
α_{s1}	8, occasionally 9
α_{s2}	10, 11, 12 or 13
β	5, occasionally 4
κ	1, occasionally 2 or perhaps 3

The number of phosphate groups in the molecule is indicated as α_{s1}-CN 8P or α_{s1}-CN 9P, etc. (CN = casein).

Disulphide bonding. The two principal caseins, α_{s1}- and β-, contain no cysteine or cystine but the two minor caseins, α_{s2}- and κ-, each contain two cysteine residues per mole which normally exist as intermolecular disulphide bonds. Under non-reducing

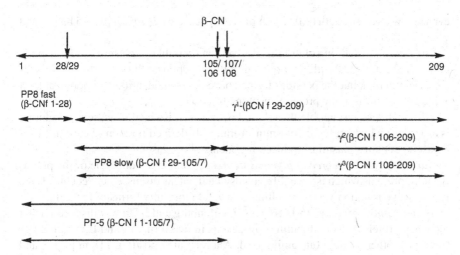

Fig. 4.9 Principal products produced from β-casein by plasmin

Table 4.3 Old and new nomenclature of the γ-caseins

Old	New
γ-CN	β-CN-A^1, A^2, A^3; B (f29–209)
TS-A^2-CN	β-CN A^2 (f106–209)
S-CN	β-CN B (f106–209)
R-CN	β-CN A^2 (f108–209)
TS-B-CN	β-CN B (f108–209)

conditions, α_{s2}-casein exists as a disulphide-linked dimer (previously known as α_{s5}-casein) while κ-casein exists as a series of disulphide-linked molecules ranging from dimers to decamers.

Hydrolysis of primary caseins by plasmin. In 1969, M. L. Groves and co-workers showed that the γ-casein fraction, as isolated by Hipp et al., is very heterogeneous, containing at least four distinct proteins which were named: γ-casein, temperature-sensitive casein (TS, which is soluble in the cold but precipitates above 20 °C), R-casein and S-casein. These four proteins were shown to be C-terminal fragments of β-casein. In 1976, the nomenclature of the γ-casein group was revised, as shown in Fig. 4.9 and Table 4.3.

γ-Caseins are produced from β-casein by proteolysis by plasmin, an indigenous proteinase in milk (Chap. 10). The corresponding N-terminal fragments are the principal components of the proteose-peptone (PP) fraction, i.e., PP5 (β-*CN* f1-105/107), PP8 slow (β-*CN* f29–105/107) and PP8 fast (β-*CN* f1-28). Normally, the γ-caseins represent only about 3 % of whole casein but levels may be much higher (up to 10 %) in late lactation or mastitic milk. Because of its high isoelectric point (~6), some γ-casein may be lost on isoelectric precipitation. γ-Caseins can be prepared readily by chromatography on DEAE-cellulose since they do not adsorb even at low ionic strength (0.02 M) at pH 6.5; γ^1-casein adsorbs at pH 8.5 but γ^2- and γ^3-caseins do not.

Isolated α_{s2}-casein in solution is also very susceptible to plasmin; eight peptide bonds are hydrolysed with the production of 14 peptides. Plasmin also hydrolyses α_{s2}-casein in milk but the peptides formed have not been identified, although at least some are included in the proteose-peptone fraction.

Although less susceptible than β- and α_{s2}-caseins, isolated α_{s1}-casein in solution is also readily hydrolysed by plasmin. A minor ill-defined fraction of casein, called λ-casein, contains plasmin-produced fragments of α_{s1}-casein.

Variations in the degree of glycosylation. κ-Casein is the only one of the principal milk proteins which is normally glycosylated but, as discussed in Sect. 4.5.1, the level of glycosylation varies, resulting in several molecular forms of κ-casein.

Genetic polymorphism. In 1956, R. Aschaffenburg and J. Drewry discovered that the whey protein, β-lactoglobulin (β-lg), exists in two forms, A and B, which differ from each other by only one amino acid, Ala for Val at position 118 in β-lg A and B, respectively. The milk of any individual animal may contain β-lg A or B or both, and the milk is indicated as AA, BB or AB with respect to β-lg. This phenomenon was referred to as genetic polymorphism and has since been shown to occur in all milk proteins; a total of about 60 variants of casein and whey proteins in bovine milk have been demonstrated by PAGE. Since PAGE differentiates on the basis of charge, only polymorphs which differ in charge, i.e., in which a charged residue is replaced by an uncharged one or *vice versa*, will be detected; therefore, it is very likely that many more than 30 polymorphs exist. The genetic variant present is indicated by a Latin letter, e.g., α_{s1}-CN A-8P, α_{s1}-CN B-8P, α_{s1}-CN B-9P, etc.

The frequency with which certain genetic variants occurs is breed-specific, and hence genetic polymorphism has been useful in the phylogenetic classification of cattle and other species. Various technologically important properties of the milk proteins, e.g., cheesemaking properties and the concentration of protein in milk, are correlated (linked) with certain polymorphs and significant research is ongoing on this subject. The genetic polymorphism of milk proteins has been comprehensively reviewed by Jakob and Puhan (1992), Ng-Kwai-Hang and Grosclaude (1992, 2003) and Martin et al. (2013a, b). Extensive polymorphism of the milk proteins of ovine and caprine milk also occurs (see Martin et al. 2013a, b) and probably of other species.

4.4.3   Nomenclature of the Caseins

During studies on casein fractionation, especially during the 1960s, various designations (Greek letters) were assigned to isolated proteins. To rationalize the nomenclature of milk proteins, the American Dairy Science Association established a Nomenclature Committee which published its first report in 1956 (Jenness et al. 1956); the report has been revised regularly (Brunner et al. 1960; Thompson et al. 1965; Rose et al. 1970; Whitney et al. 1976; Eigel et al. 1984; Farrell et al. 2004). An example of the recommended nomenclature is α_{s1}-CN A-8P, where α_{s1}-CN is the gene product, A is the genetic variant and 8P is the number of phosphate residues.

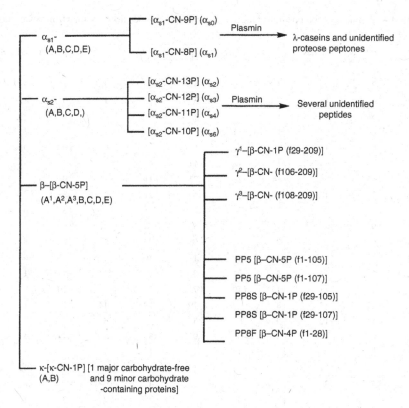

Fig. 4.10 Heterogeneity of bovine casein

The Committee recommends that in situations where confusion may arise through the use of a Greek letter alone, the relative electrophoretic mobility be given in brackets, thus α_{s2}-CN A-12P (1.00). The heterogeneity and nomenclature of the caseins in bovine milk is summarized in Fig. 4.10.

In addition to simplifying and standardizing the nomenclature of the milk proteins, the characteristics of the various caseins and whey proteins are summarized in the above articles, which are very valuable references.

4.5   Some Important Properties of the Caseins

4.5.1   Chemical Composition

The principal chemical and physicochemical properties of the principal milk proteins are summarized in Table 4.2. Some of the properties of the caseins are discussed in more detail below (see Swaisgood 1992, 2003; Huppertz 2013, for reviews).

Table 4.4 Amino acid composition of the major proteins in bovine milk (modified from Swaisgood 1982)

	α_{s1}-CN B	α_{s2}-CN A	κ-CN B	β-CN A^2	γ_1 CN A^2	γ_2 CN A^2	γ_3 CN A	β-Lg-A	α-La-B
Amino acid									
Asp	7	4	4	4	4	2	2	11	9
Asn	8	14	7	5	3	1	1	5	12
Thr	5	15	14	9	8	4	4	8	7
Ser	8	6	12	11	10	7	7	7	7
SerP	8	11	1	5	I	0	0	0	0
Glu	24	25	12	18	11	4	4	16	8
Gln	15	15	14	21	21	11	11	9	5
Pro	17	10	20	35	34	21	21	8	2
Gly	9	2	2	5	4	2	2	3	6
Ala	9	8	15	5	5	2	2	14	3
½ Cys	0	2	2	0	0	0	0	5	8
Val	11	14	11	19	17	10	10	10	6
Met	5	4	2	6	6	4	4	4	1
Ile	11	11	13	10	7	3	3	10	8
Leu	17	13	8	22	19	14	14	22	13
Tyr	10	12	9	4	4	3	3	4	4
Phe	8	6	4	9	9	5	5	4	4
Trp	2	2	1	1	1	1	1	2	4
Lys	14	24	9	11	10	4	3	15	12
His	5	3	3	5	5	4	3	2	3
Arg	6	6	5	4	2	2	2	3	1
PyroGlu	0	0	1	0	0	0	0	0	0
Total residues	199	207	169	209	181	104	102	162	123
MW	23,612	25,228	19,005	23,980	20,520	11,822	11,557	18,362	14,174
HΦ	4.89	4.64	5.12	5.58	5.85	6.23	6.29	5.03	4.68

CN casein, *β-Lg* β-lactoglobulin, *α-La* α-lactalbumin, *MW* molecular weight (Da), *HΦ* hydrophobicity (kJ/residue)

Amino acid composition. The approximate amino acid composition of the main caseins is shown in Table 4.4. Amino acid substitutions in the principal genetic variants can be deduced from the primary structures (Figs. 4.11, 4.12, 4.13, and 4.14). Four features of the amino acid profile are noteworthy:

1. All the caseins have a high content (35–45 %) of apolar amino acids (Val, Leu, Ile, Phe, Tyr, Pro) and would be expected to be poorly soluble in aqueous systems, but the high content of phosphate groups, low level of sulphur-containing amino acids and high carbohydrate content in the case of κ-casein offset the influence of apolar amino acids. The caseins are, in fact, quite soluble: solutions containing up to 20 % protein can be prepared in water at 80–90 °C. A high tem-

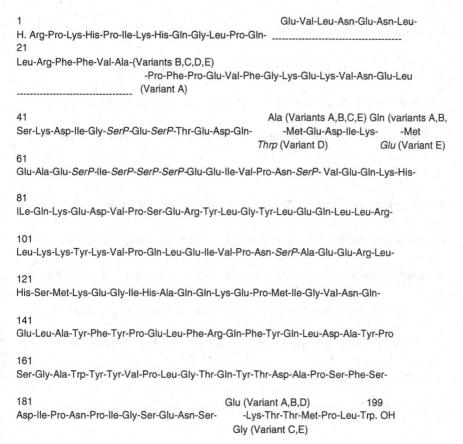

1   Glu-Val-Leu-Asn-Glu-Asn-Leu-
H. Arg-Pro-Lys-His-Pro-Ile-Lys-His-Gln-Gly-Leu-Pro-Gln- ---------------------------------------
21
Leu-Arg-Phe-Phe-Val-Ala-(Variants B,C,D,E)
                                        -Pro-Phe-Pro-Glu-Val-Phe-Gly-Lys-Glu-Lys-Val-Asn-Glu-Leu
--------------------------------- (Variant A)

41   Ala (Variants A,B,C,E) Gln (variants A,B,
Ser-Lys-Asp-Ile-Gly-*SerP*-Glu-*SerP*-Thr-Glu-Asp-Gln-   -Met-Glu-Asp-Ile-Lys-   -Met
                                                                              Thrp (Variant D)   *Glu* (Variant E)
61
Glu-Ala-Glu-*SerP*-Ile-*SerP*-*SerP*-*SerP*-Glu-Glu-Ile-Val-Pro-Asn-*SerP*- Val-Glu-Gln-Lys-His-

81
ILe-Gln-Lys-Glu-Asp-Val-Pro-Ser-Glu-Arg-Tyr-Leu-Gly-Tyr-Leu-Glu-Gln-Leu-Leu-Arg-

101
Leu-Lys-Lys-Tyr-Lys-Val-Pro-Gln-Leu-Glu-Ile-Val-Pro-Asn-*SerP*-Ala-Glu-Glu-Arg-Leu-

121
His-Ser-Met-Lys-Glu-Gly-Ile-His-Ala-Gln-Gln-Lys-Glu-Pro-Met-Ile-Gly-Val-Asn-Gln-

141
Glu-Leu-Ala-Tyr-Phe-Tyr-Pro-Glu-Leu-Phe-Arg-Gln-Phe-Tyr-Gln-Leu-Asp-Ala-Tyr-Pro

161
Ser-Gly-Ala-Trp-Tyr-Tyr-Val-Pro-Leu-Gly-Thr-Gln-Tyr-Thr-Asp-Ala-Pro-Ser-Phe-Ser-

181   Glu (Variant A,B,D)   199
Asp-Ile-Pro-Asn-Pro-Ile-Gly-Ser-Glu-Asn-Ser-   -Lys-Thr-Thr-Met-Pro-Leu-Trp. OH
                                                    Gly (Variant C,E)

Fig. 4.11 Amino acid sequence of bovine α_{s1}-casein, showing the amino acid substitutions or deletions in the principal genetic variants (from Swaisgood 1992)

perature is necessary to offset high viscosity, which is the limiting factor in pre-paring caseinate solutions. The high viscosity is a reflection of the high water-binding capacity (WBC) of casein, i.e., about 2.5 g H_2O g^{-1} protein. Such high WBC gives casein very desirable functional properties for incorporation into various foods, e.g., sausage and other comminuted meat products, instant desserts, synthetic whipping creams, etc., and large quantities of casein are used commercially for these purposes.

2. All the caseins have a very high proline content: 17, 10, 35 and 20 Pro residues per mole of α_{s1}-, α_{s2}-, β- and κ-caseins, respectively (out of a total of 199, 207, 209 and 169 residues, respectively). Such high levels of proline result in a very low content of α-helix or β-sheet structures in the caseins. The caseins are, therefore, readily susceptible to proteolysis without prior denaturation by, for example, acid or heat. Perhaps this is an important characteristic in neonatal nutrition.

1
H. Lys-Asn-Thr-Met-Glu-His-Val-*SerP*-*SerP*-*SerP*-Glu-Glu-Ser-Ile-Ile-*SerP*-Gln-Glu-Thr-Tyr-

21
Lys-Gln-Glu-Lys-Asn-Met-Ala-Ile-Asn-Pro-Ser-Lys-Glu-Asn-Leu-Cys-Ser-Thr-Phe-Cys-

41
Lys-Glu-Val-Val-Arg-Asn-Ala-Asn-Glu-Glu-Glu-Tyr-Ser-Ile-Gly-*SerP*-*SerP*-*SerP*-Glu-Glu-

61
SerP-Ala-Glu-Val-Ala-Thr-Glu-Glu-Val-Lys-Ile-Thr-Val-Asp-Asp-Lys-His-Tyr-Gln-Lys-

81
Ala-Leu-Asn-Glu-Ile-Asn-Gln-Phe-Tyr-Gln-Lys-Phe-Pro-Gln-Tyr-Leu-Gln-Tyr-Leu-Tyr-

101
Gln-Gly-Pro-Ile-Val-Leu-Asn-Pro-Trp-Asp-Gln-Val-Lys-Arg-Asn-Ala-Val-Pro-Ile-Thr-

121
Pro-Thr-Leu-Asn-Arg-Glu-Gln-Leu-*SerP*-Thr-*SerP*-Glu-Glu-Asn-Ser-Lys-Lys-Thr-Val-Asp-

141
Met-Glu-*SerP*-Thr-Glu-Val-Phe-Thr-Lys-Lys-Thr-Lys-Leu-Thr-Glu-Glu-Glu-Lys-Asn-Arg-

161
Leu-Asn-Phe-Leu-Lys-Lys-Ile-Ser-Gln-Arg-Tyr-Gln-Lys-Phe-Ala-Leu-Pro-Gln-Tyr-Leu-

181
Lys-Thr-Val-Tyr-Gln-His-Gln-Lys-Ala-Met-Lys-Pro-Trp-Ile-Gln-Pro-Lys-Thr-Lys-Val-
                                                                                            (Leu)
201   207
Ile-Pro-Tyr-Val-Arg-Tyr-leu. OH

Fig. 4.12   Amino acid sequence of bovine α_{s2}-casein A, showing 9 of the 10–13 phosphorylation sites (from Swaisgood 1992)

3. As a group, the caseins are deficient in sulphur amino acids which limits their biological value (80; egg albumen = 100). α_{s1}- and β-caseins contain no cysteine or cystine while α_{s2}- and κ-caseins have two cysteine residues per mole, which normally exist as intermolecular disulphides. The principal sulphydryl-containing protein in bovine milk is the whey protein β-lactoglobulin (β-lg), which contains two intramolecular disulphides and one sulphydryl group; normally, the sulphydryl group is buried within the molecule and is unreactive. Following denaturation, e.g., by heat above 75 °C, the -SH group of β-lg becomes exposed and reactive and undergoes a sulphydryl-disulphide interchange with κ-casein (and possibly with α_{s2}-casein and α-lactalbumin also) with very significant effects on some of the technologically important physicochemical properties of milk, e.g. heat stability and rennet coagulability (Chaps. 9 and 12).

4. The caseins, especially α_{s2}-casein, are rich in lysine, an essential amino acid in which many plant proteins are deficient. Consequently, casein and skim-milk powder are very good nutritional supplements for cereal proteins which are deficient in lysine. Owing to the high lysine content, casein and products containing

1
H.Arg-Glu-Leu-Glu-Glu-Leu-Asn-Val-Pro-Gly-Glu-Ile-Val-Glu-*SerP*-Leu*SerP*-*SerP*-*SerP*-Glu-

21   → γ1-caseins (Variant C)
Glu-Ser-Ile-Thr-Arg-Ile-Asn-Lys⌐Lys-Ile-Glu-Lys-Phe-Gln-^{Ser}-Glu-^{Lys}-Gln-Gln-Gln-
               SerP Glu
            (Variants A, B)

41
Thr-Glu-Asp-Glu-Leu-Gln-Asp-Lys-Ile-His-Pro-Phe-Ala-Gln-Thr-Gln-Ser-Leu-Val-Tyr-

61   Pro (Variants A^2, A^3)
Pro-Phe-Pro-Gly-Pro-Ile- -Asn-Ser-Leu-Pro-Gln-Asn-Ile-Pro-Pro-Leu-Thr-Gln-Thr-
      His (Variants C $A^{1,}$ and B)

81
Pro-Val-Val-Val-Pro-Pro-Phe-Leu-Gln-Pro-Glu-Val-Met-Gly-Val-Ser-Lys-Val-Lys-Glu-

       → γ3-caseins
101(Variants A^1, A^2, B, C) His
Ala-Met-Ala-Pro-Lys⌐ -Lys-Glu-Met-Pro-Phe-Pro-Lys-Tyr-Pro-Val-Glu-Pro-Phe-Thr-
 (Variant A^3) ⌐Gln

121 Ser(Variants A, C) → γ2-caseins
Glu- -Glu-Ser-Leu-Thr-Leu-Thr-Asp-Val-Glu-Asn-Leu-His-Leu-Pro-Leu-Pro-Leu-Leu-
 Arg (Variant B)

141
Gln-Ser-Trp-Met-His-Gln-Pro-His-Gln-Pro-Leu-Pro-Pro-Thr-Val-Met-Phe-Pro-Pro-Gln-

161
Ser-Val-Leu-Ser-Leu-Ser-Gln-Ser-Lys-Val-Leu-Pro-Val-Pro-Gln-Lys-Ala-Val-Pro-Tyr-

181
Pro-Gln-Arg-Asp-Met-Pro-Ile-Gln-Ala-Phe-Leu-Leu-Tyr-Gln-Glu-Pro-Val-Leu-Gly-Pro-

201   209
Val-Arg-Gly-Pro-Phe-Pro-Ile-Ile-Val.OH

Fig. 4.13 Amino acid sequence of bovine β-casein, showing the amino acid substitutions in the genetic variants and the principal plasmin cleavage sites (*inverted triangles*) (from Swaisgood 1992)

it may undergo extensive non-enzymatic Maillard browning on heating in the presence of a reducing sugar (Chap. 2).

At pH values on the acid side of their isoelectric point, proteins carry a net positive charge and react with anionic dyes (e.g., amido black or orange G), forming an insoluble protein-dye complex. This is the principle of the rapid dye-binding methods for quantifying proteins in milk and milk products and for visualizing protein bands in gel electrophoretograms; dye-binding is normally performed at pH 2.5–3.5. Another commonly-used dye for staining electrophoretograms or quantifying protein is Coomassie Brilliant Blue (R250 or G250) which forms strong non-covalent complexes with proteins, probably based on a combination of van der Waals forces and electrostatic interactions. Formation of the protein/dye complex stabilises the

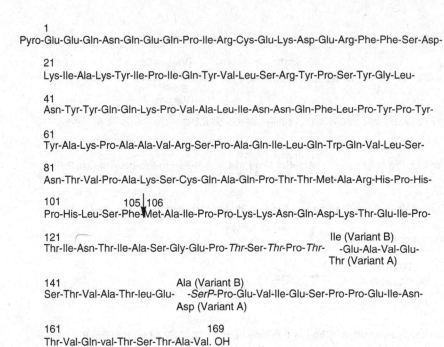

1

Pyro-Glu-Glu-Gln-Asn-Gln-Glu-Gln-Pro-Ile-Arg-Cys-Glu-Lys-Asp-Glu-Arg-Phe-Phe-Ser-Asp-

21

Lys-Ile-Ala-Lys-Tyr-Ile-Pro-Ile-Gln-Tyr-Val-Leu-Ser-Arg-Tyr-Pro-Ser-Tyr-Gly-Leu-

41

Asn-Tyr-Tyr-Gln-Gln-Lys-Pro-Val-Ala-Leu-Ile-Asn-Asn-Gln-Phe-Leu-Pro-Tyr-Pro-Tyr-

61

Tyr-Ala-Lys-Pro-Ala-Ala-Val-Arg-Ser-Pro-Ala-Gln-Ile-Leu-Gln-Trp-Gln-Val-Leu-Ser-

81

Asn-Thr-Val-Pro-Ala-Lys-Ser-Cys-Gln-Ala-Gln-Pro-Thr-Thr-Met-Ala-Arg-His-Pro-His-

101   105⎮106

Pro-His-Leu-Ser-Phe-Met-Ala-Ile-Pro-Pro-Lys-Lys-Asn-Gln-Asp-Lys-Thr-Glu-Ile-Pro-

121 ⌒   Ile (Variant B)

Thr-Ile-Asn-Thr-Ile-Ala-Ser-Gly-Glu-Pro-*Thr*-Ser-*Thr*-Pro-*Thr*-   -Glu-Ala-Val-Glu-

                                                                    Thr (Variant A)

141   Ala (Variant B)

Ser-Thr-Val-Ala-Thr-leu-Glu-   -*SerP*-Pro-Glu-Val-Ile-Glu-Ser-Pro-Pro-Glu-Ile-Asn-

                                    Asp (Variant A)

161   169

Thr-Val-Gln-val-Thr-Ser-Thr-Ala-Val. OH

Fig. 4.14 Amino acid sequence of bovine κ-casein, showing the amino acid substitutions in genetic polymorphs A and B and the chymosin cleavage site, ↓. Sites of post-translational phosphorylation or glycosylation are *italicized* (from Swaisgood 1992)

negatively charged anionic form of the dye, producing the blue colour which may then be seen on the membrane or in the gel. The bound number of dye molecules is approx. proportional to the amount of protein present per band. Lysine is the principal cationic residue in caseins, with lesser amounts of arginine and histidine ($pK_a \sim 6$). Since the caseins differ in lysine content (14, 24, 11 and 9 residues for α_{s1}-, α_{s2}-, β- and κ-caseins, respectively) they have different dye-binding capacities. This feature may be of some commercial significance in connection with dye-binding methods for protein analysis if the ratio of the caseins in the milk of individual animals varies (as it probably does). It should also be considered when calculating the protein concentration of zones on electrophoretograms stained with these dyes.

The absorbance of 1 % solutions of α_{s1}-, α_{s2}-, β- and κ-caseins at 280 nm in a 1 cm light path is 10.1, 14.0, 4.4 and 10.5, respectively. Since the protein concentration in eluates from chromatography columns is usually monitored by absorbance at 280 nm, cognisance should be taken of the differences in specific absorbance when calculating the concentrations of individual caseins in samples.

Primary structure. The primary structures of the four caseins of bovine milk are shown in Figs. 4.11, 4.12, 4.13, and 4.14. The sequences of some non-bovine caseins have been established also (see Martin et al. 2013a, b). An interesting feature of the primary structures of all caseins is that polar and apolar residues are not uniformly distributed but occur in clusters, giving hydrophobic and hydrophilic regions (Figs. 4.15 and 4.16). This feature makes the caseins good emulsifiers.

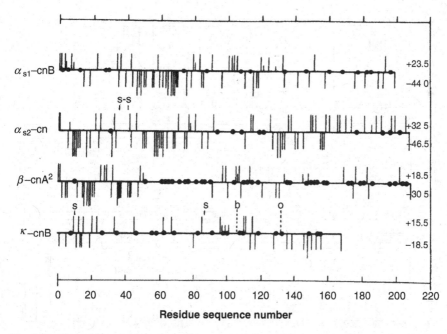

Fig. 4.15 Distribution of charged residues (pH 6–7), proline (*filled circle*) and cysteine (S) in α_{s1}-, α_{s2}-, β- and κ-caseins. *a*, Location of oligosaccharide moieties; and *b*, chymosin cleavage site in κ-casein (from Walstra and Jenness 1984)

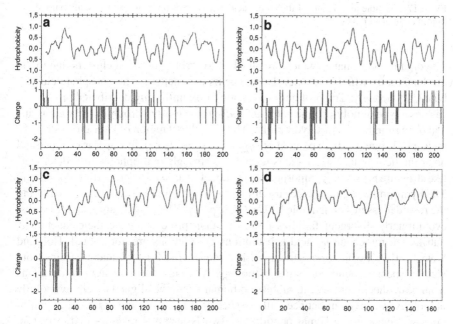

Fig. 4.16 Distribution of hydrophobicity (*top*) and charged residues (*bottom*) along the amino acid chain of α_{s1}-CN B-8P (**a**), α_{s2}-CN A-11P (**b**), β-CN A^2-5P (**c**) and κ-CN A-1P (**d**) (modified from Huppertz 2013)

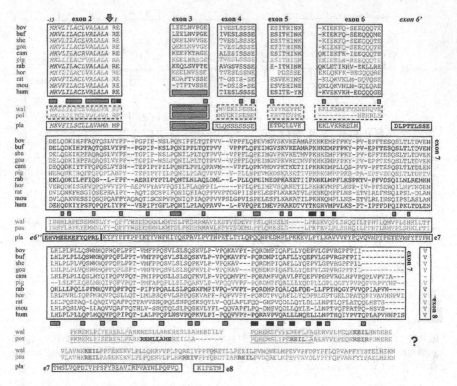

Fig. 4.17 Multiple alignments of the amino acid sequence of β-casein from 11 eutherians (from Martin et al. 2013a, b)

The organic phosphates, which are attached to serines, occur in clusters due to the mechanism by which phosphorylation occurs (see below). The phosphate clusters bind Ca^{2+} strongly. The proline residues are fairly uniformly distributed, giving the caseins a type of poly-proline helix. β-Casein is the most hydrophobic of the caseins and α_{s2}-casein is the most hydrophilic. The C-terminal region of κ-casein is strongly hydrophilic due to a high content of sugars (in some cases), few apolar residues and no aromatic residues, while the N terminus is strongly hydrophobic; this detergent-like structure is probably important for the micelle-stabilising properties of κ-casein. The hydrophilic segment of κ-casein is cleaved off during rennet action, rendering the residual caseins coagulable by Ca^{2+} (Chap. 12). The caseins are one of the most evolutionarily divergent families of mammalian proteins. Since their function is nutritional, minor amino acid substitutions or deletions are not critical. Holt and Sawyer (1993), who aligned the published sequences of α_{s1}-, β- and κ-caseins from various species found very little homology. Although the sequences of β-caseins from cow, sheep, mouse, rat, rabbit and human could be aligned readily, very little homology was evident between all six species (Fig. 4.17): the only long homologous sequence is the signal peptide, the two N-terminal residues of the mature protein and the sequence S.S.E.E (residues 18–21 of the mature protein, which is the principal phosphorylation site). The sequence of the signal peptides of α_{s1}- and κ-caseins also show a high degree of interspecies homology but several long insertions are required to obtain even a moderate degree of alignment of the sequences of the mature proteins.

Casein phosphorus. Milk contains about 900 mg phosphorus L^{-1}, which occurs in six types of phosphate-containing compounds, as will be discussed in Chap. 5:

- Inorganic: soluble and colloidal phosphates;
- Organic: phospholipids, casein and sugar phosphates, nucleotides (ATP, UTP, etc.).

Whole casein contains about 0.85 % phosphorus; α_{s1}-, β- and κ-caseins contain 1.1, 0.6 and 0.16 % P, respectively; on a molar basis, α_{s1}-, α_{s2}-, β- and κ-caseins contain 8(9), 10–13, 5(4) and 1(2, 3) moles P per mole. The phosphate groups are very important:

- Nutritionally, *per se*, and because they can bind large amounts of Ca^{2+}, Zn^{2+} and probably other polyvalent metals;
- They increase the solubility of caseins;
- They probably contribute to the high heat stability of casein; and
- They are significant in the coagulation of rennet-altered casein micelles during the secondary phase of rennet action (Chap. 12).

The phosphorus is covalently bound to the protein and is removed only by very severe heat treatment, high pH or some phosphatases. The phosphate is esterified mainly to serine (possibly a little to threonine) as a monoester:

$$Ser-O-\overset{\displaystyle O}{\underset{\displaystyle O^-}{\overset{\|}{P}}}-OH$$

Phosphorylation occurs in the Golgi membranes of the mammary cell, catalysed by two serine-specific casein kinases. Only certain serine residues are phosphorylated; the principal recognition site is Ser/Thr.X.Y, where Y is a glutamyl and occasionally an aspartyl residue; once a serine residue has been phosphorylated, SerP can serve as a recognition site. X may be any amino acid but a basic or a very bulky residue may reduce the extent of phosphorylation. However, not all serine residues in a suitable sequence are phosphorylated, suggesting that there may be a further topological requirement, e.g., a surface location in the protein conformation.

Casein carbohydrate. α_{s1}-, α_{s2}- and β-caseins contain no carbohydrate but κ-casein contains about 5 %, consisting of *N*-acetylneuraminic acid (sialic acid), galactose and N-acetylgalactosamine. The carbohydrate exists as tri- or tetrasaccharides, located toward the C-terminal of the molecule, attached through an *O*-threonyl linkage, mainly to Thr_{131} of κ-casein (Fig. 4.18). The number of

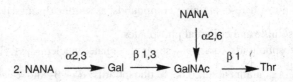

1. NANA $\xrightarrow{\alpha2,3}$ Gal $\xrightarrow{\beta1,3}$ GalNAc $\xrightarrow{\beta1}$ Thr

NANA

2. NANA $\xrightarrow{\alpha2,3}$ Gal $\xrightarrow{\beta1,3}$ GalNAc $\downarrow\alpha2,6$ $\xrightarrow{\beta1}$ Thr

NANA

3. GlcNAc $\xrightarrow{\beta1,3}$ Gal $\xrightarrow{\beta1,3}$ GalNAc $\xrightarrow{\beta2,6}$

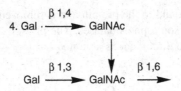

4. Gal $\xrightarrow{\beta1,4}$ GalNAc

Gal $\xrightarrow{\beta1,3}$ GalNAc $\xrightarrow{\beta1,6}$

5.   Gal $\xrightarrow{\beta1,4}$ GlcNAc

NANA $\xrightarrow{\alpha2,3}$ Gal $\xrightarrow{\beta1,3}$ GalNAc $\xrightarrow{\beta1,6}$

6. NANA $\xrightarrow{\alpha2,3}$ Gal $\xrightarrow{\beta1,4}$ GlcNAc

NANA $\xrightarrow{\alpha2,3}$ Gal $\xrightarrow{\beta1,3}$ GalNAc $\xrightarrow{\beta1,6}$

Fig. 4.18 Oligosaccharides attached to casein isolated from bovine milk (*1* and *2*) or colostrum (*1–6*) (from Eigel et al. 1984)

Table 4.5 Variability of bovine κ-casein with respect to sugars and phosphates

Fraction	Galactose	N-Acetyl-galactosamine	N-Acetyl-neuraminic acid	Phosphate
B-1	0	0	0	1
B-2	1	1	1	1
B-3	1	1	2	1
B-4	0	0	0	2
B-5	2	2	3	1
B-6	0	0	0(4)	3(1)
B-7	3	3	6	1
B-8	4	4	8	1
B-9	5	5	10	1

oligosaccharides per κ-casein molecule varies from 0 to 4. The variability of glyco-sylation results in at least nine molecular forms of κ-casein (Table 4.5). The κ-casein in colostrum is even more highly glycosylated; more sugars are present and the structures are more complex and uncertain.

The carbohydrate is attached to the (glyco)macropeptides which are produced from κ-casein on hydrolysis by rennets. The carbohydrate bestows on κ-casein quite high solubility and hydrophilicity. It is also responsible for the solubility of the gly-comacropeptides in 12 % TCA (see Chap. 12). Although the sugars increase the hydrophilicity of casein, they are not responsible for the micelle-stabilizing proper-ties of κ-casein, the carbohydrate-free form being as effective in this respect as the glycosylated forms.

4.5.2   Secondary and Tertiary Structures

Physical methods, such as optical rotary dispersion and circular dichroism, indicate that the caseins have little secondary or tertiary structure, probably due to the pres-ence of high levels of proline residues, especially in β-casein, which disrupt α-helices and β-sheets. However, theoretical calculations (Kumosinski et al. 1993a, b; Kumosinski and Farrell 1994; Huppertz 2013; Farrell et al. 2013) indicate that while α_{s1}-casein has little α-helix, it probably contains some β-sheets and β-turns. The C-terminal half of α_{s2}-casein probably has a globular conformation (i.e., a compact structure containing some α-helix and β-sheet) while the N-terminal region proba-bly forms a randomly structured hydrophilic tail. Theoretical calculations suggest that β-casein could have 10 % of its residues in α-helices, 17 % in β-sheets- and 70 % in unordered structures. κ-Casein appears to be the most highly structured of the caseins, with perhaps 23 % of its residues in α-helices, 31 % in β-sheets and 24 % in β-turns. Energy-minimized models of α_{s1}-, β- and κ-caseins are shown in Fig. 4.19a–c (Farrell et al. 2013). Holt and Sawyer (1993) coined the term 'rheomorphic' to describe the caseins as proteins with an open, flexible, mobile conformation in order to avoid using the 'demeaning' term, 'random coil'.

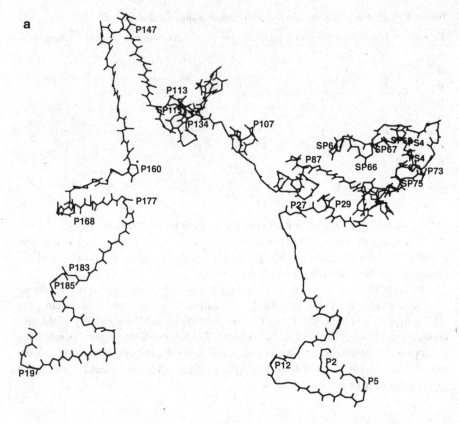

Fig. 4.19 Energy-minimized models of the tertiary structures of bovine α_{s1}- (**a**), β- (**b**) and κ-(**c**) caseins (from Kumosinski et al. 1993a, b; Kumosinski and Farrell 1994)

The lack of secondary and tertiary structures is probably significant for the following reasons:

1. The caseins are readily susceptible to proteolysis, in contrast to globular proteins, e.g., whey proteins, which are usually very resistant to proteolysis in their native state. This has obvious advantages for the digestibility of the caseins, the natural function of which is nutritional and hence easy digestibility in the 'native' state is important. The caseins are also readily hydrolysed in cheese, which is important for the development of cheese flavour and texture (Chap. 12). However, casein hydrolysates may be bitter due to a high content of hydrophobic amino acids (small hydrophobic peptides tend to be bitter). The caseins are readily hydrolysed by proteinases secreted by spoilage micro-organisms.
2. The caseins adsorb readily at air-water and oil-water interfaces due to their open structure, relatively high content of apolar amino acid residues and the uneven distribution of amino acids. This gives the caseins very good emulsifying and foaming properties, which are widely exploited in the food industry.

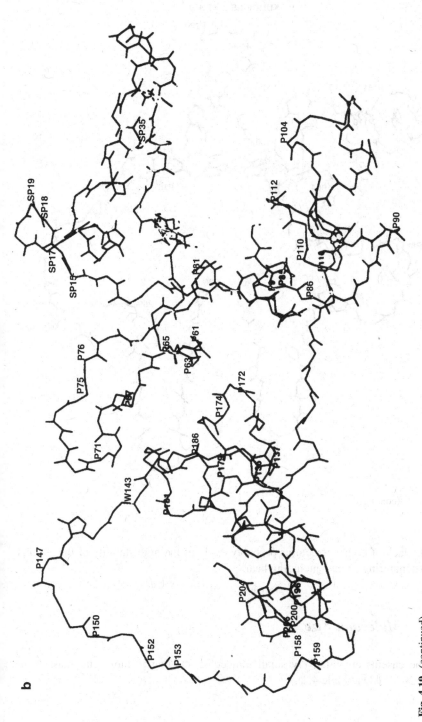

Fig. 4.19 (continued)

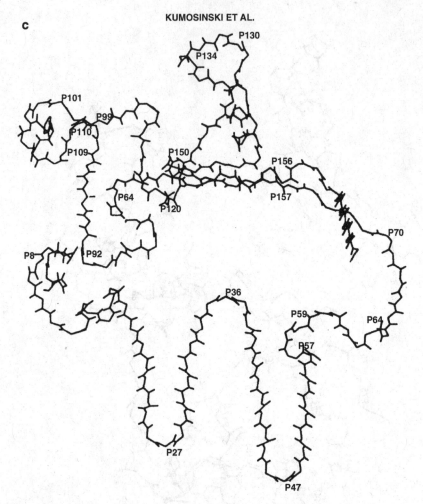

Fig. 4.19 (continued)

3. The lack of higher structures probably explains the high stability of the caseins to denaturing agents, including heat.

4.5.3 Molecular Size

All the caseins are relatively small molecules, ranging in molecular mass from about 20–25 kDa (Table 4.2).

4.5.4 Hydrophobicity

The caseins are often considered to be rather hydrophobic molecules. However, consideration of the amino acid composition indicates that they are not particularly so; in fact, some are more hydrophilic than the whey protein, β-lactoglobulin (Table 4.2). However, the caseins do have high surface hydrophobicity; in contrast to the globular whey proteins, in which the hydrophobic residues are buried as much as possible within the molecule, with most of the hydrophilic residues exposed on the surface, owing to the relative lack of secondary and tertiary structures in the caseins, such an arrangement is not possible, and hence the hydrophobic residues are rather exposed.

Thus, the caseins are relatively small, relatively hydrophobic, amphipathic, randomly or flexibly structured molecules, with relatively low levels of secondary and tertiary structures.

4.5.5 Influence of Ca^{2+} on Caseins

At all temperatures, α_{s1}-CN B and C are insoluble in calcium-containing solutions and form a coarse precipitate at Ca^{2+} concentrations greater than about 4 mM. α_{s1}-CN A, which lacks the very hydrophobic sequence, residues 13–26, is soluble at $[Ca^{2+}]$ up to 0.4 M in the temperature range 1–33 °C. Above 33 °C, it precipitates but redissolves on cooling to 28 °C. The presence of α_{s1}-CN A modifies the behaviour of α_{s1}-CN B so that an equimolar mixture of the two is soluble in 0.4 M Ca^{2+} at 1 °C; α_{s1}-CN B precipitates from the mixture at 18 °C and both α_{s1}-CN A and B precipitate at 33 °C. α_{s1}-CN A does not form normal micelles with κ-casein. Since α_{s1}-CN A occurs at very low frequency, these abnormalities are of little consequence in dairy processing but may become important if the frequency of α_{s1}-CN A increases as a result of breeding practices.

The α_{s2}-caseins are also insoluble in Ca^{2+} (above about 4 mM) at all temperatures, but their behaviour has not been studied in detail.

β-Casein is soluble at high concentrations of Ca^{2+} (0.4 M) at temperatures below 18 °C, but above 18 °C β-casein is very insoluble, even in the presence of low concentrations of Ca^{2+} (4 mM). Ca-precipitated β-casein redissolves readily on cooling to below 18 °C. About 20 °C is also the critical temperature for the temperature-dependent polymerization of β-casein and the two phenomena may be related.

κ-Casein is soluble in Ca^{2+} at all concentrations up to those at which general salting-out occurs. Solubility is independent of temperature and pH (outside the pH range at which isoelectric precipitation occurs). Not only is κ-casein soluble in the presence of Ca^{2+} but it is capable of stabilizing α_{s1}-, α_{s2}- and β-caseins against precipitation by Ca^{2+} (Sect. 4.6).

4.5.6   Action of Rennets on Casein

This subject is dealt with in Chap. 12. Suffice it to say here that κ-casein is the only casein hydrolysed by rennets during the primary phase of milk coagulation, which is the first step in the manufacture of most cheese varieties.

4.5.7   Casein Association

All the major caseins associate with themselves and with each other. In unreduced form, κ-casein is present largely as disulphide-linked polymers. κ-Casein also forms hydrogen and hydrophobic bonds with itself and other caseins but these associations have not been studied in detail.

At 4 °C, β-casein exists in solution as monomers of molecular mass ~25 kDa. As the temperature is increased, the monomers polymerize to form long thread-like chains of about 20 units at 8.5 °C and to still larger aggregates at higher temperatures. The degree of association depends on protein concentration. The ability to form thread-like polymers may be important in micelle structure. β-Casein also undergoes a temperature-dependent conformational change in which the content of poly-L-proline helix decreases with increasing temperature. The transition temperature is about 20 °C, i.e., very close to the temperature at which β-casein becomes insoluble in Ca^{2+}.

α_{s1}-Casein polymerizes to form tetramers of molecular mass ~113 kDa; the degree of polymerization increases with increasing protein concentration and increasing temperature.

The major caseins interact with each other and, in the presence of Ca^{2+}, these associations lead to the formation of **casein micelles**.

4.6   Casein Micelles

4.6.1   Composition and General Features

About 95 % of the casein exists in milk as large colloidal particles, known as micelles. On a dry matter basis, casein micelles contain ~94 % protein and 6 % low molecular mass species referred to as colloidal calcium phosphate, consisting mainly of calcium, magnesium, phosphate and citrate. The micelles are highly hydrated, binding about 2.0 g H_2O g^{-1} protein. Some of the principal properties of casein micelles are summarized in Table 4.6.

It has been known since the late nineteenth century that the caseins exist as large colloidal particles which are retained by Pasteur-Chamberland porcelain filters (roughly equivalent to modern ceramic microfiltration membranes). Electron

Table 4.6 Characteristics of bovine casein micelles (modified from McMahon and Brown 1984)

Characteristic	Value
Diameter	120 nm (range: 50–500)
Surface area	8×10^{-10} cm^2
Volume	2×10^{-15} cm^3
Density (hydrated)	1.0632 g cm^3
Mass	2.2×10^{-15} g
Water content	63 %
Hydration	3.7 g H$_2$O g^{-1} protein
Voluminosity	4.4 cm^3 g^{-1}
Molecular mass, hydrated	1.3×10^9 Da
Molecular mass, dehydrated	5×10^8 Da
Number of peptide chains	10^4
Number of micelles mL^{-1} milk	10^{14}–10^{16}
Surface area of micelles per mL milk	5×10^4 cm^2
Mean distance between micelles	240 nm

microscopy shows that casein micelles are generally spherical in shape, with a diameter ranging from 50 to 500 nm (average ~120 nm) and a mass ranging from 10^6 to 10^9 Da (average about 10^8 Da). There are very many small micelles but these represent only a small proportion of the volume or mass (Fig. 4.20). There are 10^{14}–10^{16} micelles mL^{-1} milk; they are roughly two micelle diameters (240 nm) apart, i.e., they are quite tightly packed. The surface (interfacial) area of the micelles is very large, 5×10^4 cm^2 mL^{-1}; hence, the surface properties of the micelles are critical to their behaviour.

Since the micelles are of colloidal dimensions, they are capable of scattering light and the white colour of milk is due largely to light scattering by the casein micelles; the white colour is lost if the micelles are disrupted, e.g., by removing colloidal calcium phosphate [by citrate, ethylene diaminetetraacetic acid (EDTA) or oxalate], by increasing pH (to greater than 9), or by the addition of urea or SDS or ethanol to 70 % at 70 °C.

4.6.2 Stability

1. The micelles are stable to the principal processes to which milk is normally subjected (except those in which it is intended to destabilize the micelles, e.g., rennet- and acid-induced coagulation). They are very stable at high temperatures, coagulating only after heating at 140 °C for 15–20 min at the normal pH of milk. Such coagulation is not due to denaturation in the narrow sense of the word but to major changes which occur in milk exposed to such a severe heat treatment, including a decrease in pH due to the pyrolysis of lactose to various

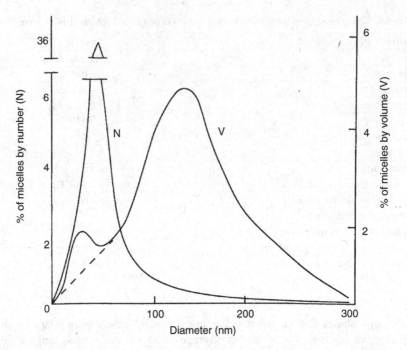

Fig. 4.20 Number and volume frequency distribution of casein micelles in bovine milk (from Walstra and Jenness 1984)

acids (principally formic), dephosphorylation of the casein, cleavage of κ-casein, denaturation of the whey proteins and their attachment to the casein micelles, precipitation of soluble calcium phosphate on the micelles and a decrease in hydration (Chap. 9).

2. They are stable to compaction, e.g., they can be sedimented by ultracentrifugation, e.g., at $100,000 \times g$ for 1 h, and redispersed readily by crushing and mild agitation.
3. They are stable to commercial homogenization but are changed slightly at very high pressure (500 MPa).
4. They are stable to high $[Ca^{2+}]$, up to at least 200 mM at a temperature up to 50 °C.
5. They aggregate and precipitate from solution when the pH is adjusted to the isoelectric point of caseins (~pH 4.6). Precipitation at this pH, which is temperature-dependent (i.e. does not occur at temperatures below 5–8 °C and occurs over a wide pH range, perhaps 3.0–5.5, at a higher temperature, e.g., 70 °C), occurs owing to the loss of net positive or negative charge as the pH approaches 4.6.
6. Concentration of milk by ultrafiltration, evaporation and spray drying can cause destabilization of casein micelles, with the extent of destabilization generally increasing with increasing concentration factor. The close packing of casein

micelles, increases in Ca^{2+} concentration and decreases in pH, caused by precipitation of CaH_2PO_4 and $CaHPO_4$ as $Ca_3(PO_4)_2$ (releasing H^+) are the main factors responsible for destabilization of casein micelles on concentration.

7. As the pH of milk is reduced, the colloidal calcium phosphate (CCP) dissolves and is completely soluble at pH 4.9 (Chap. 5). pH adjustment, followed by dialysis against bulk milk, is a convenient and widely-used technique for varying the CCP content of milk. As the concentration of CCP is reduced, the properties of the micelles are altered but they retain some of their structure even after removing 70 % of the CCP and reform on restoring the original pH. Removal of more than 70 % of the CCP results in disintegration of the micelles into smaller particles (aggregates).

8. Many proteinases hydrolyse a specific bond in κ-casein, as a consequence of which the micelles aggregate or gel in the presence of Ca^{2+}. This is the key step in the manufacture of most cheese varieties (Chap. 12).

9. On cooling of skim milk to temperatures in the range 0–5 °C, up to ~20 % of total β-casein and lesser amounts of other caseins dissociate from the micelles, presumably due to weakening of hydrophobic interactions between β-casein molecules or other caseins.

10. At room temperature, the micelles are destabilized by ~40 % ethanol at pH 6.7 and by lower concentrations if the pH is reduced. However, if the system is heated to ~70 °C, the precipitate redissolves, and the system becomes translucent. When the system is recooled, the white appearance of milk is restored, and a gel is formed if the ethanol-milk mixture is held at 4 °C, especially if a concentrated (>2×) milk was used. If the ethanol is removed by evaporation, very large aggregates (average diameter ~3,000 nm) are formed which have very different properties from those of natural micelles. The aggregates can be dispersed to particles of average diameter ~500 nm. The dissociating effect of ethanol is promoted by increasing the pH (35 % ethanol causes dissociation at 20 °C at pH 7.3) or adding NaCl. Methanol and acetone have a dissociating effect similar to ethanol, but propanol causes dissociation at ~25 °C. The mechanism by which ethanol and similar compounds cause the dissociation of casein micelles has not been established, but it is not due to the solution of CCP, which is unchanged.

11. They are destabilized by freezing (cryodestabilization) due to a decrease in pH and an increase in the $[Ca^{2+}]$ in the unfrozen phase of milk (Chaps. 2 and 5).

4.6.3 Principal Micelle Characteristics

The structure of the casein micelles has attracted the attention of scientists for a considerable time. Knowledge of micelle structure is important because the stability and behaviour of the micelles are central to many dairy processing operations, e.g., cheese manufacture, stability of sterilized, sweetened-condensed and reconstituted milks and frozen products. Without knowledge of the structure and properties of the casein micelle, attempts to solve many technological problems faced in the dairy industry are empirical and not generally applicable. From the academic viewpoint, the casein micelle presents an interesting and complex problem in the quaternary structure of proteins.

Since the pioneering work of Waugh in 1958, a considerable amount of research effort has been devoted to elucidating the structure of the casein micelle, and several models have been proposed. This work has been reviewed in the references cited in the next section. The principal properties of the casein micelles are listed below and the models which best meet these requirements discussed briefly in the next section.

1. κ-Casein, which represents about 12 % of total casein, is a critical feature of micelle structure and stability and must be located so as to be able to stabilize the calcium-sensitive α_{s1}-, α_{s2}- and β-caseins, which represent about 85 % of total casein.

2. The κ-casein content of casein micelles is inversely proportional to their size, while the content of colloidal calcium phosphate is directly related to size.

3. Ultracentrifugally sedimented micelles have a hydration of 1.6–2.7 g H_2O g^{-1} protein but a voluminosiy of 3–7 mL g^{-1} have been found by viscosity measurements and calculation of the specific hydrodynamic volume. These values suggest that the micelle has a porous structure in which the protein occupies about 25 % of the total volume.

4. Chymosin and similar proteinases, which are relatively large molecules (~36 kDa), very rapidly and specifically hydrolyse most of the micellar κ-casein,

5. When heated in the presence of whey proteins, as in normal milk, κ-casein and β-lactoglobulin interact to form a disulphide-linked complex which modifies many properties of the micelles, including rennet coagulability and heat stability.

6. Cheryan et al. (1975) reported that insolubilized (by treatment with formaldehyde) carboxypeptidase released the C-terminal amino acid from the three caseins, suggesting that there is some of all caseins on the micelle surface, However, using conjugates of dextran and pepsin or carboxypeptidase, Chaplin and Green (1982) concluded that κ-casein has a predominantly surface location and that immobilized rennets are unable to coagulate milk.

7. Removal of colloidal calcium phosphate (CCP) results in disintegration of the micelles into particles of mass ~3×10^6 Da. The properties of the CCP-free system are very different from those of normal milk, e.g., it is sensitive to and precipitated by relatively low concentrations of Ca^{2+}, it is more stable to high temperatures, e.g., 140 °C, and is not coagulable by rennets. Many of these properties can be restored, at least partially, by increased concentrations of calcium.

8. The micelles are also dissociated by urea (5 M) or SDS (suggesting the involvement of hydrogen and hydrophobic bonds in micelle integrity) or by raising the pH to >9. Under these conditions, the CCP is not dissolved; in fact, increasing the pH increases the level of CCP. If the urea is removed by dialysis against a

large excess of bulk milk, micelles reform, but these have not been characterized adequately.

9. The micelles can be destabilized by alcohols, acetone and similar solvent, suggesting an important role for electrostatic interactions in micelle structure.

10. As the temperature is lowered, caseins, especially β-casein, dissociate from the micelles; depending on the method of measurement, 10–50 % of β-casein is non-micellar at 4 °C.

11. Electron microscopy shows that the interior of the micelles are not uniformly electron dense.

12. The micelles have a surface (zeta) potential of about −20 mV at pH 6.7.

4.6.4 Micelle Structure

The structure of the casein micelles has attracted the attention of scientists for many years. Knowledge of micelle structure is important because reactions undergone by the micelles are central to many dairy processing operations (e.g., cheese manufacture; stability of sterilized, sweetened-condensed and reconstituted milks and frozen products). From the academic viewpoint, the casein micelle presents an interesting and complex problem in protein quaternary structure.

It was recognized early that the caseins in milk exist as large colloidal particles, and there was some speculation on the structure of these particles and how they were stabilized (see Fox and Brodkorp 2008). No significant progress was possible until the isolation and characterization of κ-casein (Waugh and von Hippel 1956). The first attempt to describe the structure of the casein micelle was that of D.F. Waugh, in 1958, and since then, a considerable amount of research effort has been devoted to elucidating the structure of the casein micelle. This work is summarized here.

The principal features which must be met by any micelle model are:

*   κ-Casein, which represents ~12 % of total casein, must be located so as to be able to stabilize the calcium-sensitive α_{s1}-, α_{s2}-, and β-caseins, which represent ~85 % of total casein.
*   Chymosin and similar proteases, which are relatively large molecules (~35 kDa), very rapidly and specifically hydrolyze most of the κ-casein.
*   When heated in the presence of whey proteins, as in milk, κ-casein and β-lactoglobulin (MW = 36 kDa in milk) interact to form a complex which modifies the properties of the micelles, e.g., rennet and heat coagulation.

The arrangement that would best meet these requirements is a surface layer of κ-casein surrounding the Ca-sensitive caseins, somewhat analogous to a lipid emulsion in which the triglycerides are surrounded by a thin layer of emulsifier. Removal of CCP results in disintegration of the micelles into particles of MW ~10^6 Da, suggesting that CCP is a major integrating factor in the micelles. The properties of the CCP-free system are very different from those of normal milk (e.g., it is sensitive to and precipitated by relatively low levels of Ca^{2+}, it is more stable to heat-induced coagulation, and it is not coagulable by rennets). Many of these properties can be restored, at least partially, by increased concentrations of calcium. However, CCP is not the only integrating factor, as indicated by the dissociating effect of temperature, urea, SDS, ethanol or alkaline pH. As the temperature is lowered, casein, especially, β-casein dissociates from the micelles; the amount of β-casein which dissociates varies from 10 to 50 % depending on the method of measurement; it increases to a maximum at ~pH 5.2.

Various models of casein micelle structure have been proposed over the last 50 years. They have been refined progressively as more information has become available. Progress has been reviewed regularly (see Walstra 1999; Horne 2006, 2011; de Kruif and Holt 2003; Fox and Brodkorb 2008; McMahon and Oommen 2008, 2013; Dalgleish 2011; O'Mahony and Fox 2013; de Kruif 2014, for references).

The proposed models fall into three general categories, although there is some overlap:

1. Core-coat;
2. Internal structure;
3. Subunit (submicelles); in many of the models in this category, it is proposed that the submicelles have a core-coat structure.

For many years there has been strong support for the view that the micelles are composed of submicelles of mass $~10^6$ Da and diameter 10–15 nm. This model was introduced in 1967 by Morr who proposed that the submicelles are linked together by CCP, giving the micelle an open porous structure. On removal of CCP, e.g., by acidification/dialysis, EDTA, citrate or oxalate, the micelles disintegrate. Disintegration may also be achieved by treatment with urea, SDS, 35 % ethanol at 70 °C or at pH greater than 9; these treatments do not solubilize CCP, suggesting that other forces, e.g. hydrophobic and hydrogen bonds, contribute to micelle structure.

The submicellar model has undergone several refinements (see Schmidt 1982; Walstra and Jenness 1984). The current view is that the κ-casein content of the submicelles varies and that the κ-casein-deficient submicelles are located in the interior of the micelles with the κ-casein-rich submicelles concentrated at the surface, giving the micelles a κ-casein-rich layer but with some α_{s1}-, α_{s2}- and β-caseins also exposed on the surface. It is proposed that the hydrophilic C-terminal region of κ-casein protrudes from the surface, forming a layer 5–10 nm thick and giving the micelles a hairy appearance (Fig. 4.21). This hairy layer is responsible for micelle stability through a major contribution to zeta potential (~20 mV) and steric stabilization. If the hairy layer is removed, e.g. specific hydrolysis of κ-casein, or collapsed, e.g. by ethanol, the colloidal stability of the micelles is destroyed and they coagulate or precipitate.

Although the submicellar model of the casein micelle readily explains many of the principal features and physicochemical reactions undergone by the micelles and has been widely supported, it has never enjoyed unanimous support and two alternative models have been proposed recently. Visser (1992) proposed that the micelles

Fig. 4.21 Submicelle model
of the casein micelle (from
Walstra and Jenness 1984)

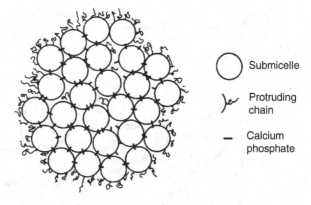

Fig. 4.22 Model of the
casein micelle (modified from
Holt 1994)

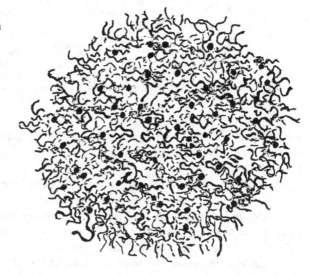

are spherical conglomerates of individual casein molecules randomly aggregated
and held together partly by salt bridges in the form of amorphous calcium phosphate
and partly by other forces, e.g., hydrophobic bonds, with a surface layer of κ-casein.
Holt (1992, 1994) depicted the casein micelle as a tangled web of flexible casein
molecules forming a gel-like structure in which microgranules of colloidal calcium
phosphate are an integral feature and from the surface of which the C-terminal
region of κ-casein extends, forming a hairy layer (Fig. 4.22). These models retain
two of the central features of the submicellar model, i.e. the cementing role of CCP
and the predominantly surface location of κ-casein.

Dalgleish (1998) agreed that the micellar surface is only partially covered with
κ-casein, which is distributed non-uniformly on the surface. This surface coverage
provides steric stabilization against the approach of large particles, such as other

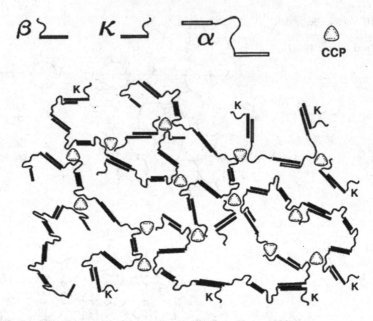

Fig. 4.23 Dual-bonding model of the casein micelle (from Horne 1998)

micelles, but the small-scale heterogeneities and the gaps between κ-casein molecules provide relatively easy access for molecules with dimensions of individual proteins or smaller.

Much of the evidence for a sub-micellar structure came from electron microscopy studies, which appeared to show variations in electron density, which was interpreted as indicating sub-micelles, i.e., a raspberry-like structure. However, artifacts may arise in electron microscopy owing to fixation, exchanging water for ethanol, air drying or metal coating. Using a new cryopreparation electron microscopy stereo-imaging technique, McMahon and McManus (1998) found no evidence to support the sub-micellar model and concluded that if the micelles do consist of sub-micelles, these must be smaller than 2 nm or less densely packed than previously presumed. The TEM micrographs appear very similar to the model prepared by Holt (1994). Cryo-transmission electron tomography also failed to show a sub-micellar structure (Marchin et al. 2007; Trejo et al. 2011). Holt (1998) concluded that none of the sub-micelle models of casein micelle structure explained the results of gel permeation chromatography of micelles dissociated by removal of CCP or by urea. de Kruif (1998) supported the structure of the casein micelle as depicted by Holt (1992, 1994) and describes the behaviour and properties of the micelles in terms of adhesive hard spheres.

A more recent model for casein micelle structure is the 'dual-bonding' model put forward by Horne (1998, 2002, 2006, 2011, 2014). This model suggests that micelle structure is governed by a balance of hydrophobic interactions and colloidal calcium phosphate-mediated cross-linking of hydrophilic regions (Fig. 4.23).

The study of casein micelle structure continues to be an active and exciting area of research with developments in analytical approaches contributing new information about casein micelle structure and stability (see Bouchoux et al. 2010). In comparison to the micelles of bovine milk, the casein micelles in the milk of other species has been studied only superficially; it is presumed that the micelles of all species are basically similar but their detailed study is a vast area for further research.

Holt (1992, 1994) also proposed that, in addition to supplying amino acids, caseins should be considered to have a biological function, i.e. to enable a high concentration of calcium to be carried in stable form in milk; without the stabilizing effect of casein, calcium phosphate would precipitate in the mammary cells, resulting in ectopic mineralization, which might lead to the death of the mammary gland or of the whole animal. A similar situation occurs with kidney stones, gallstones and calcified synovial and salivary fluid (see Chap. 5).

Since the micelles are closely packed, inter-micellar collisions are frequent; however, the micelles do not normally remain together after collisions. The micelles are stabilized by two principal factors: (1) a surface (zeta) potential of ~20 mV at pH 6.7, which, alone, is probably too small for colloidal stability, and (2) steric stabilization due to the protruding κ-casein hairs.

4.7 Whey Proteins

About 20 % of the total protein of bovine milk belongs to a group of proteins referred to as whey or serum proteins or non-casein nitrogen. Acid and rennet wheys also contain casein-derived peptides; both contain proteose-peptones, produced by plasmin, mainly from β-casein, and the latter also contains (glyco)macropeptides produced by rennets from κ-casein, These peptides are excluded from the present discussion.

4.7.1 Preparation

The whey proteins, as a group, are readily prepared from milk by any of the methods described in Sect. 4.3, i.e.

1. The proteins that remain soluble at pH 4.6;
2. Protein soluble in saturated NaCl;
3. Protein soluble after rennet coagulation of the caseins;
4. By gel permeation chromatography;
5. By ultracentrifugation, with or without added Ca^{2+};
6. Microfiltration.

The whey prepared by any of the above methods, except 4, contains lactose and soluble salts. Total whey proteins may be prepared from the whey by dialysis and drying the retentate. The products prepared by these various methods differ: acid whey contains some γ-casein and proteose-peptones; immunoglobulins are co-precipitated with the caseins by saturated NaCl; rennet whey contains the κ-CN macropeptides produced by rennet action, plus, perhaps, very small amounts of other caseins; small casein micelles remain in the ultracentrifugal supernatant, especially if Ca is not added. The salt composition of the serum differs very considerably in wheys produced by various methods.

On a commercial scale, whey protein-rich products are prepared by:

1. Ultrafiltration/diafiltration of acid or rennet whey to remove a variable amount of lactose, and spray-drying to produce whey protein concentrate (30–85 % protein).
2. Ion-exchange chromatography: proteins are adsorbed on an ion exchanger, washed free of lactose and salts and then eluted by pH adjustment. The eluate is desalted by ultrafiltration and spray-dried to yield whey protein isolate, containing about 95 % protein.
3. Demineralization by electrodialysis and/or ion exchange, thermal evaporation of water and crystallization of lactose.
4. Thermal denaturation, recovery of precipitated protein by filtration/centrifugation and spray-drying, to yield "lactalbumin" which has very low solubility and limited functionality.

Several other methods are available for the removal of whey proteins from whey but are not used commercially. Several methods for the purification of the major and minor whey proteins on a commercial scale have also been developed and will be discussed briefly in Sect. 4.18.6.

4.7.2 Heterogeneity of Whey Proteins

It was recognized about 1890 that whey prepared by any of the above methods contained two well-defined groups of proteins which could be fractionated by saturated $MgSO_4$ or half saturated $(NH_4)_2SO_4$; the precipitate (roughly 20 % of total N) was referred to as lactoglobulin and the soluble protein as lactalbumin.

The lactoglobulin fraction consists mainly of immunoglobulins (Ig), especially IgG_1, with lesser amounts of IgG_2, IgA and IgM (Sect. 4.12). The lactalbumin fraction of bovine milk contains three main proteins, β-lactoglobulin (β-lg), α-lactalbumin (α-la) and blood serum albumin (BSA), which represent approximately 50, 20 and 10 % of total whey protein, respectively, and trace amounts of several other proteins, notably lactoferrin, serotransferrin and several enzymes. The whey proteins of sheep, goat and buffalo milk are roughly similar to those in bovine milk. Human milk contains no β-lg and the milk of some species contains α-la and whey acidic protein (WAP). β-lg, α-la and WAP are synthesized in the mammary gland and are milk-specific; most of the other proteins in whey originate from blood or mammary tissue.

Since the 1930s, several methods have been developed for the isolation of homogeneous whey proteins, which have been crystallized (McKenzie 1970, 1971). Today, homogeneous whey proteins are usually prepared by ion-exchange chromatography on DEAE cellulose.

4.8 β-Lactoglobulin

4.8.1 Occurrence and Microheterogeneity

β-Lactoglobulin is a major protein in bovine milk, representing about 50 % of total whey protein or 12 % of the total protein of milk. It was among the first proteins to be crystallized, and since crystallizability was long considered to be a good criterion of homogeneity, β-lg, which is a typical globular protein, has been studied extensively and is very well characterized (reviewed by McKenzie 1971; Hambling et al. 1992; Sawyer 2003, 2013).

β-Lg is the principal whey protein (WP) in bovine, ovine, caprine and buffalo milks, although there are slight interspecies differences. Some years ago, it was believed that β-lg occurs only in the milk of ruminants but it is now known that it occurs in the milk of the sow, mare, kangaroo, dolphin, manatee and other species. However, β-lg does not occur in human, rat, mouse or guinea-pig milk, in which α-la is the principal WP.

The two principal genetic variants of bovine β-lg, are A and B with 11 other variants occurring less frequently. A variant, which contains carbohydrate, has been identified in the Australian breed, Droughtmaster (Dr). Further variants occur in the milk of yak and Bali cattle. Genetic polymorphism also occurs in β-lg of other species (see Sawyer 2013; Martin et al. 2013a, b).

4.8.2 Amino Acid Composition

The amino acid composition of some β-lg variants is shown in Table 4.4. It is rich in sulphur-containing amino acids which give it a high biological value of 110. It contains 2 mol of cystine and 1 mol of cysteine per monomer of 18 kDa. The cysteine is especially important since it reacts, following heat denaturation, with the disulphide of κ-casein and significantly affects rennet coagulation and the heat stability properties of milk; it is also responsible for the cooked flavour of heated milk. Some β-lgs, e.g. porcine, do not contain a free sulphydryl group. The isoionic point of bovine β-lg is ~pH 5.2.

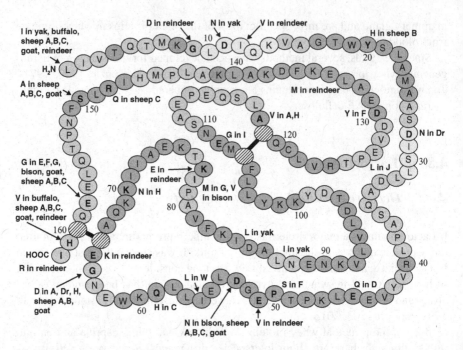

Fig. 4.24 Amino acid sequence variation within ruminant β-lactoglobulins relative to bovine genetic variant B (from Sawyer 2013)

4.8.3   Primary Structure

The amino acid sequence of β-lg consists of 162 residues per monomer; the sequence of bovine β-lg B and the substitutions in other variants of bovine β-lg and that of other ruminants is shown in Fig. 4.24.

4.8.4   Secondary Structure

β-Lg is a highly structured protein: optical rotary dispersion and circular dichroism measurements show that in the pH range 2–6, β-lg exists as 10–15 % α-helix, 43 % β-sheet and 47 % unordered structure, including β-turns.

4.8.5   Tertiary Structure

The tertiary structure of β-lg has been studied in considerable detail using X-ray crystallography (see Sawyer 2013). It has a very compact globular structure in which the β-sheets run anti-parallel to form a β-barrel-type structure or calyx (Figs. 4.25 and 4.26). Each monomer exists almost as a sphere with a diameter of about 3.6 nm.

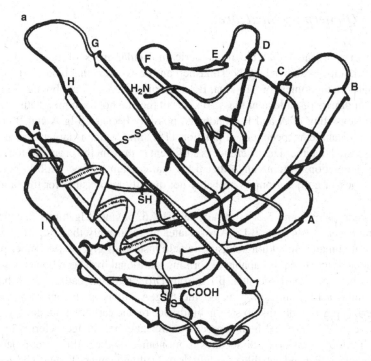

Fig. 4.25 Schematic representation of the tertiary structure of bovine β-lactoglobulin, showing the binding of retinol; *arrows* indicate anti parallel β-sheet structures (from Papiz et al. 1986)

Fig. 4.26 Structure of bovine β-lactoglobulin viewed into the central ligand-binding calyx at the bottom of which is Trp$_{19}$ (from Sawyer 2013)

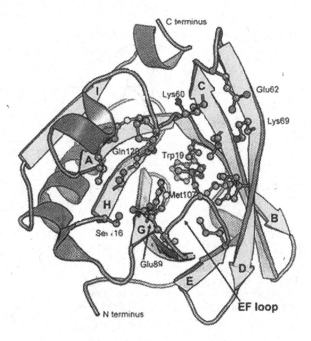

4.8.6   Quaternary Structure

Early work indicated that the monomeric molecular mass of bovine β-lg was ~36 kDa but it was shown by S.N. Timasheff and co-workers that below pH 3.5, β-lg dissociates to monomers of ~18 kDa. Between pH 5.5 and 7.5, all bovine β-lg variants form dimers of molecular mass 36 kDa but they do not form mixed dimers, i.e., a dimer consisting of A and B monomers, possibly because β-lg A and B contain valine and alanine, respectively, at position 118. Since valine is larger than alanine, it is suggested that the size difference is sufficient to prevent the proper fit for hydrophobic interaction. Porcine and other β-lgs that contain no free thiol do not form dimers; lack of a thiol group is probably not directly responsible for the failure to dimerize.

Between pH 3.5 and 5.2, especially at pH 4.6, bovine β-lg forms octamers of molecular mass ~144 kDa. β-Lg A associates more strongly than β-lg B, possibly because it contains an additional aspartic acid instead of glycine (in B) per monomer; the additional Asp is capable of forming additional hydrogen bonds in the pH region where it is undissociated. β-Lg from Droughtmaster cattle, which has the same amino acid composition as bovine β-lg A but is a glycoprotein, fails to octamerize, presumably due to stearic hindrance by the carbohydrate moiety.

Above pH 7.5, bovine β-lg undergoes a conformational change (referred to as the N↔R, Tanford, transition), dissociates to monomers and the thiol group becomes exposed and active and capable of sulphydryl-disulphide inter-change. The association of β-lg is summarized in Fig. 4.27.

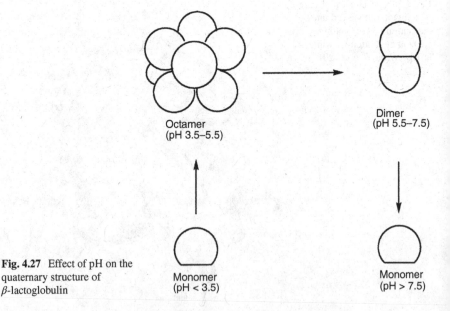

Octamer
(pH 3.5–5.5)

Dimer
(pH 5.5–7.5)

Fig. 4.27 Effect of pH on the quaternary structure of β-lactoglobulin

Monomer
(pH < 3.5)

Monomer
(pH > 7.5)

4.8.7 Physiological Function

Since the other principal whey proteins have a biological function, it has long been felt that β-lg might have a biological role; it appears that this role may be to act as a carrier for retinol (vitamin A). β-Lg can bind retinol in a hydrophobic pocket (see Fig. 4.25), protect it from oxidation and transport it through the stomach to the small intestine where the retinol is transferred to a retinol-binding protein, which has a similar structure to β-lg. β-Lg is capable of binding many hydrophobic molecules and hence its ability to bind retinol may be incidental. Unanswered questions are how retinol is transferred from the core of the fat globules, where it occurs in milk, to β-lg and how humans and rodents have evolved without β-lg.

β-Lg also binds free fatty acids and thus it stimulates lipolysis (lipases are inhibited by free fatty acids); perhaps this is its physiological function. BSA also binds hydrophobic molecules, including fatty acids; perhaps BSA serves a similar function to β-lg in those species lacking β-lg.

4.8.8 Denaturation

Denaturation of whey proteins is of major technological significance and is discussed in Chap. 9.

4.9 Whey Acidic Protein

Whey acidic protein (WAP) was identified first in the milk of mouse and has since been found also in the milk of rat, rabbit, pig, camel, wallaby, possum, echidna and platypus. Since the milk of all of these species lacks β-Lg, it was thought that these proteins were mutually exclusive. However, porcine milk, which contains β-lg, was recently found to contain WAP also (see Simpson et al. 1998). The MW of WAP is 14–30 kDa (the variation may be due to differences in glycosylation) and it contains two (in eutherians) or three (in monotrenes and marsupials) 4-disulphide domains. Since human milk lacks β-lg it might be expected to contain WAP but there are no reports to this effect. In humans and ruminants, the WAP gene is frame-shifted and is a pseudogene. WAP functions as a proteinase inhibitor, is involved in terminal differentiation in the mammary gland and has antibacterial activity (for reviews see Simpson and Nicholas 2002; Hajjoubi et al. 2006; Martin et al. 2011).

4.10   α-Lactalbumin

α-Lactalbumin (α-la) represents about 20 % of the proteins of bovine whey (3.5 % of total milk protein); it is the principal protein in human milk. α-La is a small protein with a molecular mass of c ~14 kDa. Recent reviews of the literature on this protein include Kronman (1989), Brew and Grobler (1992) and Brew (2003, 2013).

4.10.1   Amino Acid Composition

The amino acid composition is shown in Table 4.4. α-La is relatively rich in tryptophan (four residues per mole). It is also rich in sulphur (1.9 %) which is present in cystine (four intramolecular disulphides per mole) and methionine; it contains no cysteine (sulphydryl group). α-La contains no phosphorus or carbohydrate, although some minor forms may contain either or both. The isoionic point is ~pH 4.8 and minimum solubility in 0.5 M NaCl is also at pH 4.8.

4.10.2   Genetic Variants

The milk of Western cattle contains mainly α-la B but Zebu and Droughtmaster cattle secrete two variants, A and B. α-La A contains no arginine, the one Arg residue of α-la B being replaced by glutamic acid. Two rare variants, C and D, have been reported in Bali cattle.

4.10.3   Primary Structure

The primary structure of α-la is shown in Fig. 4.28. There is considerable homology between the sequence of α-la and lysozyme from many sources. The primary structures of α-la and chicken egg white lysozyme are very similar. Out of a total of 123 residues in α-la, 54 are identical to corresponding residues in lysozyme and a further 23 residues are structurally similar (e.g., Ser/Thr, Asp/Glu).

4.10.4   Secondary and Tertiary Structure

α-La is a compact globular protein, which exists in solution as a prolate ellipsoid with dimensions of 2.3×2.6×4.0 nm. It exists as 26 % α-helix, 14 % β-structure and 60 % unordered structure. The metal-binding (Sect. 4.10.8) and molecular conformational properties of α-la were discussed in detail by Kronman (1989). The tertiary structure of α-la is very similar to that of lysozyme. X-ray crystallography of α-la from several species has been reported (Brew 2013); the 3-D structure of the molecule is shown in Fig. 4.29.

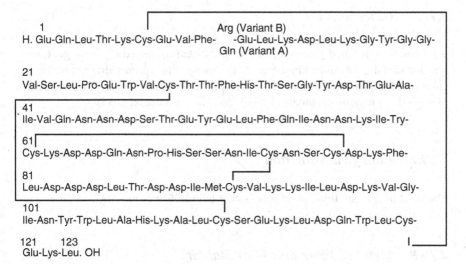

```
1                                      Arg (Variant B)
H. Glu-Gln-Leu-Thr-Lys-Cys-Glu-Val-Phe-   -Glu-Leu-Lys-Asp-Leu-Lys-Gly-Tyr-Gly-Gly-
                                       Gln (Variant A)

21
Val-Ser-Leu-Pro-Glu-Trp-Val-Cys-Thr-Thr-Phe-His-Thr-Ser-Gly-Tyr-Asp-Thr-Glu-Ala-

41
Ile-Val-Gln-Asn-Asn-Asp-Ser-Thr-Glu-Tyr-Glu-Leu-Phe-Gln-Ile-Asn-Asn-Lys-Ile-Try-

61
Cys-Lys-Asp-Asp-Gln-Asn-Pro-His-Ser-Ser-Asn-Ile-Cys-Asn-Ser-Cys-Asp-Lys-Phe-

81
Leu-Asp-Asp-Asp-Leu-Thr-Asp-Asp-Ile-Met-Cys-Val-Lys-Lys-Ile-Leu-Asp-Lys-Val-Gly-

101
Ile-Asn-Tyr-Trp-Leu-Ala-His-Lys-Ala-Leu-Cys-Ser-Glu-Lys-Leu-Asp-Gln-Trp-Leu-Cys-

121    123
Glu-Lys-Leu. OH
```

Fig. 4.28 Amino acid sequence of *α*-lactalbumin showing intramolecular disulphide bonds (*dashed lines*) and amino acid substitutions in genetic polymorphs (from Brew and Grobler 1992)

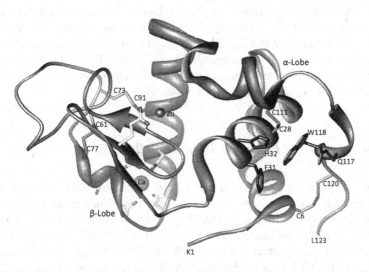

Fig. 4.29 The 3D structure of the Ca/Zn complex of human *α*-lactalbumin (from Brew 2013)

4.10.5   Quaternary Structure

α-La associates under a variety of environmental conditions but the association process has not been well studied.

4.10.6   Other Species

α-La has been isolated from several species, including the cow, sheep, goat, sow, human, buffalo, rat and guinea-pig. Some minor interspecies differences in the amino acid sequence and properties have been reported (see Brew 2013). The milk of several sea mammals contains little or no α-la (see Oftedal 2011).

4.10.7   Biological Function

The most interesting function of α-la is its role in lactose synthesis (see Chap. 2)

4.10.8   Metal Binding and Heat Stability

α-La is a metallo-protein; it binds one Ca^{2+} per mole in a pocket containing four Asp residues (Figs. 4.29 and 4.30); these residues are highly conserved in all α-la's and in lysozyme. The Ca-containing protein is quite heat stable (it is the most heat stable whey protein) or more correctly, the protein renatures following heat denaturation (denaturation occurs at a relatively low temperature, as indicated by differential scanning calorimetry). When the pH is reduced to below about 5, the Asp residues become protonated and lose their ability to bind Ca^{2+}. The metal-free protein is denatured at quite a low temperature and does not renature on cooling; this characteristic has been exploited to isolate α-la from whey.

4.10.9   Apoptosis Effect on Tumour Cells

Recently, an interesting non-native state of apo-α-la, stabilized by complex formation with oleic acid, has been found to selectively induce apoptosis in tumour cells — this complex is known as HAMLET (Human α-la Made Lethal to Tumour cells). The complex can be generated from apo-α-la by chromatography on an ion-exchange column, pre-conditioned with oleic acid. The complex can be formed from human (i.e., HAMLET) or bovine (i.e., BAMLET) apo-α-la (Liskova et al. 2010),

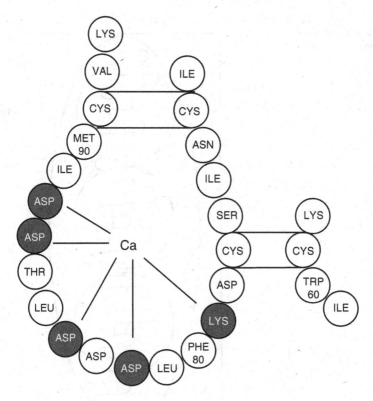

Fig. 4.30 Calcium-binding loop in bovine α-lactalbumin (modified from Berliner et al. 1991)

with both forms reported to have comparable cytotoxic activity against three different cancer cell lines (Brinkmann et al. 2011). This complex may offer potential as a premium functional food ingredient (see Brew 2013 and also Chap. 11).

4.11 Blood Serum Albumin

Normal bovine milk contains a low level of blood serum albumin (BSA) (0.1–0.4 g L^{-1}; 0.3–1.0 % of total N), presumably as a result of leakage from blood. The molecular mass of BSA is ~65 kDa; it contains 582 amino acid residues, 17 disulphides and one sulphydryl. All the disulphides involve cysteines that are relatively close in the polypeptide chain, which is therefore organized in a series of relatively short loops (Fig. 4.31). The molecule is elliptical in shape and is divided into three domains. Owing to its biological function, BSA has been studied extensively; reviews include Carter and Ho (1994) and Nicholson et al. (2000). In blood, BSA serves various functions: it controls the osmotic pressure of blood (and thus

Fig. 4.31 Model of the
bovine serum albumin
molecule

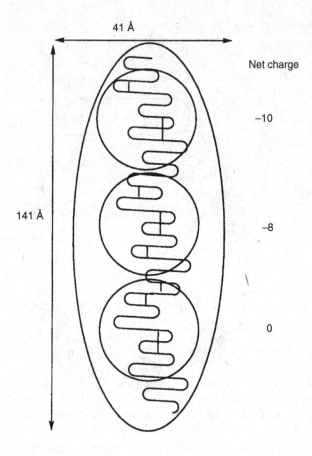

regulates the uptake of fluids from tissues), transports thyroid and other hormones,
fatty acids and many drugs, it binds Ca^{2+} and buffers pH. It probably has little or no
significance in milk, although by binding metals and fatty acids, it may enable it to
stimulate lipase activity.

4.12 Immunoglobulins (Ig)

Mature milk contains 0.6–1 g Ig L^{-1} (~3 % of total N) but colostrum contains up to
100 g L^{-1}, the level of which decreases rapidly post-partum (Fig. 4.2).

Igs are very complex proteins which will not be reviewed here (see Hurley and
Theil 2013 and textbooks on Biochemistry, Physiology or Immunology). There are
five classes of Ig: IgA, IgG, IgD, IgE and IgM; IgA, IgG and IgM are present in
milk. These occur as subclasses, e.g., IgG occurs as IgG_1 and IgG_2. IgG consists of
two long (heavy) and two shorter (light) polypeptide chains linked by disulphides
(Fig. 4.32). IgA consists of two such units (i.e., eight chains) linked together by

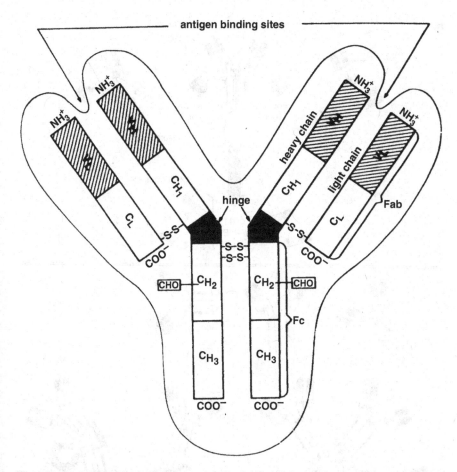

Fig. 4.32 Model of the basic 7S immunoglobulin (Ig) molecule showing two heavy and two light chains joined by disulphide bonds: *V* variable region, *C* constant region; *L* light chain, *H* heavy chain, 1, 2 and 3 subscripts refer to the three constant regions of the heavy chains, *CHO* carbohydrate groups, Fab refers to the (*top*) antigen-specific portion of the Ig molecule, Fe refers to the cell-binding effector portion of the Ig molecule (from Larson 1992)

secretory component (SC) and a junction (j) component, while IgM consists of five linked four-chain units (Fig. 4.33). The heavy and light chains are specific to each type of Ig. For reviews of immunoglobulins in milk, see Larson (1992), Hurley (2003) and Hurley and Theil (2011, 2013).

The function of Ig is to provide various types of immunity in the body. The principal Ig in bovine milk is IgG_1 while in human milk it is IgA. The calf (and the young of other ruminants) is born without Ig in its blood serum and hence is very susceptible to infection. However, the intestine of the calf is permeable to large molecules for about 3 days post-partum and therefore Ig is absorbed intact and active from its mother's colostrum; Igs from colostrum appear in the calf's blood

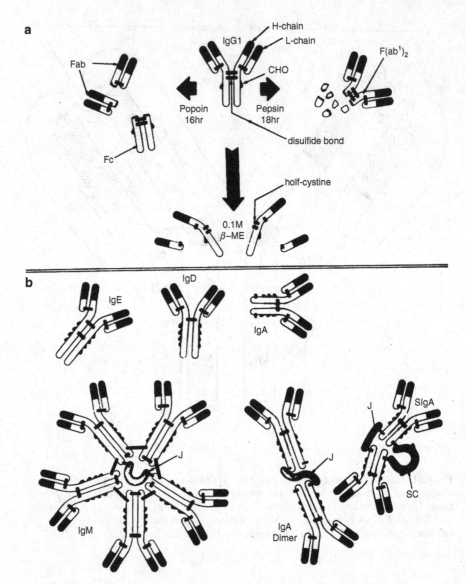

Fig. 4.33 Models of IgG, IgA, IgD, IgE and IgM. (**a**) Structural model of IgG₁ before and after fragmentation by pepsin and papain and reduction with a sulphydryl reagent. *Solid black chain portion* = variable regions; *light chain portion* = constant regions. *Small black lines* represent disulphide and half-cystine (–SH) groups. *Small black dots* in Fc regions represent attached carbohydrate groups. The various parts of the model are labelled. (**b**) The structure of four classes of immunoglobulins are shown with monomeric IgA, dimeric IgA and secretory IgA. Location of the J-chain, secretory component (SC) and carbohydrate is approximate. (From Larson 1992)

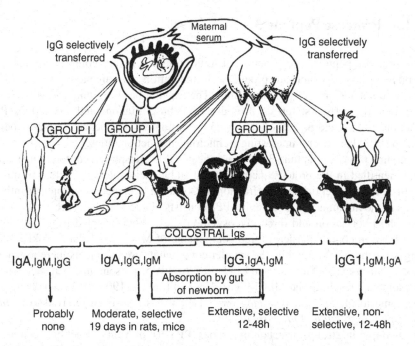

Fig. 4.34   Transfer of maternal immunoglobulins to the foetus and neonate of representative mammalian species. Group I species transfer Ig *in utero* before birth. Group II species transfer Ig both *in utero* before birth and via colostrum after birth. Group III species transfer Ig only via colostrum after birth. The size of the immunoglobulin notation (IgA, IgM, IgG, IgG$_1$) indicates the relative percentage composition of the immunoglobulins in colostrum. Species in group II may have IgG as the predominant Ig in colostrum. Significant IgG$_2$ also may be present in the colostrum of some Group III species. The relative absorption of immunoglobulins in the gut of the neonate is also shown. (From Larson 1992)

within about 3 h of suckling and persist for about 3 months, although the calf is able to synthesize its own Ig within about 2 weeks. It is, therefore, essential that a calf should receive colostrum within a few hours of birth. The human baby obtains Ig *in utero* and hence, unlike the calf, is not as dependent on Ig from milk (in fact its intestine is impermeable to Ig). However, the Ig in human colostrum is beneficial to the baby, e.g. it reduces the risk of intestinal infections.

As regards the type and function of Ig in colostrum, mammals fall into three groups (Fig. 4.34) — (I) the cow, other ruminants pig and horse, (II) humans and rabbit and (III) mouse, rat and dog which have features of the other two groups (Larson 1992).

Colostrum differs markedly from mature milk; its composition and properties have been reviewed by McGrath (2014). The modern dairy cow produces much more colostrum than its calf requires and even consumes. Colostrum is excluded from the milk supply for about 6 days *post partum*; the excess colostrum is frequently fed to older calves or pigs; some is processed for human consumption as a liquid or as cheese-like products.

4.13 Proteose Peptone 3

The proteose-peptone (PP) fraction of milk protein is a very complex mixture of peptides, most of which are produced by the action of indigenous plasmin (see above) but some are indigenous to milk. The fraction has been only partially characterised; the current status has been described by O'Mahony and Fox (2013). PP fractions 5, 8_{slow} and 8_{fast} have little or no technological significance, proteose peptone 3 (PP3) has several interesting technological functionalities.

Bovine proteose peptone 3 (PP3) is a heat-stable phosphoglycoprotein that was first identified in the proteose-peptone (heat-stable, acid-soluble) fraction of milk. Unlike the other peptides in this fraction, PP3 is an indigenous milk protein, synthesized in the mammary gland. Bovine PP3 consists of 135 amino acid residues, with five phosphorylation and three glycosylation sites. When isolated from milk, the PP3 fraction contains at least three components of $MW \approx 28$, 18 and 11 kDa; the largest of these is PP3, while the smaller components are fragments thereof generated by plasmin (see Girardet and Linden 1996). PP3 is present mainly in acid whey but some is present in the MFGM. Girardet and Linden (1996) proposed changing the name to *lactophorin* or *lactoglycophorin*; it has also been referred to as *the hydrophobic fraction of proteose peptone*.

Owing to its strong surfactant properties, PP3 can prevent contact between milk lipase and its substrates, thus preventing spontaneous lipolysis and its emulsifying properties have also been evaluated in dairy products such as ice cream and recombined dairy cream (Vanderghem et al. 2007; Innocente et al. 2011). Although its amino acid composition suggests that PP3 is not a hydrophobic protein, it behaves hydrophobically, possibly because it forms an amphiphilic α-helix, one side of which contains hydrophilic residues while the other side is hydrophobic. The biological role of PP3 is unknown.

4.14 Minor Milk Proteins

Milk contains numerous minor proteins, including about 60 indigenous enzymes, some of which, e.g., lipoprotein lipase, proteinase, phosphatases, lactoperoxidase and xanthine oxidoreductase, are technologically important (Chap. 10). Most of the minor proteins have biological functions and probably play very significant roles (Wynn and Sheehy 2013 and Chap. 11)

4.15 Non-protein Nitrogen

Nitrogen soluble in 12 % TCA is referred to as non-protein nitrogen (NPN), of which milk contains 250–300 mg L^{-1}, i.e. 5–6 % of total milk nitrogen. The NPN is a very heterogeneous fraction (Table 4.7).

Table 4.7 Non-protein nitrogen of cow's milk

Component	N (mg L^{-1})
Ammonia	6.7
Urea	83.8
Creatinine	4.9
Creatine	39.3
Uric acid	22.8
α-Amino nitrogen	37.4
Unaccounted	88.1

The 'unaccounted' N includes some phospholipids, amino sugars, nucleotides, hippuric acid and orotic acid. The α-amino N includes free amino acids and small peptides; almost a complete range of amino acids, including ornithine, has been identified in milk, but glutamic acid predominates.

All the components of NPN are present in blood, from which they are probably transferred into milk. The technological and nutritional significance of NPN is not known but the amino acids are likely to be important for the nutrition of starter micro-organisms, especially of weakly proteolytic strains. Urea, which is the principal component of the NPN (6 mmol L^{-1}), is strongly correlated with the heat stability of milk; the urea content of milk from cows on pasture is twice as high as that from cows on dry feed and hence the heat stability of the former is considerably higher. The level of NPN in freshly drawn milk is fairly constant but it does increase on ageing, especially if significant growth of psychrophilic bacteria, which may be strongly proteolytic, occurs.

4.16   Interspecies Comparison of Milk Proteins

This chapter has been concerned mainly with the protein system of bovine milk, which is by far the most important commercially. However, there are ~4,500 species of mammal, each of which produces milk, the composition and properties of which are more or less species-specific. Unfortunately, the milk of most species has not been studied at all; some information is available on the milk of ~180 species. However, the data on the milk of only about 50 species are considered to be reliable, in that a sufficient number of samples were analyzed and that these samples were reliable, properly taken, and covering the lactation period adequately. Milk from the commercially important species, cow, goat, sheep, buffalo, yak, horse and pig are quite well characterized. For medical and nutritional reasons, human milk is also well characterized, as is that of experimental laboratory animals, especially rats and mice. For reviews on non-bovine milks see O'Mahony and Fox (2013) and a set of articles in Fuquay et al. (2011).

The milk of the species for which data are available show considerable differences in protein content, i.e., from ~1 to 20 %. The protein content reflects the growth rate of the neonate of the species, i.e., its requirements for essential amino acids. The milk of all species for which data are available contain two groups of protein, caseins and whey proteins. Both groups show genus- and even species-specific characteristics which presumably reflect some unique nutritional or physiological requirements of the neonate of the species. Interestingly, and perhaps significantly, of the milks that have been characterized, human and bovine milks are more or less at opposite ends of the spectrum.

There is considerably more and better information available on the interspecies comparison of individual milk proteins than of overall milk composition; this is not surprising since only one sample of milk from one animal is sufficient to yield a particular protein for characterization in addition to advances in DNA homology studies. The two principal milk-specific whey proteins, α-la and β-lg, from quite a wide range of species have been characterized, and, in general, show a high degree of homology (see Sawyer 2013; Brew 2013). However, the caseins show much greater inter-species diversity, especially in the α-casein fraction — all species that have been studied appear to contain a protein that has an electrophoretic mobility similar to that of bovine β-casein (Fig. 4.7), but the β-caseins that have been sequenced show a low level of homology (Holt and Sawyer 1993; Martin et al. 2003, 2013a, b). Human β-casein occurs in multi-phosphorylated form (0–5 mol P per mol protein; see Atkinson and Lonnerdal 1989), as does equine β-casein (Ochirkuyag et al. 2000). Considering the critical role played by κ-casein, it would be expected that all casein systems contain this protein. Human κ-casein is very highly glycosylated, containing 40 – 60 % carbohydrate (compared with approximately 10 % for bovine κ-casein), which occurs as oligosaccharides which are much more diverse and complex than those in bovine milk (see Atkinson and Lonnerdal 1989).

The α_s-casein fraction differs markedly between species (Fig. 4.7); human milk lacks an α_s-casein while the α-casein fractions in horse and donkey milk are very heterogeneous. The caseins of only about ten species have been studied in some detail. In addition to the references cited earlier in this section, the literature has been reviewed by Martin et al. (2003, 2013a, b). Martin et al. (2013a, b) includes numerous references on the proteins and milk protein genes from several species. There are very considerable inter-species differences in the minor proteins of milk. The milks of those species which have been studied in sufficient depth contain approximately the same profile of minor proteins, but there are very marked quantitative differences. Most of the minor proteins in milk have some biochemical or physiological function, and the quantitative inter-species differences presumably reflect the requirements of the neonate of the species. Many of the minor milk proteins are considered in Chap. 11.

In the milk of all species, the caseins probably exist as micelles (at least the milks appear white) but the properties of the micelles in the milk of only a few species have been studied. The micelles in caprine milks were studied by Ono and Creamer (1986). The water buffalo is the second most important dairy animal and is particularly important in India. The composition and many of the physico-chemical properties

Table 4.8 Some important differences between bovine and human milk proteins

Constituent	Bovine	Human
Protein concentration (%)	3.5	1.0
Casein:NCN	80:20	40:60
Casein types[a]	$\alpha_{s1} = \beta > \alpha_{s2} = \kappa$	$\beta > \kappa$; no α_{s1}
β-Lactoglobulin	50 % of NCN	None
Lactoferrin	Trace	20 % of total N
Lysozyme	Trace	Very high (6 % TN; 3,000×bovine)
Glycopeptides	Trace	High
NPN (as % TN)	3	20
Taurine	Trace	High
Lactoperoxidase	High	Low
Immunoglobulins (Ig) (colostrum)	Very high	Lower
Ig type	$IgG_1 > IgG_2 > IgA$	$IgA > IgG > IgG_2$

NCN non-casein nitrogen, *NPN* non-protein nitrogen, *TN* total nitrogen
[a]A low level of α_{s1}-casein was reported by (Martin et al. 1996) in human milk

of buffalo milk differ considerably from those of bovine milk (see Patel and Mistry 1997). Other properties of buffalo milk will be mentioned for comparative purposes in other chapters. Some properties of the casein micelles in camel milk have been described by Attia et al. (2000). Possibly because porcine milk is relatively easily obtained, but also because it has interesting properties, the physico-chemical behaviour of porcine milk has been studied fairly thoroughly and the literature reviewed by Gallagher et al. (1997). Equine and asinine milks have also been the subject of some detailed characterization over the last 20 years or so (Oftedal and Jenness 1988; Salimei et al. 2004; Uniacke-Lowe et al. 2010; Uniacke-Lowe and Fox 2011).

Some of the more important differences between human and bovine milk are summarized in Table 4.8. At least some of these differences are probably nutritionally and physiologically important. It is perhaps ironic that human babies are the least likely of all species to receive the milk intended for them.

4.17   Synthesis and Secretion of Milk Proteins

The synthesis and secretion of milk proteins have been studied in considerable detail; reviews include Mercier and Gaye (1983), Mepham (1987), Mepham et al. (1982, 1992), Violette et al. (2003, 2013).

4.17.1   Sources of Amino Acids

Arteriovenous (AV) difference studies and mammary blood flow measurements (Chap. 1) have shown that in both ruminants and non-ruminants, amino acids for milk protein synthesis are obtained from blood plasma but that some inter-conversions occur. The amino acids can be divided into two major groups:

1. Those for which uptake from blood is adequate to supply the requirements for milk protein synthesis and which correspond roughly to the essential amino acids (EAA); and
2. Those for which uptake is inadequate, i.e., the non-essential amino acids (NEAA).

Studies involving AV difference measurements, isotopes and perfused gland preparations indicate that the EAA may be subdivided into those for which uptake from blood and output in milk proteins are almost exactly balanced (Group I) and those for which uptake significantly exceeds output (Group 11). Group 11 amino acids are metabolized in the mammary gland and provide amino groups, via transamination, for the biosynthesis of those amino acids for which uptake from blood is inadequate (Group Ill), their carbon skeletons are oxidized to CO_2. Considered as a whole, total uptake and output of amino acids from blood are the major, or sole, precursors of the milk-specific proteins (i.e., the caseins, β-lactoglobulin and α-lactalbumin).

• Group I amino acids: methionine, phenylalanine, tyrosine, histidine and tryptophan.
• Group II amino acids: valine, leucine, isoleucine, lysine, arginine and threonine.
• Group III amino acids: aspartic acid, glutamic acid, glycine, alanine, serine, cysteine/cystine, proline.

The interrelationships between the carbon and nitrogen of amino acids are summarized in Fig. 4.35.

4.17.2   Amino Acid Transport into the Mammary Cell

Since the cell membranes are composed predominantly of lipids, amino acids (which are hydrophilic) cannot enter by diffusion and are transported by special carrier systems (see Violette et al. 2013).

4.17.3   Synthesis of Milk Proteins

Synthesis of the major milk proteins occurs in the mammary gland; the principal exceptions are serum albumin and some of the immunoglobulins, which are transferred from the blood. Polymerization of the amino acids occurs on ribosomes fixed on the rough endoplasmic reticulum of the secretory cells, by the method common to all cells.

Fig. 4.35 Summary
diagrams of amino acid
metabolism in mammary
tissue. (**a**) Amino acid carbon
interrelationships, (**b**) amino
acid nitrogen
interrelationships (from
Mepham et al. 1982)

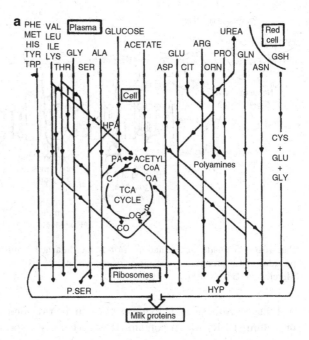

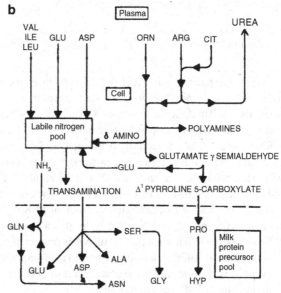

The primary blueprint for the amino acid sequence of proteins is contained in deoxyribonucleic acid (DNA) within the cell nucleus. The requisite information is transcribed in the nucleus to ribonucleic acid (RNA) of which there are three types: messenger RNA (mRNA), transfer RNA (tRNA) and ribosomal RNA (rRNA); these are transferred to the cytoplasm where each plays a specific role in protein synthesis.

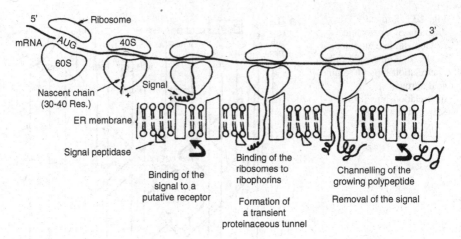

Fig. 4.36 Schematic representation of ribosomes attached to mRNA showing the growing polypeptides and a proposed mechanism for cotranslational crossing of the RER membrane (from Mercier and Gaye 1983)

Protein synthesis actually takes place in the ribosomes of the rough endoplasmic reticulum (RER) which contain rRNA. There is a specific tRNA for each amino acid, with which it forms an acyl complex:

$$\text{Amino acid} + \text{tRNA} + \text{ATP} \xrightarrow[\text{Mg}^+]{\text{amino acyl-tRNA synthetase}} \text{amino acyl-tRNA} + \text{AMP} + \text{PPi}$$

There is a specific amino acyl-tRNA synthetase for each amino acid; these enzymes have two specific binding sites, one for the amino acid and the second for the appropriate tRNA. The specificity of the tRNAs is determined by the sequence of the anticodon which recognizes and hydrogen bonds with the complementary codon of the mRNA. Interaction between the tRNA and the appropriate amino acid occurs in the cytoplasm but the remaining reactions in protein synthesis occur in the ribosomes, which are complex structures of rRNA and a number of proteins (including enzymes, initiators and controlling factors). The ribosomes of animal cells have a diameter of about 22 nm and a sedimentation coefficient of 80S; they consist of two principal subunits: 60S and 40S. mRNA passes through a groove or tunnel between the 60S and 40S subunits; while in the groove, mRNA is protected from the action of ribonuclease (Fig. 4.36). The information for the amino acid sequence is contained in the mRNA.

Synthesis commences at the correct codon of the mRNA because a special amino acid derivative, N-formyl methionine:

$$
\begin{array}{c}
\text{H} \\
| \\
\text{C}=\text{O} \\
| \\
\text{NH} \\
| \\
\text{H}_3\text{CSCH}_2\text{CH}_2\text{C-COOH} \\
| \\
\text{H}
\end{array}
$$

is bound to a specific special codon and forms the temporary N-terminal residue of the protein; N-formyl methionine is later hydrolysed off, together with a short hydrophobic signal peptide, exposing the permanent *N*-terminal residue. The acyl amino acid-tRNA is bound to the mRNA just outside the ribosome by becoming attached to its corresponding codon; presumably, a full range of amino acid-tRNAs are available in the environment but only the tRNA with the appropriate anticodon is bound. GTP and a number of specific cytoplasmic protein factors are required for binding.

In the ribosome, the amino group of the newly-bound amino acid reacts through nucleophilic substitution with the C-terminal carbonyl carbon of the existing peptide, and in the process the peptide is transferred to the newly-bound tRNA, releasing the tRNA just vacated. Condensation is catalysed by a peptidyl transferase, which is part of the ribosomal subunit.

For the next cycle, a new acyl amino acid-tRNA is bound to the mRNA, the ribosome tracks along the mRNA and the vacated tRNA is ejected. As the polypeptide is elongated it assumes its secondary and tertiary structure (Fig. 4.36).

Termination of synthesis is controlled by a special ribosomally-bound protein, TB 3–1, the **protein release factor** (RF) which recognises any of three "stop codons" UAG, UAA and UGA. RF promotes the hydrolysis of the ester bond linking the polypeptide with the tRNA.

A strand of mRNA is long enough to accommodate several ribosomes along its length, e.g., the mRNA for haemoglobin (150 amino acid residues/molecule) contains 450 nucleotides and is ~150 nm long; since each ribosome is about 20 nm in diameter, five to six ribosomes can be accommodated. The ribosomes are connected to each other by the mRNA strand, forming a polysome (polyribosome) which can be isolated intact if adequate care is taken. Each ribosome in a polysome is at a different stage in the synthesis of a protein molecule, thereby utilizing the mRNA more efficiently (Fig. 4.36).

Milk proteins are destined to be exported from the cell. Like other exported proteins, translocation through cell membranes is facilitated by a **signal sequence**, a sequence of 15–29 amino acids at the amino terminal of the growing polypeptide chain. This sequence causes the ribosome to bind to the ER membrane, in which a 'channel' forms, allowing the growing chain to enter the ER lumen (Fig. 4.36). Subsequently, the signal sequence is cleaved from the polypeptide by a **signal peptidase**, an enzyme located on the luminal side of the ER membrane.

4.17.4   Modifications of the Polypeptide Chain

In addition to proteolytic processing (i.e., removal of the signal peptide), the polypeptide is subject to other covalent modifications: N- and O-glycosylation and O-phosphorylation. After synthesis and transportation across the ER lumen, the proteins pass to the Golgi apparatus and thence, *via* secretory vesicles, to the apical membrane. Covalent modification must therefore occur at some point(s) along this route. Such modifications may be either co-translational (occurring when chain elongation is in progress) or post-translational. Proteolytic cleavage of the signal peptide is co-translational and this seems to be the case also for *N*-glycosylation, in which dolichol-linked oligosaccharides are enzymatically transferred to asparaginyl residues of the chain when these are present in the sequence code, Asn-X-Thr/Ser (where X is any amino acid except proline). The large oligosaccharide component may be 'trimmed' as it traverses the secretory pathway. Formation of disulphide bonds between adjacent sections of the chain, or between adjacent chains (as in κ-casein), may also be partly co-translational,

In contrast, *O*-glycosylation and *O*-phosphorylation appear to be post-translational events. Glycosylation of the principal milk-specific glycoprotein, casein, is believed to be effected by membrane-bound glycosyltransferases (three such enzymes have been described) located in the Golgi apparatus. *O*-Phosphorylation involves transfer of the γ-phosphate of ATP to serine (or, less frequently, threonine) residues, occurring in the sequence, Ser/Thr-X-A (where X is any amino acid residue and A is an acidic residue, such as aspartic or glutamic acid or a phosphorylated amino acid). Phosphorylation is effected by casein kinases which are located chiefly in the Golgi membranes. In addition to the correct triplets, the local conformation of the protein is also important for phosphorylation of Ser since not all serines in caseins in the correct sequence are phosphorylated. Some serine residues in β-lg occur in a Ser-X-A sequence but are not phosphorylated, probably due to extensive folding of this protein.

The Golgi complex is also the locus of casein micelle formation. In association with calcium, which is actively accumulated by Golgi vesicles, the polypeptide chains associate to form submicelles, and then micelles, prior to secretion.

4.17.5   Structure and Expression of Milk Protein Genes

The structure, organization and expression of milk protein genes are now understood in considerable detail. This subject is considered to be outside the scope of this book and the interested reader is referred to Mepham et al. (1992), Martin et al. (2013a, b), Oftedal 2013 and Singh et al. (2014). Such knowledge permits the genetic engineering of milk proteins with respect to the transfer of genes from one species to another, the over-expression of a particular desirable protein(s), the elimination of certain undesirable proteins, changing the amino acid sequence by point mutations to modify the functional properties of the protein or transfer of a milk protein gene to a plant or microbial host. This topic is also considered to be outside the scope of this text and the interested reader is referred to Richardson et al. (1992) and Leaver and Law (2003).

4.17.6   Secretion of Milk-Specific Proteins

Following synthesis in the ribosomes and passage into the ER lumen, the polypeptides are transferred to Golgi lumina. The route of transfer from ER to the Golgi has not been established with certainty. It is possible that lumina of the ER and Golgi

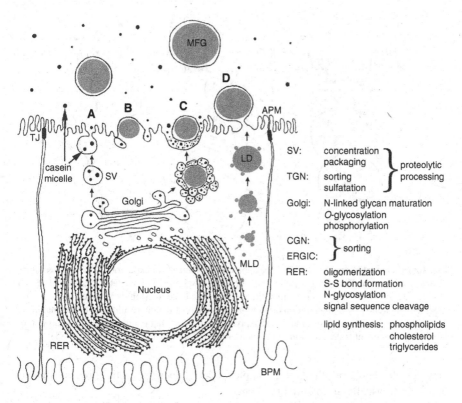

Fig. 4.37 Schematic representation of a mammary secretory cell: *BPM* basal plasma membrane, *RER* rough endoplasmic reticulum, *SV* secretory vesicles, *APM* apical plasma membrane, *MLD's* microlipid droplets, *CLD's* cytoplasmic lipid droplets, *CGN* cis-Golgi network, *ERGIC* ER-Golgi intermediate compartment, *LD* lipid droplet, *MFG* milk fat globule, *TJ* tight junction, *TGN* trans-Golgi network (from Violette et al. 2013)

apparatus are connected, or that small vesicles bud from the ER and subsequently fuse with the Golgi membranes. In either case, casein molecules aggregate in the Golgi cisternal lumina in the form of micelles (Violette et al. 2013).

Lumina at the nuclear face of the Golgi apparatus (Fig. 4.37) are termed *cis* cisternae; those at the apical face *trans* cisternae. Proteins appear to enter the complex at the *cis* face and progress, undergoing post-translational modification, towards the *trans* face. Transfer between adjacent Golgi cisternae is thought to be achieved by budding and subsequent fusion of vesicles.

In the apical cytosol there are numerous protein-containing secretory vesicles (Fig. 4.37). EM studies suggest that they move to the apical plasmalemma and fuse with it, releasing their contents by exocytosis. Current ideas on intracellular transport of vesicles suggest participation of cytoskeletal elements - microtubules and microfilaments. In mammary cells, these structures are orientated from the basal to the apical membrane suggesting that they may act as 'guides' for vesicular movement. Alternatively, vesicle transport may involve simple physical displacement as

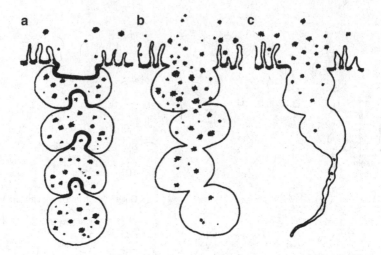

Fig. 4.38 Schematic representation of one apparent mechanism for exocytotic release of secretory vesicle contents. (**a**) Vesicles assemble into a chain through ball-and-socket interaction. The exit vesicle interacts with apical plasma membrane via a vesicle depression. (**b**) Linked vesicles fuse together, apparently by disintegration of membrane in areas of fusion, resulting in the formation of a continuum with the alveolar lumen. (**c**) Emptying of the vesicular chain appears to result in collapse and subsequent fragmentation of the membrane. (From Keenan and Dylewski 1985)

new vesicles bud from the Golgi complex, or an 'electrophoretic' process, dependent on a transcellular potential gradient.

Secretory vesicles seem to become attached to the cytoplasmic face of the apical plasmalemma. The vesicles have a distinctive coat on their outer surface which appears to react with appropriate receptors on the apical membrane, forming a series of regularly-spaced bridges. Presumably, these bridges, and the contiguous vesicle and apical membrane material, are subsequently eliminated and the vesicular contents released, but the process seems to be very rapid and it has proved difficult to visualize the details of the sequence by EM. However, secretory vesicle membrane becomes incorporated, however briefly, into the apical membrane as a consequence of exocytosis.

Alternatively, protein granules are transported through the lumina of a contiguous sequence of vesicles, so that only the most apical vesicle fuses with the apical membrane (Fig. 4.38). The process has been called **compound** exocytosis.

Thus, the synthesis and secretion of milk proteins involves eight steps: transcription, translation, segregation, modification, concentration, packaging, storage and exocytosis, as summarized schematically in Fig. 4.39.

4.17.7   Secretion of Immunoglobulins

Interspecies differences in the relative importance of colostral Igs are discussed in Sect. 4.12. The IgG of bovine colostrum is derived exclusively from blood plasma. It is presumed that cellular uptake involves binding of IgG molecules, *via* the Fc

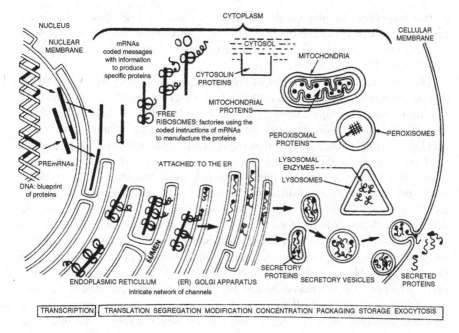

Fig. 4.39 Schematic representation of the intra cellular transport of proteins in mammary cells (from Mepham et al. 1982)

fragment (Fig. 4.32), to receptors situated in the basal membranes; just prior to parturition, there is a sharp increase in the number of such receptors showing a high affinity for IgG$_1$, which is selectively transported into bovine colostrum. The intra-cellular transport route has not been described with any degree of certainty, but the most likely scheme appears to involve vesicular transport, followed by exocytosis at the apical membrane.

IgA in colostrum is derived partly from intra-mammary synthesis and partly by accumulation in the gland after being transported in the blood from other sites of synthesis. In either case, IgA molecules are transported into the secretory cells across the basal membrane by means of a large, membrane-bound form of secretory component, which acts as a recognition site. It is presumed that, following **endocytosis**, the sIgA complex (Fig. 4.33) is transported to the apical membrane of the secretory cell where, following cleavage of a portion of the complex, the mature sIgA complex is secreted by exocytosis.

4.18   Functional Milk Protein Products

Although modified by fat, the physico-chemical properties of most dairy products depend on the proteins, the only exceptions are anhydrous milk fat (or ghee) and butter, and to some extent cream. The existence of many dairy products is due to certain properties of milk proteins. The physico-chemical properties of cheese and fermented milk products and the heat stability of milk are described in Chaps. 9, 12 and 13. In the following, the production, properties and applications of "Functional Milk Protein Products" will be described.

The term 'functional properties of proteins' in relation to foods refers to those physico-chemical properties of a protein which affect the functionality of the food, i.e., its texture (rheology), colour, flavour, water sorption/binding and stability. Probably the most important physico-chemical properties are solubility, hydration, rheology, surface activity and gelation, the relative importance of which depends on the food in question; these properties are, at least to some extent, interdependent. The physical properties of many foods, especially those of animal origin, are determined primarily by their constituent proteins, but those properties are not the subject of this section. Rather, we are concerned with isolated, more or less pure, proteins which are added to foods for specific purposes. The importance of such proteins has increased greatly in recent years, partly because suitable technology for the production of such proteins on a commercial scale has been developed to a level where it is possible to implement industrially and partly because a market for functional proteins has been created through the growth of formulated foods, i.e., foods manufactured from enriched/pure ingredients (proteins, fats/oils, sugars/polysaccharides, flavours, colours). Perhaps one should view the subject the other way round, i.e., formulated foods developed because suitable functional proteins were available. Some functional proteins have been used in food applications for a very long time, e.g., egg white in various types of foamed products or gelatine in gelled products. The principal functional food proteins are derived from milk (caseins and whey proteins) or soybeans; other important sources are egg white, blood, connective tissue (gelatine) and wheat (gluten). Probably because of the ease with which casein can be produced from skim milk, essentially free of lipids, lactose and salts, by rennet or isoelectric coagulation and washing of the curd, acid and rennet caseins have been produced commercially since the beginning of the twentieth century. However, in the early years, they were used for industrial applications, e.g., for the production of glues, plastics or fibres or as dye-binders for paper glazing. Although some casein is still used for industrial applications, the vast majority of world production is now used in food formulation applications (e.g., cream liqueurs, processed cheese and analogue cheese). This change has occurred partly because cheaper, and possibly better, materials have replaced casein for industrial applications while growth in the production of formulated foods has created a demand for functional proteins at higher prices than those available for industrial-grade products. Obviously, the production of a food-grade protein requires more stringent hygienic standards than industrial proteins; the pioneering work in this area was done mainly in Australia and to a lesser extent in New Zealand in the 1960s. Although heat-denatured whey protein, referred to as lactalbumin, has been available for many years for food applications, it was of little significance, mainly because the product is insoluble (similar to acid casein) and therefore has limited functionality.

The commercial production of functional whey protein became possible with the development of ultrafiltration in the US in the 1960s. Whey protein concentrates

(WPCs), containing 30–85 % protein produced by ultrafiltration and diafiltration are now of major commercial importance, with many specific food applications (e.g., ice cream, infant formula, protein-enriched drinks). The extent of whey protein enrichment can be increased further by using higher volume concentration factors, diafiltration and microfiltration (MF; to remove fat) to produce whey protein-enriched ingredients with protein content ≥90 % (i.e., whey protein isolates, WPIs). WPI ingredients have increased in commercial importance in recent years as they are an excellent source of high nutritional quality protein (i.e., high in branched-chain amino acids and low in non-protein nitrogen), with very low carbohydrate (i.e., lactose) contribution, making them attractive for formulation of premium protein-enriched formulated food products such as nutritional beverages (e.g., "protein fortified" water, i.e., containing up to 5 % whey protein). WPI may also be produced by ion exchange chromatography but the production thereof is limited due to the higher processing costs, compared with combined UF/DF/MF. As discussed in Chap. 11, many of the whey proteins have interesting biological properties. It is now possible to isolate individual whey proteins on a commercial scale in a relatively pure form for specific applications.

4.18.1   Industrial Production of Caseins

There are two principal established methods for the production of casein on an industrial scale: isoelectric precipitation and enzymatic (rennet) coagulation. There are a number of comprehensive reviews on the subject (e.g., Muller 1982; Mulvihill 1989, 1992; Fox and Mulvihill 1992; Mulvihill and Ennis 2003; O'Regan et al. 2009; O'Regan and Mulvihill 2011) which should be consulted for references. Acid casein is produced from skim milk by direct acidification, usually with HCl, or by fermentation with a *Lactococcus* culture, to about pH 4.6. The curds/whey are cooked to about 50 °C, separated using an inclined perforated screen or decanting centrifuge, washed thoroughly with water (usually in counter-current flow mode), dewatered by pressing, dried (fluidized bed, attrition or ring dryers) and milled. A flow diagram of the process and a line diagram of the plant are shown in Figs. 4.40 and 4.41. Acid casein is insoluble in water but soluble caseinate can be formed by dispersing the acid casein in water and adjusting the pH to 6.5–7.0 with NaOH (usually), KOH, Ca(OH)$_2$ or NH$_3$ to produce sodium, potassium, calcium or ammonium caseinate, respectively (Fig. 4.42). The caseinates are usually spray dried. Caseinates form very viscous solutions and solutions containing only up to about 20 % casein can be prepared; this low concentration of protein increases drying costs and leads to a low-density powder. Calcium caseinate forms highly aggregated colloidal dispersions.

A relatively recent development in the production of acid casein is the use of ion exchangers for acidification. In one such method, a portion of the milk is acidified to approximately pH 2 at 10 °C by treatment with a strong acid ion exchanger and then mixed with unacidified milk in proportions so that the mixture has a pH of 4.6.

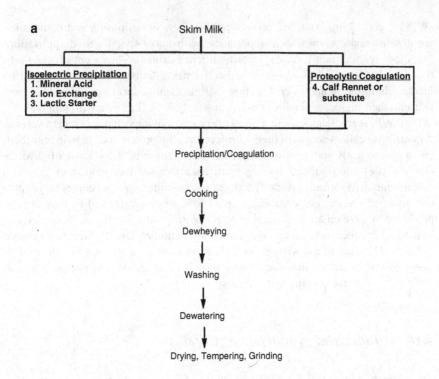

a   Skim Milk

Isoelectric Precipitation	Proteolytic Coagulation
1. Mineral Acid	4. Calf Rennet or
2. Ion Exchange	substitute
3. Lactic Starter	

Precipitation/Coagulation

Cooking

Dewheying

Washing

Dewatering

Drying, Tempering, Grinding

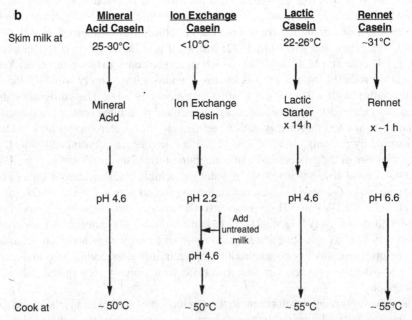

b   Mineral   Ion Exchange   Lactic   Rennet
             Acid Casein   Casein   Casein   Casein
Skim milk at
             25-30°C   <10°C   22-26°C   ~31°C

             Mineral   Ion Exchange   Lactic   Rennet
             Acid   Resin   Starter
                                             x 14 h   x ~1 h

             pH 4.6   pH 2.2   pH 4.6   pH 6.6
                                  Add
                              untreated
                                milk
                              pH 4.6

Cook at   ~ 50°C   ~ 50°C   ~ 55°C   ~ 55°C

Fig. 4.40 (a) Line diagram of industrial processes for the manufacture of acid and rennet casein. The conditions (time, temperature and pH) of precipitation are shown in (b). (Modified from Mulvihill 1992)

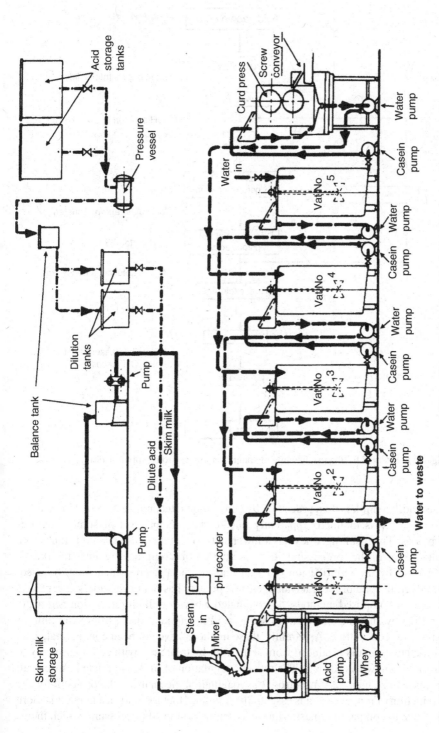

Fig. 4.41 Line diagram of an acid casein manufacturing plant: *solid line* milk/casein flow lines; *thick dashed lines* water flow lines; and *dashed dotted lines* acid flow lines (from Muller 1982)

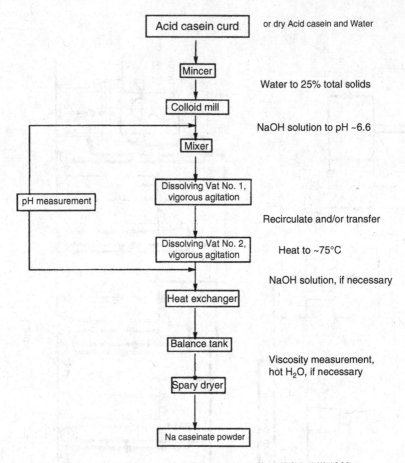

Fig. 4.42 Protocol for the manufacture of sodium caseinate (from Mulvihill 1992)

The acidified milk is then processed by conventional techniques. A yield increase of about 3.5 % is claimed, apparently due to the precipitation of some proteose-peptones. The resulting whey has a lower salt content than normal and is thus more suitable for further processing. The elimination of strong acid reduces the risk of corrosion by the chloride ion (Cl⁻) and hence cheaper equipment may be used. However, in spite of these advantages, this process has not been widely accepted.

In other proposed methods, deproteinated whey or milk ultrafiltration permeate, acidified by an ion exchanger, is used to acid-precipitate casein from skim milk or skim-milk concentrate. Apparently, these methods have not been commercialized.

Rennet casein is produced from skim milk by treatment with certain proteolytic enzymes, known as rennets. The rennet coagulation of milk and related aspects are discussed in Chap. 12. Apart from the coagulation mechanism, the protocol for the production of rennet casein is essentially similar to that for acid casein but it is more sensitive to compaction/plasticization by temperature and/or pressure, which therefore must be controlled carefully. Rennet casein is insoluble in water or alkali but can be solubilized by treatment with a range of calcium binding salts such as citrate- and phosphate-based salts, which when combined with heat energy and shear, bind calcium, allowing the protein to hydrate, binding water and emulsifying oil — a process which is widely used in the production of cheese analogues (e.g., for pizza topping application). Caseins and caseinates are used is a wide range of formulated food products including analogue cheese, coffee creamers, milk beverages, ice cream, frozen desserts, comminuted meat products, soups, sauces and edible films/ coatings, due to their excellent water binding, texture enhancing, emulsifying, foaming and nutritional properties. Current annual global production of casein and caseinate ingredients is approximately 250,000 tonnes but total production has been decreasing steadily over the last decade or so, due mainly to the displacement of casein and caseinate as functional ingredients in formulated foods by newer ingre- dients such as milk protein concentrates/isolates.

4.18.2 Novel Methods for Casein Production

Cryoprecipitation. When milk is frozen and stored at about −10 °C, the ionic strength of the liquid phase increases with a concomitant increase in [Ca^{2+}] and a decrease in pH (to approximately 5.8) due to precipitation of calcium phosphate with the release of hydrogen ions (H^+) (Chap. 5). These changes destabilize the casein micelles which precipitate when the milk is thawed. Cryodestabilization of casein limits the commercial feasibility of frozen milk, which may be attractive in certain circumstances. However, cryodestabilized casein might be commercially viable, especially if applied to milk concentrated by ultrafiltration, which is less stable than normal milk. Cryodestabilized casein may be processed in the usual way. The product is dispersible in water and can be reconstituted as micelles in water at 40 °C. The heat stability and rennet coagulability of these micelles are generally similar to those of normal micelles and casein produced by cryodestabili- zation may be suitable for the production of fast-ripening cheeses, e.g. Mozzarella or Camembert, when the supply of fresh milk is inadequate. However, as far as we are aware, casein is not produced commercially by cryodestabilization.

Precipitation with ethanol. The casein in milk coagulates at pH 6.6 on addition of ethanol to about 40 %; stability decreases sharply as the pH is reduced and only 10 – 15 % ethanol is required at pH 6. Ethanol-precipitated casein may be dispersed in a micellar form and has very good emulsifying properties. Ethanol-precipitated casein is probably economically viable but the process is not being used commercially.

Membrane processing. The use of ultrafiltration (UF) for the production of whey protein concentrates/isolates (WPC's/WPIs) is well established, with such technol- ogy being commercially adopted more recently for the production of total milk protein-based ingredients such as milk protein concentrates/isolates (MPCs/MPIs). Microfiltration is also used to separate casein micelles from the serum-phase con- stituents (i.e., whey proteins, minerals and lactose) of skim milk for the production of a range of native casein micelle-enriched ingredients such as phosphocaseinate and micellar casein isolate (see earlier section).

High-speed centrifugation. The casein micelles may be sedimented by centrifugation at greater than $100,000 \times g$ for 1 h, which is widely used on a laboratory scale. A combination of ultrafiltration and ultracentrifugation has been proposed for the industrial production of 'native' phosphocaseinate. Almost all the casein in skim milk or UF retentate can be sedimented by centrifugation at greater than $75,000 \times g$ for 1 h at 50 °C.

'Native' casein. An exciting new development is the production of 'native' casein. Few details on the process are available at present but it involves electrodialysis of skim milk at 10 °C against acidified whey to reduce the pH to about 5; the acidified milk is centrifuged and the sedimented casein dispersed in water, concentrated by UF and dried. The product disperses readily in water and is claimed to have properties approaching those of native casein micelles.

4.18.3   Fractionation of Casein

As discussed in Sect. 4.4, individual caseins may be isolated on a laboratory scale by methods based on differences in solubility in urea solutions at around pH 4.6, by selective precipitation with $CaCl_2$ or by various forms of chromatography, especially ion-exchange or reverse-phase high performance liquid chromatography (RP-HPLC). Obviously, these methods are not amenable to scale-up for industrial application. There is considerable interest in developing techniques for the fractionation of caseins on an industrial scale for special applications. For example:

- β-Casein has very high surface activity and may find special applications as an emulsifier or foaming agent.
- Human milk contains β- and κ-caseins no α_{s1}-casein; hence β-casein should be an attractive ingredient for bovine milk-based infant formulae.
- β-Casein is reported to increase the strength of rennet-induced milk gels.
- Reduction of β-casein levels in milk have been shown to improve the melt and flow properties of cheese made therefrom.
- κ-Casein, which is responsible for the stability of casein micelles, might be a useful additive for certain milk products.
- As discussed in Chap. 11, all the principal milk proteins contain sequences which have biological properties when released by proteolysis; the best studied of these are the β-casomorphins. The preparation of biologically active peptides requires purified proteins.

Methods with the potential for the isolation of β-casein on a large scale, leaving a residue enriched in α_{s1}-, α_{s2}- and κ-caseins, have been published. The methods largely exploit the temperature-dependent association characteristics of β-casein, the most hydrophobic of the caseins. Up to 80 % of the β-casein may be recovered from sodium caseinate by UF at 2 °C; the β-casein may be recovered from the per-

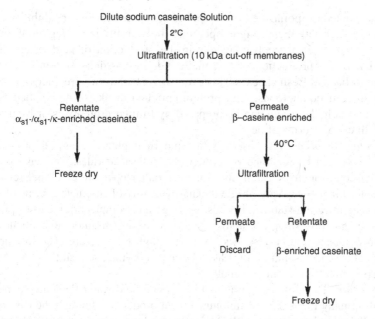

Fig. 4.43 Method for preparing α_{s1}-/α_{s2}-/κ- and β-casein-enriched fractions by ultrafiltration (from Murphy and Fox 1991)

meate by UF at 40 °C (Fig. 4.43). MF at 2 °C has been used to isolate casein from milk or sodium caseinate. Industrial uptake of many of these earlier methods was hampered by the fact that large proportions of insoluble β-casein depleted material was generated as a side-stream or β-casein needed to be thermally precipitated from the enriched stream, leading to impairment of techno-functional properties in the final ingredient. More recently, a process was developed whereby cold microfiltration of skim milk at 4–6 °C was shown to be successful in removing up to 20 % of the total β-casein with thermo-reversible aggregation of β-casein in the resultant permeate allowing the enrichment of β-casein from whey proteins, minerals and lactose. This method has the advantages of using standard membrane processing technology and maintaining solubility of β-casein with no production of aggregated/precipitated streams. β-casein-enriched products are being produced commercially for use in infant formulae.

4.18.4   Functional (Physicochemical) Properties of Caseins

Solubility. Solubility is an important functional property *per se*, i.e., in fluid products, and is essential for other functionalities since insoluble proteins cannot perform useful physical functions in foods. The caseins are, by definition, insoluble at their isoelectric point, i.e., in the pH range ~3.5–5.5; the insolubility range becomes wider with increasing temperature. Insolubility in the region of the isoelectric point is clearly advantageous in the production of acid casein and is exploited in the production of two major families of dairy products, i.e., fermented milks and fresh cheeses. However, such insolubility precludes the use of casein in acid liquid foods, e.g., protein-enriched drinks or carbonated beverages. Acid-soluble casein can be prepared by limited proteolysis or by interaction with certain forms of pectin.

Rheological properties. Viscosity, an important physicochemical property of many foods, can be modified by proteins or polysaccharides. The caseins form rather viscous solutions, a reflection of their rather open structure and relatively high water-binding capacity. While the high viscosity of caseinate may be of some importance in casein-stabilized emulsions (e.g., cream liqueurs), it causes production problems; for example, due to very high viscosity, not more than about 20 % sodium caseinate can be dissolved even at a high temperature. The low protein content of caseinate solutions increases the cost of drying and results in low-density powders which are difficult to handle.

Hydration. The ability of proteins to bind and hold water without syneresis is critical in many foods, e.g., comminuted meat products. Although the caseins are relatively hydrophobic, they bind ~2 g H_2O g^{-1} protein, which is typical of proteins. Hydration increases with increasing pH and is relatively independent of NaCl concentration, which is especially important in the efficacy of casein in meat-based products. The water-holding capacity of sodium caseinate is higher than that of calcium caseinate or micellar casein.

Gelation. One of the principal functional applications of proteins is the formation of gels. In milk, caseins undergo gelation when the environment is changed in one of several ways, but the most important are rennet-induced coagulation for cheese or rennet casein manufacture (which is discussed in Chap. 12) or on acidification to the isoelectric point (pH 4.6), which is exploited in the preparation of fermented milk products and isoelectric casein. In addition, casein may be gelled or coagulated by organic solvents, prolonged heat treatment or during storage of heat-sterilized products; these changes are usually undesirable. Heat-induced gelation is used in the preparation of many food products but, as discussed in Chap. 9, the caseins are remarkably heat stable and do not undergo thermally-induced gelation except under extremely severe conditions; their stability is a major advantage in milk processing.

Surface activity. Probably the outstanding property of caseins, as far as their functionality in foods is concerned, is their surface activity, which makes them good foaming agents and especially good emulsifiers. Surface-active agents are molecules with hydrophilic and hydrophobic regions which can interact with the aqueous and non-aqueous (air or lipid) phases of emulsions and foams, thus reducing the interfacial or surface tension. Caseins are among the most surface-active proteins available to food technologists, β-casein being particularly effective. To exhibit good surface activity, a protein must possess three structural features:

1. It should be relatively small, since the rate of migration to the interface is inversely proportional to the molecular mass. In actual food processing operations, the rate of diffusion is not particularly important since the production of emulsions and foams involves a large input of work with vigorous agitation which moves the protein rapidly to the interface.
2. The molecule must be capable of adsorbing at the oil-water or air-water interface and hence must have relatively high surface hydrophobicity; the caseins, especially β-casein, meet this requirement very well.
3. Once adsorbed, the molecule must open and spread over the interface; thus, an open, flexible structure is important. The caseins, which have relatively low levels of secondary and tertiary structures and have no intramolecular disulphide bonds, can open and spread readily.

In practice, while the caseins are very good emulsifiers and foam readily, the resultant foams are not very stable, possibly because the lamella of the foam bubbles are thin and drain rapidly in contrast to the thicker foams formed by egg albumin.

4.18.5 Applications of Caseins and Whey Proteins

Casein/caseinates and whey proteins have a wide range of applications in the food industry, as summarized in Table 4.9.

4.18.6 Casein-Whey Protein Co-precipitates

Following denaturation, the whey proteins in skim milk coprecipitate with the caseins on acidification to pH 4.6 or addition of $CaCl_2$ at 90 °C, to yield a range of products known as casein-whey protein co-precipitates (Fig. 4.44). The main attraction of such products is an increase in yield of about 15 %, but the products also have interesting functional properties. However, they have not been commercially successful.

New forms of co-precipitate, referred to as soluble lactoprotein or total milk protein, with improved solubility, have been developed recently (Fig. 4.45). By adjusting the milk to an alkaline pH before denaturing the whey proteins and co-precipitating them with the caseins at pH 4.6, the functionality of the caseins is not adversely affected; probably, the denatured whey proteins do not complex with the casein micelles at the elevated pH.

Table 4.9 Applications of milk proteins in food products (modified from Mulvihill 1992)

Bakery products	
Caseins/caseinates/co-precipitates	
Used in	Bread, biscuits/cookies, breakfast cereals, cake mixes, pastries, frozen cakes and pastries, pastry glaze
Effect	Nutritional, sensory, emulsifier, dough consistency, texture, volume/yield
Whey proteins	
Used in	Bread, cakes, muffins, croissants
Effect	Nutritional, emulsifier, egg replacer
Dairy products	
Caseins/caseinates/co-precipitates	
Used in	Imitation cheeses (vegetable oil, caseins/caseinates, salts and water)
Effect	Fat and water binding, texture enhancing, melting properties, stringiness and shredding properties
Used in	Coffee creamers (vegetable fat, carbohydrate, sodium caseinate, stabilizers and emulsifiers)
Effect	Emulsifier, whitener, gives body and texture, promotes resistance to feathering, sensory properties
Used in	Cultured milk products, e.g., yoghurt
Effect	Increased gel firmness, reduced syneresis
Used in	Milk beverages, imitation milk, liquid milk fortification, milk shakes
Effect	Nutritional, emulsifier, foaming properties
Used in	High-fat powders, shortening, whipped toppings and butter-like spreads
Effect	Emulsifier, texture enhancing, sensory properties
Whey proteins	
Used in	Yoghurt, Quarg, Ricotta cheese
Effect	Yield, nutritional, consistency, curd cohesiveness
Used in	Cream cheeses, cream cheese spreads, sliceable/squeezable cheeses, cheese fillings and dips
Effect	Emulsifier, gelling, sensory properties
Beverages	
Caseins/caseinates/co-precipitates	
Used in	Drinking chocolate, fizzy drinks and fruit beverages
Effect	Stabilizer, whipping and foaming properties
Used in	Cream liqueurs, wine aperitifs
Effect	Emulsifier
Used in	Wine and beer industry
Effect	Fines removal, clarification, reduce colour and astringency
Whey proteins	
Used in	Soft drinks, fruit juices, powdered or frozen orange beverages
Effect	Nutritional
Used in	Milk-based flavoured beverages
Effect	Viscosity, colloidal stability

(continued)

Table 4.9   (continued)

Dessert products	
Caseins/caseinates/co-precipitates	
Used in	Ice-cream, frozen desserts
Effect	Whipping properties, body and texture
Used in	Mousses, instant puddings, whipped toppings
Effect	Whipping properties, film former, emulsifier, imparts body and flavour
Whey proteins	
Used in	Ice-cream, frozen juice bars, frozen dessert coatings
Effect	Skim-milk solids replacement, whipping properties, emulsifying, body/texture
Confectionary	
Caseins/caseinates/co-precipitates	
Used in	Toffee, caramel, fudges
Effect	Confers firm, resilient, chewy texture; water binding, emulsifier
Used in	Marshmallow and nougat
Effect	Foaming, high temperature stability, improve flavour and brown colour
Whey proteins	
Used in	Aerated candy mixes, meringues, sponge cakes
Effect	Whipping properties, emulsifier
Pasta products	
Used in	Macaroni, pasta and imitation pasta
Effect	Nutritional, texture, freeze-thaw stability, microwaveable
Meat products	
Caseins/caseinates/co-precipitates	
Used in	Comminuted meat products
Effect	Emulsifier, water binding, improves consistency, releases meat proteins for gel formation and water binding
Whey proteins	
Used in	Frankfurters, luncheon meats
Effect	Pre-emulsion, gelatin
Used in	Injection brine for fortification of whole meat products
Effect	Gelation, yield
Convenience foods	
Used in	Gravy mixes, soup mixes, sauces, canned cream soups and sauces, dehydrated cream soups and sauces, salad dressings, microwaveable foods, low lipid convenience foods
Effect	Whitening agents, dairy flavour, flavour enhancer, emulsifier, stabilizer, viscosity controller, freeze-thaw stability, egg yolk replacement, lipid replacement
Textured products	
Used in	Puffed snack foods, protein-enriched snack-type products, meat extenders
Effect	Structuring, texturing, nutritional
Pharmaceutical and medical products	
Special dietary preparations for	
Ill or convalescent patients	
Dieting patients/people	

(continued)

Table 4.9 (continued)

Athletes
Astronauts
Infant foods
Nutritional fortification
'Humanized' infant formulae
Low-lactose infant formulae
Specific mineral balance infant foods
Casein hydrolysates: used for infants suffering from diarrhoea, gastroenteritis, galactosaemia, malabsorption, phenylketonuria
Whey protein hydrolysates used in hypoallergenic formulae preparations
Nutritional fortification
Intravenous feeds
Patients suffering from metabolic disorders, intestinal disorders for postoperative patients
Special food preparations
Patients suffering from cancer, pancreatic disorders of anaemia
Specific drug preparations
β-Caseinomorphins used in sleep or hunger regulation or insulin secretion
Sulphonated glycopeptides used in treatment of gastric ulcers
Miscellaneous products
Toothpastes
Cosmetics
Wound treatment preparations

4.19   Methods for Quantitation of Proteins in Foods

4.19.1   Kjeldahl Method

The first successful method for determination of the protein content of foods and tissues was developed by Johan Kjeldahl at the Carlsberg Laboratories, Copenhagen, in 1883 for determination of the protein content of barley (a high concentration of protein in malting barley is undesirable because it causes a haze in beer). The Kjeldahl method is still the standard reference method for determining the protein content of foods.

The Kjeldahl method involves digesting (wet ashing) the organic matter in the sample (lipids, proteins, carbohydrates, etc.) with concentrated H_2SO_4 at 370 to 400 °C. K_2SO_4 is added to increase the boiling point of the H_2SO_4 and a catalyst [Cu (CuSO$_4$), Hg (undesirable because it is very toxic) or Se] is added. During digestion, C in the sample is converted to CO_2, oxygen to CO_2 and H_2O, H to H_2O and N to $(NH_4)_2SO_4$. Some H_2SO_4 is degraded to the irritating and toxic gas, SO_2. When digestion is complete, as indicated by clearing of the sample (light blue colour if $CuSO_4$ is used as catalyst; may require several hours), the solution is made strongly

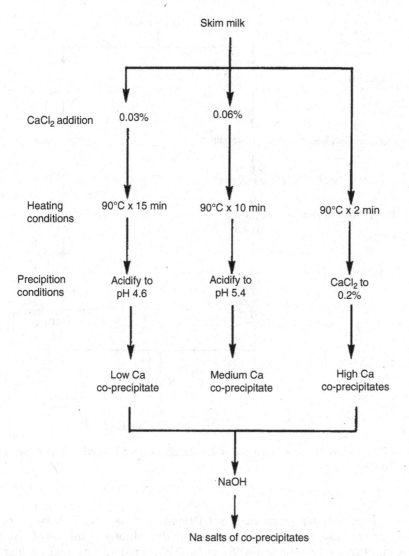

Fig. 4.44 Protocols for the manufacture of conventional casein—whey protein co-precipitates (from Mulvihill 1992)

alkaline by addition of concentrated NaOH. Under these conditions, the $(NH_4)_2SO_4$ is converted to NH_3 which is steam distilled by heating the system and trapped in an acid. A measured volume of standard HCl or H_2SO_4 may be used and back-titrated with standard NaOH, enabling the amount of acid neutralized by the NH_3 to be calculated. Alternatively, and more commonly, 2–4 % boric acid, H_3BO_3 may be used with 0.2 % methyl red—0.1 % methylene blue as indicator (colour change blue

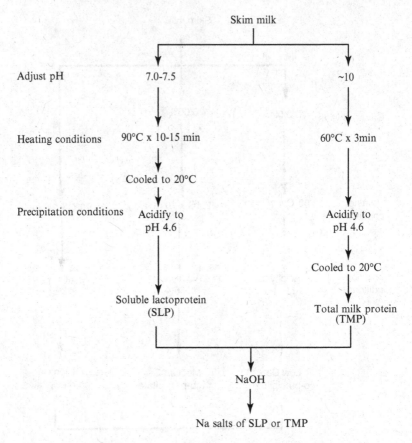

Fig. 4.45 Protocols for the manufacture of soluble lactoprotein and total milk proteins (from Mulvihill 1992)

to green); the NH_3 trapped raises the pH abruptly (boric acid has little buffering at acidic pH) and causes a change in the colour of the indicator; at the end of distillation, the H_3BO_3 is back-titrated with standard HCl to restore the original colour of the indicator. Only one standard solution, HCl, is required.

Since a typical protein contains 16 % N, the protein content of the sample can be determined by multiplying % N by 100/16 [i.e., %N×6.25]. The N content of milk proteins is 15.7 % (rather than 16 %) and the conversion factor for N to protein is 6.38 (rather than 6.25).

Sample calculation:

Weight of sample = 10 g
Volume of 0.1 M HCl required for back titration = 15 mL
1 mol HCl ≡ 1 mol NH_3 ≡ 1 mol N

1 L of 1 M HCl $\equiv$ 17 g NH$_3$ $\equiv$ 14 g N
1 mL 0.1 M HCl = 0.0014 g N
15 mL 0.1 M HCl $\equiv$ 0.0014 × 15 g N
% protein = 0.0014 × 15 × 100/10 × 6.38 = 1.34 %

The Kjeldahl method is slow and potentially dangerous and requires specialized equipment, including a good fume hood to remove the irritating and toxic SO$_2$; it is not suitable for analysis of large numbers of samples and been automated successfully. Consequently, easy, rapid, but still accurate, alternatives were sought and a series of possible methods developed, some specifically for milk, including:

4.19.2   The Formol Titration

A sample of milk is titrated to the phenolphthalein end-point (pH 8.4) with 0.1 M NaOH, formaldehye is then added which converts the primary, R–NH$_2$, group of lysine, the pK$_a$ of which is ~9.5, i.e. above the phenolphthalein end-point, to a tertiary amine, R–N (CH$_3$)$_2$, which has a pK$_a$ of about 6.5 and is below the indicator end-point; the red colour fades and the sample is re-titrated with NaOH—the volume of NaOH required for the second titration is proportional to the concentration of lysine, and hence protein, in the sample. The formol titration (FT) is the volume of 1 M NaOH required per 100 mL milk for the second titration. The slope of a plot of FT against protein (Kjeldahl) is 1.74; hence

$$\% \text{ protein} = FT \times 1.74$$

The formol titration method has been used in research laboratories but not by the dairy industry; it is not sufficiently accurate and sensitive.

4.19.3   Absorbance of UV Light

The aromatic amino acids tyrosine and tryptophan, absorb UV light strongly at 280 nm, which may be used to determine the concentration of protein in a sample. The A$_{280}$ of a 0.1 % solution of a typical protein in a 1 cm cuvette, is 1.0.

If the absorbance of a 1:10 dilution of a protein solution is 0.8, the concentration of protein in the sample is:

$$8 \div 10 = 0.8 \text{ %}$$

UV absorbance is widely used in biochemistry to determine the protein content of eluates from chromatography columns but interference from light scattering by casein micelles and fat globules makes it unsuitable for milk and, obviously the method is not suitable for solid samples.

The peptide bond absorbs strongly at 180 nm but there are technical difficulties making measurements at this low wavelength and absorbance at 210 or 220 nm is used instead. A_{220} is about 20 times as sensitive as A_{280}. This method is not used in the food industry but is used in food research laboratories.

4.19.4   Biuret Method

The peptide bonds of proteins react with Cu^{2+} at alkaline pH to give a blue-coloured complex with λ_{max} 540 nm, the intensity of which is proportional to the concentration of protein in the sample. This reaction is called the biuret reaction because biuret, $H_2NC(O)NHC(O)NH_2$, undergoes the same reaction. Various modifications of the method have been developed; the Lowry method is commonly used. This method is used in dairy research laboratories but not by the dairy industry.

4.19.5   Folin-Ciocalteau (F-C) Method

Phenolic groups (e.g., tyrosine) react with phosphotungstate-phosphomolybdate to produce a blue-coloured complex, with a λ_{max} at 660 nm, the intensity of which is proportional to the concentration of Tyr, and hence of protein, in a sample. In the Lowry method, the F-C reagent is combined with Cu^{2+} for increased sensitivity. The method is 50–100 times more sensitive than the biuret method and 10–20 times more sensitive than absorbance at 280 nm. The method is used in dairy research laboratories but not by the dairy industry.

4.19.6   Dye-Binding Methods

As discussed in Sect. 4.5, proteins carry a positive charge at acidic pH and bind anionic dyes. In the early 1960s analytical methods were developed for quantitation of the proteins in milk. Amido Black 10B (λ_{max}, 615 nm), Orange G (λ_{max} 475) or Acid Orange 12 (λ_{max}, 475) were usually used. The protein-dye complex precipitates and is removed by centrifugation or filtration. The amount of dye bound is proportional to the protein concentration in the sample and is calculated from the difference between the absorbance of the original dye solution and that of the supernatant/filtrate, The analysis is performed under standard conditions [a sample of milk, the protein content of which had been determined by the Kjeldahl method, should be included] and the protein concentration determined from a standard curve of % protein (by Kjeldahl) plotted against the amount of dye bound.

Dye-binding methods were widely used by the dairy industry in the 1960s but were soon displaced by IR methods.

4.19.7 Bradford Method

When the dye, Coomassie Brilliant Blue G250, binds to a protein, the dye changes from a reddish to a blue colour and its absorbance maximum is shifted from 465 to 595 nm. The change in A595 is proportional to the concentration of protein in the sample.

The method is used in dairy research laboratories but not by the dairy industry.

4.19.8 Infra-Red Spectroscopy

As discussed in Chaps. 2 and 3, the peptide bond absorbs IR radiation at 6.46 μm, permitting the concentration of protein to be determined readily, simultaneously with the concentrations of fat and lactose.

Analysis by IR absorbance is now the routine method for determination of the protein content of milk and many dairy products. The equipment is relatively expensive and must be maintained carefully.

4.19.9 Dumas Method

Actually, the first method developed for determination of the nitrogen content of a specimen, including foods, was developed in 1826 by Jean-Baptiste Dumas but lack of appropriate equipment meant that this method was not suitable for routine use until recently.

The method involves heating a sample at 900 °C in an atmosphere of oxygen, i.e., pyrolysis. The sample is converted to CO_2, H_2O and N_2. The CO_2 is absorbed by KOH and the remaining gas passed over a thermal conductivity detector and the concentration of N_2 quantified. The instrument must be carefully standardized using a compound of known N content.

The N content of a sample is converted to protein as for the Kjeldahl method. The method is fast, requiring only a few minutes per sample; it best suited for solid samples, including milk powders and is used fairly widely by the dairy industry.

Appendix 4A   Structures of Amino Acids Occurring in Proteins

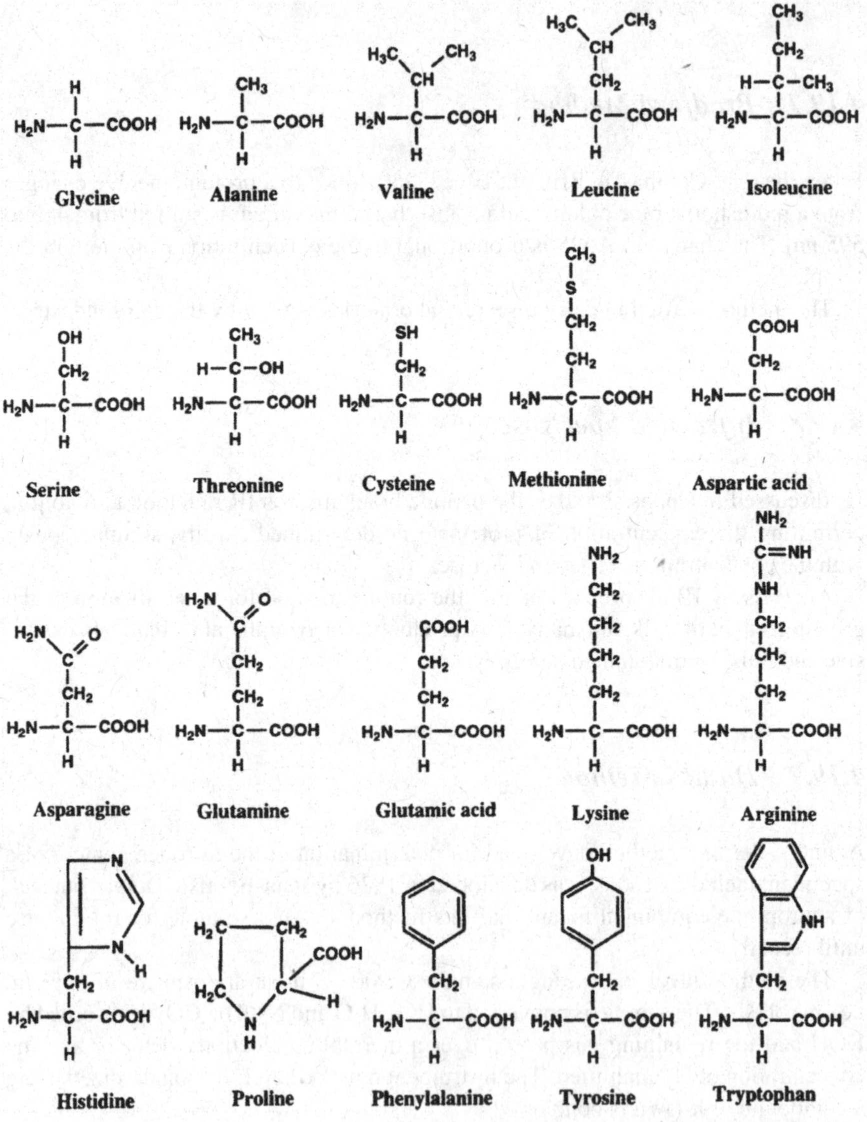

Fig. 4A.1 Protocols for the manufacture of conventional casein—whey protein co-precipitates (from Mulvihill 1992)

References

Atkinson, S. A., & Lonnerdal, B. (1989a). *Protein and non-protein nitrogen in human milk.* Boca Raton, FL: CRC Press.

Attia, H., Kherouatou, N., Nasri, M., & Khorcheni, T. (2000). Characterization of the dromedary milk casein micelle and study of its changes during acidification. *Le Lait, 80,* 503–515.

Berliner, L. J., Meinholtz, D. C., Hirai, Y., Musci, G., & Thompson, M. P. (1991). Functional implications resulting from disruption of the calcium binding loop in bovine α-lactalbumin. *Journal of Dairy Science, 74,* 2394–2402.

Bernhart, F. W. (1961). Correlation between growth-rate of the suckling of various species and the percentage of total calories from protein in the milk. *Nature, 191,* 358–360.

Bouchoux, A., Gesan-Guizou, G., Perez, J., & Cabane, B. (2010). How to squeeze a sponge; casein micelles under osmotic stress, a SAXS study. *Biophysical Journal, 99,* 3754–3762.

Brew, K. (2003). α-Lactalbumin. In P. F. Fox & P. L. H. McSweeney (Eds.), *Advanced dairy chemistry* (Protein: Part A 3rd ed., Vol. 1, pp. 387–419). New York, NY: Kluwer Academic/Plenum.

Brew, K. (2013). α-Lactalbumin. In P. L. H. McSweeney & P. F. Fox (Eds.), *Advanced dairy chemistry* (Protein: Basic aspects 4th ed., Vol. 1A, pp. 261–273). New York, NY: Springer.

Brew, K., & Grobler, I. A. (1992). α-Lactalbumin. In P. F. Fox (Ed.), *Advanced dairy chemistry* (Proteins, Vol. 1, pp. 191–229). London, UK: Elsevier Applied Science.

Brinkmann, C. R., Hergaard, C. W., Petersen, T. E., Jensenius, J. C., & Thiel, S. (2011). The toxicity of BAMLET is highly dependent on oleic acid and induces killing in cancer cell lines and non-cancer derived primary cells. *FEBS Journal, 278,* 1955–1967.

Brunner, J. R., Ernstrom, C. A., Hollis, R. A., Larson, B. L., Whitney, R. M. L., & Zittle, C. A. (1960). Nomenclature of the proteins of bovine milk—First revision. *Journal of Dairy Science, 43,* 901–911.

Carter, D. C., & Ho, J. X. (1994). Structure of serum albumin. *Advances in Protein Chemistry, 45,* 153–203.

Chaplin, B., & Green, M. L. (1982). Probing the location of casein fractions in the casein micelle using enzymes and enzyme-dextran complexes. *Journal of Dairy Research, 49,* 631–643.

Cheryan, M., Richardson, T., & Olson, N. F. (1975). Structure of bovine casein micelles elucidated with immobilized carboxypeptidase. *Journal of Dairy Science, 58,* 651–659.

Chevalier, F. (2011a). Analytical methods: Electrophoresis. In J. W. Fuquay, P. F. Fox, & P. L. H. McSweeney (Eds.), *Encyclopedia of dairy sciences* (2nd ed., Vol. 1, pp. 185–192). Oxford, UK: Academic Press.

Chevalier, F. (2011b). Milk proteins: Proteomics. In J. W. Fuquay, P. F. Fox, & P. L. H. McSweeney (Eds.), *Encyclopedia of dairy sciences* (2nd ed., Vol. 3, pp. 843–847). Oxford, UK: Academic Press.

Dalgleish, D. G. (1998). Casein micelles as colloids: Surface structures and stabilities. *Journal of Dairy Science, 81,* 3013–3018.

Dalgleish, D. G. (2011). On the structural models of bovine casein micelles—Review and possible improvements. *Soft Matter, 7,* 2265–2272.

De Kruif, C. G. (1998). Supra-aggregates of casein micelles as a prelude to coagulation. *Journal of Dairy Science, 81,* 3019–3028.

De Kruif, C. G. (2014). The structure of casein micelles: A review of small-angle scattering data. *Journal of Applied Crystallography, 47,* 1479–1489.

De Kruif, C. G., & Holt, C. (2003). Casein micelle structure, functions and interactions. In P. F. Fox & P. L. H. McSweeney (Eds.), *Advanced dairy chemistry* (Proteins: Part A 3rd ed., Vol. 1, pp. 233–276). New York, NY: Kluwer Academic/Plenum.

Eigel, W. N., Butler, J. E., Emstrom, C. A., Farrell, H. M., Jr., Harwalkar, V. R., Jenness, R., et al. (1984). Nomenclature of proteins of cow's milk: Fifth revision. *Journal of Dairy Science, 67,* 1599–1631.

Farrell, H. M., Jr., Brown, E. M., & Malin, E. L. (2013). Higher order structures of the caseins; a paradox. In P. F. Fox & P. L. H. McSweeney (Eds.), *Advanced dairy chemistry* (Proteins: Basic aspects 4th ed., Vol. 1A, pp. 161–184). New York, NY: Springer.

Farrell, H. M., Jimenez-Flores, R., Bloch, G. T., Brown, E. M., Butler, J. E., Creamer, L. K., et al. (2004). Nomenclature of proteins of cow's milk: Sixth revision. *Journal of Dairy Science, 87,* 1641–1674.

Fox, P. F., & Brodkorp, A. (2008). The casein micelle: Historical aspects, current concepts and significance. *International Dairy Journal, 18,* 677–684.

Fox, P. F., & Mulvihill, D. M. (1992). Developments in milk protein processing. *Food Science and Technology Today, 7,* 152–161.

Fuquay, J. W., Fox, P. F., & McSweeney, P. L. H. (Eds.). (2011). *Encyclopedia of dairy sciences* (Vol. 3, pp. 458–590). Oxford, UK: Academic Press.

Gallagher, D. P., Cotter, P. F., & Mulvihill, D. M. (1997). Porcine milk proteins: A review. *International Dairy Journal, 7,* 99–118.

Girardet, J.-M., & Linden, G. (1996). PP3 component of bovine milk: A phosphorylated glycoprotein. *Journal of Dairy Research, 63,* 333–350.

Hajjoubi, S., Rival-Gervier, S., Hayes, H., Floriot, S., Eggen, A., Pivini, F., et al. (2006). Ruminants genome no longer contains acidic whey protein gene but only a pseudo-gene. *Gene, 370,* 104–112.

Hambling, S. G., McAlpine, A. S., & Sawyer, L. (1992). β-Lactoglobulin. In P. F. Fox (Ed.), *Advanced dairy chemistry* (Proteins, Vol. 1, pp. 141–190). London, UK: Elsevier Applied Science.

Holt, C. (1992). Structure and stability of bovine casein micelles. *Advances in Protein Chemistry, 43,* 63–151.

Holt, C. (1994). The biological function of casein. *Yearbook 1994, The Hannah Institute,* Ayr, Scotland, pp. 60–68.

Holt, C. (1998). Casein micelle substructure and calcium phosphate interactions studied by Sephacryl column chromatography. *Journal of Dairy Science, 81,* 2994–3003.

Holt, C., & Sawyer, L. (1993). Caseins as rheomorphic proteins: Interpretation of primary and secondary structures of α_{s1}-, β- and κ-caseins. *Journal of the Chemical Society, Faraday Transactions, 89,* 2683–2692.

Horne, D. S. (1998). Casein interactions: Casting light on the black boxes, the structure of dairy products. *International Dairy Journal, 8,* 171–177.

Horne, D. S. (2002). Casein structure, self assembly and gelation. *Current Opinion in Colloid and Interface Science, 7,* 456–461.

Horne, D. S. (2006). Casein micelle structure: Models and muddles. *Current Opinion in Colloid and Interface Science, 11,* 148–153.

Horne, D. S. (2011). Casein, molecular structure. In J. W. Fuquay, P. F. Fox, & P. L. H. McSweeney (Eds.), *Encyclopedia of dairy sciences* (2nd ed., Vol. 3, pp. 772–779). Oxford, UK: Academic Press.

Horne, D. S. (2014). Casein micelle structure and stability. In H. Singh, M. Boland, & A. Thompson (Eds.), *Milk proteins: From expression to food* (2nd ed., pp. 169–200). Amsterdam, Netherlands: Elsevier.

Huppertz, T. (2013). Chemistry of the caseins. In P. L. H. McSweeney & P. F. Fox (Eds.), *Advanced dairy chemistry* (Proteins: Basic aspects 4th ed., Vol. 1A, pp. 135–160). New York, NY: Springer.

Hurley, W. L. (2003). Immunoglobulins in mammary secretions. In P. F. Fox & P. L. H. McSweeney (Eds.), *Advanced dairy chemistry* (Proteins, Part A 3rd ed., Vol. 1, pp. 421–447). New York, NY: Kluwer Academic/Plenum.

Hurley, W. L., & Theil, P. K. (2011). Perspectives on immunoglobulins in colostrum and milk. *Nutrients, 3,* 442–474.

Hurley, W. L., & Theil, P. K. (2013). Immunoglobulins in mammary secretions. In P. L. H. McSweeney & P. F. Fox (Eds.), *Advanced dairy chemistry* (Protein: Basic aspects 4th ed., Vol. 1A, pp. 275–294). New York, NY: Springer.

Innocente, N., Biasutti, M., & Blecker, C. (2011). HPLC profile and dynamic surface properties of the proteose peptone fraction from bovine milk and from whey protein concentrate. *International Dairy Journal, 21,* 222–228.

Jakob, E., & Puhan, Z. (1992). Technological properties of milk as influenced by genetic polymorphism of milk proteins—A review. *International Dairy Journal, 2*, 157–178.

Jenness, R., Larson, B. L., McMeekin, T. L., Swanson, A. M., White, C. H., & Whitnay, R. M. L. (1956). Nomenclature of the proteins of bovine milk. *Journal of Dairy Science, 39*, 536–541.

Keenan, T. W., & Dylewski, D. P. (1985). Aspects of intracellular transit of serum and lipid phases of milk. *Journal of Dairy Science, 68*, 1025–1040.

Kronman, M. J. (1989). Metal-ion binding and the molecular conformational properties of α-lactalbumin. *Critical Reviews in Biochemistry and Molecular Biology, 24*, 565–667.

Kumosinski, T. F., Brown, E. M., & Farrell, H. M., Jr. (1993a). Three-dimensional molecular modeling of bovine caseins: An energy-minimized β-casein structure. *Journal of Dairy Science, 76*, 931–945.

Kumosinski, T. F., Brown, E. M., & Farrell, H. M., Jr. (1993b). Three-dimensional molecular modeling of bovine caseins: A refined, energy-minimized κ-casein structure. *Journal of Dairy Science, 76*, 2507–2520.

Kumosinski, T. F., & Farrell, H. M., Jr. (1994). Solubility of proteins: Salt-water interactions. In N. S. Hettiarachchy & G. R. Ziegler (Eds.), *Protein functionality in food systems* (pp. 39–77). New York, NY: Marcel Dekker.

Larson, B. L. (1992). Immunoglobulins of the mammary secretions. In P. F. Fox (Ed.), *Advanced dairy chemistry* (Proteins, Vol. 1, pp. 231–354). London, UK: Elsevier Applied Science.

Leaver, J., & Law, A. J. R. (2003). Genetic engineering of milk proteins. In P. F. Fox & P. L. H. McSweeney (Eds.), *Advanced dairy chemistry* (Proteins, Part B 3rd ed., Vol. 1, pp. 817–837). New York, NY: Kluwer Academic/Plenum.

Liskova, K., Kelly, A.L., O'Brien, N., & Brodkorb, A. (2010). Effect of denaturation of α-lactalbumin on the formation of BAMLET (bovine α-lactalbumin made lethal to tumor cells). *Journal of Agricultural and Food Chemistry, 58*, 4421–4427.

Marchin, S., Putaux, J. L., Pignon, E., & Lenoil, J. (2007). Effects of environmental factors on the casein micelle structure studied by cryotransmission electron microscopy and small-angle X-ray scattering/ultrasmall-angle X-ray scattering. *Journal of Chemical Physics, 126*(4), 1–10.

Martin, P., Blanchi, L., Cebo, C., & Miranda, G. (2013a). Genetic polymorphism of milk proteins. In P. L. H. McSweeney & P. F. Fox (Eds.), *Advanced dairy chemistry* (Proteins: Basic aspects 4th ed., Vol. 1A, pp. 463–514). New York, NY: Springer.

Martin, P., Cebo, G., & Miranda, G. (2013b). Inter-species comparison of milk proteins: Quantitative variability and molecular diversity. In P. L. H. McSweeney & P. F. Fox (Eds.), *Advanced dairy chemistry* (Proteins: Basic aspects 4th ed., Vol. 1A, pp. 387–429). New York, NY: Springer.

Martin, P., Brignon, G., Furet, J. P., & Leroux, C. (1996). The gene encoding α_{s1}-casein is expressed in human mammary epithelial cells during lactation. *Le Lait, 76*, 523–535.

Martin, P., Cebo, G., & Miranda, G. (2011). Inter-species comparison of milk proteins: Quantitative variability and molecular diversity. In J. W. Fuquay, P. F. Fox, & P. L. H. McSweeney (Eds.), *Encyclopedia of dairy sciences* (2nd ed., Vol. 3, pp. 821–842). Oxford, UK: Academic Press.

Martin, P., Ferranti, P., Leroux, C., & Addeo, F. (2003). Non-bovine caseins: Quantitative variability and molecular diversity. In P. F. Fox & P. L. H. McSweeney (Eds.), *Advanced dairy chemistry* (Proteins: Part A 3rd ed., Vol. 1, pp. 277–317). New York, NY: Kluwer Academic/Plenum.

McGrath, B. A. (2014). *Generation and characterisation of biologically active milk-derived protein and peptide fractions*. Ph.D. Thesis, National University of Ireland, Cork.

McKenzie, H. A. (Ed.). (1970a). *Milk proteins: Chemistry and molecular biology* (Vol. 1). New York, NY: Academic Press.

McKenzie, H. A. (1971a). β-Lactoglobulin. In H. A. McKenzie (Ed.), *Milk proteins. Chemistry and molecular biology* (Vol. 11, pp. 257–330). New York, NY: Academic Press.

McMahon, D. J., & Brown, R. J. (1984). Composition, structure and integrity of casein micelles: A review. *Journal of Dairy Science, 67*, 499–512.

McMahon, D. J., & McManus, W. R. (1998). Rethinking casein micelle structure using electron microscopy. *Journal of Dairy Science, 81*, 2985–2993.

McMahon, D. J., & Oommen, B. S. (2008). Supramolecular structure of the casein micelle. *Journal of Dairy Science, 91*, 1709–1721.

McMahon, D. J., & Oommen, B. S. (2013). Casein micelle structure, functions and interactions. In P. L. H. McSweeney & P. F. Fox (Eds.), *Advanced dairy chemistry* (Proteins: Basic aspects 4th ed., Vol. 1A, pp. 185–209). New York, NY: Springer.

Mepham, T. B. (1987). *Physiology of lactation*. Milton, Keynes, UK: Open University Press.

Mepham, T. B., Gaye, P., Martin, P., & Mercier, J.-C. (1992). Biosynthesis of milk proteins. In P. F. Fox (Ed.), *Advanced dairy chemistry* (Proteins, Vol. 1, pp. 491–543). London, UK: Elsevier Applied Science.

Mepham, T. B., Gaye, P., & Mercier, J.-C. (1982). Biosynthesis of milk proteins. In P. F. Fox (Ed.), *Developments in dairy chemistry* (Proteins, Vol. 1, pp. 115–156). London, UK: Elsevier Applied Science.

Mercier, J.-C., & Gaye, P.-C. (1983). Milk protein syntheses. In T. B. Mepham (Ed.), *Biochemistry of lactation* (pp. 177–227). Amsterdam, Netherlands: Elsevier.

Muller, L. L. (1982). Manufacture of casein. Caseinates and casein co-precipitates. In P. F. Fox (Ed.), *Developments in dairy chemistry* (Proteins, Vol. 1, pp. 315–337). London, UK: Applied Science.

Mulvihill, D. M. (1989). Casein and caseinates: Manufacture. In P. F. Fox (Ed.), *Developments in dairy chemistry* (Functional milk proteins, Vol. 4, pp. 97–130). London, UK: Elsevier Applied Science.

Mulvihill, D. M. (1992). Production, functional properties and utilization of milk proteins. In P. F. Fox (Ed.), *Advanced dairy chemistry* (Proteins, Vol. 1, pp. 369–404). London, UK: Elsevier Applied Science.

Mulvihill, D. M., & Ennis, M. P. (2003). Functional milk proteins: Production and utilization. In P. F. Fox & P. L. H. McSweeney (Eds.), *Advanced dairy chemistry* (Proteins, Part B 3rd ed., Vol. 1, pp. 1175–1228). New York, NY: Kluwer Academic/Plenum.

Murphy, J. F., & Fox, P. F. (1991). Fractionation of sodium caseinate by ultrafiltration. *Food Chemistry, 39*, 27–38.

Ng-Kwai-Hang, K. F., & Grosclaude, F. (1992). Genetic polymorphism of milk proteins. In P. F. Fox (Ed.), *Advanced dairy chemistry* (Proteins 2nd ed., Vol. 1, pp. 405–455). London, UK: Elsevier Applied Science.

Ng-Kwai-Hang, K. F., & Grosclaude, F. (2003). Genetic polymorphism of milk proteins. In P. F. Fox & P. L. H. McSweeney (Eds.), *Advanced dairy chemistry* (Part B: Proteins 3rd ed., Vol. 1, pp. 739–816). New York, NY: Kluwer Academic/Plenum.

Nicholson, J. P., Wolmarans, M. R., & Park, G. R. (2000). The role of albumin in critical illness. *British Journal of Anaesthesia, 84*, 599–610.

O'Donnell, R., Holland, J. W., Deeth, H. C., & Alewood, P. (2004). Milk proteomics. *International Dairy Journal, 14*(1013), 1023.

O'Mahony, J. A., & Fox, P. F. (2013). Milk proteins: Introduction and historical aspects. In P. L. H. McSweeney & P. F. Fox (Eds.), *Advanced dairy chemistry* (Proteins: Basic aspects 4th ed., Vol. 1A, pp. 43–85). New York, NY: Springer.

O'Regan, J., Ennis, M. P., & Mulvihill, D. M. (2009). Milk proteins. In G. O. Phillips & P. A. Williams (Eds.), *Handbook of hydrocolloids* (2nd ed., pp. 298–358). Cambridge, UK: Woodhead.

O'Regan, J., & Mulvihill, D. M. (2011). Casein and caseinates, industrial production, compositional standards, specifications, and regulatory aspects. In J. W. Fuquay, P. F. Fox, & P. L. H. McSweeney (Eds.), *Encyclopedia of dairy sciences* (2nd ed., Vol. 3, pp. 855–863). Oxford, UK: Academic Press.

Ochirkuyag, B., Chobert, J. M., Dalgalarrondo, M., & Haertle, T. (2000). Characterization of mare casein: Identification of α_{s1}- and α_{s2}-caseins. *Le Lait, 80*, 223–235.

Oftedal, O. T. (2011). Milk of marine mammals. In J. W. Fuquay, P. F. Fox, & P. L. H. McSweeney (Eds.), *Encyclopedia of dairy sciences* (2nd ed., Vol. 3, pp. 563–580). Oxford, UK: Academic Press.

Oftedal, O. T. (2013). Origin and evolution of the major milk constituents of milk. In P. L. H. McSweeney & P. F. Fox (Eds.), *Advanced dairy chemistry* (Proteins: Basic aspects 4th ed., Vol. 1A, pp. 1–42). New York, NY: Springer.

Oftedal, O. T., & Jenness, R. (1988). Interspecies variation in milk composition among horses, zebras and asses (*Perissodactyla: Equidae*). *Journal of Dairy Research, 55,* 57–66.

Ono, T., & Creamer, L. K. (1986). Structure of goat casein micelles. *New Zealand Journal of Dairy Science and Technology, 21,* 57–64.

Papiz, M. Z., Sawyer, L., Eliopoulos, E. E., North, A. C. T., Finlay, J. B. C., Sivaprasadaro, R., et al. (1986). The structure of β-lactoglobulin and its similarity to plasma retinol-binding protein. *Nature, 324,* 383–385.

Patel, R. J., & Mistry, V. V. (1997). Physicochemical properties of ultrafiltered buffalo milk. *Journal of Dairy Science, 80,* 812–817.

Richardson, T., Oh, S., Jimenez-Flores, R., Kumosinski, T. F., Brown, E. M., & Farrell, H. M., Jr. (1992). Molecular modeling and genetic engineering of milk proteins. In P. F. Fox (Ed.), *Advanced dairy chemistry* (Proteins, Vol. 1, pp. 545–577). London, UK: Elsevier Applied Science.

Rose, D., Brunner, J. R., Kalan, E. B., Larson, B. L., Melnychyn, P., Swaisgood, H. E., et al. (1970). Nomenclature of the proteins of cow's milk: Third revision. *Journal of Dairy Science, 53,* 1–17.

Salimei, E., Fantuz, F., Coppola, R., Chiolfalo, B., Polidori, P., & Varisco, G. (2004). Composition and characteristics of ass' milk. *Animal Research, 53,* 67–78.

Sawyer, L. (2003). β-Lactoglobulin. In P. F. Fox & P. L. H. McSweeney (Eds.), *Advanced dairy chemistry* (Protein: Part A 3rd ed., Vol. 1, pp. 319–386). New York, NY: Kluwer Academic/ Plenum.

Sawyer, L. (2013). β-Lactoglobulin. In P. L. H. McSweeney & P. F. Fox (Eds.), *Advanced dairy chemistry* (Proteins: Basic aspects 4th ed., Vol. 1A, pp. 211–259). New York, NY: Springer.

Schmidt, D. G. (1982). Association of caseins and casein micelle structure. In P. F. Fox (Ed.), *Developments in dairy chemistry* (Proteins, Vol. 1, pp. 61–86). London, UK: Applied Science.

Shalabi, S. I., & Fox, P. F. (1987). Electrophoretic analysis of cheese: Comparison of methods. *Irish Journal of Food Science and Technology, 11,* 135–151.

Simpson, K. J., Bird, P., Shaw, D., & Nicholas, K. (1998). Molecular characterisation and hormone-dependent expression of the porcine whey acidic protein gene. *Journal of Molecular Endocrinology, 20,* 27–34.

Simpson, K. J., & Nicholas, K. (2002). Comparative biology of whey proteins. *Journal of Mammary Gland Biology and Neoplasia, 7,* 313–326.

Singh, H., Boland, M., & Thompson, A. (2014a). *Milk proteins: From expression to food* (2nd ed.). Amsterdam, Netherlands: Academic Press.

Strange, D. E., Malin, E. L., van Hekken, D. L., & Basch, J. J. (1992). Chromatographic and electrophoretic methods used for analysis of milk proteins. *Journal of Chromatography, 624,* 81–102.

Swaisgood, H. E. (Ed.). (1975). *Methods of gel electrophoresis of milk proteins.* Champaign, IL: American Dairy Science Association. 33 pp.

Swaisgood, H. E. (1982). Chemistry of milk proteins. In P. F. Fox (Ed.), *Developments in dairy chemistry* (Proteins, Vol. 1, pp. 1–59). London, UK: Applied Science.

Swaisgood, H. E. (1992). Chemistry of the caseins. In P. F. Fox (Ed.), *Advanced dairy chemistry* (Proteins 2nd ed., Vol. 1, pp. 63–110). London, UK: Elsevier Applied Science.

Swaisgood, H. E. (2003). Chemistry of the caseins. In P. F. Fox (Ed.), *Advanced dairy chemistry* (Proteins: Part A 3rd ed., Vol. 1, pp. 139–201). London, UK: Elsevier Applied Science.

Thompson, M. P., Tarassuk, N. P., Jenness, R., Lillevik, H. A., Ashworth, U. S., & Rose, D. (1965). Nomenclature of the proteins of cow's milk—Second revision. *Journal of Dairy Science, 48,* 159–169.

Trejo, R., Dokland, T., Jurat-Fuentes, J., & Harte, F. (2011). Cryo-transmission electron tomography of native casein micelles from bovine milk. *Journal of Dairy Science, 94,* 5770–5775.

Uniacke-Lowe, T., & Fox, P. F. (2011). Equid milk. In J. W. Fuquay, P. F. Fox, & P. L. H. McSweeney (Eds.), *Encyclopedia of dairy sciences* (2nd ed., Vol. 3, pp. 518–529). Oxford, UK: Academic Press.

Uniacke-Lowe, T., Huppertz, T., & Fox, P. F. (2010). Equine milk proteins: Chemistry, structure and nutritional significance. *International Dairy Journal, 20*, 609–629.

Vanderghem, D., Danthine, S., Blecker, C., & Deroanne, C. (2007). Effect of proteose peptone addition on some physico-chemical characteristics of recombined dairy creams. *International Dairy Journal, 17*, 889–895.

Violette, J.-L., Chanat, E., Le Provost, F., Whitelaw, C. B. A., Kolb, A., & Shennan, D. B. (2013). Genetics and biosynthesis of milk proteins. In P. L. H. McSweeney & P. F. Fox (Eds.), *Advanced dairy chemistry* (Proteins: Basic aspects 4th ed., Vol. 1A, pp. 431–461). New York, NY: Springer.

Violette, J.-L., Whitelaw, C. B. A., Ollivier-Bousquet, M., & Shennan, D. B. (2003). Biosynthesis of milk proteins. In P. F. Fox & P. L. H. McSweeney (Eds.), *Advanced dairy chemistry* (Proteins: Part A 3rd ed., Vol. 1, pp. 698–738). New York, NY: Kluwer Academic/Plenum.

Visser, H. (1992). A new casein micelle model and its consequences for pH and temperature effects on the properties of milk. In H. Visser (Ed.), *Protein interactions* (pp. 135–165). Weinheim, Germany: VCH.

Walstra, P. (1999). Casein sub-micelles: Do they exist? *International Dairy Journal, 9*, 189–192.

Walstra, P., & Jenness, R. (1984a). *Dairy chemistry and physics*. New York, NY: John Wiley & Sons.

Waugh, D. F., & von Hippel, P. H. (1956). κ-Casein and the stabilisation of casein micelles. *Journal of the American Chemical Society, 78*, 4576–4582.

Whitney, R. M. L., Brunner, J. R., Ebner, K. E., Farrell, H. M., Jr., Josephson, R. U., Morr, C. V., et al. (1976). Nomenclature of cow's milk: Fourth revision. *Journal of Dairy Science, 59*, 795–815.

Wynn, P. C., & Sheehy, P. A. (2013). Minor proteins, including growth factors. In P. L. H. McSweeney & P. F. Fox (Eds.), *Advanced dairy chemistry* (Proteins: Basic aspects 4th ed., Vol. 1A, pp. 317–335). New York, NY: Springer.

Suggested Reading

Atkinson, S. A., & Lonnerdal, B. (1989b). *Protein and non-protein nitrogen in human milk*. Boca Raton, FL: CRC Press.

Fox, P. F. (1982). *Developments in dairy chemistry* (Proteins, Vol. 1). London, UK: Applied Science.

Fox, P. F. (Ed.). (1989). *Developments in dairy chemistry* (Functional milk proteins, Vol. 4). London, UK: Applied Science.

Fox, P. F. (Ed.). (1992). *Advanced dairy chemistry* (Milk proteins, Vol. 1). London, UK: Elsevier Applied Science.

Fox, P. F., & Mc Sweeney, P. L. H. (Eds.). (2003). *Advanced dairy chemistry* (Proteins 3rd ed., Vol. 1A & B). New York, NY: Kluwer Academic/Plenum.

Kinsella, J. E. (1984). Milk proteins: Physicochemical and functional properties. *CRC Critical Reviews in Food Science and Nutrition, 21*, 197–262.

McKenzie, H. A. (Ed.). (1970b). *Milk proteins: Chemistry and molecular biology* (Vol. 1). New York, NY: Academic Press.

McKenzie, H. A. (Ed.). (1971b). *Milk proteins: Chemistry and molecular biology* (Vol. 2). New York, NY: Academic Press.

McSweeney, P. L. H., & Fox, P. F. (Eds.). (2013). *Advanced dairy chemistry* (Proteins: Basic aspects 4th ed., Vol. 1A, pp. 317–335). New York, NY: Springer.

Mepham, T. B. (Ed.). (1983). *Biochemistry of lactation*. Amsterdam, Netherlands: Elsevier.

Singh, H., Boland, M., & Thompson, A. (2014b). *Milk proteins: From expression to food* (2nd ed.). Amsterdam, Netherlands: Academic Press.

Walstra, P., & Jenness, R. (1984b). *Dairy chemistry and physics*. New York, NY: John Wiley & Sons.

Walstra, P., Geurts, T. J., Noomen, A., Jellema, A., & von Boekel, M. A. J. S. (1999). *Dairy technology: Principles of milk properties and processes*. New York, NY: Marcel Dekker.

Walstra, P., Wouters, J. T. M., & Geurts, T. J. (2005). *Dairy science and technology*. Oxford, UK: CRC/Taylor and Francis.

Webb, B. H., Johnson, A. H., & Alford, J. A. (Eds.). (1974). *Fundamentals of dairy chemistry* (2nd ed.). Westport, CT: AVI.

Wong, N. P., Jenness, R., Keeney, M., & Marth, E. H. (Eds.). (1988). *Fundamentals of dairy chemistry* (3rd ed.). Westport, CT: AVI.

Chapter 5
Salts of Milk

5.1 Introduction

The salts of milk are mainly the phosphates, citrates, chlorides, sulphates, carbonates and bicarbonates of sodium, potassium, calcium and magnesium. Approximately 20 other elements are found in milk in trace quantities, including copper, iron, lead, boron, manganese, zinc, iodine, etc. Strictly speaking, the proteins of milk should be included as part of the salt system since these carry positively and negatively charged groups and can form salts with counter-ions; however, they are not normally treated as such. There is no lactate in freshly drawn milk but may be present in stored milk and in milk products. Many of the inorganic elements are of importance in nutrition, in the preparation, processing and storage of milk products due to their marked influence on the conformation and stability of milk proteins, especially caseins, in the activity of some indigenous enzymes and to a lesser extent in the stability of lipids.

5.2 Methods of Analysis

The mineral content of foods is usually determined from the ash prepared by heating a sample at 500–600 °C in a muffle furnace for ~4 h to oxidize organic matter. The ash content of milk is not truly representative of the salt system because: (1) the ash is a mixture of the carbonates and oxides of the inorganic elements present in the food but not of the original salts; (2) phosphorus and sulphur from proteins and lipids are present in the ash, while organic ions such as citrate are lost during ashing; (3) the temperature usually used in ashing may also vaporize certain volatile elements, e.g., sodium and potassium. Therefore, it is difficult to estimate accurately the ash content of a food, and low values are obtained for certain mineral elements by ash analysis compared to direct analysis of the intact food. The various mineral

© Springer International Publishing Switzerland 2015
P.F. Fox et al., *Dairy Chemistry and Biochemistry*,
DOI 10.1007/978-3-319-14892-2_5

constituents may be determined by titration, colorimetric, polarographic, flame photometric, atomic absorption spectrometry, inductively coupled plasma atomic emission spectrometry and inductively coupled plasma mass spectrometry techniques; however, the quantitative estimation of each ion in a mixture is frequently complicated by interfering ions. The major elements/ions may be determined by the following specific methods:

(a) Inorganic phosphate is usually determined by the method of Fiske and Stubbarow, as described in the official methods of the AOAC or IDF.
(b) Calcium by titration with EDTA or by atomic absorption spectroscopy on 12 % TCA filtrates.
(c) Magnesium by titration with EDTA (Davies and White 1962).
(d) Citrate by the colorimetric method of Marier and Boulet (1958), as modified by White and Davies (1963), by complexation of citrate with copper ions (Pierre and Brule 1983) or by enzymatic assay involving the use of citrate lyase, malate dehydrogenase, lactate dehydrogenase and NADH (Mutzelburg 1979).
(e) Ionized calcium is usually determined by the method of Smeets (1955), as modified by Tessier and Rose (1958), or using a calcium ion (Ca^{2+})-selective electrode (Demott 1968).
(f) Sodium and potassium may be determined by flame photometry, atomic absorption spectroscopy or ion selective electrodes.
(g) Chloride by titration with $AgNO_3$ using potentiometric or indicator end-point detection.
(h) Sulphate is usually precipitated by $BaCl_2$ and determined gravimetrically.
(i) Lactate may be determined spectrophotometrically after reaction with $FeCl_2$ or enzymatically (using lactate dehydrogenase) which can detect D- and L-isomers or by HPLC.

A comprehensive overview of the methods used to analyse the minerals in milk can be found in Gaucheron (2010).

5.3   Composition of Milk Salts

The ash content of bovine milk is relatively constant at 0.7–0.8 %, but the relative concentrations of the various ions can vary considerably. Table 5.1 shows the average concentration of the principal ions in milk, the usual range and the extreme ranges encountered. The latter undoubtedly includes abnormal milk, e.g., colostrum, very late lactation milk or milk from cows with mastitic infection.

The concentration of ash in human milk is only ~0.2 %; the concentration of all principal and several minor ions is higher in bovine than in human milk (Table 5.2). Consumption of unmodified bovine milk by human babies is not advised/practiced, due, at least in part, to the higher salts content in bovine milk compared with human milk and the negative health implications of increased renal solute load. The introduction of electrodialysis as a unit processing operation suitable for industrial

Table 5.1 Variation of content of milk salt constituents in mg per L milk (from various sources)

Constituent	Average content	Usual range	Extremes reported
Sodium	500	350–600	110–1,150
Potassium	1,450	1,350–1,550	1,150–2,000
Calcium	1,200	1,000–1,400	650–2,650
Magnesium	130	100–150	20–230
Phosphorus (total)[a]	950	750–1,100	470–1,440
Phosphorus (inorganic)[b]	750		
Chloride	1,000	800–1,400	540–2,420
Sulphate	100		
Carbonate (as CO_2)	200		
Citrate (as citric acid)	1,750		

[a]Total phosphorus includes colloidal inorganic phosphate, casein (organic) phosphate, soluble inorganic phosphate, ester phosphate and phospholipids
[b]Inorganic phosphorus includes colloidal inorganic phosphate and soluble inorganic phosphate

Table 5.2 Nutritionally important macro elements (mmol/L) in milks of selected species (A) and differences in composition of micro elements (µg/L) in mature human or bovine milk (B)

(A)	Cow	Human	Sheep	Goat	Sow	Mare
Calcium	29.4	7.8	56.8	23.1	104.1	16.5
Magnesium	5.1	1.1	9.0	5.0	9.6	1.6
Sodium	24.2	5.0	20.5	20.5	14.4	5.7
Potassium	34.7	16.5	31.7	46.6	31.4	11.9
Phosphate	20.9	2.5	39.7	15.6	51.2	6.7
Citrate	9.8	2.2	4.9	5.4	8.4	3.1
Chloride	30.2	6.2	17.0	34.2	28.7	6.6

(B)	Mature human milk		Cow's milk	
Constituent	Mean	Range	Mean	Range
Iron (µg)	760	620–930	500	300–600
Zinc (µg)	2,950	2,600–3,300	3,500	2,000–6,000
Copper (µg)	390	370–430	200	100–600
Manganese (µg)	12	7–15	30	20–50
Iodine (µg)	70	20–120	260	–
Fluoride (µg)	77	21–155	–	30–220
Selenium (µg)	14	8–19	–	5–67
Cobalt (µg)	12	1–27	1	0.5–1.3
Chromium (µg)	40	6–100	10	8–13
Molybdenum (µg)	8	4–16	73	18–120
Nickel (µg)	25	8–85	25	0–50
Silicon (µg)	700	150–1,200	2,600	750–7,000
Vanadium (µg)	7	tr-15	–	tr-310
Tin (µg)	–	–	170	40–500
Arsenic (µg)	50	–	45	20–60

demineralisation (specifically removal of sodium and chloride) from liquid whey made the commercial introduction of whey protein-dominant infant nutritional products possible in the 1960s. Nowadays, other technologies such as nanofiltration and ion exchange chromatography are also used industrially for the production of demineralised milk-based ingredients, most of which are used in the formulation of infant nutritional products.

5.4 Secretion of Milk Salts

The secretion of milk salts, which is not well understood, has been reviewed and summarized by Holt (1985). Despite the importance of milk salts in determining the processing characteristics of milk, relatively little interest has been shown in the nutritional manipulation of milk salts composition.

Three principles must be considered when discussing the milk salts system:

1. the need to maintain electrical neutrality,
2. the need to maintain milk isotonic with blood; as a result of this, a set of correlations exist between the concentrations of lactose, Na, K and Cl,
3. the need to form casein micelles which puts constraints on the pH and ionic calcium concentration ($[Ca^{2+}]$) and requires the complexation of calcium phosphate with casein.

Skim milk can be considered as a two-phase system of casein colloidal calcium phosphate in quasi-equilibrium with an aqueous solution of salts and proteins; the phase boundary is ill defined because of the intimate association between the calcium phosphate and the caseins (phosphoproteins).

A fat-free primary secretion is formed within vesicles by blebbing-off of the Golgi dictyosomes; the vesicles pass through the cytoplasm to the apical membrane where exocytosis occurs. The vesicles contain casein (synthesized in the rough endoplasmic reticulum toward the base of the mammocyte); fully-formed casein micelles are present within the Golgi vesicles. The vesicles also contain lactose synthetase (UDP: galactosyl transferase and α-lactalbumin) and there is good evidence showing that lactose synthesis occurs within the vesicles from glucose and UDP-galactose transported from the cytosal.

The intracellular concentrations of Na and K are established by a Na/K-activated ATPase and Na and K can permeate across the vesicle membranes. Ca is probably necessary to activate the UDP: galactosyl transferase and is transported by a Ca/Mg ATPase which concentrates Ca against an electrical potential gradient from μM concentrations in the cytosol to mM concentrations in the vesicles. Inorganic phosphorus (P_i) can be formed within the vesicles from UDP formed during the synthesis of lactose from UDP-galactose and glucose. UDP, which cannot cross the membrane, is hydrolyzed to UMP and P_i, both of which can re-enter the cytosol (to avoid product inhibition); however, some of the P_i is complexed by Ca^{2+}. Calcium ions are also chelated by citrate to form largely soluble, undissociated complexes and by casein to form casein micelles.

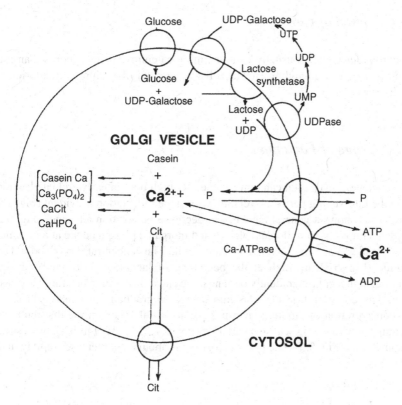

Fig. 5.1 Summary of some transport mechanisms for calcium, phosphate and citrate from the cytosol of the secretory cell to the inside of Golgi vesicles (from Holt 1985)

Water movement across the vesicle membranes is controlled by osmotic pressure considerations, e.g., lactose synthesis. Thus, the concentrations of both soluble and colloidal salts in milk are strongly influenced by lactose concentration and the mechanism by which it is synthesized.

Inter-relationships in the biosynthesis of the principal milk salts are summarized in Fig. 5.1. Para-cellular transport of several ionic species occurs during early and late lactation and during mastitic infection when the cell membranes are relatively more permeable.

5.5 Factors Influencing Variation in Salt Composition

The composition of milk salts is influenced by a number of factors including breed, individuality of the cow, stage of lactation, feed, mastitic infection and season of the year. The more important factors are discussed below.

5.5.1   Breed of Cow

Milk from Jersey cows usually contains more calcium and phosphorus than milk from other breeds, including Holstein, but its content of sodium and chloride are usually lower.

5.5.2   Stage of Lactation

The concentration of total calcium is generally high both in early and late lactation but in the intervening period no relation with stage of lactation is evident (Fig. 5.2). Phosphorus shows a general tendency to decrease as lactation advances (Fig. 5.2). The concentrations of colloidal calcium and inorganic phosphate are at a minimum in early and at a maximum in late lactation milk. The concentrations of sodium and chloride (Fig. 5.3) are high at the beginning of lactation, followed by a rapid decrease, then increase gradually until near the end of lactation when rapid increases occur. The concentration of potassium decreases gradually throughout lactation. The concentration of citrate, which has a marked influence on the distribution of calcium, shows a strong seasonal variation (Fig. 5.4). The pH of milk shows a strong seasonal trend (Fig. 5.5). The pH of colostrum is about 6 but increases rapidly in the

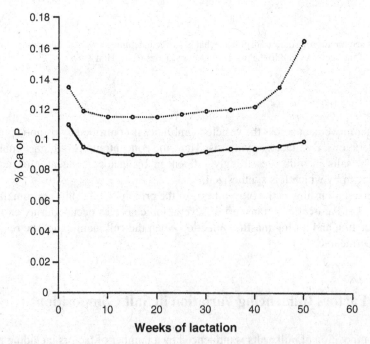

Fig. 5.2 Changes in the concentrations of calcium (*dashed line*) and phosphorus (*solid line*) in bovine milk during lactation

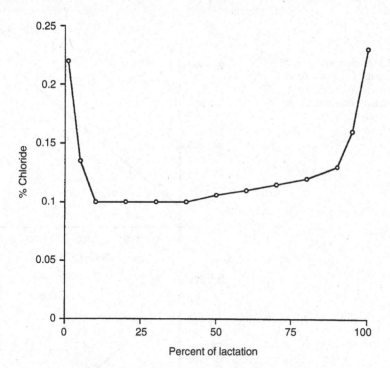

Fig. 5.3 Changes in the concentration of chloride in bovine milk during lactation

Fig. 5.4 Seasonality of the concentration of citric acid in bovine milk

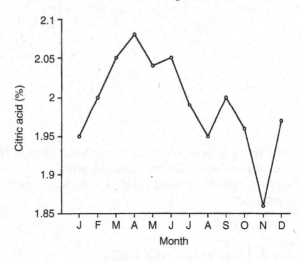

Fig. 5.5 Schematic representation of the pH of milk during lactation

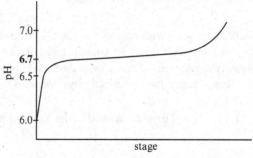

Fig. 5.6 Correlations between the concentration of sodium and chloride (**a**) and sodium and potassium (**b**) in bovine milk

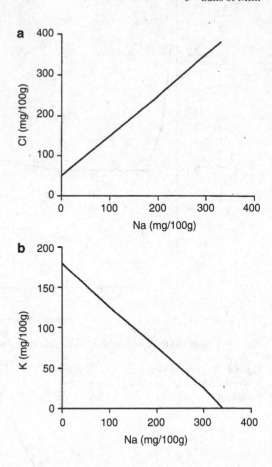

early stages of lactation, to reach the normal value of ~6.7 shortly after parturition and changes little until late lactation, when the pH reaches as high as 7.2, i.e., approaches that of blood (pH 7.4) due to degeneration of the mammary cell membrane.

5.5.3   Infection of the Udder

Milk from cows with mastitic infections has a low level of total solids, especially lactose, and high levels of sodium and chloride, the concentrations of which are directly related (Fig. 5.6a). The sodium and chloride ions come from the blood and are compensated for osmotically by depressed lactose synthesis.

$$\text{These are related in the Koestler number} = \frac{100 \times \% \, Cl}{\% \, lactose}$$

which is normally 1.5–3.0 but increases on mastitic infection and has been used as an index of such (better methods are now available, e.g., somatic cell count, activity of certain enzymes, especially catalase and N-acetylglucosamidase). The pH of milk increases to approach that of blood during mastitic infection. The concentrations of potassium and sodium are inversely related (Fig. 5.6b).

5.5.4   Feed

Feed has relatively little effect on the concentration of most elements in milk because the animal's skeleton acts as a reservoir of calcium (and other minerals). Milk fever occurs when the cow depletes its skeleton of Ca to maintain the level of Ca in its milk. Milk fever occurs mainly when cows are on fresh pasture which may have a low level of minerals and can be avoided by supplementation with magnesium spread on pasture. The level of citrate in milk decreases on diets very deficient in roughage and results in the "Utrecht phenomenon"—i.e., milk of very low heat stability due to a high concentration of Ca^{2+} arising from low citrate content. Relatively small changes in the concentrations of milk salts, especially of Ca, P_i and citrate, can have very significant effects on the processing characteristics of milk and hence these can be altered by the level and type of feed but definitive studies on this are lacking.

5.6   Interrelations of Milk Salt Constituents

Various milk salts are interrelated and the inter-relationships are affected by pH (Table 5.3).

Those constituents, the concentrations of which are related to pH in the same way, are also directly related to each other, e.g., the concentrations of total soluble calcium and ionized calcium, while those related to pH in opposite ways are inversely related, e.g., the concentrations of potassium and sodium.

Relationships between some of the more important ions/molecules are shown in Fig. 5.7. Three families of correlations of milk salt concentrations can be identified:

1. Correlations between the concentrations of lactose, K^+, Na^+ and Cl^- (Fig. 5.7a) have been recognised for many years and result from the requirement that milk must be isotonic with blood, i.e., [lactose] is negatively correlated with [K^+], and the operation of a para-cellular pathway, i.e., [Na^+] is positively correlated with [Cl^-].
2. Arising from the correlation of the concentrations of various salts with pH, certain other correlations arise. There is a direct correlation between [diffusible Ca] (and diffusible Mg) and [diffusible citrate] (Fig. 5.7b); this correlation, which is very good at constant pH, exists because chelation of Ca^{2+} by citrate is much stronger than that by phosphate.

Table 5.3 Relationships between the pH of milk and the concentrations of certain milk salt constituents

Inversely related to pH	Directly related to pH
Titratable acidity	Colloidal inorganic calcium
Total soluble calcium	Caseinate calcium
Soluble un-ionized calcium	Colloidal inorganic phosphorus
Ionized calcium	Colloidal calcium phosphate
Soluble magnesium	Sodium
Soluble citrate	Chloride
Total phosphorus	
Soluble inorganic phosphorus	
Ester phosphorus	
Potassium	

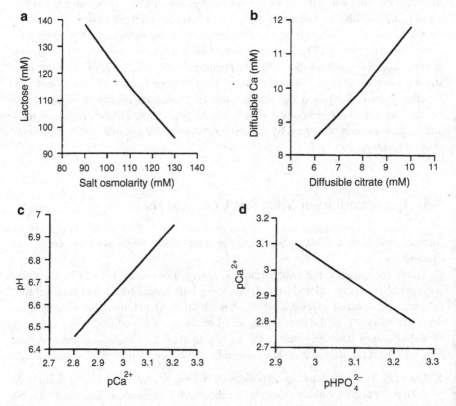

Fig. 5.7 Inter-relationships between lactose and soluble salts (osmolarity) and between some soluble salts in bovine milk (from Holt 1985)

3. The ratio $HPO_4^{2-}/H_2PO_4^-$ is strongly pH dependent, as is the solubility of $Ca_3(PO_4)_2$ (see below). As the pH is reduced, colloidal CaP_i is converted to soluble CaP_i but $HPO_4^{2-} \rightarrow H_2PO_4^-$ as the pH is reduced and hence both $[Ca^{2+}]$ and soluble P_i are directly related to pH (Fig. 5.7c) the $[HPO_4^{2-}]$ is inversely related to $[Ca^{2+}]$ (Fig. 5.7d).

5.6.1 Partition of Milk Salts Between Colloidal and Soluble Phases

Certain of the milk salts, e.g., chlorides, and the salts of sodium and potassium are sufficiently soluble to be present almost entirely in the dissolved phase. The concentration of others, in particular calcium phosphate, is higher than can be maintained in solution at the normal pH of milk. Consequently, these exist partly in soluble form and partly in an insoluble or colloidal form associated with casein. The state and distribution of these salts has been extensively reviewed by Pyne (1962), Holt (1985) and Gaucheron (2010, 2011).

The dividing line between soluble and colloidal is somewhat arbitrary, its exact position depending very much on the method used to separate the phases. However, a fairly sharp separation between the two phases is not difficult since the insoluble salts occur mainly associated with the colloidal casein micelles.

5.6.2 Methods Used to Separate the Colloidal and Soluble Phases

The methods used include dialysis, ultrafiltration, high-speed centrifugation, Donnan membrane technique and rennet-induced coagulation. The method used must not cause changes in the equilibrium between the two phases. The two most important precautions are to avoid changes in pH (lowering the pH dissolves colloidal calcium phosphate, see below) and temperature (reducing the temperature dissolves colloidal calcium phosphate and *vice versa*). Since milk comes from the cow at ~40 °C, working at 20 °C and certainly at 4 °C, will cause significant shifts in calcium phosphate equilibrium.

Ultrafiltrates obtained by filtering through cellophane or more modern polysulfone membranes at 20 °C and a pressure of 103.5 kPa (1034.6 mBar) are satisfactory, but the concentrations of citrate and calcium are slightly low due to a sieving effect accentuated by high pressures. Dialysis of a small volume of water (1:50) against milk (to which a little chloroform or azide has been added as preservative) at 20 °C for 48 h is the most satisfactory separation procedure and agrees closely with results obtained by ultrafiltration and renneting techniques, although the latter tends to be slightly high in calcium. As mentioned above, the temperature at which dialysis is performed is important, e.g. diffusate prepared from milk at 3 °C contains more total calcium, ionized calcium and phosphate than a diffusate prepared at 20 °C (Table 5.4). The partition of salts between the soluble and colloidal phases are summarized in Table 5.5.

In general, most or all the sodium, potassium and chloride are in solution, nearly all citrate, 1/3 of the calcium and 2/3 of the magnesium and about 40 % of the inorganic phosphate are present in true solution.

Table 5.4 Effect of temperature on the composition of diffusate obtained by dialysis

	mg per 100 g milk	
Constituent	20 °C	3 °C
Total calcium	37.8	41.2
Ionized calcium	12.0	12.9
Magnesium	7.7	7.9
Inorganic phosphorus	32.0	32.6
Citrate (as citric acid)	177.0	175.0
Sodium	58.0	60.0
Potassium	133.0	133.0

Table 5.5 Distribution of salts between the soluble and colloidal phases of milk

	Total in milk	Diffusate	Colloidal
Constituent	mg/l milk		
Total calcium	1,142	381 (33.5 %)	761 (66.5 %)
Ionized calcium	117		
Magnesium	110	74 (67 %)	36 (33 %)
Sodium	500	460 (92 %)	40 (8 %)
Potassium	1,480	1,370 (92 %)	110 (8 %)
Total phosphorus	848	377 (43 %)	471 (57 %)
Inorganic phosphorus	318		
Citrate (as citric acid)	1,660	1,560 (94 %)	100 (6 %)
Chloride	1,063	1,065 (100 %)	0 (0 %)

The phosphorus of milk occurs in five classes of compounds:

Organic	Inorganic
1. Lipid	4. Soluble
2. Casein	5. Colloidal
3. Small soluble esters	

The distribution of total phosphorus between these classes is shown schematically in Fig. 5.8.

5.6.3   Soluble Salts

The soluble salts are present in various ionic forms, complex ions and unionized complexes. Sodium and potassium are present totally as cations. Similarly, chloride and sulphate, i.e., anions of strong acids, are present as anions at the pH of milk. The salts of weak acids (phosphates, citrates and carbonates) are distributed between various ionic forms which can be calculated approximately from the analytical composition of milk serum and the dissociation constants of phosphoric, citric and carbonic acid, after

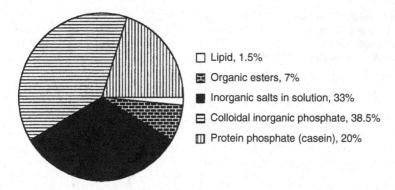

Fig. 5.8 Distribution of phosphorus among various classes of compounds in bovine milk

allowance has been made for binding of calcium and magnesium to citrate as anionic complexes and to phosphate as undissociated salts. The distribution of the various ionic forms can be calculated according to the Henderson-Hasselbalch equation:

$$pH = pK_a + \log\frac{[salt]}{[acid]}$$

Phosphoric acid (H_3PO_4) dissociates as follows:

$$H_3PO_4 \leftrightarrow H^+ + H_2PO_4^- \leftrightarrow H^+ + HPO_4^{2-} \leftrightarrow H^+ + PO_4^{3-}$$

$$pK_{a1} = 1.96 \quad pK_{a2} = 6.83 \quad pK_{a3} = 12.32$$

The titration curve for H_3PO_4 using NaOH is shown in Fig. 5.9. $H_2PO_4^-$, HPO_4^{2-} and PO_4^{3-} are referred to as primary, secondary and tertiary phosphate, respectively.

Citric acid is also triprotic:

$$
\begin{array}{c}
H \\
| \\
H\text{--}C\text{--}COOH \\
H\text{--}C\text{--}COOH \\
H\text{--}C\text{--}COOH \\
| \\
H
\end{array}
$$

and carbonic acid is diprotic ($H_2C\text{-}(COOH)_2$). The exact value of the dissociation constant to use depends on the total ionic concentration. Consequently, the constants used are only an approximation of the situation in milk. The following values are generally used:

Acid	pK_1	pK_2	pK_3
Citric	3.08	4.74	5.4
Phosphoric	1.96	6.83	12.32
Carbonic	6.37	10.25	

Fig. 5.9 Titration curve for phosphoric acid (H_3PO_4); *plus sign* indicates pK_{a1} (1.96), pK_{a2} (6.8) and pK_{a3} (12.3)

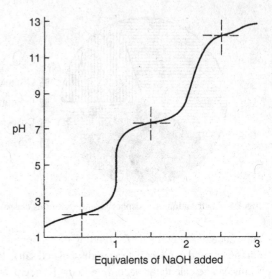

Equivalents of NaOH added

In milk, the critical dissociation constants are pK_3 of citric acid, pK_2 of phosphoric acid and pK_1 of carbonic acid. Bearing in mind the limitations and assumptions of the above data, the following calculations can be made for the distribution of the various ions in milk at pH 6.7 (average pH value):

(a) Phosphoric acid

$$pH = pK_{a1} + \log\frac{[salt]}{[acid]}$$

$$6.7 = 1.96 + \log\frac{[salt]}{[acid]}$$

$$\frac{[salt]}{[acid]}, \text{i.e.,} \frac{H_2PO_4^-}{H_3PO_4} = 43,700$$

Therefore, there is essentially no H_3PO_4 in milk.

For the second dissociation, i.e., $H_2PO_4^- \rightarrow HPO_4^{2-} + H^+, pK_{a2} = 6.83$

$$6.6 = 6.83 + \log\frac{[salt]}{[acid]}$$

$$\log\frac{[salt]}{[acid]} = -0.23$$

$$\frac{[salt]}{[acid]}, \text{i.e.,} \frac{HPO_4^{2-}}{H_2PO_4^-} = 0.59$$

Dihydrogenphosphate (primary) and monohydrogen phosphate (secondary) are the predominant forms, in the ratio of 60:40. 1.0:0.74: i.e., 57 % $H_2PO_4^-$ and 43 % HPO_4^{2-}

(b) Similarly for citrate, using pK's of 3.08, 4.74 and 5.4:

$$\frac{H_2Citrate^-}{H_3Citric\,acid}=3,300$$

$$\frac{HCitrate^{2-}}{H_2Citrate^-}=72$$

$$\frac{Citrate^{3-}}{HCitrate^{2-}}=16$$

Therefore, tricitrate and dicitrate, in the ratio 16:1, are the predominant forms. The small amount of carbonic acid present occurs mainly as bicarbonate, HCO_3^-. Some calcium and magnesium in milk exist as complex undissociated ions with citrate, phosphate or bicarbonate, e.g., Ca Citr$^-$, $CaPO_4^-$, $CaCO_3$.

Calculations by Smeets (1955) suggest the following distribution for the various ionic forms in the <u>soluble</u> phase:

- Ca+Mg: 35 % as ions, 55 % bound to citrate and 10 % bound to phosphate.
- Citrates: 14 % tertiary (citr^{3-}), 1 % secondary (H citr^{2-}) and 85 % bound to calcium and magnesium.
- Phosphates: 51 % primary ($H_2PO_4^-$), 39 % secondary (HPO_4^{2-}) and 10 % bound to calcium and magnesium.

Combining this information with the distribution of the various salts between colloidal and soluble phases (Table 5.5), gives the following quantitative distribution of the salts in milk (Table 5.6):

It is possible to determine the concentrations of anions such as Cl$^-$, PO_4^{3-} and Cit^{3-} in milk using anion exchange chromatography. Detection is normally performed using conductivity with supression of the signal due to the eluant to maximise the signal to noise ratio, allowing greater sensitivity of analysis (Buldini et al. 2002). Capillary electrophoresis has also been shown to be effective for the analysis of anions in milk and dairy products (Izco et al. 2003). It is also evident, given recent advancements in analytical and chemometric capability, that nuclear magnetic resonance (NMR) spectroscopy will play an increasingly important role in characterising interactions between minerals and proteins in milk and other dairy products (Rulliere et al. 2013; Sundekilde et al. 2013).

Making certain assumptions and approximations as to the state of various ionic species in milk, Lyster (1981) and Holt et al. (1981) have developed computer programmes that permit calculation of the concentrations of various ions and soluble complexes in typical milk diffusate. The outcome of both sets of calculations are in fairly good agreement and are also in good agreement with those species for which experimentally determined values are available. The ionic strength of milk is ~0.08 M.

Table 5.6 Distribution of milk salts

Species	Concentration (mg/l)	Soluble %	Form	Colloidal %
Sodium	500	92	Completely ionized	8
Potassium	1,450	92	Completely ionized	8
Chloride	1,200	100	Completely ionized	–
Sulphate	100	100	Completely ionized	–
Phosphate	750	43	10 % bound to Ca and Mg	57
			51 % H_2PO^-	
			39 % HPO_4^{2-}	
Citrate	1,750	94	85 % bound to Ca and Mg	
			14 % $Citr^{3-}$	
			1 % $HCitr^{2-}$	
Calcium	1,200	34	35 % Ca^{2+}	66
			55 % bound to citrate	
			10 % bound to phosphate	
Magnesium	130	67	Probably similar to calcium	33

5.6.4   Measurement of Calcium and Magnesium Ions

Ca^{2+}, along with H^+, play especially important roles in the stability of the caseinate system and its behaviour during milk processing, especially in the coagulation of milk by rennet, heat or ethanol. The concentration of these ions is also related to the solubility of the colloidal calcium phosphate. Consequently, there is considerable interest in determining their concentrations; three methods are available:

1. An ion-exchange method in which Ca^{2+} and Mg^{2+} are adsorbed onto an ion-exchange resin added to milk, the resin is removed and the Ca^{2+} and Mg^{2+} desorbed. It is assumed that the treatment does not alter the ionic equilibrium in milk.
2. The murexide method, which depends on the formation of a complex between Ca^{2+} and ammonium purpurate (murexide).

$$Ca^{2+} + M \leftrightarrow CaM$$

The free dye (M) has an absorption maximum at 520 nm while Ca M absorbs maximally at 480 nm. The concentration of Ca^{2+} can be calculated from a standard curve in which E_{480} is plotted as a function of $[Ca^{2+}]$ or preferably from a standard curve of $(E_{520} - E_{480})$ as a function of $[Ca^{2+}]$ which is less curved and more sensitive (Fig. 5.10). Using this method, the Ca^{2+} concentration in milk was found to be 2.5–3.4 mM. The murexide method measures calcium ions only; magnesium, at the concentration in milk, does not affect the indicator appreciably. Calculation of Mg^{2+} concentration is possible when the total calcium and magnesium (obtained by EDTA titration) is known. This is based on the assumption that the same proportion of each cation is present in the ionic form, which is justifiable since the dissociation constants of their citrate and phosphate salts are

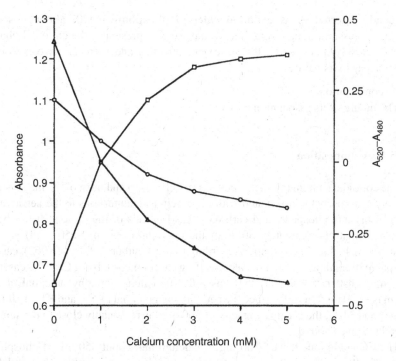

Fig. 5.10 Standard curve for the absorbance of murexide at 520 nm (*open circle*) and of Ca-murexide at 480 nm (*open square*) and $A_{520} - A_{480}$ (*open triangle*)

virtually identical. In contrast to the above, $[Ca^{2+}]$ determined colorimetrically appears to be 0.8 mM higher than that determined by the other methods.

3. Ca^{2+} activity (rather than concentration) can be determined rapidly and accurately using a Ca ion-selective electrode. Care must be exercised to ensure that the potentiometer is properly standardized using solutions that simulate the composition of milk serum, i.e., with the background ionic strength of the medium in which Ca^{2+} (e.g., milk) is to be measured, must be simulated in the preparation of suitable standard solutions (Crowley et al. 2014b). The Ca^{2+} activity is lower than the Ca^{2+} concentration, by about 2 mM, as measured by murexide titration. For further information please see a comprehensive review completed by Lewis (2011) on the measurement and significance of ionic calcium in milk.

5.6.5   Colloidal Milk Salts

As shown in Table 5.5, all the major ionic species in milk, with the exception of Cl⁻, are distributed between the soluble and colloidal phases but the principal colloidal salt is calcium phosphate; about 67 % and 57 %, respectively, of the total calcium and phosphate are in the colloidal phase. The colloidal inorganic salts are, therefore, frequently referred to as **colloidal calcium phosphate** (CCP) although some sodium, potassium, magnesium and citrate are also present in the colloidal phase. CCP is closely associated with the casein micelles and there are two principal questions as to its nature:

1. Its composition
2. The nature of its association with casein

5.6.5.1 Composition

All the colloidal sodium (40 mg/l), potassium (110 mg/l) and most of the magnesium (30 mg/l) are probably associated with the casein as counter-ions to the negatively-charged organic phosphate and carboxylic acid groups of the protein. It has been calculated that approximately 30 % of the colloidal calcium (~250 mg/l) is also directly attached to these groups. According to most authors (cf. Pyne 1962), casein is capable of binding 25–30 mol calcium/10^5 g casein (i.e. ~1,160 g Ca/10^5 g casein). Assuming that milk contains 25 g casein/L, the calcium-binding potential of the casein is ~300 mg/l of milk. Since the neutralizing potential of Na^+ and K^+ is half that of Ca^{2+} and Mg^{2+}, the binding capacity of 300 mg/l is reasonably close to the sum of the values given above.

These calculations leave ~500 mg of calcium and about 350 mg of phosphate present in the colloidal phase per litre of milk to be accounted for. The available evidence suggests that the excess CCP is present largely as tricalcium phosphate, $Ca_3(PO_4)_2$, or similar type of salt.

The so-called Ling oxalate titration indicates that CCP consists of 80 % $Ca_3(PO_4)_2$ and 20 % $CaHPO_4$, with an overall Ca:P ratio of 1.4:1 (see Pyne 1962). However, the oxalate titration procedure has been criticised because of the authenticity of many of the assumptions made. Pyne and McGann (1960) developed a new technique to study the composition of CCP. Milk at ~2 °C, is acidified to ~pH 4.9, followed by exhaustive dialysis of the acidified milk against a large excess of bulk milk; this procedure restores the acidified milk to normality in all respects except that CCP is not reformed. Analysis of milk and CCP-free milk (assumed to differ from milk only in respect of CCP) shows that the ratio of Ca:P in CCP was 1.7:1. The difference between this value and that obtained by the oxalate titration (i.e., 1.4:1) was attributed to the presence of citrate in CCP; citrate is not measured by the oxalate method. Pyne and McGann (1960) suggested that CCP has an apatite structure with the formula:

$$3Ca_3\left(PO_4\right)_2 \cdot CaHCitr^- \quad or \quad 2.5Ca_3\left(PO_4\right)_2 \cdot CaHPO_4 \cdot 0.5Ca_3Citr_2^-$$

Based on the assumption that the amount of Ca bound directly to casein is equivalent to the number of ester phosphate groups present, Schmidt (1982) argued that CCP is most likely to be amorphous tricalcium phosphate [$Ca_3(PO_4)_2$]. The argument is as follows: It is likely that the phosphoserine residues of the caseins are potential sites for interaction with CCP. The importance of these residues in calcium binding has been demonstrated also for dentine and salivary phosphoproteins. In a casein micelle of particle weight 10^8 Da, consisting of 93.3 % casein, with an ester-bound phosphorus content of 0.83 %, there are 25,000 ester phosphate groups. In such a micelle, 70,600 calcium atoms and 30,100 inorganic phosphate residues are present from which 5,000 $Ca_9(PO_4)_6$ clusters might be formed, leaving 25,500 calcium atoms. This means that there is approximately one calcium atom for each ester phosphate group and that about 40 % of these ester phosphate groups can be linked in pairs via $Ca_9(PO_4)_6$ clusters as shown in Fig. 5.11. The electrostatic interaction between casein and negatively charged ester phosphate groups of casein and $Ca_9(PO_4)_6$ clusters, which, by adsorption of two calcium atoms easily fit into the crystal grid, are positively changed. The proposed structure and association with the casein micelles is shown in Fig. 5.11.

Holt and collaborators (see for example Holt 1985; Holt et al. 1998, 2013, 2014), used X-ray absorption spectroscopy, IR spectroscopy, small-angle neutron scattering and NMR to study the structure of CCP. From these studies, they concluded that

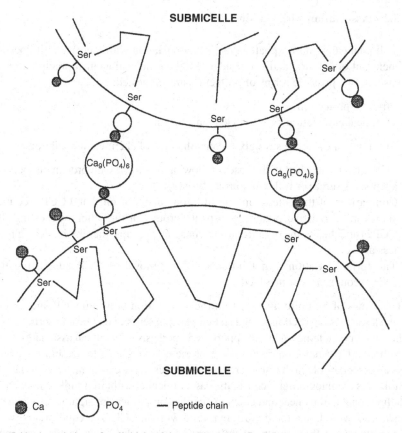

Fig. 5.11 Association of colloidal calcium phosphate ($Ca_3(PO_4)_2$) with the serine phosphate groups of casein (from Schmidt 1982)

it had a brushite-type structure, $CaHPO_4 \cdot 2H_2O$, the structure of which and its interaction with the phosphoserine residues of casein was described by Holt et al. (1998). Holt explains the difference between the Ca:P ratio found by analysis, i.e. 1.5–1.6 and the Ca:P ratio of $CaHPO_4$, i.e., 1.0, as being due to the ability of the phosphate moiety of phosphoserine to substitute in surface sites of a brushite-type lattice. The size and composition of the nanoclusters have been investigated using a number of different approaches over the years, including extensive hydrolysis of casein using the enzyme pronase to release Ca, PO_4, citrate, Mg and Zn from CCP (McGann et al. 1983), studying CCP attached to casein phosphopeptides, and using β-casein f1-25 to stabilize CCP nanoclusters (Holt et al. 1998). The most recent research aimed at elucidating the molecular mass (MW) of CCP used size exclusion chromatography coupled with a multi-angle laser light scattering detector of casein digests (Choi et al. 2011); using this approach and assuming that four casein phosphopeptides stabilise each CCP and if the MW of each of these phosphopeptides was about 2,500 Da, then the MW of CCP would be around 7,450 Da.

5.6.5.2 Association with Casein

The colloidal calcium phosphate is in close association with the casein; it does not sediment out of solution and is considered to be protected against precipitation by the casein. Two possible forms of protection are suggested:

(i) Physical protection
(ii) Chemical association between CCP and casein

Experimental evidence strongly favours the idea of chemical association:

(a) CCP remains attached to the casein following treatment with protein dissociating agents, e.g. urea, or following proteolysis,
(b) Comparison of the potentiometric titration curves of milk and CCP-free milk shows more reactive organic phosphate groups in the latter, suggesting that CCP is attached to the organic casein phosphate groups, thereby rendering them less active,
(c) The formol titration is not influenced by removal of CCP, suggesting that εNH_2-groups are not involved.

The views of Schmidt and Holt on the association between CCP and casein, i.e., via a shared Ca^{2+} (Schmidt) or a shared phosphoserine, i.e., phosphoserine is part of the CCP crystal lattice (Holt), support the hypothesis of chemical association.

Although CCP represents only ~6 % of the dry weight of the casein micelle, it plays an essential role in the structure and properties of the casein micelles and hence of milk; it is the integrating factor in the casein micelle, without it milk is not coagulable by rennet and its heat and calcium stability properties are significantly altered. In fact, milk would be a totally different fluid without CCP. The equilibria between the soluble and colloidal salts of milk are influenced by many factors, the more important of which are discussed below, and which consequently modify the processing properties of milk.

5.7   Changes in Milk Salts Equilibrium Induced by Various Treatments

Milk serum is supersaturated with calcium phosphate, the excess of salts being present in the colloidal phase, as described above. The balance between the colloidal and soluble phases may be upset by various factors, including changes in temperature, dilution or concentration, addition of acid, alkali, salts, thermal processing or other processing treatments such as high hydrostatic pressure.

5.7.1   Addition of Acid or Alkali

Acidification of milk is accompanied by a progressive solubilization of CCP and other colloidal salts from casein solubilization (Fig. 5.12) and is complete below ~pH 4.9 (Fig. 5.13). Alkalization has the opposite effect and at about pH 11 almost all the soluble calcium phosphate exists as the colloidal form. The changes on alkalisation are not reversible on subsequent dialysis against untreated milk. Milk has strong buffering capacity over a wide pH range (Lucey et al. 1993; Salaun et al. 2005).

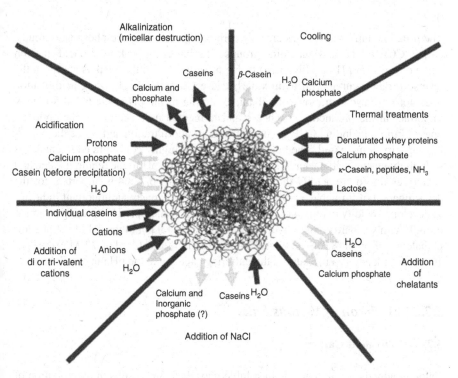

Fig. 5.12 Schematic representation of the changes that occur in distribution of salts in milk between the colloidal and serum phases in response to additions of selected salts, changes in pH and changes in temperature (i.e., cooling and thermal treatment) (from Gaucheron 2011)

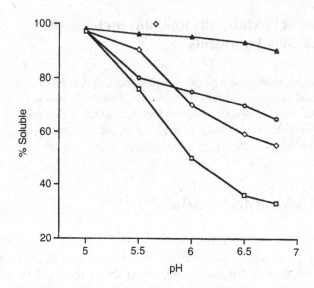

Fig. 5.13 Effect of pH on the distribution of calcium (*open square*), inorganic phosphorus (*open rhombus*), magnesium (*open circle*) and citrate (*open triangle*) between the colloidal and soluble phases in bovine milk

The principal buffering groups are: β/γ carboxyl groups, organic phosphate groups, citrate, CCP, ε-NH$_2$ and guanidino groups. The buffering peak in the acid titration curve of milk at ~pH 4.8 (Fig. 5.14) is due mainly to CCP; this peak is absent in the reverse titration curve with hydroxide because CCP dissolved during acidification does not reform on titration with alkali. Buffering in the pH range 6.7–4.5 is very important in the production of cheese and fermented milk products. The colloidal phase of milk, including CCP, is concentrated by ultrafiltration and hence the buffering capacity is increased. The buffering capacity of highly concentrated retentates is so high that lactic acid bacteria are unable to reduce the pH to the required value and it is necessary to pre-acidify the milk. Buffering at pH 6.7–8.4 is important for the determination of titratable acidity (TA) which was previously an important indicator of developed acidity in milk, i.e., of milk quality. TA is measured by titrating a sample of milk with standard sodium hydroxide from its natural pH to the pH of the phenolphthalein end-point (8.4). The typical TA of fresh milk is 0.14 ml of 0.1 M NaOH per 10 ml of milk and is a measure of the buffering capacity in the pH range 6.7–8.4.

5.7.2 Addition of Various Salts

5.7.2.1 Divalent Cations

Calcium added to milk reacts with soluble phosphate and results in precipitation of colloidal calcium phosphate (Fig. 5.12), an increase in ionized calcium, a decrease in the concentration of soluble phosphate and a decrease in pH.

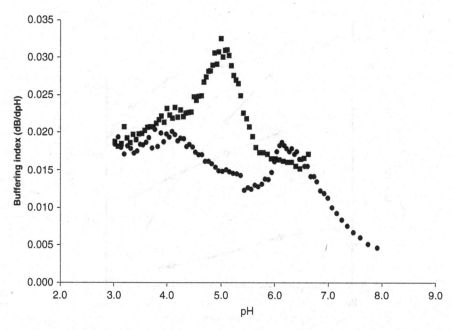

Fig. 5.14 Buffering curves of milk titrated from the initial pH (pH 6.7) to pH 3.0 with 0.1 N HCl (*filled square*) followed by back titration to pH 8.0 with 0.1 N NaOH (*filled circle*) (O'Mahony, unpublished data)

5.7.2.2   Phosphate

Addition of secondary Na or K phosphate (i.e., Na_2HPO_4) causes the precipitation of colloidal calcium phosphate, with concomitant decreases in the concentrations of soluble calcium and calcium ions. Polyphosphates, e.g., Na-hexametaphosphate, chelate Ca strongly and dissolve CCP.

5.7.2.3   Citrate

Addition of citrate reduces the concentrations of calcium ions and colloidal calcium phosphate and increases the soluble calcium, soluble phosphate and pH.

5.7.3   Effect of Changes in Temperature

The solubility of calcium phosphate is markedly temperature-dependent; unlike most compounds, the solubility of calcium phosphate decreases with increasing temperature—therefore, heating causes precipitation of calcium phosphate while cooling increases the concentrations of soluble calcium and phosphate at the expense of CCP

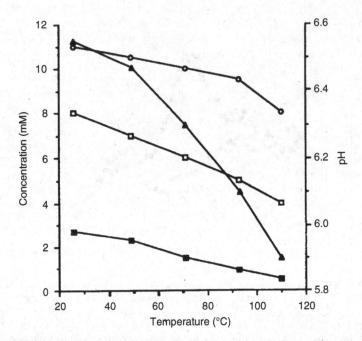

Fig. 5.15 Concentration of total calcium (*open square*), calcium ions (*filled square*), phosphate (*open circle*) and pH (*open triangle*) of ultrafiltrates prepared from milk at various temperatures (from Rose and Tessier 1959)

(Fig. 5.12). At low temperatures, shifts in the ionic balance are readily reversible, but after heating at a high temperature, reversibility becomes more sluggish and incomplete. Comparatively slight changes (20–3 °C) cause substantial changes in equilibrium (Table 5.4) which are completely reversible. The effects of high temperature treatments were studied by Rose and Tessier (1959) using hot ultrafiltration of milk heated to various temperatures. Calcium and phosphate precipitated on heating, to an extent dependent on temperature and time (Fig. 5.15), but the distribution of Na, K, Mg or citrate were not affected. On cooling, these changes were partly reversible. An extension of this approach has been used recently in studies on thermal processing of milk, whereby ultrafiltration permeate was removed from milk in the holding section of an ultra high temperature (UHT) plant to better understand the role of changes in mineral equilibrium (and consequently pH) in stability to heating of milk (On-Nom et al. 2010).

5.7.4   Changes in pH Induced by Temperature

The pH of milk is changed following heating due to changes in two salt systems. Fresh milk contains about 200 mg CO_2/L (i.e., 10 % by volume); about 50 % of this is lost on standing, with additional losses on heating. This results in a decrease in

Table 5.7 The pH of milk at various temperatures

Temperature (°C)	pH of milk
20	6.64
30	6.55
40	6.45
50	6.34
60	6.23

titratable acidity and an increase in pH. The formation of colloidal calcium phosphate during heating more than compensates for the loss of CO_2. The effect of temperature on pH is shown in Table 5.7 and Fig. 5.15.

The change in pH can be described as follows:

$$3Ca^{2+} + 2HPO_4^{2-} \xleftrightarrow[\text{cooling}]{\text{heating}} Ca_3(PO_4)_2 + 2H^+$$

The reaction is reversible on cooling after heating to a moderate temperature but becomes only partially reversible following more severe heating. The shifts in calcium phosphate equilibrium and pH increase when milk is concentrated.

5.7.5 Effect of Dilution and Concentration

Since milk is saturated with respect to calcium and phosphate, dilution reduces the concentration of Ca^{2+} and HPO_4^{2-} and causes solubilisation of some colloidal calcium phosphate, making the milk more alkaline. Concentration of milk causes precipitation of colloidal phosphate and shifts the PM of milk to the acid side, e.g., concentration by a factor of 2.1:1 reduces the pH to ~6.2.

1.   Dilution $Ca_3(PO_4)_2 \xrightarrow{H_2O} 3Ca^{2+} + 2HPO_4^{2-} + 2OH^-$

2.   Concentration $3Ca^{2+} + 2HPO_4^{2-} \rightarrow Ca_3(PO_4)_2 + 2H^+$

5.7.6 Effect of Freezing

Freezing milk causes crystallization of pure water and the unfrozen liquid becomes more saturated with respect to various salts. Some calcium phosphate precipitates as Ca_3PO_4, with the release of H^+ and a decrease in pH (e.g., to 5.8 at −20 °C).

As discussed in Chap. 2, crystallization of lactose as α-monohydrate exacerbates the situation. The combination of increased concentrations of Ca^{2+} and reduced pH causes destabilization of the casein micelles (see Sect. 4.3.9).

5.7.7   Effect of Ultrafiltration

Ultrafiltration is widely using in the dairy processing industry for the manufacture of ingredients such as milk protein concentrates (MPC) and milk protein isolates (MPI). Ultrafiltration of milk is designed to permit the permeation of serum phase/ diffusible salts through the semi-permeable membrane (in addition to lactose and other soluble solutes), to facilitate enrichment of protein in the retentate fraction. Acidification or addition of calcium-binding salts (e.g., citrates) to the milk feed material is often practiced as it has the effect of solubilising more of the calcium, magnesium and phosphate salts which can then permeate the membrane, with important implications for technological functionality of the retentate materials (e.g., viscosity, gelation and heat stability).

5.7.8   Effect of High Pressure Processing

High pressure processing (HPP) has beneficial effects on the microbiological quality (i.e., microbial inhibition) and some technological properties of milk (e.g., faster rennet-induced coagulation and increased cheese yield). HPP of bovine milk at pressures up to approximately 300 MPa causes increased levels of minerals in the serum phase of milk, caused mainly by partial disruption of hydrophobic and ionic interactions stabilising the casein micelles. These changes in mineral distribution between the colloidal and serum phases of milk on HPP are species-specific (the milk of at least cows, goats, sheep and humans has been studied), with the differences between species resulting mainly from the differences in susceptibility of casein micelles to HPP-induced dissociation.

5.8   Fortification of Milk with Inorganic Elements

Milk is an important nutritional source of salts for the population in many countries and there is a well-established association between milk and salts for the consumer, hence, salt-fortified milks are becomming increasingly common. The most common salt that is fortified in milk is calcium. Milk typically contains 1,100–1,200 mg calcium/L and a number of calcium-fortified products are available commercially with a calcium fortification level as high as 800 mg/l. Calcium can be added to such products in various forms, including, but not limited to, milk calcium (typically prepared by heat-acid precipitation from milk/whey permeate), calcium phosphate, calcium citrate, calcium carbonate, calcium chloride, calcium gluconate, calcium malate, calcium oxalate, calcium hydroxide and calcium glycerol-phosphate. Each of these salts containing calcium have different calcium potency, solubility and pH in solution/dispersion, and as a consequence, have different implications for techno-logical properties such as heat and physical stability during processing and shelf-life (Crowley et al. 2014a). Gaps in current knowledge of protein-mineral interactions make it difficult to develop a calcium-enriched milk containing more than 2 g/L of calcium, while retaining good consumer acceptability and being stable to processing and storage (Gaucheron 2010). Other salts can impact the sensory quality of mineral-fortified milk-based products, with specific examples including the addition of ferrous sulphate for iron enrichment causing oxidation and the addition of calcium carbonate for calcium enrichment causing sedimentation and coarse mouthfeel. Mineral fortification of dairy products is a complex area involving successful integration of nutrition, dairy chemistry and dairy technology to successfully develop cost effective, nutritionally-balanced, stable, good tasting products.

5.9 Synthetic Milk Ultrafiltrate

The diffusate or ultrafiltrate of skim milk contains all of the serum-phase constituents of milk (Table 5.5) and is a very appropriate dispersant for various studies on the chemistry and technological aspects of milk proteins. However, for use in such studies, the preparation of milk diffusate/ultrafiltrate is not always possible to obtain fresh and is time-consuming to prepare. Jenness and Koops (1962) developed a salt solution which is designed to replicate the serum/salt system of milk diffusate/ultrafiltrate, which is referred to as simulated milk ultrafiltrate (SMUF). SMUF has been used extensively by researchers for work involving the solubilisation and dispersion of milk proteins to study several factors, including, but not limited to, rehydration of milk protein powders, changes in protein particle size in response to various treatments, as a medium/buffer for chemical reactions and in studies of heat stability of milk proteins. SMUF typically contains phosphate, citrate, carbonate and sulphate salts in addition to calcium chloride and lactose, if required. It is important to note that SMUF replicates the salts/serum phase of bovine milk only and further work is required to develop versions of SMUF for use in replication of salts/serum phases of other species (e.g., human in particular for studies on infant formula ingredients).

References

Buldini, P. L., Cavalli, S., & Sharma, J. L. (2002). Matrix removal for the ion chromatographic determination of some trace elements in milk. *Microchemical Journal, 72*, 277–284.

Choi, J., Horne, D. S., & Lucey, J. A. (2011). Determination of molecular weight of a purified fraction of colloidal calcium phosphate derived from the casein micelles of bovine milk. *Journal of Dairy Science, 94*, 3250–3261.

Crowley, S. V., Kelly, A. L., & O'Mahony, J. A. (2014a). Fortification of reconstituted skim milk powder with different calcium salts: Impact of physicochemical changes on stability to processing. *International Journal of Dairy Technology, 67*, 474–482.

Crowley, S. V., Megemont, M., Gazi, I., Kelly, A. L., Huppertz, T., & O'Mahony, J. A. (2014b). Heat stability of reconstituted milk protein concentrate powders. *International Dairy Journal, 37*, 104–110.

Davies, D. T., & White, J. C. D. (1962). The determination of calcium and magnesium in milk and milk diffusate. *Journal of Dairy Research, 29*, 285–296.

Demott, B. J. (1968). Ionic calcium in milk and whey. *Journal of Dairy Science, 51*, 1008–1012.

Gaucheron, F. (2010). *Analysing and improving the mineral content of milk.* Cambridge: Woodhead Publishing.

Gaucheron, F. (2011). Milk salts: Distribution and analysis. In *Encyclopedia of dairy sciences* (2nd ed., pp. 908–916). Academic Press, Oxford, UK.

Holt, C. (1985). The milk salts: Their secretion, concentration and physical chemistry. In P. F. Fox (Ed.), *Developments in dairy chemistry, volume 3, lactose and minor constituents* (pp. 143–181). London: Elsevier Applied Science.

Holt C., Dalgleish, D.G. and Jenness, R. (1981). Calculation of the ion equilibria in milk diffusate and comparison with experiment. *Analytical Biochemistry*, 113, 154–163.

Holt, C., Timmins, P. A., Errington, N., & Leaver, J. (1998). A core-shell model of calcium phosphate nanoclusters stabilised by β-casein phosphopeptides, derived from sedimentation equilibrium and small-angle X-ray and neutron-scattering measurements. *European Journal of Biochemistry, 252*, 73–78.

Holt, C., Carver, J. A., Ecroyd, H., & Thorn, D. C. (2013). Caseins and the casein micelle: Their biological functions, structures, and behaviour in foods. *Journal of Dairy Science, 96*, 6127–6146.

Holt, C., Lenton, S., Nylander, T., Sørensen, E. S., & Teixeira, S. C. M. (2014). Mineralisation of soft and hard tissues and the stability of biofluids. *Journal of Structural Biology, 185*, 383–396.

Izco, J. M., Tormo, M., Harris, A., Tong, P. S., & Jimenez-Flores, R. (2003). Optimisation and validation of a rapid method to determine citrate and inorganic phosphate in milk by capillary electrophoresis. *Journal of Dairy Science, 86*, 86–95.

Jenness, R., & Koops, J. (1962). Preparation and properties of salt solution which simulates milk ultrafiltrate. *Netherlands Milk and Dairy Journal, 16*, 153–164.

Lewis, M. J. (2011). The measurement and significance of ionic calcium in milk – a review. *International Journal of Dairy Technology, 1*, 1–13.

Lucey, J. A., Hauth, B., Gorry, C., & Fox, P. F. (1993). The acid-base buffering properties of milk. *Milchwissenschaft, 48*, 268–272.

Lyster, R. L. J. (1981). Calculation by computer of individual concentrations in a simulated milk salt solution. II. An extension to the previous model. *Journal of Dairy Research*, 48, 85–89.

McGann, T. C. A., Buchheim, W., Kearney, R. D., & Richardson, T. (1983). Composition and ultrastructure of calcium phosphate - citrate complexes in bovine milk systems. *Biochimica et Biophysica Acta, 760*, 415–420.

Mutzelburg, I. D. (1979). An enzymatic method for the determination of citrate in milk. *Australian Journal of Dairy Technology, 34*, 82–84.

Pierre, A., & Brule, G. (1983). Dosage rapide du citrate dans l'ultrafiltrat de lait par complexation cuivrique. *Le Lait, 63*, 66–74.

Pyne, G. T. (1962). A review on the progress of dairy science. Some aspects of the physical chemistry of the salts in milk. *Journal of Dairy Research, 29*, 101–130.

Pyne, G. T., & McGann, T. C. A. (1960). The colloidal phosphate of milk. II. Influence of citrate. *Journal of Dairy Research, 27*, 9–17.

Rose, D., & Tessier, H. (1959). Composition of ultra-filtrates from milk heated at 80 to 220 °F in relation to heat stability. *Journal of Dairy Science, 42*, 969–980.

Rulliere, C., Rondeau-Mouro, C., Raouche, S., Dufrechou, M., & Marchesseau, S. (2013). Studies of polyphosphate composition and their interaction with dairy matrices by ion chromatography and ^{31}P NMR spectroscopy. *International Dairy Journal, 28*, 102–108.

Salaun, F., Mietton, B., & Gaucheron, F. (2005). Buffering capacity of dairy products. *International Dairy Journal, 15*, 95–109.

Schmidt, D. G. (1982). Association of caseins and casein micelle structure. In P. F. Fox (Ed.), Developments in Dairy Chemistry, Vol. 1. Protein (pp. 61–86). London: Elsevier Applied Science.

Smeets, W. J. G. M. (1955). The determination of the concentration of calcium ions in milk ultrafiltrate. *Netherland Milk and Dairy Journal, 9*, 249–260.

Sundekilde, U. K., Poulsen, N. A., Larsen, L. B., & Bertram, H. C. (2013). Nuclear magnetic resonance metabonomics reveals strong association between milk metabolites and somatic cell count in bovine milk. *Journal of Dairy Science, 96*, 290–299.

Tessier, H., & Rose, D. (1958). Calcium ion concentration in milk. *Journal of Dairy Science, 41*, 351–359.

White, J. C. D., & Davies, D. T. (1963). The determination of citric acid in milk and milk sera. *Journal of Dairy Research, 30*, 171–189.

Suggested Reading

Considine, T., Flanagan, J. and Loveday, S,M. (2014). Interations between Milk Proteins and Micronutrients. Milk Proteins: From Expression to Food. 2nd Edition. H. Singh, M. Boland, A. Thompson, eds. Academic Press, London. pp. 421–449.

Davies, D. T., & White, J. C. D. (1960). The use of ultrafiltration and dialysis in isolating the aqueous phase of milk and in determining the partition of milk constituents between the aqueous and disperse phases. *Journal of Dairy Research, 27*, 171–190.

de la Fuente, M. A. (1998). Changes in the mineral balance of milk submitted to technological treatments. *Trends in Food Science and Technology, 9*, 281–288.

Edmonson, L. F., & Tarassuk, N. P. (1956). Studies on the colloidal proteins of skim milk. II. The effect of heat and disodium phosphate on the composition of the casein complex. *Journal of Dairy Science, 49*, 123–128.

Fiske, C. H., & Stubbarow, J. J. (1925). The colorimetric determination of phosphorus. *Journal of Biological Chemistry, 66*, 375–400.

Gao, R., Temminghoff, E. J. M., van Leeuwen, H. P., van Valenberg, H. J. F., Eisner, M. D., & van Boekel, M. A. J. S. (2009). Simultaneous determination of free calcium, magnesium, sodium and potassium ion concentrations in simulated milk ultrafiltrate and reconstituted skim milk using the Donnan Membrane Technique. *International Dairy Journal, 19*, 431–436.

Gaucheron, F. (2000). Iron fortification in dairy industry. *Trends in Food Science and Technology, 11*, 403–409.

Greenwald, I., Redish, J., & Kibrick, A. (1940). The dissociation of calcium phosphates. *Journal of Biological Chemistry, 135*, 65–76.

Hastings, A. B., McLean, F. C., Eichelberger, L., Hall, J. L., & DaCosta, E. (1934). The ionization of calcium, magnesium, and strontium citrates. *Journal of Biological Chemistry, 107*, 351–370.

Jenness, R., & Patton, S. (1959). The effects of heat on milk. In *Principles of dairy chemistry* (pp. 329–334). New York: John Wiley & Sons.

Marier, J. R., & Boulet, M. (1958). Direct determination of citric acid in milk with an improved pyridine-acetic anhydride method. *Journal of Dairy Science, 41*, 1683–1692.

McMeckin, J. L., & Groves, M. L. (1964). In B. H. Webb, A. H. Johnson, & J. A. Alford (Eds.), *Fundamentals of dairy chemistry* (2nd ed.). Westport, CT: AVI Publication Corporation.

Mekmene, O., Le Graet, Y. L., & Gaucheron, F. (2009). A model for predicting salt equilibria in milk and mineral-enriched milks. *Food Chemistry, 116*, 233–239.

Miller, P. G., & Sommer, H. H. (1940). The coagulation temperature of milk as affected by pH, salts, evaporation and previous heat treatment. *Journal of Dairy Science, 23*, 405–421.

On-Nom, N., Grandison, A. S., & Lewis, M. J. (2010). Measurement of ionic calcium, pH and soluble divalent cations in milk at high temperature. *Journal of Dairy Science, 93*, 515–523.

Pyne, G. T., & Ryan, J. J. (1950). The colloidal phosphate of milk. I. Composition and titrimetric estimation. *Journal of Dairy Research, 17*, 200–205.

Rose, D. (1965). Protein stability problems. *Journal of Dairy Science, 48*, 139–146.

Tabor, H., & Hastings, A. B. (1943). The ionization constant of secondary magnesium phosphate. *Journal of Biological Chemistry, 148*, 627–632.

Verma, T. S., & Sommer, H. H. (1957a). Study of the naturally occurring salts in milk. *Journal of Dairy Science, 40*, 331.

Verma, T. S., & Sommer, H. H. (1957b). Study of the naturally occurring salts in milk. *Journal of Dairy Science, 40*, 331.

White, J. C. D., & Davies, D. T. (1958). The relation between the chemical composition of milk and the stability of the caseinate complex. I. General introduction, description of samples, methods, and chemical composition of samples. *Journal of Dairy Research, 25*, 236–255.

Chapter 6
Vitamins in Milk and Dairy Products

6.1 Introduction

Vitamins are organic chemicals required by the body in trace amounts but which cannot be synthesized by the body. The vitamins required for growth and maintenance of health differ between species; compounds regarded as vitamins for one species may be synthesized at adequate rates by other species. For example, only primates and guinea pigs require ascorbic acid (vitamin C; see Sect. 6.4) from their diet; other species possess the enzyme gluconolactone oxidase which is necessary for the synthesis of vitamin C. The chemical structures of the vitamins have no relationship with each other. Vitamins may be classified based on their solubility in water. Water-soluble vitamins are the B group [thiamine, riboflavin, niacin, biotin, pantothenate, folate, pyridoxine (and related substances, vitamin B_6)] and cobalamin (and its derivatives, vitamin B_{12}) and ascorbic acid (vitamin C) while the fat-soluble vitamins are retinol (vitamin A), calciferols (vitamin D), tocopherols (and related compounds, vitamin E) and phylloquinone (and related compounds, vitamin K). The water-soluble vitamins and vitamin K function as co-enzymes while vitamin A is important in the vision process, vitamin D functions like a hormone and vitamin E is primarily an antioxidant.

Milk is the only source of nutrients for the neonatal mammal during the early stage of life until weaning. Thus, in addition to providing macronutrients (protein, carbohydrate and lipid) and water, milk must also supply sufficient vitamins and minerals to support the growth of the neonate. Human beings continue to consume milk into adulthood and thus milk and dairy products continue to be important sources of nutrients in the diet of many people worldwide. The concentrations of macronutrients and minerals in milk have been discussed in Chaps. 1 and 5; vitamin levels in milk and dairy products will be considered here. Milk is normally processed to a lesser or greater extent before consumption. Thus, it is important to consider the influence of processing on the vitamin status of milk and dairy products.

© Springer International Publishing Switzerland 2015
P.F. Fox et al., *Dairy Chemistry and Biochemistry*,
DOI 10.1007/978-3-319-14892-2_6

The recommended dietary allowance (RDA) for a vitamin is the intake of that vitamin required to ensure the good health of a high proportion (97.5 %) of healthy individuals. Nutrient intakes equal to the RDA pose only a very small risk of deficiency.

6.2   Fat-Soluble Vitamins

6.2.1   *Retinol (Vitamin A)*

Vitamin A (retinol, Fig. 6.1) is the parent of a range of compounds known as retinoids, which possess the biological activity of vitamin A. In general, foods of animal origin provide preformed vitamin A as retinyl esters (e.g., Fig. 6.5, which are easily hydrolysed in the gastrointestinal tract) while plant-derived foods provide precursors of vitamin A, i.e., carotenoids. Only carotenoids with a β-ionone ring (e.g., β-carotene) can serve as vitamin A precursors. β-Carotene (Fig. 6.6) is cleaved at its centre by the enzyme β-carotene-15,15′-monooxygenase (present in the intestinal mucosa) to yield 2 mol retinol per mol. However, cleavage of other bonds results in the formation of only 1 molecule of retinol per molecule of β-carotene. The extent of conversion of β-carotene to vitamin A in humans is between 60 and 75 % with some 15 % β-carotene absorbed intact. Due to variable absorption of carotenoids, 1 μg retinol activity equivalent (RE) is defined as 1 μg retinol or 12 μg all *trans* β-carotene from food.

Retinol can be oxidized to retinal (Fig. 6.2) and further to retinoic acid (Fig. 6.3). *Cis-trans* isomerization can also occur, e.g., the conversion of all *trans* retinal to 11-*cis*-retinal (Fig. 6.4), which is important for vision.

Vitamin A has a number of roles in the body: it is involved in the vision process, in cell differentiation, embryogenesis, reproduction and growth and in the immune system. The RDA for vitamin A is 900 μg RE day^{-1} for men and 700 μg RE day^{-1} for

Fig. 6.1 Retinol

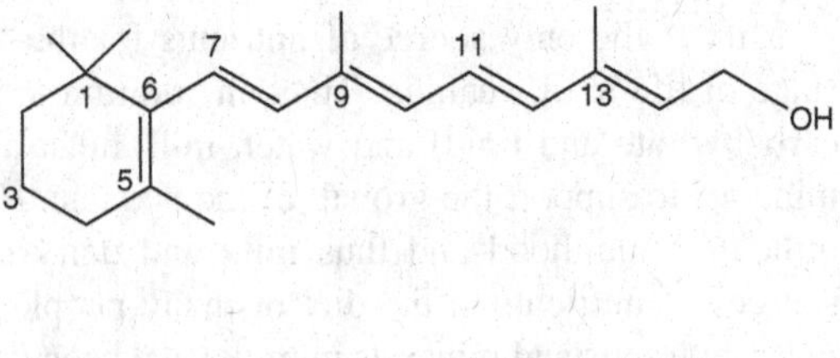

Fig. 6.2 Retinal

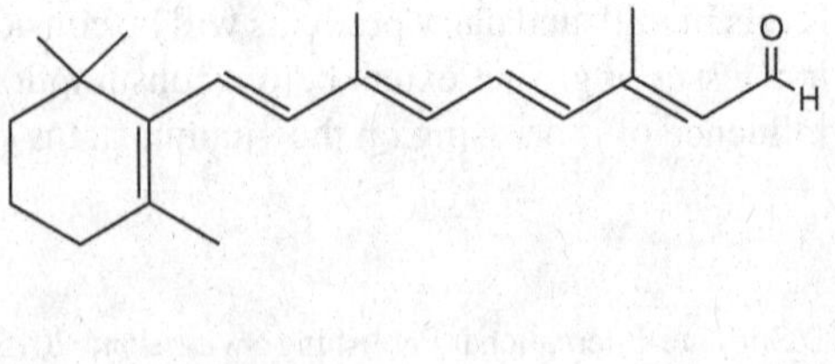

Fig. 6.3 Retinoic acid

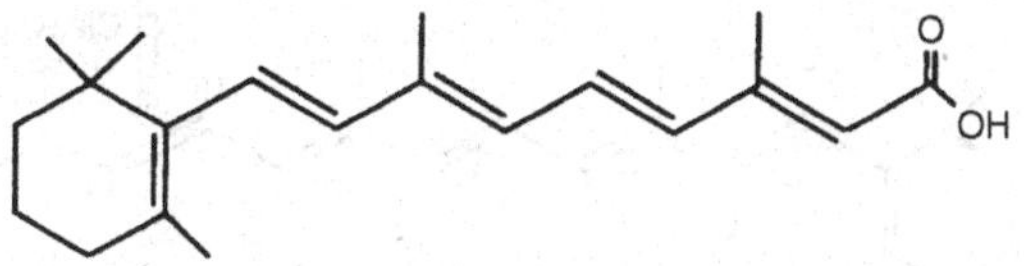

Fig. 6.4 11-*cis*-retinal

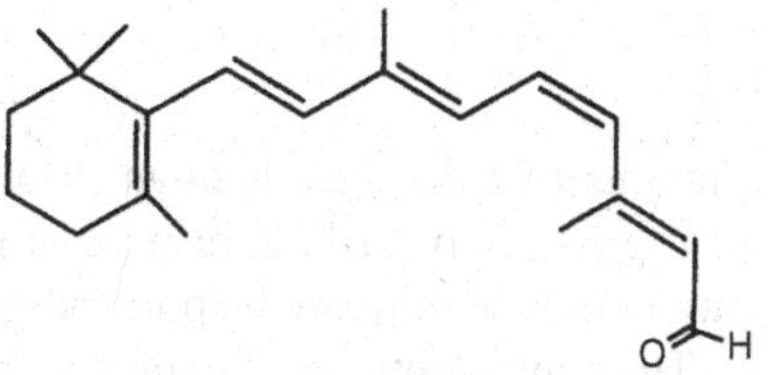

Fig. 6.5 Retinyl palmitate

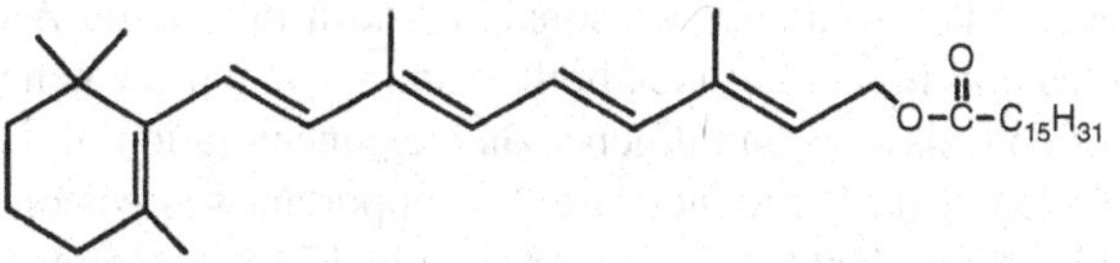

women. European population reference intake (PRI) value for vitamin A is 700 and 600 µg RE day^{-1} for adult men and women, respectively. The body will tolerate a wide range of vitamin A intakes (500–15,000 µg RE day^{-1}) but insufficient or excessive intakes result in illness. Vitamin A deficiency (<500 µg RE day^{-1}) results in night blindness, xerophthalmia (progressive blindness caused by drying of the cornea of the eye), keratinization (accumulation of keratin in digestive, respiratory and urinary-genital tract tissues) and finally exhaustion and death. At excessive intake levels (>100 times recommended intake for adults and >20 times the recommended intake for children), vitamin A is toxic. Symptoms of hypervitaminosis A include headache, vomiting, alopecia, cracking of lips, ataxia and anorexia in addition to bone and liver damage; 13-*cis*-retinoic acid is also a human teratogen.

The major dietary sources of retinol are dairy products, eggs and liver, while important sources of β-carotene are spinach and other dark-green leafy vegetables, deep orange fruits (apricots, cantaloupe) and vegetables (squash, carrots, sweet potatoes, pumpkin). The richest natural sources of vitamin A are fish liver oils, particularly halibut and shark.

Vitamin A activity is present in milk as retinol, retinyl esters and carotenes (Figs. 6.1–6.6). Whole cows' milk contains an average of 40 µg retinol and 20 µg carotene per 100 g (Morrissey and Hill 2009). The concentration of retinol in raw sheeps' and pasteurized goats' milk is 83 and 44 µg per 100 g, respectively, although the milk of these species are reported (Holland et al. 1991) to contain only trace

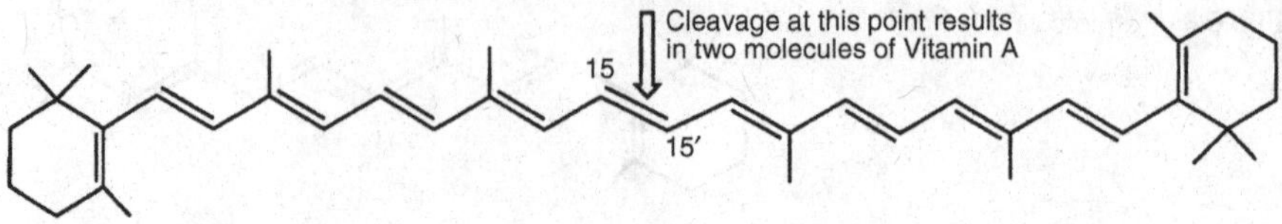

Fig. 6.6 β-Carotene

amounts of carotenes. Human milk and colostrum contain an average of 57 and 155 µg retinol per 100 g, respectively. In addition to their role as provitamin A, the carotenoids in milk are responsible for the colour of milk fat (see Chap. 8).

The concentration of vitamin A and carotenoids in milk is strongly influenced by the carotenoid content of the animal's feed. Milk from animals fed on pasture contains higher levels of carotenes than that from animals fed on concentrate feeds. There is also a large seasonal variation in vitamin A concentration; summer milk contains higher levels of both vitamin A and β-carotene than winter milk. The breed of cow also has an influence on the concentration of vitamin A in milk.

Other dairy products are also important sources of vitamin A. Whipping cream (39 % fat) contains ~565 µg retinol and 265 µg carotene per 100 g. The level of vitamin A in cheese varies with the fat content (Table 6.1). Camembert (23.7 % fat) contains 230 µg retinol and 315 µg carotene per 100 g, while Cheddar (34.4 % fat) contains 325 µg retinol and 225 µg carotene per 100 g. Whole milk yogurt (3 % fat; unflavoured) contains ~28 µg retinol and 21 µg carotene per 100 g while the corresponding values for ice-cream (9.8 % fat) are 115 and 195 µg per 100 g, respectively.

Vitamin A is relatively stable to most dairy processing operations and loss of vitamin A activity happens principally through autoxidation or geometric isomerisations (Morrissey and Hill 2009). Heating at <100 °C (e.g., pasteurization) has little effect on the vitamin A content of milk although some loss may occur at temperatures >100 °C (e.g., when frying using butter). Losses of vitamin A can occur in UHT milk during its long shelf-life at ambient temperature. Vitamin A is stable in pasteurized milk at refrigeration temperatures provided the milk is protected from light but substantial losses can occur in milk depending on packaging materials and storage under fluorescent light. Low-fat milks are often fortified with vitamin A for nutritional reasons. Added vitamin A is less stable to light than the indigenous vitamin; the composition of the lipid used as a carrier for the exogenous vitamin influences its stability. Protective compounds (e.g., ascorbyl palmitate or β-carotene) will reduce the rate at which exogenous vitamin A is lost during exposure to light. Yogurts containing fruit often contain higher concentrations of vitamin A precursor carotenoids than natural yogurts. The manufacture of dairy products which involves concentration of the milk fat (e.g., cheese, butter) results in a *pro rata* increase in the concentration of vitamin A. The increased surface area of dried milk products accelerates the loss of vitamin A; supplementation of milk powders with vitamin A and storage at a low temperature minimizes these losses.

Table 6.1 Concentrations of vitamin A and carotene and of vitamins E, D and C (per 100 g) in Dairy Products (modified from Holland et al. 1991)

Product	Retinol (µg)	Carotene (µg)	Vitamin D (µg)	Vitamin E (µg)	Vitamin C (mg)
Skimmed milk					
Pasteurized	1	Tr	Tr	Tr	1
UHT, fortified	61	18	0.1	0.02	35[a]
Whole milk					
Pasteurized	52	21	0.03	0.09	1
Summer	62	31	0.03	0.10	1
Winter	41	11	0.03	0.07	1
Sterilized, in container	52	21	0.03	0.09	Tr
Dried skimmed milk[b] (fortified)	350	5	2.10	0.27	13
With vegetable fat (fortified)	395	15	10.50	1.32	11
Evaporated milk, whole	105	100	3.95[c]	0.19	1
Goat's milk, pasteurized	44	Tr	0.11	0.03	1
Human milk, colostrum	155	(135)	N	1.3	7
Transitional	85	(37)	N	0.48	6
Mature	58	(24)	0.04	0.34	4
Sheep's milk, raw	83	Tr	0.18	0.11	5
Fresh whipping cream, pasteurized (39.3 % fat)	565	265	0.22	0.86	1
Cheeses					
Brie	285	210	0.20	0.84	Tr
Camembert	230	315	(0.18)	0.65	Tr
Cheddar, average	325	225	0.26	0.53	Tr
Cottage cheese					
Plain	44	10	0.03	0.08	Tr
Reduced fat (1.4 % fat)	16	4	0.01	0.03	Tr
Cream cheese	385	220	0.27	1.00	Tr
Danish Blue	280	250	(0.23)	0.76	Tr
Edam	175	150	(0.19)	0.48	Tr
Feta	220	33	0.50	0.37	Tr
Parmesan	345	210	(0.25)	0.70	Tr
Processed cheese, plain	270	95	0.21	0.55	Tr
Stilton, blue	355	185	0.27	0.61	Tr
Whole milk yogurt					
Plain	28	21	0.04	0.05	1
Fruit	39	16	(0.04)	(0.05)	1
Ice cream					
Dairy, vanilla	115	195	0.12	0.21	1

Tr trace, *N* nutrient present in significant quantities but there is no reliable information on amount, ( )=estimated value

[a]Unfortified milk would contain only traces of vitamin C

[b]Unfortified skimmed milk powder contains approximately 8 µg retinol, 3 µg carotene, Tr vitamin D and 0.01 mg vitamin E per 100 g. Some brands contain as much as 755 µg retinol, 10 µg carotene and 4.6 µg vitamin D per 100 g

[c]This is for fortified product. Unfortified evaporated milk contains approximately 0.09 µg vitamin D per 100 g

6.2.2 Calciferols (Vitamin D)

The term "vitamin D" is used for group of closely related secosteroids with anti-rachitic (anti-rickets) properties. The major dietary form of vitamin D obtained from foods of animal origin is cholecalciferol (vitamin D_3; Fig. 6.8); ergocalciferol (vitamin D_2) is obtained from fungi and protozoa. Cholecalciferol can be formed from a steroid precursor, 7-dehydrocholesterol (Fig. 6.7), by the skin when exposed to sunlight; with sufficient exposure to the sun, no preformed vitamin D is required from the diet.

UV light (290–320 nm) causes the photoconversion of 7-dehydrocholesterol to pre-vitamin D_3. This pre-vitamin can undergo further photoconversion to tachysterol and lumisterol or can undergo a temperature-dependent isomerization to cholecalciferol (vitamin D_3; Fig. 6.8). At body temperature, this conversion requires ~28 h to convert 50 % of pre-vitamin D_3 to vitamin D_3. Thus, production of vitamin D_3 in the skin can take a number of days. Preformed vitamin D_3 is obtained from the diet. Vitamin D_3 is stored in various fat deposits around the body. Regardless of the source of vitamin D_3, it must undergo two hydroxylations to become fully active. Vitamin D_3 is transported by a specific binding protein through the circulatory system to the liver where the enzyme, 25-hydroxylase, converts it to 25-hydroxycholecalciferol [25(OH)D_3; Fig. 6.9] which is the principal circulating form of vitamin D and a commonly used index of vitamin D status. 25(OH)D_3 is converted to 1,25-dihydroxycholcalciferol [1,25(OH)$_2$$D_3$ or calcitriol; Fig. 6.10] by

Fig. 6.7
7-Dehydrocholesterol

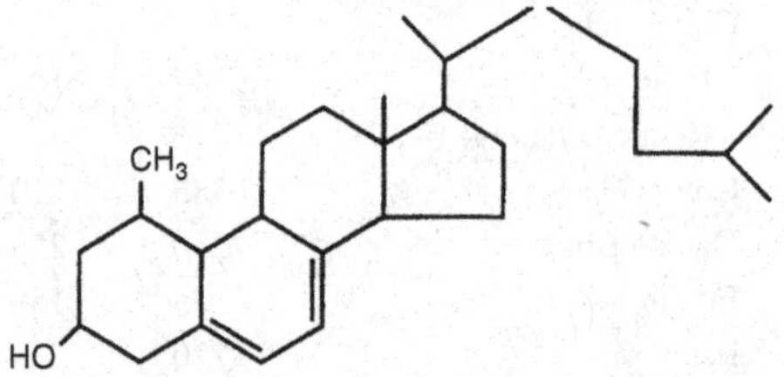

Fig. 6.8 Cholecalciferol, Vitamin D_3

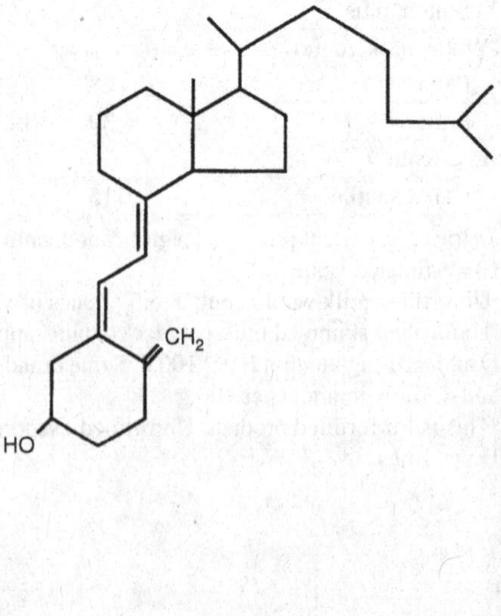

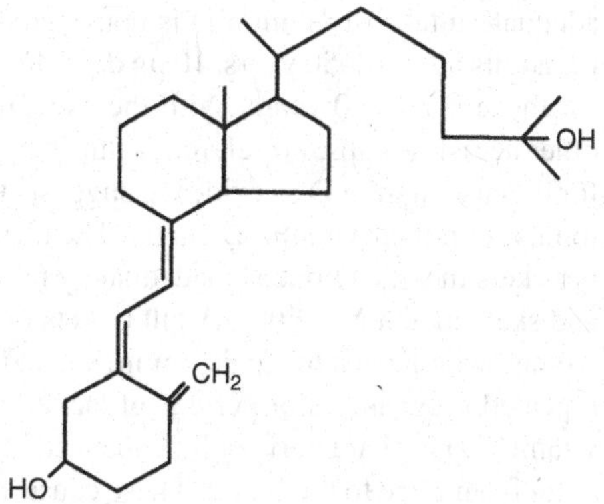

Fig. 6.9
25-Hydroxycholecalciferol

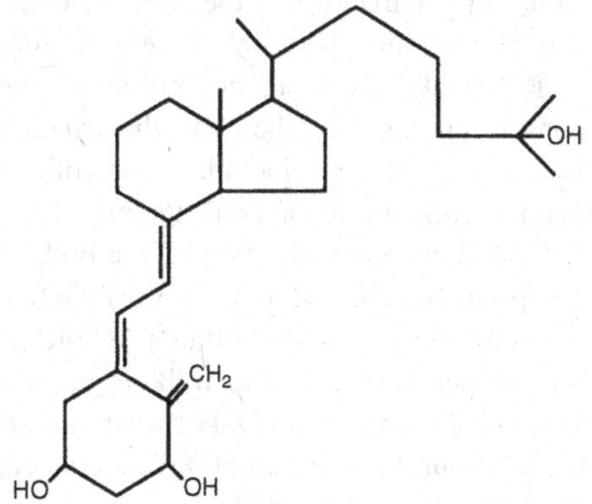

Fig. 6.10 1,25-
Dihydroxycholecalciferol

the enzyme, 1-hydroxylase, in the kidney; $1,25(OH)_2D_3$ is the major active metabolite of vitamin D. Alternatively, $25(OH)D_3$ can be hydroxylated at position 24 to form 24,25-dihydroxycholecalciferol [$24,25(OH)_2D_3$] which is metabolically inactive. About 50 different metabolites of vitamin D_2 and D_3 have been studied.

Vitamin D_2 (ergocalciferol) is formed by the photoconversion of ergosterol, a sterol present in certain fungi and yeasts, and differs from cholecalciferol in having an extra methyl group at carbon 24 and an extra double bond between C_{22} and C_{23}. Ergocalciferol was widely used for many years as a therapeutic agent.

The principal physiological role of vitamin D in the body is to maintain the level of plasma calcium by stimulating its absorption from the gastrointestinal tract, its retention by the kidney and by promoting its transfer from bone to the blood. Vitamin D acts in association with other vitamins, hormones and nutrients in the bone mineralization process. In addition, Vitamin D has a wider physiological role in other tissues in the body, including the brain and nervous system, muscles and cartilage, pancreas, skin, reproductive organs and immune cells.

Establishing dietary requirements for vitamin D is difficult because exposure to sunlight has a major influence on levels of $25(OH)D_2$ and $25(OH)D_3$ in the bloodstream which are used as indices of vitamin D status. In the United States, adequate intake of vitamin D is considered 5 µg day^{-1} for children, 0–10 µg day^{-1} for adults aged 18–50 years, 10 µg day^{-1} for adults aged 51–70 years and 15 µg day^{-1} for those aged >70 years. With the exception of the very young and very old and other at-risk groups, no reference nutrient intake (RNI) values are given in the UK for dietary vitamin D as evidence suggests that most individuals do not rely on food to maintain their vitamin D status. The classical syndrome of vitamin D deficiency is rickets in which bone is inadequately mineralized, resulting in growth retardation and skeletal abnormalities. Adult rickets or osteomalacia occurs most commonly in women who have a low calcium intake and little exposure to sunlight and have had repeated pregnancies or periods of lactation. Hypervitaminosis D (excess intake of vitamin D) is characterized by enhanced absorption of calcium and transfer of calcium from bone to the blood. These cause excessively high concentrations of serum calcium which can precipitate at various locations in the body, causing kidney stones or calcification of the arteries. Vitamin D can exert these toxic effects if consumed continuously at only relatively small amounts in excess of the RDA.

Relatively few foods contain significant amounts of vitamin D. In addition to conversion *in situ* by the body, the principal sources of vitamin D are foods derived from animal sources, including egg yolk, fatty fish and liver. Unfortified cows' milk is not an important source of vitamin D.

The major form of vitamin D in both cows' and human milk is 25(OH)D$_3$. This compound is reported to be responsible for most of the vitamin D in the blood serum of exclusively breast-fed infants. Whole cows' milk contains only ~0.1–1.5 µg vitamin D per L. Therefore, milk is often fortified (at the level of approximately 10 µg L^{-1}) with vitamin D. Fortified milk, dairy products or margarine are important dietary sources of vitamin D. The concentration of vitamin D in unfortified dairy products is usually quite low.

As with other fat-soluble vitamins, the concentration of vitamin D in dairy products is increased *pro rata* by concentration of the fat (e.g., in the production of butter or cheese). Vitamin D is relatively stable during storage and to most dairy processing operations. Studies on the degradation of vitamin D in fortified milk have shown that the vitamin may be degraded by exposure to light. However, the conditions necessary to cause significant losses are unlikely to be encountered in practice. Extended exposure to light and oxygen are needed to cause significant losses of vitamin D.

6.2.3 Tocopherols and Related Compounds (Vitamin E)

Eight compounds have vitamin E activity, four of which are derivatives of tocopherol (Fig. 6.11) and four of tocotrienol (Fig. 6.12); all are derivatives of 6-chromanol. Tocotrienols differ from tocopherols in having three carbon-carbon double bonds in their hydrocarbon side chain. α-, β-, γ- or δ-tocopherols and tocotrienols differ with respect to the number and position of methyl groups on the chromanol ring. The biological activity of the different forms of the tocopherols and tocotrienols varies to their structure. Enantiomers of vitamin E also occur which differ in biological

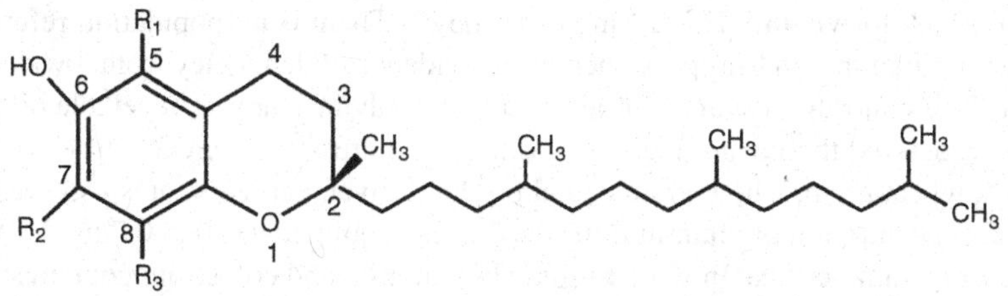

Fig. 6.11 Tocopherols

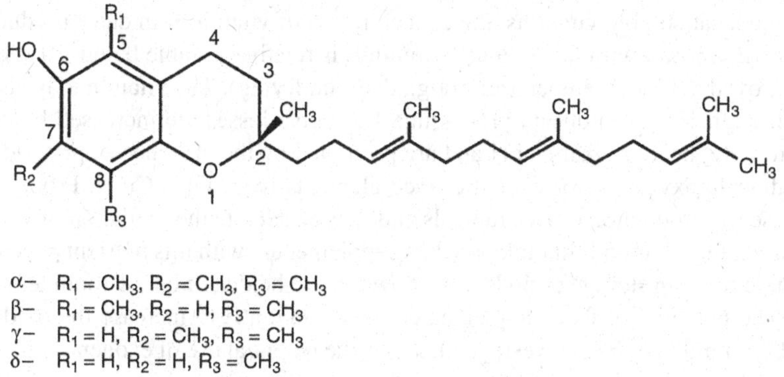

α– $R_1 = CH_3$, $R_2 = CH_3$, $R_3 = CH_3$
β– $R_1 = CH_3$, $R_2 = H$, $R_3 = CH_3$
γ– $R_1 = H$, $R_2 = CH_3$, $R_3 = CH_3$
δ– $R_1 = H$, $R_2 = H$, $R_3 = CH_3$

Fig. 6.12 Tocotrienols: α—$R_1=CH_3$, $R_2=CH_3$, $R_3=CH_3$; β—$R_1=CH_3$, $R_2=H$, $R_3=CH_3$; γ—$R_1=H$, $R_2=CH_3$, $R_3=CH_3$; δ—$R_1=H$, $R_2=H$, $R_3=CH_3$

activity. Vitamin E activity can be expressed as tocopherol equivalents (TE), where 1 TE is equivalent to the vitamin E activity of 1 mg α-tocopherol. The biological activity of β- and γ-tocopherols and δ-tocotrienol is 50, 10 and 33 % of the activity of α-tocopherol, respectively.

Vitamin E is a very effective antioxidant. It can easily donate a hydrogen from the phenolic -OH group on the chromanol ring to free radicals. The resulting vitamin E radical is quite unreactive as it is stabilized by delocalization of its unpaired electron into the aromatic ring. Vitamin E thus protects the lipids (particularly polyunsaturated fatty acids) and membranes in the body against damage caused by free radicals. The role of vitamin E is of particular importance in the lungs where exposure of cells to oxygen is greatest. Vitamin E also exerts a protective effect on red and white blood cells. It has been suggested that the body has a system to regenerate active vitamin E (perhaps involving vitamin C) once it has acted as an antioxidant.

Vitamin E deficiency is normally associated with diseases of fat malabsorption and is rare in humans. Deficiency is characterized by erythrocyte haemolysis and prolonged deficiency can cause neuromuscular dysfunction. Hypervitaminosis E is not common, despite an increased intake of vitamin E supplements. Extremely high doses of the vitamin may interfere with the blood clotting process.

The RDA for vitamin E is 15 mg α-TE day^{-1}. There is no population reference intake for vitamin E in Europe as there is no evidence of deficiency from low dietary intake. The major food sources of vitamin E are polyunsaturated vegetable oils and products derived therefrom (e.g., margarine, salad dressings), green and leafy vegetables, wheat germ, whole grain cereal products, liver, egg yolk, nuts and seeds.

The concentration of vitamin E in cows' milk is quite low (~0.2–0.7 mg L^{-1}) and is higher in summer than in winter milk. Human milk and colostrum contain somewhat higher concentrations (~3–8 to ~14–22 mg L^{-1}, respectively). Most dairy products contain a low level of vitamin E (Table 6.1) and thus are not important sources of this nutrient. However, levels are higher in dairy products supplemented with vegetable fat (e.g., some ice creams, imitation creams, fat-filled dried skim milk). Like other fat-soluble vitamins, the concentration of vitamin E in dairy products is increased *pro rata* with fat content. Vitamin E is relatively stable below 100 °C but is destroyed at higher temperatures (e.g., deep-fat frying). The vitamin may also be lost through oxidation during processing. Oxidative losses are increased by exposure to light, heat or alkaline pH and are promoted by the presence of prooxidants, including lipoxygenase or catalytic trace elements (e.g., Fe^{3+}, Cu^{2+}). Prooxidants increase the production of free radicals and thus accelerate the oxidation of vitamin E. Exogenous vitamin E in milk powders supplemented with this nutrient appears to be stable for long storage periods if the powders are held at or below room temperature. The potential of feed supplemented with vitamin E to increase the oxidative stability of milk has been investigated, as has the potential use of exogenous tocopherols added directly to the milk fat.

6.2.4 Phylloquinone and Related Compounds (Vitamin K)

The structure of vitamin K is characterized by 2-methyl-1,4-naphthoquinone rings. It exists naturally in two forms: phylloquinone (vitamin K$_1$, the major dietary source of vitamin K in western diets; Fig. 6.13) occurs only in plants while menaquinones (vitamin K$_2$, Fig. 6.14) is a family of compounds with a side chain consisting of 1–14 isoprene units. Menaquinones are synthesized only by bacteria (which inhabit the human gastrointestinal tract and thus provide some of the vitamin K required by the body). Menadione (vitamin K$_3$; Fig. 6.15) is a synthetic compound with vitamin K activity. Unlike K$_1$ and K$_2$, menadione is water-soluble and is not active until it is alkylated *in vivo*.

Fig. 6.13 Phylloquinone, Vitamin K$_1$

Fig. 6.14 Menaquinone, vitamin K_2

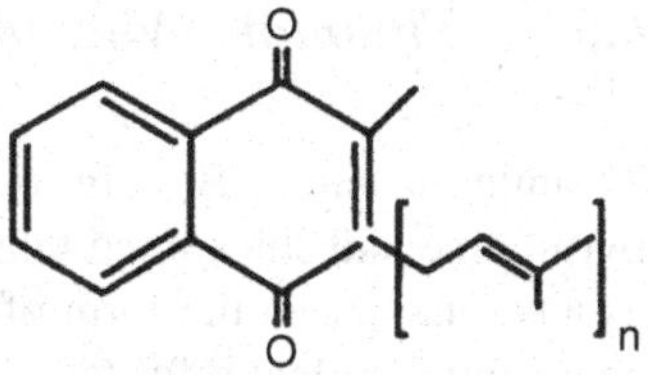

Fig. 6.15 Menadione, vitamin K_3

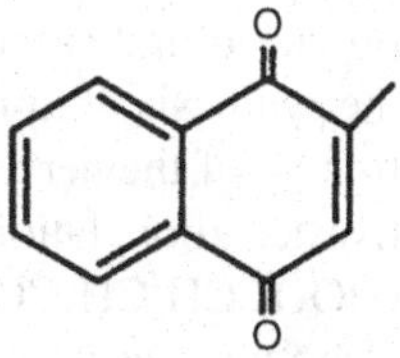

The physiological role of vitamin K is in blood clotting and is essential for the synthesis of at least four of the proteins (including prothrombin) involved in this process. Vitamin K also plays a role in the synthesis of proteins (osteocalcin and matrix Gla protein) in bone. Vitamin K deficiency is rare but can result from impaired absorption of fat. Vitamin K level in the body is also reduced if the intestinal flora is killed (e.g., by antibiotics). Vitamin K toxicity is also rare but can be caused by excessive intake of vitamin K supplements. Symptoms include erythrocyte haemolysis, jaundice, brain damage and reduced effectiveness of anticoagulants.

The RDA for vitamin K for people aged 19–24 years is 70 and 60 μg day^{-1} for men and women, respectively. Corresponding values for adults aged 25 years and over are 80 and 65 μg day^{-1}. The Department of Health (1991) suggested that a vitamin K intake of 1 μg kg^{-1} bodyweight per day is safe and adequate. The principal food sources of vitamin K are liver, green leafy vegetables and milk.

Whole cows' milk contains 3.5–18 μg vitamin K per L while human milk contains ~0.25 μg L^{-1}. Human colostrum contains a higher concentration of vitamin K than mature milk which is necessary since bacteria capable of synthesizing vitamin K take time to become established in the intestine of the neonate. Irradiation under anerobic and apolar conditions can result in *cis/trans* isomerization, resulting in loss of activity since only the *trans* isomer has vitamin K activity. However, unit operations in dairy processing are unlikely to affect the stability of this nutrient.

6.3 B-Group Vitamins

The B-group is a heterogeneous collection of water-soluble vitamins, most of which function as co-enzymes or are precursors of co-enzymes. The B-group vitamins are thiamine, riboflavin, niacin, biotin, pantothenic acid, pyridoxine (and related substances, vitamin B_6), folate and cobalamin (and its derivatives, vitamin B_{12}).

6.3.1   Thiamine (Vitamin B₁)

Thiamine (vitamin B_1; Fig. 6.16) consists of two heterocyclic rings (substituted pyrimidine and substituted thiazole), linked by a methylene bridge. Thiamine acts as a co-enzyme in the form of thiamine pyrophosphate (TPP; Fig. 6.17) which is an essential co-factor for many enzyme-catalyzed reactions in carbohydrate metabolism. TPP-dependent pyruvate dehydrogenase catalyses the conversion of pyruvate ($CH_3COCOOH$) to acetyl-CoA (CH_3CO-CoA) in mitochondria. The acetyl-CoA produced in this reaction enters the Kreb's cycle and also serves as a substrate for the synthesis of lipids and acetylcholine (and thus is important for the normal functioning of the nervous system). TPP is necessary in the Kreb's cycle for the oxidative decarboxylation of α-ketoglutarate ($HOOCCH_2CH_2COCOOH$) to succinyl-CoA ($HOOCCH_2CH_2CO$-CoA) by the α-ketoglutarate dehydrogenase complex. TPP also functions in reactions involving the decarboxylation of ketoacids derived from branched-chain amino acids and in transketolase reactions in the hexose monophosphate pathway for glucose metabolism.

The characteristic disease caused by prolonged thiamine deficiency is beriberi, the symptoms of which include oedema, enlarged heart, abnormal heart rhythms, heart failure, wasting, weakness, muscular problems, mental confusion and paralysis.

Thiamine is widespread in many nutritious foods but pigmeat, liver, whole-grain cereals, legumes and nuts are particularly rich sources. The RDA for thiamine is approximately equivalent to 1.2 mg and 1.0 mg day^{-1} for men and women, respectively.

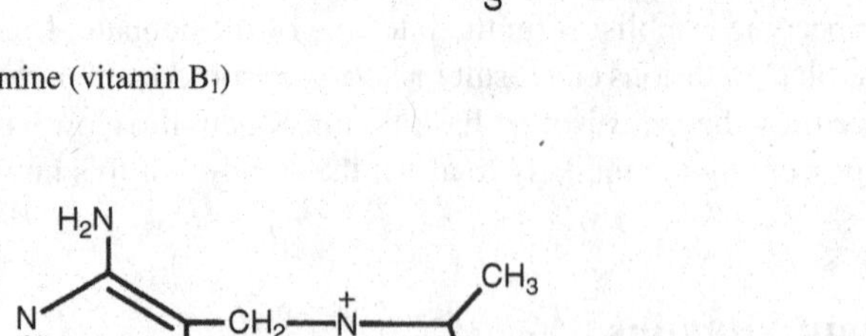

Fig. 6.16 Thiamine (vitamin B_1)

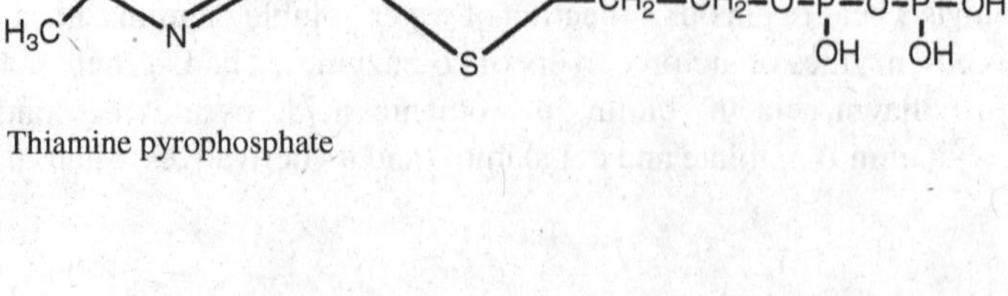

Fig. 6.17 Thiamine pyrophosphate

Milk contains, on average, 37 µg thiamine per 100 g. Most (50–70 %) of the thiamine in bovine milk is in the free form; lesser amounts are phosphorylated (18–45 %) or protein-bound (7–17 %). The concentration in mature human milk is somewhat lower (~15 µg per 100 g). Human colostrum contains only trace amounts of thiamine which increase during lactation. Most of the thiamine in bovine milk is produced by microorganisms in the rumen and, therefore, feed, breed of the cow or season have relatively little effect on its concentration in milk.

Thiamine levels in milk products (Table 6.2) are generally 20–50 µg per 100 g. As a result of the growth of *Penicillium* mould, the rind of Brie and Camembert cheese is relatively rich in thiamine.

Table 6.2 Concentrations of vitamins B1, 2, 3, 5 in milk, dairy products and cheese (in alphabetical order) (from Nohr and Biesalski 2009)

Food	B-Vitamin			
	Thiamine (µg per 100 g)	Riboflavin (µg per 100 g)	Niacin (µg per 100 g)	Pantothenic acid (µg per 100 g)
Blue cheese (50 % fat in dry matter)		500	870	2,000
Brie (50 % in dry matter)			1,100	690
Buttermilk	34	160	100	300
Camembert (45 % fat in dry matter)	45	600	1,100	800
Condensed milk (min. 10 % fat)	88	480	260	840
Consumer milk (3.5 % fat)	37	180	90	350
Cottage cheese	29			
Cream (min. 30 % fat)	25	150	80	300
Cream cheese (min 60 % fat in dry matter)	45	230	110	440
Dried whole milk	270	1,400	700	2,700
Gouda	30			
Limburger (40 % fat in dry matter)		350	1,200	1,200
Parmesan cheese	20	620		530
Quark/fresh cheese (from skim milk)	43	300		740
Skim milk	38	170	95	280
Sterilized milk	24	140	90	350
Sweet whey	37	150	190	340
UHT milk	33	180	90	350
Yoghurt (min. 3.5 % fat)	37	180	90	350
Milk from				
Buffalo	50	100	80	370
Cow	37	180	90	350
Donkey	41	64	74	
Goat	49	150	320	310
Horse	30		140	300
Human	15	38	170	270
Sheep	48	230	450	350

Thiamine is relatively unstable and is easily cleaved by a nucleophilic displacement reaction at its methylene carbon. The hydroxide ion (OH⁻) is a common nucelophile which can cause this reaction in foods. Thiamine is thus more stable under slightly acid conditions. Dairy processing can cause losses of thiamine: minimum pasteurisation results in 3–4 % loss, boiling milk to 4–8 % loss, spray drying 10 % loss, sterilization 20–45 % loss and evaporation 20–60 % loss. The light sensitivity of thiamine is less than that of other light-sensitive vitamins.

6.3.2   Riboflavin (Vitamin B₂)

Riboflavin (vitamin B_2; Fig. 6.18) consists of an isoalloxazine ring linked to an alcohol derived from ribose. The ribose side chain of riboflavin can be modified by the formation of a phosphoester (forming flavin mononucleotide, FMN, Fig. 6.19). FMN can be joined to adenine monophosphate to form flavin adenine dinucelotide (FAD, Fig. 6.20). FMN and FAD act as co-enzymes by accepting or donating two hydrogen atoms and thus are involved in redox reactions. Flavoprotein enzymes are involved in many metabolic pathways. Riboflavin is a yellow-green fluorescent compound and in addition to its role as a vitamin, it is responsible for the colour of milk serum (see Chap. 8).

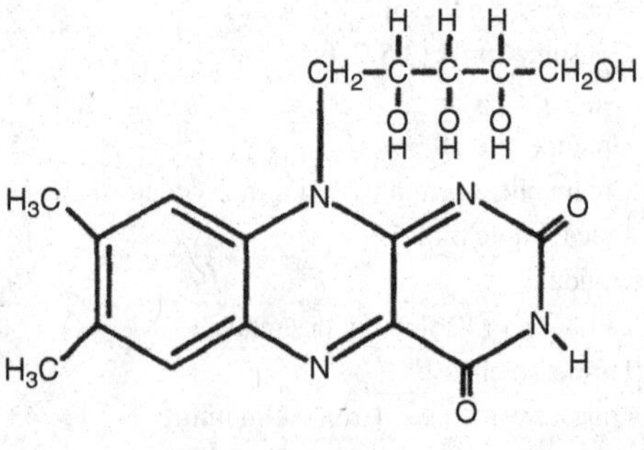

Fig. 6.18 Riboflavin

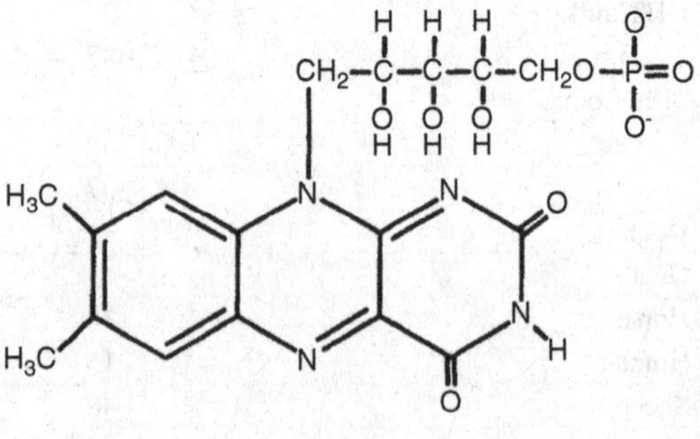

Fig. 6.19 Flavin mononucleotide

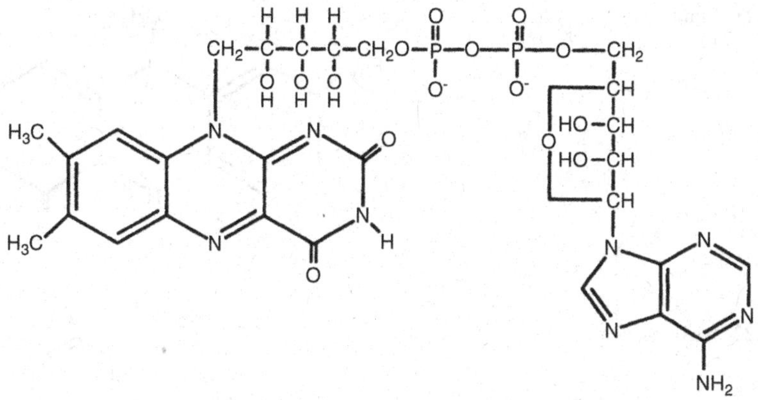

Fig. 6.20 Flavin adenine dinucleotide

Symptoms of riboflavin deficiency include cheilosis (cracks and redness at the corners of the mouth), glossitis (painful, smooth tongue), inflamed eyelids, sensitivity of the eyes to light, reddening of the cornea, skin rash and brain dysfunction. The recommended daily uptake for riboflavin is about 1.4 and 1.2 mg day^{-1} for men and women, respectively. Important dietary sources of riboflavin include milk and dairy products, meat and leafy green vegetables. Cereals are poor sources of riboflavin, unless fortified. There is no evidence for riboflavin toxicity.

Milk is a good source of riboflavin; whole milk contains ~0.18 mg per 100 g. Most (65–95 %) of the riboflavin in milk is present in the free form; the remainder is present as FMN or FAD. Milk also contains small amounts (~11 % of total flavins) of a related compound, 10-(2′-hydroxyethyl) flavin, which acts as an antivitamin. The concentration of this compound must be considered when evaluating the riboflavin activity in milk. The concentration of riboflavin in milk is influenced by the breed of cow (milk from Jersey and Gurnsey cows contains more riboflavin than Holstein milk). Summer milk generally contains slightly higher levels of riboflavin than winter milk. Interspecies variations in concentration are also apparent. Raw sheep's milk contains ~0.23 mg per 100 g while the mean value for goats' milk (0.15 mg per 100 g) is lower; human milk contains 0.015 mg per 100 g. Dairy products also contain significant amounts of riboflavin (Table 6.2). Cheese contains 0.3–0.5 mg per 100 g and yogurt about 0.3 mg per 100 g. The whey protein fraction of milk contains a riboflavin-binding protein (RfBP) which probably originates from blood plasma, although its function in milk is unclear (see Chap. 11).

Riboflavin is stable in the presence of oxygen, heat and at acid pH. However, it is labile to thermal decomposition under alkaline conditions. The concentration of riboflavin in milk is unaffected by pasteurization and little loss is reported for UHT-treated milks. The most important parameter affecting the stability of riboflavin in dairy products is exposure to light (particularly wavelengths in the range 415–455 nm). At alkaline pH, irradiation cleaves the ribitol portion of the molecule, leaving a strong oxidizing agent, lumiflavin (Fig. 6.21). Irradiation under acidic conditions results in the formation of lumiflavin and a blue fluorescent compound,

Fig. 6.21 Lumiflavin

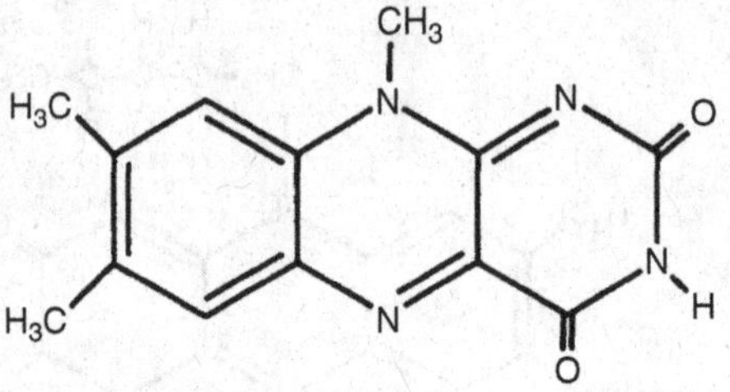

Fig. 6.22 Nicotinic acid

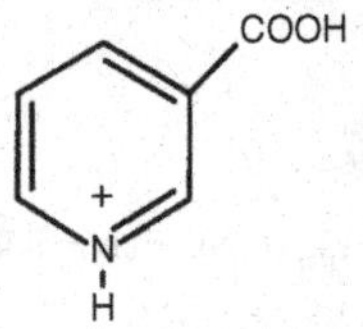

lumichrome. Lumiflavin is capable of oxidizing other vitamins, particularly ascorbate (see Sect. 6.4, and Chap. 11). Loss of riboflavin in milk packaged in materials that do not protect against light can be caused by either sunlight or by lights in retail outlets. Packaging in paperboard containers is the most efficient method for minimising this loss, although glass containing a suitable pigment has also been used. Riboflavin is more stable in high-fat than in low-fat or skim milk, presumably as a result of the presence of antioxidants (e.g., vitamin E) in the milk fat which protect riboflavin against photooxidation.

6.3.3   Niacin

Niacin is a generic term which refers to two related chemical compounds, nicotinic acid (Fig. 6.22) and its amide, nicotinamide (Fig. 6.23); both are derivatives of pyridine. Nicotinic acid is synthesized chemically and can be easily converted to the amide in which form it is found in the body. Niacin is obtained from food or can be synthesized from tryptophan (60 mg of dietary tryptophan have the same metabolic effect as 1 mg niacin). Niacin forms part of two important co-enzymes, nicotinamide adenine dinucleotide (NAD) and nicotinamide adenine dinucleotide phosphate (NADP), which are co-factors for many enzymes that participate in various metabolic pathways and function in electron transport.

The classical niacin deficiency disease is pellagra which is characterized by symptoms including diarrhoea, dermatitis, dementia and eventually death. High-protein diets are rarely deficient in niacin since, in addition to the pre-formed vitamin, such diets supply sufficient tryptophan to meet dietary requirements. Large doses of niacin can cause the dilation of capillaries, resulting in a painful tingling sensation.

The RDA for niacin is approximately equivalent to 15 and 13 mg niacin equivalents (NE) per day for men and women, respectively. The richest dietary sources of niacin are meat, poultry, fish and whole-grain cereals.

Fig. 6.23 Nicotinamide

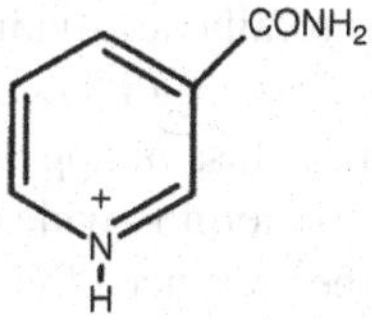

Fig. 6.24 Biotin

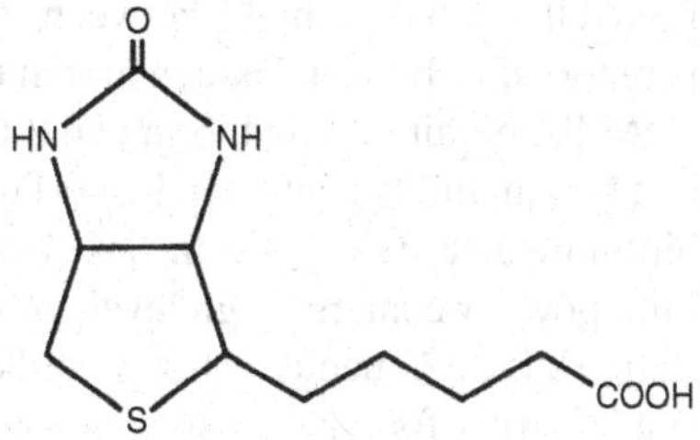

Milk contains ~0.09 mg niacin per 100 g and thus is not a rich source of the pre-formed vitamin. Tryptophan contributes ~0.7 mg NE per 100 g milk. In milk, niacin exists primarily as nicotinamide and its concentration appears not to be affected greatly by breed of cow, feed, season or stage of lactation. Pasteurized goats' (~0.3 mg niacin and 0.7 mg NE from tryptophan per 100 g) and raw sheep's (~0.45 mg niacin and 1.3 mg NE from tryptophan per 100 g) milk are somewhat richer than cows' milk. Niacin levels in human milk are 0.17 mg niacin and 0.5 mg NE from tryptophan per 100 g. The concentration of niacin in most dairy products is low (Table 6.2) but is compensated somewhat by tryptophan released on hydroly-sis of the proteins.

Niacin is relatively stable to most food processing operations. It is stable to expo-sure to air and resistant to autoclaving (and is therefore stable to pasteurization and UHT treatments). The amide linkage in nicotinamide can be hydrolyzed to the free carboxylic acid (nicotinic acid) by treatment with acid but the vitamin activity is unaffected. Like other water-soluble vitamins, niacin can be lost by leaching.

6.3.4   Biotin

Biotin (vitamin B_7; Fig. 6.24) consists of an imidazole ring fused to a tetrahydro-thiophene ring with a valeric acid side chain. Biotin acts as a co-enzyme for carboxylases involved in the synthesis and catabolism of fatty acids and for branched-chain amino acids and gluconeogenesis.

Biotin deficiency is rare but under laboratory conditions it can be induced by feeding subjects with large amounts of raw egg white which contains the protein, avidin, which has a binding site for the imidazole moiety of biotin, thus making it unavailable. Avidin is denatured by heat and, therefore, biotin binding occurs only in raw egg albumen. Symptoms of biotin deficiency include scaly dermatitis, hair loss, loss of appetite, nausea, hallucinations and depression.

Biotin is widespread in foods, although its availability is affected somewhat by the presence of binding proteins. Biotin is required in only small amounts. Although US RDA values have not been established, the estimated safe and adequate intake of biotin is 30–60 µg day⁻¹ for adults. The UK Department of Health (1991) suggested that a biotin intake between 10 and 200 µg day⁻¹ is safe and adequate. Biotin is reported to be non-toxic in amounts up to at least 10 mg day⁻¹.

Milk contains ~4 µg biotin per 100 g, apparently in the free form. Caprine, ovine and human milks contain 4, 9 and 1 µg per 100 g, respectively. The concentration of biotin in cheese ranges from 1.4 (Gouda) to 7.6 (Camembert) µg per 100 g. Whole milk powder contains high levels of biotin (~24 µg per 100 g) owing to the concentration of the aqueous phase of milk during its manufacture (Table 6.3). Biotin is stable during food processing and storage and is unaffected by pasteurization.

6.3.5 Pantothenic Acid

Pantothenic acid (Fig. 6.25) is a dimethyl derivative of butyric acid linked to β-alanine. Pantothenate is part of the structure of co-enzyme A (CoA), and as such is vital as a co-factor for numerous enzyme-catalyzed reactions in lipid and carbohydrate metabolism.

Pantothenate deficiency is rare, occurring only in cases of severe malnutrition; characteristic symptoms include vomiting, intestinal distress, insomnia, fatigue and occasional diarrhoea. Pantothenate is widespread in foods; meat, fish, poultry, whole-grain cereals and legumes are particularly good sources. Although no RDA or RNI values have been established for panthothenate, safe and adequate intake of this vitamin for adults is estimated to be 3–7 mg day⁻¹. Pantothenate is non-toxic at doses up to 10 g day⁻¹.

Milk contains, on average, 0.35 mg panthothenate per 100 g. Pantothenate exists partly free and partly bound in milk and its concentration is influenced by breed, feed and season. Ovine and caprine milk contains 0.35 and 0.31 mg per 100 g, respectively. The value for pantothenate in human milk is approximately 0.27 mg per 100 g. Mean concentration of pantothenate in cheese varies from ~0.3 (Cream cheese) to ~2 (Blue cheese) mg per 100 g (Table 6.2). Pantothenate is stable at neutral pH but is easily hydrolyzed by acid or alkali at high temperatures. Pantothenate is reported to be stable to pasteurization.

6.3.6 Pyridoxine and Related Compounds (Vitamin B₆)

Vitamin B₆ occurs naturally in three related forms: pyridoxine (Fig. 6.26; the alcohol form), pyridoxal (Fig. 6.27; aldehyde) and pyridoxamine (Fig. 6.28; amine). All are structurally related to pyridine. The active co-enzyme form of this vitamin is

Table 6.3 Concentrations of vitamins B6, 7, 9, 12 in milk, dairy products and cheese (in alphabetical order) (from Nohr and Biesalski 2009)

Food	B-Vitamin			
	Pyridoxine (μg per 100 g)	Biotin (μg per 100 g)	Folic acid (μg per 100 g)	Cobalamin (μg per 100 g)
Blue cheese (50 % fat in dry matter)			40	
Brie (50 % in dry matter)	230	6	65	2
Buttermilk	40	2	5	<1
Camembert (45 % fat in dry matter)	250	5	44	3
Condensed milk (min. 10 % fat)	77	8	6	<1
Consumer milk (3.5 % fat)	36	4	6	<1
Cream (min. 30 % fat)	36	3	4	<1
Cream cheese (min 60 % fat in dry matter)	60	4		<1
Dried whole milk	200	24	40	1
Emmental cheese	111			3
Limburger (40 % fat in dry matter)		9	60	
Quark/fresh cheese (from skim milk)		7	16	
Skim milk	50	2	5	<1
Sterilized milk	23	4	3	<1
Sweet whey	42	1		<1
UHT milk	41	4	5	<1
Yoghurt (min. 3.5 % fat)	46	4	13	<1
Milk from				
Buffalo	25	11		<1
Cow	36	4	7	<1
Donkey				<1
Goat	27	4	1	<1
Horse	30			<1
Human	14	1	8	<1
Sheep		9		<1

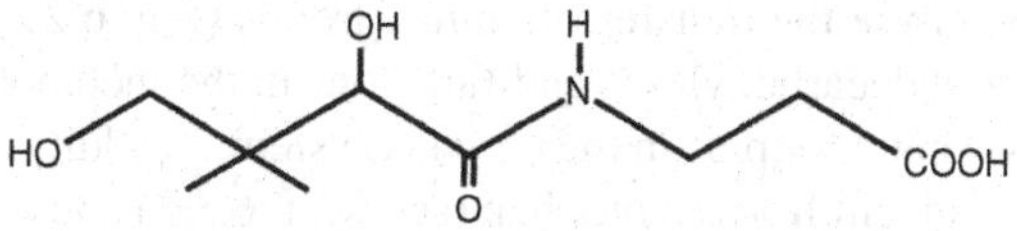

Fig. 6.25 Pantothenic acid

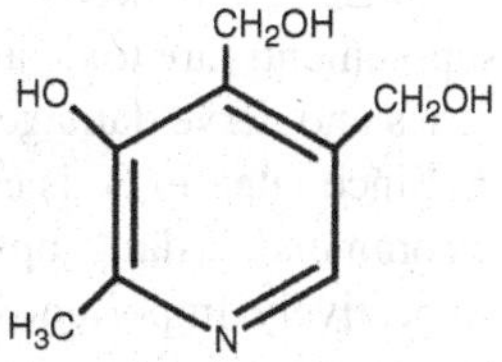

Fig. 6.26 Pyridoxine

Fig. 6.27 Pyridoxal

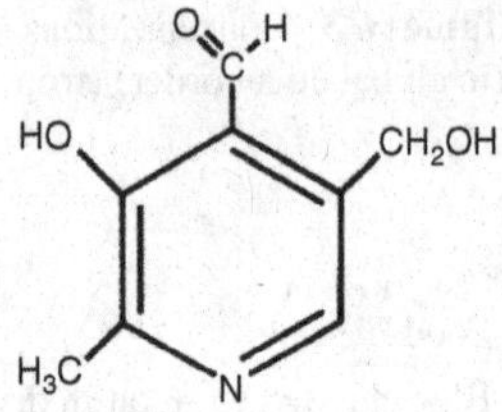

Fig. 6.28 Pyridoxamine

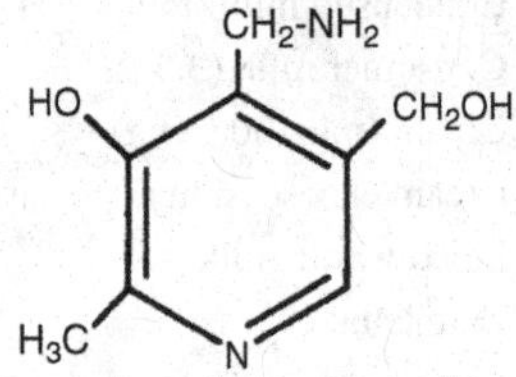

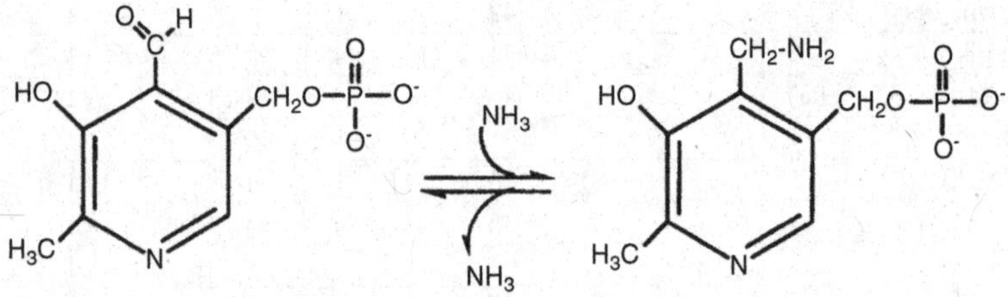

Fig. 6.29 Pyridoxal phosphate and Pyridoxamine phosphate

pyridoxal phosphate (PLP; Fig. 6.29), which is a co-factor for transaminases which catalyse the transfer of amino groups (Fig. 6.29). PLP is also important for amino acid decarboxylases and functions in the metabolism of glycogen and the synthesis of sphingolipids in the nervous system. In addition, PLP is involved in the formation of niacin from tryptophan (see Sect. 6.3.3) and in the initial synthesis of heme.

Deficiency of vitamin B_6 is characterized by weakness, irritability and insomnia and later by convulsions and impairment of growth, motor functions and immune response. High doses of vitamin B_6, often associated with excessive intake of supplements, are toxic and can cause bloating, depression, fatigue, irritability, headaches and nerve damage.

Since vitamin B_6 is essential for amino acid (and hence protein) metabolism, its recommended daily uptake is about 1.5 and 1.2 mg day^{-1} for men and women, respectively. Important sources of B_6 include green, leafy vegetables, meat, fish and poultry, shellfish, legumes, fruits and whole grains.

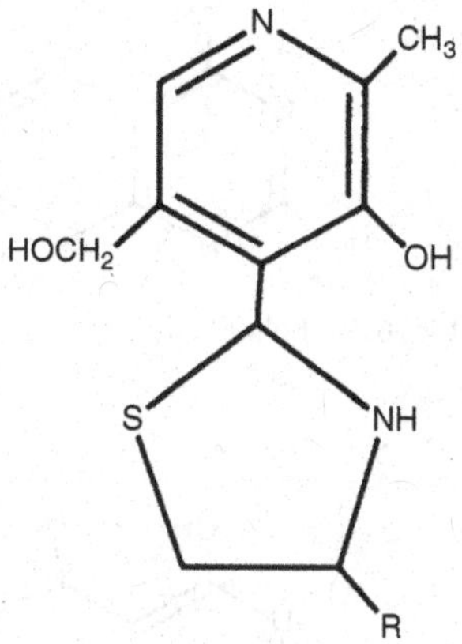

Fig. 6.30 Thiazolidine
derivative of pyridoxal

Whole milk contains, on average, 36 μg B_6 per 100 g, mainly in the form of pyridoxal (80 %); the balance is mainly pyridoxamine (20 %), with trace amounts of pyridoxamine phosphate. Concentrations in raw ovine and pasteurized caprine milks are similar to those in cows' milk. The concentration of B_6 varies during lactation; colostrum contains lower levels than mature milk. Seasonal variation in the concentration of vitamin B_6 has been reported in Finnish milk; levels were higher (14 %) when cattle were fed outdoors than when they were fed indoors. Mature human milk contains about 14 μg B_6 per 100 g.

In general, dairy products are not major sources of B_6 in the diet. Concentrations in cheeses and related products vary from ~40 (Fromage frais, Cream cheese) to ~250 (Camembert) μg per 100 g (Table 6.3). Whole milk yogurt contains ~46 μg per 100 g and the concentration in whole milk powder is ~200 μg per 100 g.

All forms of B_6 are sensitive to UV light and may be decomposed to biologically inactive compounds. Vitamin B_6 may also be decomposed by heat. The aldehyde group of pyridoxal and the amine group of pyridoxamine shows some reactivity under conditions that may be encountered during milk processing. An outbreak of B_6 deficiency in 1952 was attributed to the consumption of heated milk products. Pyridoxal and/or its phosphate can react directly with the sulphydryl group of cysteine residues in proteins, forming an inactive thiazolidine derivative (Fig. 6.30). Losses during pasteurization and UHT treatments are relatively small, although losses of up to 50 % can occur in UHT milk during its shelf-life. (Losses of 0–8 %, <10 %, 20–50 % and 35–50 % have been reported for pasteurisation, UHT treatment, sterilization and evaporation, respectively.)

6.3.7 Folate

Folate (vitamin B_9) consists of a substituted pteridine ring linked through a methylene bridge to *p*-aminobenzoic acid and glutamic acid (Fig. 6.31). Up to seven glutamic acid residues can be attached by γ-carboxyl linkages, producing polyglutamyl folate

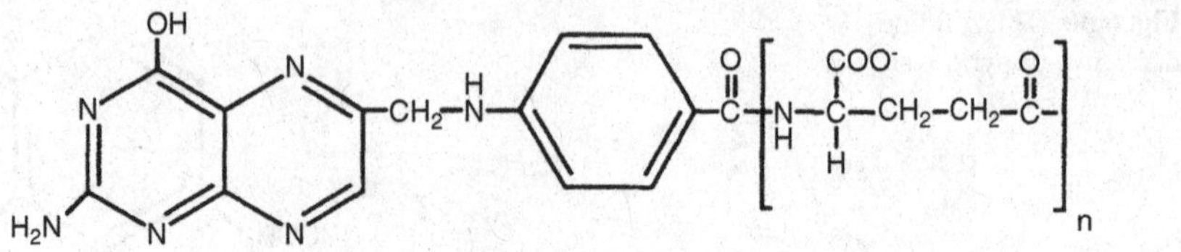

Fig. 6.31 Folate

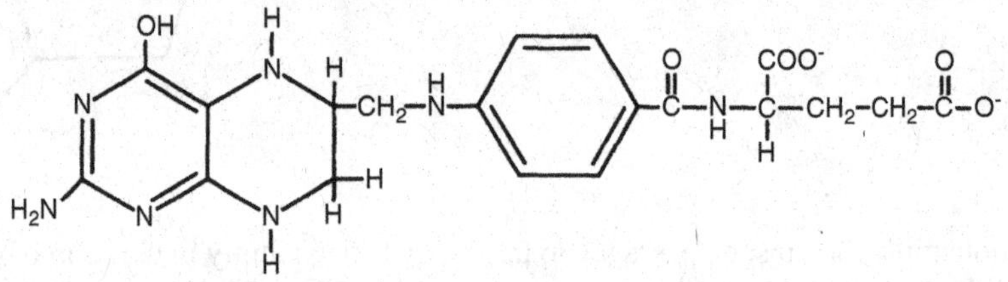

Fig. 6.32 Tetrahydrofolate

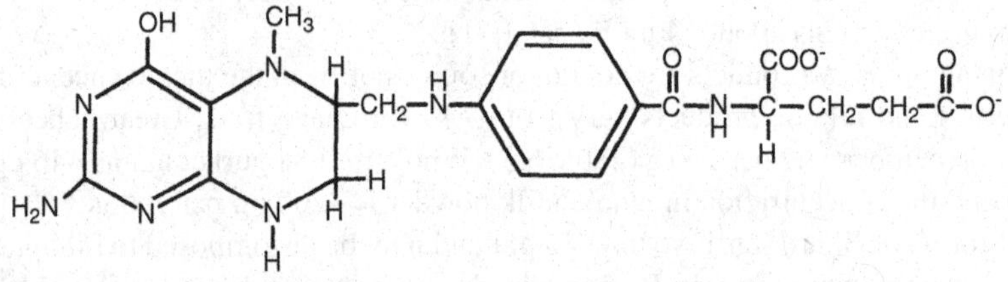

Fig. 6.33 5-Methyl tetrahydrofolate

(Fig. 6.31) which is the major dietary and intracellular form of the vitamin. Reductions and substitutions on the pteridine ring result in tetrahydrofolate (H_4 folate; Fig. 6.32) and 5-methyl tetrahydrofolate (5-methyl-H_4 folate; Fig. 6.33). Folate is a co-factor in the enzyme-catalyzed transfer of carbon atoms in many metabolic pathways, including the biosynthesis of purines and pyrimidines (essential for DNA and RNA) and interconversions of amino acids. Folate interacts with vitamin B_{12} (see Sect. 6.3.8) in the enzyme-catalyzed synthesis of methionine and in the activation of 5-methyl-H_4 folate to H_4 folate. H_4 Folate is involved in a complex and inter-linked series of metabolic reactions (see Nohr and Biesalski 2009).

Folate deficiency impairs cell division and protein biosynthesis; symptoms include megaloblastic anaemia, digestive system problems (heartburn, diarrhoea and constipation), suppression of the immune system, glossitis and problems with the nervous system (depression, fainting, fatigue, mental confusion). The recommended daily uptake for folate is 39 and 51 µg MJ^{-1} for men and women, respectively (equivalent to ~400 µg day^{-1}). Higher intakes of folate have been suggested for women of child-bearing age to prevent the development of neural tube defects in the developing foetus.

Rich dietary sources of folate include leafy green vegetables, legumes, seeds and liver. Milk contains ~7 µg folate per 100 g. The dominant form of folate in milk is 5-methyl-H_4 folate. Folate in milk is mainly bound to the folate-binding protein and ~40 % occurs as conjugated polyglutamate forms. The folate binding proteins of milks of various species have been characterized (see Nohr and Biesalski 2009). The folate level in human milk is approximately 8 µg per 100 g. Folate levels in some dairy products are shown in Table 6.3. Cream contains ~4 µg per 100 g while the value for cheese varies widely up to 65 µg per 100 g; the high concentration found in mould- and smear-ripened varieties presumably reflects biosynthesis of folate. The concentration of folate in yogurt is ~13 µg per 100 g. The higher level of folate in yogurt than in milk is due to bacterial biosynthesis, particularly by *Streptococcus thermophilus* and perhaps to some added ingredients.

Folate is a relatively unstable nutrient; processing and storage conditions that promote oxidation are of particular concern since some of the forms of folate found in foods are easily oxidized. The reduced forms of folate (dihydro- and tetrahydrofolate) are oxidized to *p*-aminobenzoylglutamic acid and pterin-6-carboxylic acid, with a concomitant loss in vitamin activity. 5-Methyl-H_4 folate can also be oxidized. Antioxidants (particularly ascorbic acid in the context of milk) can protect folate against destruction. The rate of the oxidative degradation of folate in foods depends on the derivative present and the food itself, particularly its pH, buffering capacity and concentration of catalytic trace elements and antioxidants.

Folate is sensitive to light and may be subject to photodecomposition. Heat-treatment influences folate level in milk. Pasteurization and the storage of pasteurized milks have relatively little effect on the stability of folate but UHT treatments can cause substantial losses. The concentration of oxygen in UHT milk (from the headspace above the milk or by diffusion through the packaging material) has an important influence on the stability of folate during the storage of UHT milk, as have the concentrations of ascorbate in the milk and of O_2 in the milk prior to heat treatment. Folate and ascorbic acid (see Sect. 6.4) are the least stable vitamins in powdered milks.

The heat stability of folate-binding proteins in milk should also be considered in the context of folate in dairy foods. Breast-fed babies require less dietary folate (55 µg folate day^{-1}) to maintain their folate status than bottle-fed infants (78 µg day^{-1}). The difference has been attributed to the presence of active folate-binding proteins in breast milk; folate-binding proteins originally present in milk formulae are heat-denatured during processing. However, a study involving feeding radiolabelled folate to rats together with dried milks, prepared using different heat treatments, showed no differences in folate bioavailability (see Öste et al. 1997).

6.3.8   Cobalamin and Its Derivatives (Vitamin B_{12})

Vitamin B_{12} consists of several cobalt-containing corriods; a corrin ring is a porphyrin-like structure with four reduced pyrrole rings, with an atom of Co chelated at its centre, linked to a nucleotide base, ribose and phosphoric acid (Fig. 6.34). A number of different groups can be attached to the free ligand site on the cobalt. Cyanocobalamin has cyanide at this position and is the commercial and therapeutic form of the vitamin, although the principal dietary forms of B_{12} are 5′-deoxyadenosylcobalamin (with 5′-deoxyadenosine at the R position), methylcobalamin (-CH$_3$) and hydroxocobalamin (-OH). Vitamin B_{12} acts as a co-factor for methionine synthetase and methylmalonyl-CoA mutase. The former enzyme catalyzes the transfer of the methyl group of 5-methyl-H$_4$ folate to cobalamin and thence to homocysteine,

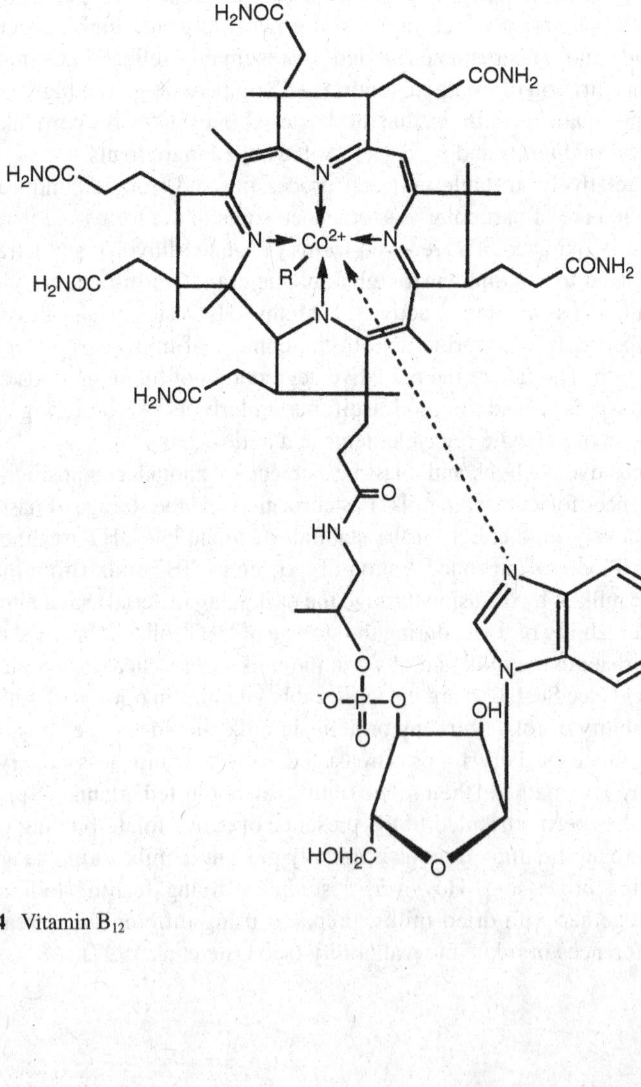

Fig. 6.34 Vitamin B_{12}

forming methionine. Methylmalonyl CoA-mutase catalyzes the conversion of methylmalonyl-CoA to succinyl-CoA in the mitochondrion.

Vitamin B_{12} deficiency normally results from inadequate absorption rather than inadequate dietary intake. Pernicious anaemia is caused by vitamin B_{12} deficiency; symptoms include anaemia, glossitis, fatigue and degeneration of the peripheral nervous system and hypersensitivity of the skin. The recommended daily uptake of B_{12} for adult (21–51 years) is $3 \mu g$ day^{-1}, respectively. Unlike other vitamins, B_{12} is obtained exclusively from animal food sources, such as meat, fish, poultry, eggs, shellfish, milk, cheese and eggs. Vitamin B_{12} in these foods is protein-bound and released by the action of HCl and pepsin in the stomach.

Bovine milk contains, on average, $<1 \mu g$ B_{12} per 100 g (Table 6.3). The predominant form is hydroxycobalamin and >95 % of this nutrient is protein-bound. The concentration of B_{12} in milk is influenced by the Co intake of the cow. The predominant source of B_{12} for the cow, and hence the ultimate origin of B_{12} in milk, is biosynthesis in the rumen. Therefore, its concentration in milk is not influenced greatly by feed, breed or season. Higher concentrations are found in colostrum than in mature milk.

The B_{12}-binding proteins of human milk have been studied in detail (see Chap. 11). The principal binding protein (R-type B_{12}-binding protein) has a molecular weight of ~63 kDa and contains ~35 % carbohydrate. Most or all of the B_{12} in human milk is bound to this protein. A second protein, transcobalamin II, is present at low concentrations.

Vitamin B_{12} is stable to pasteurization and storage of pasteurized milk (<10 % loss). UHT heat treatment, and in particular storage of UHT milk, causes greater losses. Storage temperature has a major influence on the stability of B_{12} in UHT milk. Losses during storage at 7 °C are minimal for up to 6 months but at room temperature (the normal storage conditions for UHT milk), losses can be significant after only a few weeks. Oxygen level in UHT milk does not appear to influence the stability of B_{12}.

6.4 Ascorbic Acid (Vitamin C)

Ascorbic acid (Fig. 6.35) is a carbohydrate which can be synthesized from D-glucose or D-galactose by most species with the exception of primates, guinea pigs, Indian fruit bats and certain birds and fish. Ascorbate can be oxidized

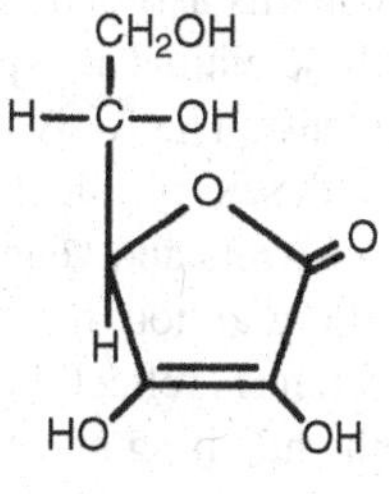

Fig. 6.35 Ascorbic acid

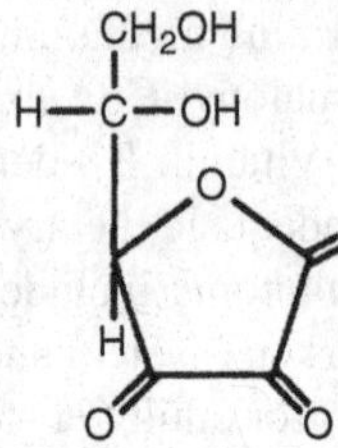

Fig. 6.36 Dehydroascorbic acid

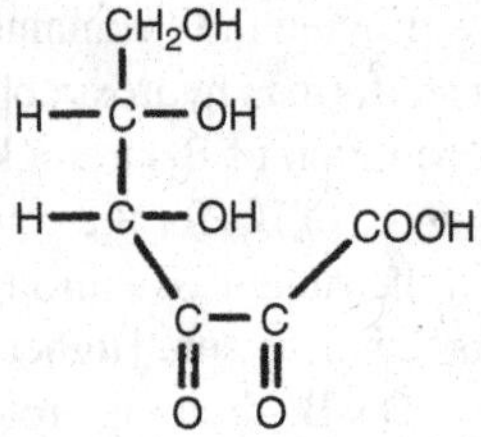

Fig. 6.37 2,3-Diketogulonic acid

reversibly to dehydroascorbate (Fig. 6.36) in the presence of transition metal ions, heat, light or mildly alkaline conditions without loss of vitamin activity. Dehydroascorbate can be oxidized irreversibly to 2,3-diketogulonic acid (Fig. 6.37) with loss of activity. 2,3-Diketogulonic acid can be broken down to oxalic and L-threonic acids and ultimately to brown pigments.

Ascorbic acid is a strong reducing agent and therefore is an important antioxidant in many biological systems. It is also necessary for the activity of the hydroxylase that catalyzes the post-translational conversion of proline to hydroxyproline and lysine to hydroxylysine. This post-translational hydroxylation is vital for the formation of collagen, the principal protein in connective tissue. Ascorbate functions to maintain iron in its correct oxidation state and aids in its absorption. Vitamin C also functions in amino acid metabolism, in the absorption of iron and increases resistance to infection. The classical vitamin C deficiency syndrome is scurvy, the symptoms of which include microcytic anaemia, bleeding gums, loose teeth, frequent infections, failure of wounds to heal, muscle degeneration, rough skin, hysteria and depression. The popular scientific literature has suggested major health benefits associated with ascorbate intakes far in excess of the RDA. While many of these claims are spurious, they have led to the widespread use of vitamin C supplements. The RDA for vitamin C is 60 day^{-1}. However, ascorbate requirement varies with sex, physical stress and perhaps with age. The richest sources of ascorbic acid are fruits and vegetables; milk is a poor source. Milk contains ~2 mg ascorbate per 100 g although reported values range from ~1.65 to 2.75 mg per 100 g. These differences reflect the fact that the levels of ascorbate can be reduced markedly during the handling and storage of milk. A ratio of ascorbate to dehydroascorbate in milk of 4 to 1 has been reported, although this ratio is strongly influenced by oxidation.

Some authors have reported seasonal differences in the concentration of vitamin C in milk (highest in winter milk) but the influence of this factor is unclear.

Human milk contains ~3–4 mg ascorbate per 100 g. Ascorbate is readily oxidized at the pH of milk. The rate of oxidation is influenced by factors including temperature, light, concentration of oxygen and the presence of catalytic trace elements. Ascorbic acid is of great importance in establishing and maintaining redox equilibria in milk (as discussed in detail in Chap. 8), the protection of folate (see Sect. 6.3.7) and in the prevention of oxidized flavour development in milk. The photochemical degradation of riboflavin (Sect. 6.3.2) catalyses the oxidation of ascorbate.

At least 75 % of the vitamin C in milk survives pasteurization and losses during storage of pasteurized milk are usually minimal. However, considerable losses of vitamin C have been reported in milk packaged in transparent containers. The extent of losses during UHT treatment depends on the amount of oxygen present during heat treatment and subsequent storage and on storage temperature. The concentration of ascorbate in creams and yogurts is similar to, or a little lower than, that in milk (Table 6.1); cheese contains only trace amounts.

References and Suggested Reading

Belitz, H.-D., & Grosch, W. (1987). *Food chemistry*. New York, NY: Springer.

Combs, G. T., Jr. (2012). *The vitamins: Fundamental aspects in nutrition and health* (4th ed.). San Diego, CA: Academic Press.

Department of Health. (1991). *Dietary reference values for food energy and nutrients for the United Kingdom: Report on health and social subjects* (Vol. 40). London, UK: HMSO.

Fouquay, J. W., Fox, P. F., & McSweeney, P. L. H. (Eds.). (2011). *Encyclopedia of dairy sciences* (2nd ed.). Oxford, UK: Academic Press.

Fox, P. F., & Flynn, A. (1992). Biological properties of milk proteins. In P. F. Fox (Ed.), *Advanced dairy chemistry* (Proteins, Vol. 1, pp. 255–284). London, UK: Elsevier Applied Science.

Garrow, J. S., & James, W. P. T. (1993). *Human nutrition and dietetics*. Edinburgh, UK: Churchill Livingstone.

Holland, B., Welch, A. A., Unmin, I. D., Buss, D. H., Paul, A. A., & Southgate, D. A. T. (1991). *McCance and Widdowson's the composition of foods* (5th ed.). Cambridge, UK: Royal Society of Chemistry and Ministry of Agriculture, Fisheries and Food.

Jensen, R. G. (Ed.). (1995). *Handbook of milk composition*. San Diego, CA: Academic Press.

Morrissey, P. A., & Hill, T. R. (2009). Fat-soluble vitamins and vitamin C in milk and dairy products. In P. L. H. McSweeney & P. F. Fox (Eds.), *Advanced dairy chemistry* (Lactose, water, salts and minor constituents 3rd ed., Vol. 3, pp. 527–589). New York, NY: Springer.

Nohr, D., & Biesalski, H. K. (2009). Vitamins in milk and dairy products: B-group vitamins. In P. L. H. McSweeney & P. F. Fox (Eds.), *Advanced dairy chemistry* (Lactose, water, salts and minor constituents 3rd ed., Vol. 3, pp. 591–630). New York, NY: Springer.

Öste, R., Jägerstad, M., & Andersson, I. (1997). Vitamins in milk and milk products. In P. F. Fox (Ed.), *Advanced dairy chemistry* (Lactose and minor constituents, Vol. 3, pp. 347–402). London, UK: Chapman & Hall.

Whitney, E. N., & Rolfes, S. R. (1996). *Understanding nutrition*. St. Paul, MN: West Publishing.

Chapter 7
Water in Milk and Dairy Products

7.1 Introduction

The water content of dairy products ranges from ~2.5 to~94 % (w/w) (Table 7.1) and is the principal component, by weight, in most dairy products, including milk, cream, ice cream, yogurt and most cheeses. The moisture content of foods (or more correctly their water activity, see Sect. 7.3), together with temperature and pH, are of great importance to food technology. As described in Sect. 7.8, water plays an extremely important role even in relatively low-moisture products such as butter (~16 % moisture) or dehydrated milk powders (~2.5 to 4 % moisture). Water, the most important diluent in foodstuffs, has an important influence on the physical, chemical and microbiological changes which occur in dairy products, and is an important plasticizer of non-fat milk solids.

7.2 General Properties of Water

Some physical properties of water are listed in Table 7.2. Water has higher melting and boiling temperatures, surface tension, dielectric constant, heat capacity, thermal conductivity and heats of phase transition than similar molecules (Table 7.3). Water has a lower density than would be expected from comparison with the above molecules and has the unusual property of expansion on solidification. The thermal conductivity of ice is approximately four times greater than that of water at the same temperature and is high compared with other non-metallic solids. Likewise, the thermal diffusivity of ice is about nine times greater than that of water.

The water molecule (HOH) is formed by covalent (σ) bonds between two of the four sp^3 bonding orbitals of oxygen (formed by hybridization of the 2s, $2p_x$, $2p_y$ and $2p_z$ orbitals) and two hydrogen atoms (Fig. 7.1a). The remaining two sp^3 orbitals of oxygen contain non-bonding electrons. The overall arrangement of the orbitals

© Springer International Publishing Switzerland 2015
P.F. Fox et al., *Dairy Chemistry and Biochemistry*,
DOI 10.1007/978-3-319-14892-2_7

Table 7.1 Approximate water content of some dairy products (modified from Holland et al. 1991)

Product	Water, g/100 g
Skimmed milk, average	91
Pasteurized	91
Fortified plus SMP	89
UHT sterilized	91
Whole bovine milk, average	88
Pasteurized[a]	88
Summer	88
Winter	88
Sterilized	88
Channel Island milk	
Whole, pasteurized	86
Summer	86
Winter	86
Semi-skimmed, UHT	89
Dried skimmed milk	3.0
With vegetable fat	2.0
Evaporated milk, whole	69
Flavoured milk	85
Goats' milk, pasteurized	89
Human colostrum	88
Mature milk	87
Sheep's milk, raw	83
Fresh cream, whipping	55
Cheeses	
Brie	49
Camembert	51
Cheddar, average	36
Vegetarian	34
Cheddar-type, reduced fat	47
Cheese spread, plain	53
Cottage cheese, plain	79
with additions	77
reduced fat	80
Cream cheese	46
Danish blue	45
Edam	44
Feta	57
Fromage frais, fruit	72
plain	78
very low fat	84
Full-fat soft cheese	58
Gouda	40
Hard cheese, average	37

(continued)

Table 7.1 (continued)

Product	Water, g/100 g
Lymeswold	41
Medium-fat soft cheese	70
Parmesan	18
Processed cheese, plain	46
Stilton, blue	39
White cheese, average	41
Whey	94
Drinking yogurt	84
Low-fat plain yogurt	85
Whole milk yogurt, plain	82
fruit	73
Ice cream, dairy, vanilla	62
non-dairy, vanilla	65

[a]The value for pasteurized milk is similar to unpasteurized milk

Table 7.2 Physical constants of water and ice (from Fennema 1985)

Molecular weight	18.01534			
Phase transition properties				
Melting point at 101.3 kPa (1 atm)	0.000 °C			
Boiling point at 101.3 kPa (1 atm)	100.00 °C			
Critical temperature	374.15 °C			
Critical pressure	22.14 MPa (218.6 atm)			
Triple point	0.0099 °C and 610.4 kPa (4.579 mmHg)			
Heat of fusion at 0 °C	6.012 kJ (1.436 kcal)/mol			
Heat of vaporization at 100 °C	40.63 kJ (9.705 kcal)/mol			
Heat of sublimation at 0 °C	50.91 kJ (12.16 kcal)/mol			
Other properties at	20 °C	0 °C	0 °C (ice)	−20 °C (ice)
Density (kg/l)	0.9998203	0.999841	0.9168	0.9193
Viscosity (Pa s)	1.002×10^{-3}	1.787×10^{-3}	–	–
Surface tension against air (N/M)	72.75×10^{-3}	75.6×10^{-3}	–	
Vapor pressure (Pa)	2.337×10^{-3}	6.104×10^{2}	6.104×10^{2}	1.034×10^{2}
Specific heat (J/kg K)	4.1819	4.2177	2.1009	1.9544
Thermal conductivity (J/m s k)	5.983×10^{2}	5.644×10^{2}	22.40×10^{2}	24.33×10^{2}
Thermal diffusivity (m²/s)	1.4×10^{-5}	1.3×10^{-5}	$\sim 1.1 \times 10^{-4}$	$\sim 1.1 \times 10^{-4}$
Dielectric constant				
static[a]	80.36	80.00	91[b]	98[b]
at 3×10^{9} Hz	76.7	80.5	–	3.2

[a]Limiting value at low frequencies
[b]Parallel to c-axis of ice; values about 15% larger if perpendicular to c-axis

Table 7.3 Properties of water and other similar compounds (from Roos 1997)

Property	Ammonia NH_3	Hydrofluoric acid HF	Hydrogen sulphide H_2S	Methane CH_4	Water H_2O
Molecular weight	17.03	20.02	34.08	16.04	18.015
Melting point (°C)	−77.7	−83.1	−85.5	−182.6	0.00
Boiling point (°C)	−33.35	19.54	−60.7	−161.4	100.00
Critical T (°C)	132.5	188.0	100.4	−82.1	374.15
p (bar)	114.0	64.8	90.1	46.4	221.5

Fig. 7.1 Schematic representations (**a–c**) of a water molecule and hydrogen bonding between water molecules (**d**)

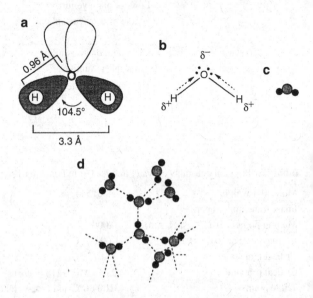

around the central oxygen atom is tetrahedral and this shape is almost perfectly retained in the water molecule. Due to electronegativity differences between oxygen and hydrogen, the O-H bond in water is strongly polar, with a vapour state dipole moment of 1.84 Debye units. The dipole moment of some other small dipolar molecules is: HF, 1.82; HCCl, 1.08; HBr, 0.82, HI, 0.44. This results in a partial negative charge on the oxygen and a partial positive charge on each hydrogen (Fig. 7.1b). Hydrogen bonding can occur between lone electron pairs in the oxygen atom and the hydrogen atoms of other molecules which, due to the above-mentioned differences in electronegativity, have some of the characteristics of bare protons. Thus, each water molecule can form four hydrogen bonds arranged in a tetrahedral fashion around the oxygen (Fig. 7.1d). The structure of water has been described as a continuous three-dimensional network of hydrogen-bonded molecules, with a local preference for tetrahedral geometry but with a large number of strained or broken hydrogen bonds. This tetrahedral geometry is usually maintained only over short distances. The structure is dynamic; molecules can rapidly exchange one hydrogen bonding partner for another and there may be some unbonded water molecules.

Fig. 7.2   Unit cell of an ice crystal at 0 °C. *Circles* represent the oxygen atoms of water molecules, *dashed lines* indicates hydrogen bonding (modified from Fennema 1985)

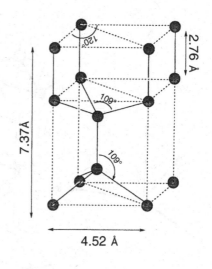

Fig. 7.3   The "basal plane" of ice (combinations of two planes of slightly different elevation) viewed from above. The *closed circles* represent oxygen atoms of water molecules in the lower plane and the *shaded circles* oxygen atoms in the upper plane, (**a**) seen from above and (**b**) from the side (from Fennema 1985)

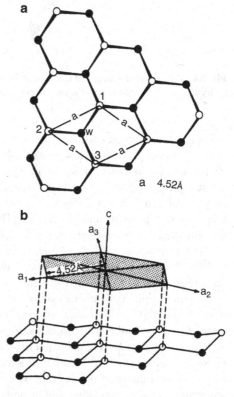

Water crystallizes to form ice. Each water molecule associates with four others in a tetrahedral fashion as is apparent from the unit cell of an ice crystal (Fig. 7.2). The combination of a number of unit cells, when viewed from above, results in a hexagonal symmetry (Fig. 7.3). Because of the tetrahedral arrangement around each

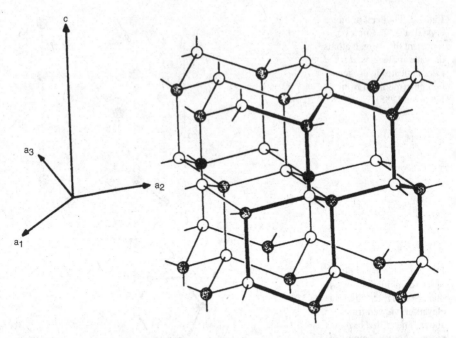

Fig. 7.4 The extended structure of ice. *Closed* and *shaded circles* represent oxygen atoms of water molecules in the lower and upper layers, respectively, of a basal plane (from Fennema 1985)

molecule, the three-dimensional structure of ice (Fig. 7.4) consists of two parallel planes of molecules lying close to each other ("basal planes"). The basal planes of ice move as a unit under pressure. The extended structure of ice is formed by stacking several basal planes. This is the only crystalline form of ice that is stable at a pressure of 1 atm at 0 °C, although ice can exist in a number of other crystalline forms, as well as in an amorphous state. The above description of ice is somewhat simplified; in practice the system is not perfect due to the presence of ionized water (H_3O^+, OH^-), isotopic variants, solutes and vibrations within the water molecules.

With the exceptions of water vapour and ice, water in dairy products contains numerous solutes. Thus, the interactions of water with solutes is very important. Hydrophilic compounds interact strongly with water by ion-dipole or dipole-dipole interactions while hydrophobic substances interact poorly with water and prefer to interact with each other ("hydrophobic interaction").

Water in food products can be described as being free or bound. "Bound" water is considered as that part of the water in a food which does not freeze at −40 °C and exists in the vicinity of solutes and other non-aqueous constituents, has reduced molecular mobility and other significantly altered properties compared with the "bulk water" of the same system (Fennema 1985; Roos 1997; Simatos et al. 2009). The actual amount of bound water varies in different products and the amount measured is often a function of the assay technique. Bound water is not permanently immobilized since interchange of bound water molecules occurs frequently.

Fig. 7.5 Arrangement
of water molecules in the
vicinity of sodium and
chloride ions (modified
from Fennema 1985)

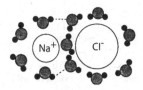

There are several types of bound water. Constitutional water is the most strongly bound and is an integral part of another molecule (e.g., within the structure of a globular protein). Constitutional water represents only a small fraction of the water in high moisture foods. "Vicinal", or monolayer, water is that bound to the first layer sites of the most hydrophilic groups. Multilayer water occupies the remaining hydrophilic sites and forms a number of layers beyond the monolayer water. There is often no clear distinction between constitutional, monolayer and multilayer water since they differ only in the length of time a water molecule remains associated with a food constituent.

The addition of dissociable solutes to water disrupts its normal tetrahedral structure. Many simple inorganic solutes do not possess hydrogen bond donors or acceptors and therefore can interact with water only by dipole interactions (e.g., Fig. 7.5 for NaCl). Multilayer water exists in a structurally-disrupted state while bulk-phase water has properties similar to those of water in a dilute aqueous salt solution. Ions in solution impose structure on the water but disrupt its normal tetrahedral structure. Concentrated solutions probably do not contain much bulk-phase water and structures caused by the ions predominate. The ability of an ion to affect the structure of water is influenced by its electric field. Some ions (principally small and/or multivalent) have strong electric fields and loss of the inherent structure of the water is more than compensated for by the new structure resulting from the presence of the ions. However, large, monovalent ions have weak electric fields and thus have a net disruptive effect on the structure of water.

In addition to hydrogen bonding with itself, water may also form hydrogen bonds with suitable donor or acceptor groups on other molecules. Water-solute hydrogen bonds are normally weaker than water-water interactions. By interacting through hydrogen bonds with the polar groups of solutes, the mobility of water is reduced and, therefore, is classified as either constitutional or monolayer water. Some solutes which are capable of hydrogen bonding with water do so in a manner that is incompatible with the normal structure of water and therefore have a disruptive effect on this structure. For this reason, solutes depress the freezing point of water (see Chap. 8). Water can potentially hydrogen bond with lactose or a number of groups on proteins (e.g., hydroxyl, amino, carboxylic acid, amide or imino; Fig. 7.6) in dairy products.

Milk contains a considerable amount of hydrophobic material, especially lipids and hydrophobic amino acid side chains. The interaction of water with such groups is thermodynamically unfavourable due to a decrease in entropy caused by increased water-water hydrogen bonding (and thus an increase in structure) adjacent to the non-polar groups.

Fig. 7.6 Schematic
representation of the
interaction of water
molecules with carboxylic
acid (**a**), alcohol (**b**), –NH
and carbonyl groups (**c**) and
amide groups (**d**)

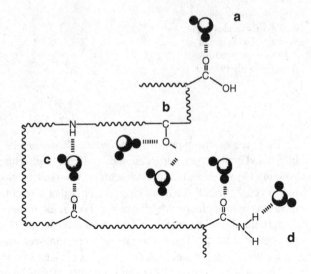

7.3 Water Activity

Water activity (a_w) is defined as the ratio between the water vapour pressure exerted
by the water in a food system (p) and that of pure water (p_o) at the same
temperature:

$$a_w = \frac{p}{p_o} \tag{7.1}$$

Due to the presence of various solutes, the vapour pressure exerted by water in a
food system is always less than that of pure water (unity). Water activity is a
temperature-dependent property of water which may be used to characterize the
equilibrium or steady state of water in a food system (Roos 1997, 2011).

For a food system in equilibrium with a gaseous atmosphere (i.e., no net gain or
loss of moisture to or from the system caused by differences in the vapour pressure
of water), the equilibrium relative humidity (ERH) is related to a_w by:

$$\text{ERH}(\%) = a_w \times 100 \tag{7.2}$$

Thus, under ideal conditions, ERH is the % relative humidity of an atmosphere
in which a foodstuff may be stored without a net loss or gain of moisture. Water
activity, together with temperature and pH, is one of the most important parameters
which determine the rates of chemical, biochemical and microbiological changes
which occur in foods. However, since a_w presupposes equilibrium conditions, its
usefulness is limited to foods in which these conditions exist.

Fig. 7.7 Clausius-Clapeyron relationship between water activity and temperature for native potato starch. *Numbers on curves* indicate water content, in g per g dry starch (from Fennema 1985)

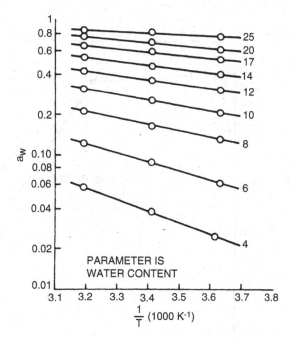

Water activity is influenced by temperature and therefore the assay temperature must be specified. The temperature dependence of a_w is described by the Clausius-Clapeyron equation in modified form:

$$\frac{d\ln(a_w)}{d(1/T)} = \frac{-\Delta H}{R} \tag{7.3}$$

where T is temperature (K), R is the universal gas constant and ΔH is the change in enthalpy. Thus, at a constant water content, there is a linear relationship between log a_w and 1/T (Fig. 7.7). This linear relationship is not obeyed at extremes of temperature or at the onset of ice formation.

The concept of a_w can be extended to cover sub-freezing temperatures. In these cases, a_w is defined (Fennema 1985) relative to the vapour pressure of supercooled water ($p_{o(SCW)}$) rather than to that of ice:

$$a_w = \frac{p_{ff}}{p_{o(SCW)}} = \frac{p_{ice}}{p_{o(SCW)}} \tag{7.4}$$

where p_{ff} is the vapour pressure of water in the partially frozen food and p_{ice} that of pure ice. There is a linear relationship between log a_w and 1/T at sub-freezing temperatures (Fig. 7.8). The influence of temperature on a_w is greater below the freezing point of the sample and there is normally a pronounced break at the freezing point.

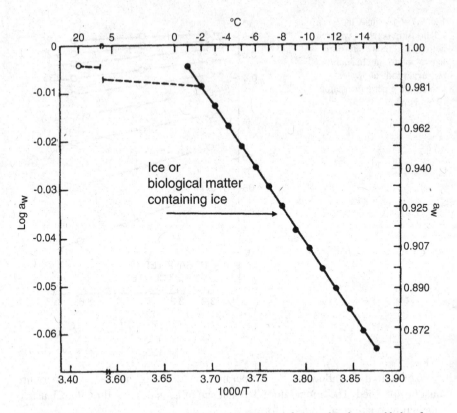

Fig. 7.8 Relationship between water activity and temperature for samples above and below freezing (from Fennema 1985)

Unlike the situation above freezing (where a_w is a function of composition and temperature), a_w below freezing is independent of sample composition and is influenced only by temperature. Thus, a_w values of foods at sub-freezing temperatures cannot be used to predict the a_w of foods above freezing. Sub-freezing a_w values are far less useful indicators of potential changes in foods than a_w values determined above the freezing point.

Water activity may be measured by a number of techniques (see Marcos 1993; Simatos et al. 2011). Comparison of manometric readings taken simultaneously on a food system and on pure water is the most direct technique. a_w can also be measured in dilute solutions and liquid foods with low solute concentrations by cryoscopy since under certain conditions, a_w can be considered as a colligative property. In these cases, the Clausius-Clapeyron equation is valid:

$$a_w = \gamma \left[n_2 / \left(n_1 + n_2 \right) \right] \tag{7.5}$$

where n_1 and n_2 are the number of moles of solute and water, respectively, and γ is the activity coefficient (approximately one for dilute solutions); n_2 can be determined by measuring the freezing point from the relation:

$$n_2 = \frac{G\Delta T_f}{1000\,K_f} \qquad (7.6)$$

where G is the grammes of solvent in the sample, ΔT_f is the freezing point depression (°C) and K_f is the molal freezing point depression constant for water, i.e., 1.86.

Water activity may also be calculated by determining the ERH for a food sample, using (7.2). ERH may be estimated by measuring the relative humidity of the headspace over a food in a small, sealed container using a hygrometer or psychrometer (a wet and dry bulb thermometer) or directly by measuring the moisture content of the air by gas chromatography. ERH can be estimated by moisture-related colour changes in paper impregnated with cobalt thiocyanate [Co(SCN)$_2$] and compared to standards of known a_w.

Differences in the hygroscopicity of various salts may be used to estimate a_w. Samples of the food are exposed to a range of crystals of known a_w; if the a_w of the sample is greater than that of a given crystal, the crystal will absorb water from the food. a_w may be measured by isopiestic equilibration. In this method, a dehydrated sorbent (e.g., microcrystalline cellulose) with a known moisture sorption isotherm (see Sect. 7.4) is exposed to the atmosphere in contact with the sample in an enclosed vessel. After the sample and sorbent have reached equilibrium, the moisture content of the sorbent can be measured gravimetrically and related to the a_w of the sample.

The a_w of a sample can also be estimated by exposing it to atmospheres with a range of known and constant relative humidities (RH). Moisture gains or losses to or from the sample may then be determined gravimetrically after equilibration. If the weight of the sample remains constant, the RH of the environment is equal to the ERH of the sample. The a_w of the food may be estimated by interpolation of data for RH values greater and less than the ERH of the sample.

For certain foodstuffs, a_w may be estimated from their chemical composition. A nomograph relating the a_w of freshly-made cheese to its content of moisture and NaCl is shown in Fig. 7.9. Likewise, various equations relating the a_w of cheese to [NaCl], [ash], [12 % trichloroacetic acid-soluble N] and pH have been developed (see Marcos 1993).

7.4   Water Sorption

Sorption of water vapour to or from a food depends on the vapour pressure exerted by the water in the food. If this vapour pressure is lower than that of the atmosphere, *absorption* occurs until vapour pressure equilibrium is reached. Conversely, *desorption* of water vapour results if the vapour pressure exerted by water in the food is

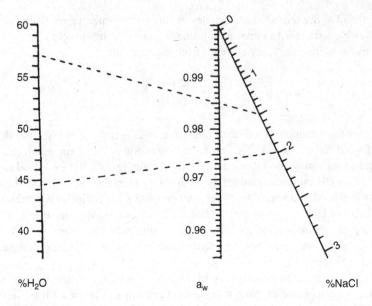

Fig. 7.9 Nomograph for direct estimation of water activity (a_w) of unripe cheeses from % H_2O and % NaCl. Examples: If % $H_2O = 57.0$, and % NaCl = 1.5, then $a_w = 0.985$; if % $H_2O = 44$, % NaCl = 2.0, then $a_w = 0.974$ (from Marcos 1993)

greater than that of the atmosphere. *Adsorption* is regarded as the sorption of water at a physical interface between a solid and its environment. *Absorption* is regarded as a process in which *adsorption* occurs in the interior of the substance (Kinsella and Fox 1986).

The water sorption characteristics of dairy products are governed by their non-fat constituents (principally lactose and proteins). However, in many milk and whey products, the situation is complicated by structural transformations and/or solute crystallization.

The relationship between the water content of a food (g H_2O/g dry matter) and a_w at a constant temperature is known as a *sorption isotherm*. Sorption isotherms are prepared by exposing a set of previously dried samples to atmospheres of high RH; desorption isotherms can be determined by a similar technique. Isotherms provide important information regarding the difficulty of removing water from a food during dehydration and on its stability since both ease of dehydration and stability are related to a_w. A typical sorption isotherm is shown in Fig. 7.10. Most sorption isotherms are sigmoidal in shape, although foods which contain large amounts of low molecular weight solutes and relatively little polymeric material generally exhibit a J-shaped isotherm. The rate of water sorption is temperature dependent and for a given vapour pressure, the amount of water lost by desorption or gained by resorption may not be equal and therefore sorption hysteresis may occur (Fig. 7.11).

The moisture present in Zone I (Fig. 7.10) is the most tightly bound and represents the monolayer water bound to accessible, highly polar groups of the dry food.

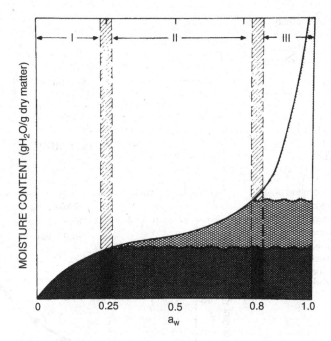

Fig. 7.10 Generalized moisture sorption isotherm for a food (from Fennema 1985)

Fig. 7.11 Hysteresis of a
moisture sorption isotherm
(from Fennema 1985)

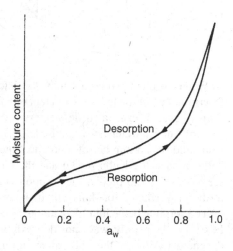

The boundary between Zones I and II represents the monolayer moisture content of the food. The moisture in Zone II consists of multilayer water in addition to the monolayer water while the extra water added in Zone III consists of the bulk-phase water.

Table 7.4 Water activity of saturated salt solutions, at 30 °C, used for determination of water sorption isotherms

Salt	Water activity at 30 °C
KOH	0.0738
$MgCl_2·6H_2O$	0.3238
K_2CO_3	0.4317
$NaNO_3$	0.7275
KCl	0.8362
$BaCl_2·2H_2O$	0.8980

Fig. 7.12 Adsorption of water by skim milk and sorption isotherms predicted by the Braunauer-Emmett-Teller (BET), Kühn and Guggenheim-Andersson-De Boer (GAB) sorption models (from Roos 1997)

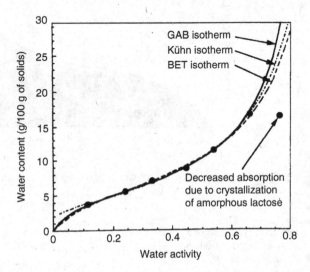

Water sorption isotherms may be determined experimentally by gravimetric determination of the moisture content of a food product after it has reached equilibrium in sealed, evacuated dessicators containing saturated solutions of different salts of known water activity (Table 7.4). Data obtained in this manner may be compared with a number of theoretical models (including the Braunauer-Emmett-Teller model, the Kühn model and the Gruggenheim-Andersson-De Boer model, see Roos 1997) to predict the sorption behaviour of foods. Examples of sorption isotherms predicted for skim milk by three such models are shown in Fig. 7.12.

The sorption behaviour of a number of dairy products is known (see Kinsella and Fox 1986). Generally, whey powders exhibit sigmoidal sorption isotherms, although the characteristics of the isotherm are influenced by the composition and history of the sample. Examples of sorption isotherms for whey protein concentrate (WPC), dialyzed WPC and its dialysate (principally lactose) are shown in Fig. 7.13. At low a_w values, sorption is due mainly to the proteins present. The sharp decrease in a_w observed in the sorption isotherm of lactose at a_w values between 0.35 and 0.50 (e.g., Fig. 7.13), is due to the crystallization of amorphous lactose in the α-form, which contains 1 mole of water of crystallization per mole (Chap. 2). Above a_w

Fig. 7.13 Water vapour
sorption by whey protein
concentrate (*A*), dialyzed
whey protein concentrate (*B*)
and dialyzate (lactose) from
whey protein concentrate (*C*)
(from Kinsella and Fox 1986)

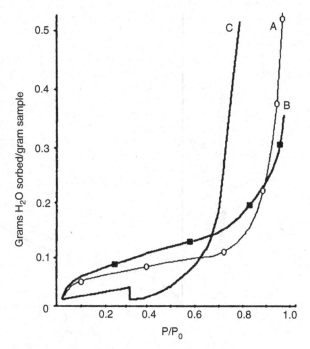

values of ~0.6, water sorption is influenced principally by small molecular weight components (Fig. 7.13).

Despite some conflicting evidence (see Kinsella and Fox 1986), it appears that denaturation has little influence on the amount of water bound by whey proteins. However, other factors which may accompany denaturation (e.g., Maillard browning, association or aggregation of proteins) may alter protein sorption behaviour. Drying technique affects the water sorption characteristics of WPC. Freeze-dried or spray-dried WPC preparations bind more water at the monolayer level than roller, air or vacuum dried samples, apparently due to larger surface areas in the former. As discussed above. temperature also influences water sorption by whey protein preparations. The sorption isotherm for β-lactoglobulin is typical of many globular proteins.

In milk powders, the caseins are the principal water sorbents at low and intermediate values of a_w. The water sorption characteristics of the caseins are influenced by their micellar state, their tendency to self-associate, their degree of phosphorylation and their ability to swell. Sorption isotherms for casein micelles and sodium caseinate (Fig. 7.14) are generally sigmoidal, but the isotherm of sodium caseinate shows a marked increase at a_w between 0.75 and 0.95. This has been attributed to the presence of certain ionic groups, bound Na^+ or the increased ability of sodium caseinate to swell.

Heating of casein influences its water sorption characteristics, as does pH. With some exceptions at low pH, the hydration of sodium caseinate increases with pH (Fig. 7.15b). Minimum water sorption occurs around the isoelectric pH (4.6). At low and intermediate values of a_w, increasing pH, and thus [Na^+], has little influence on

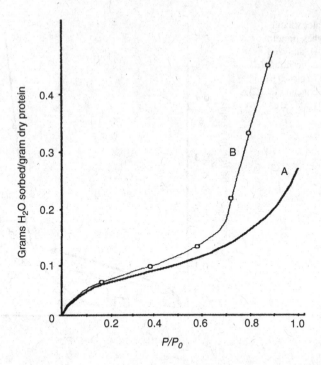

Fig. 7.14 Sorption isotherm for casein micelles (*A*) and sodium caseinate (*B*) at 24 °C, pH 7 (from Kinsella and Fox 1986)

water sorption. At low a_w values, water is bound strongly to binding sites on the protein while at higher a_w both protein and NaCl sorb available water in multilayer form. Water sorption by casein micelles (Fig. 7.15a) has a minimum at about pH 6–7 at high a_w. This difference in sorption minima between caseinate and casein micelles is because the hydration of caseinate is due mainly to ion effects (Na^+ being more effective in this respect than Cl^-). The hydration behaviour of casein micelles, on the other hand, reflects effects of pH on micelle integrity. Hydrolysis of κ-casein by rennet appears to have only a small influence on its ability to bind water, although the chemical modification of amino groups has a greater effect. Variation in the amino acid sequences of the caseins caused by genetic polymorphism also influences water sorption. The addition of NaCl to isoelectric casein greatly increases water sorption.

The greatest consequences of water sorption are in the context of dehydrated dairy products. In addition to being influenced by relative humidity, temperature and the relative amounts and intrinsic sorption properties of its constituents, the amount of water sorbed by milk powders is influenced by the method of preparation, the state of lactose, changes in protein conformation and swelling and dissolution of solutes such as salts. As discussed in Chap. 2, amorphous lactose is hygroscopic and may absorb large amounts of water at low relative humidities while water sorption by crystalline lactose is significant only at high relative humidities and thus water sorption by milk products containing crystallized lactose is due mainly to their protein fraction.

Fig. 7.15 Equilibrium water
content of (*A*) casein micelles
and (*B*) sodium caseinate and
casein hydrochloride as a
function of pH and changing
water activities (isopsychric
curves) (from Kinsella and
Fox 1986)

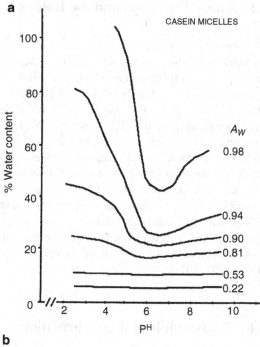

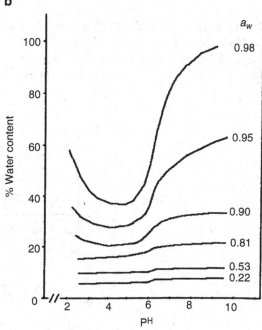

7.5   Glass Transition and the Role of Water in Plasticization

The non-fat solids in low-moisture dairy products (e.g., milk powders) or frozen milk products (since dehydration occurs on freezing) are amorphous in most dairy products (except those containing pre-crystallized lactose). The non-fat solids exist in a metastable, non-equilibrium state as a solid glass or a supercooled liquid. Phase changes can occur between these states with a phase transition temperature range called the glass transition (T_g; Roos 1997, 2011). Changes in heat capacity, dielectric properties, volume, molecular mobility and various mechanical properties occur on glass transition. The temperature of onset of the glass transition of amorphous water (i.e., the transformation of a solid, amorphous glass into a supercooled liquid and *vice versa*) is about −135 °C. T_g increases with increasing weight fraction of solids (Fig. 7.16). The addition of water causes a sharp decrease in T_g.

The stability of dairy products decreases sharply above a critical water activity (see Sect. 7.8). This decrease in stability is related to the influence of water on the glass transition and the role of water as a plasticizer of amorphous milk constituents (Roos 1997, 2011).

7.6   Non-equilibrium Ice Formation

Cooling aqueous solutions to below their freezing point results in the formation of ice. If solutions of sugars are cooled rapidly, non-equilibrium ice formation occurs. This is the most common form of ice in frozen dairy products (e.g., ice cream). Rapid freezing of ice cream mixes results in the freeze concentration of lactose and

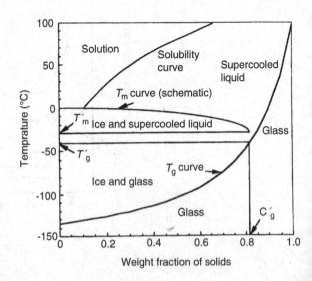

Fig. 7.16 State diagram of lactose (from Roos 1997)

other sugars, resulting in supersaturated solutions if the temperature is too low to permit crystallization. The rapid cooling of lactose results in the formation of a supersaturated, freeze-concentrated amorphous matrix.

Various thermal transitions can occur in rapidly-cooled solutions, including glass transition, devitrification (ice formation on warming a rapidly-frozen solution) and melting of ice. The relationship between temperature, weight fraction of solids, solubility and glass transition of lactose is shown in Fig. 7.16.

7.7 Role of Water in Stickiness and Caking of Powders and Crystallization of Lactose

As discussed in Chap. 2, drying of whey or other solutions containing a high concentration of lactose is difficult since the semi-dry powder may stick to the metal surfaces of the dryer. The influence of dryer temperature and other process parameters on stickiness during the drying of whey are discussed in Chap. 2. The role of agglomeration on the wetting and reconstitution of dairy powders was also discussed in Chap. 2.

The principal cause of sticking and caking is the plasticization of amorphous powders by heating or by exposure to high relative humidity. As discussed by Roos (1997), heating or the addition of water reduces surface viscosity (thus permitting adhesion) by creating an incipient liquid state of lower viscosity at the surface of the particle. If sufficient liquid is present and flows by capillary action, it may form bridges between particles strong enough to cause adhesion. Factors that affect liquid bridging include water sorption, melting of components (e.g., lipids), the production of H_2O by chemical reactions (e.g., Maillard browning), the release of water of crystallization and the direct addition of water.

The viscosity of lactose in the glassy state is extremely high and thus a long contact time is necessary to cause sticking. However, above T_g, viscosity decreases markedly and thus the contact time for sticking is reduced. Since T_g is related to sticking point, it may be used as an indicator of stability. Caking of powders at high RH occurs when the addition of water plasticizes the components of the powder and reduces T_g to below the ambient temperature. The crystallization of amorphous lactose is discussed in Chap. 2.

7.8 Water and the Stability of Dairy Products

The most important practical aspect of water in dairy products is its effect on their chemical, physical and microbiological stability. Chemical changes which are influenced by a_w include Maillard browning (including loss of lysine), lipid oxidation, loss of certain vitamins, pigment stability and the denaturation of proteins. Physical changes involve crystallization of lactose. Control of the growth of microorganisms

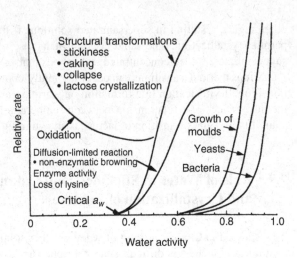

Fig. 7.17 Stability map for non-fat milk solids showing schematic rates of various deteriorative changes and growth of microorganisms as a function of water activity (from Roos 1997)

by reducing a_w is of great significance for the stability of a number of dairy products. The relationship between the stability of foods and a_w is summarized in Fig. 7.17.

Milk is the only naturally-occurring protein-rich food which contains a large amount of a reducing sugar. Maillard browning is undesirable in the context of nearly all dairy foods. Since lactose is a reducing sugar, it can participate in these browning reactions and essentially all dairy products (with the exceptions of butter oil, butter and dairy spreads) have sufficient protein to supply the necessary amino groups. Many of the stages of Maillard browning (see Chap. 2) have a high activation energy and thus the process is accelerated at high temperatures. The combination of the presence of lactose and a high temperature occurs during the production of many milk and whey powders, processed cheese and when dairy products are heated during cooking (e.g., the browning of Mozzarella cheese during baking of pizza). The loss of lysine accompanies the early stages of the Maillard reaction in which its ε-amino group participates. Loss of lysine is significant from a nutritional standpoint since it is an essential amino acid. Loss of lysine may occur without visible browning.

For a given product composition and temperature, the rate of browning is affected by a_w. The influence of water on the rate of Maillard browning depends on the relative importance of a number of factors. Water imparts mobility to reacting species (thus increasing the rate of browning) but may also dilute reactants (thus reducing the rate of browning). At low values of a_w, the increase in molecular mobility is most significant while at higher values of a_w, the dilution effect predominates. At lower a_w values, water can also dissolve new reacting species. The presence of water can retard certain steps in browning in which water is released as a product (product inhibition, e.g., the initial glycosylamine reaction) or enhance other reactions (e.g., deamination). For many foods, the rate of Maillard browning usually reaches a maximum at an intermediate moisture level ($a_w \approx 0.40$ to 0.80). However, the maximum rate is greatly influenced by the presence of other constituents in the food such as glycerol or other liquid humectants which can shift the maximum to a lower a_w value. The rate of browning of milk powders is also accelerated by the crystallization of lactose.

Lipid oxidation can cause defects in high-fat dairy products. The mechanism of lipid oxidation is discussed in Chap. 3. At low a_w, the rate of oxidation decreases with increasing a_w and reaches a minimum around the monolayer value and then increases at higher a_w. The antioxidant effect of water at low values of a_w has been attributed to bonding of hydroperoxide intermediates and the hydration of metal ions, which act as catalysts. The increased rate of oxidation at higher a_w is a consequence of increased mobility of reactants. In general, water may influence the rate of lipid oxidation by affecting the concentration of initiating radicals, the degree of contact, the mobility of reacting species and the relative importance of radical transfer versus recombination events. Side reactions associated with lipid oxidation (e.g., cross-linking of proteins, enzyme inactivation by peroxidation products, degradation of amino acids) are also influenced by a_w.

The stability of some vitamins is influenced by a_w. In general, the stability of retinol (vitamin A), thiamine (vitamin B_1) and riboflavin (vitamin B_2) decreases with increasing a_w. At low a_w (<0.40), metal ions do not have a catalytic effect on the destruction of ascorbic acid. The rate of loss of ascorbic acid increases exponentially as a_w increases. The photodegradation of riboflavin (see Chap. 6) is also accelerated by increasing a_w.

Water activity influences the rate of thermal denaturation of proteins, including enzymes. Generally, the denaturation temperature increases with decreasing a_w. The rate of nearly all enzyme-catalyzed reactions increases with increasing a_w, as a consequence of increased molecular mobility.

The emulsion state of water in butter (i.e., the water droplet size) is very important for the quality of the product. Bacteria in butter can grow only in the aqueous emulsified phase. A finely-divided aqueous phase restricts bacterial growth since the nutrients available in small droplets will quickly become limiting. Also, unless bacterial contamination is high, it is likely that most small water droplets in butter are sterile.

Together with pH and temperature, a_w has a major influence on the rate of growth of microorganisms. Indeed, reduction of a_w by drying or the addition of NaCl or sugars is one of the principal traditional techniques used to preserve food. The minimum a_w required for microbial growth is ~0.62, which permits the growth of xerophilic yeasts. As a_w increases, moulds and other yeasts can grow and, finally, bacteria (above ~0.80). a_w also controls the growth of pathogenic microorganisms. *Staphylococcus aureus* will not grow below a_w ~0.86 while the growth of *Listeria monocytogenes* does not occur below a_w ~0.92.

The significance of water activity in cheese differs from that in other dairy products because cheese is a dynamic product and many of its important characteristics, e.g., flavour and texture, develop during ripening after curd manufacture whereas the desirable characteristics of other dairy products are optimal at the end of manufacture and changes are undesirable. The a_w of cheese has a major influence on its ripening; it affects the activity of indigenous and added enzymes, especially rennets, and the survival, growth and activity of indigenous and inoculated microorganisms, the activity of which has major and characteristic effects on the flavour, texture, stability and safety of cheese. The a_w of cheese varies widely (Table 7.5), and is determined by its moisture content, Table 7.1, and added NaCl. The chemistry and biochemistry of cheese are described in Chap. 12.

Table 7.5 Typical water activity of some common cheese varieties at 25 °C (modified from Roos 2011)

Variety	Water activity
Appenzeller	0.96
Brie	0.98
Camembert	0.98
Blue	0.94
Cheddar	0.95
Cottage	0.99
Edam	0.96
Gouda	0.95
Emmentaler	0.97
Mozzarella	0.99
Parmesan	0.92

References

Fennema, O. R. (Ed.). (1985). *Food chemistry* (2nd ed.). New York: Marcel Dekker, Inc.

Holland, B., Welch, A. A., Unwin, I. D., Buss, D. H., Paul, A. A., & Southgate, D. A. T. (1991). *The composition of foods* (5th ed.). Cambridge: McCance and Widdowson's, Royal Society of Chemistry and Ministry of Agriculture, Fisheries and Food.

Kinsella, J. E., & Fox, P. F. (1986). Water sorption by proteins: Milk and whey proteins. *CRC Critical Reviews in Food Science and Nutrition, 24*, 91–139.

Marcos, A. (1993). Water activity in cheese in relation to composition, stability and safety. In P. F. Fox (Ed.), *Cheese: Chemistry, physics and microbiology* (2nd ed., Vol. 1, pp. 439–469). London: Chapman & Hall.

Roos, Y. (1997). Water in milk products. In P. F. Fox (Ed.), *Advanced dairy chemistry - 3 - lactose, water, salts and vitamins* (pp. 306–346). London: Chapman & Hall.

Roos, Y. H. (2011). Water in dairy products: Significance. In J. W. Fuquay, P. F. Fox, & P. L. H. McSweeney (Eds.), *Encyclopedia of dairy sciences* (2nd ed., Vol. 4, pp. 707–714). Oxford: Academic.

Simatos, D., Champion, D., Lorient, D., Loupiac, C., & Roudaut, G. (2009). Water in dairy products. In P. L. H. McSweeney & P. F. Fox (Eds.), *Advanced dairy chemistry, volume 3, lactose, water, salts and minor constituents* (3rd ed., pp. 467–526). New York: Springer.

Simatos, D., Roudaut, D., & Champion, D. (2011). Analysis and measurement of water activity. In J. W. Fuquay, P. F. Fox, & P. L. H. McSweeney (Eds.), *Encyclopedia of dairy sciences* (2nd ed., Vol. 4, pp. 715–726). Oxford: Academic.

Suggested Reading

Fennema, O.R., ed. (1985). Food Chemistry, 2nd edn., Marcel Dekker, Inc., New York.

Rockland, L.B. and Beuchat, L.R., eds. (1987). Water Activity: Theory and Applications to Food, Marcel Dekker, Inc., New York.

Roos, Y. (1997). Water in milk products, in, Advanced Dairy Chemistry - 3 - Lactose, Water, Salts and Vitamins, P.F. Fox, ed, Chapman & Hall, London, pp 306–346.

Simatos, D., Champion, D., Lorient, D., Loupiac, and Roudaut, G. (2009). Water in dairy products, in, Advanced Dairy Chemisry, volume 3, Lactose, Water, Salts and Minor Constituents 3rd edition, P.L.H. McSweeney and P.F. Fox, eds, Springer, New York. pp 467–526.

Chapter 8
Physical Properties of Milk

Milk is a dilute emulsion consisting of an oil/fat dispersed phase and an aqueous colloidal continuous phase. The physical properties of milk are similar to those of water but are modified by the presence of various solutes (proteins, lactose and salts) in the continuous phase and by the degree of dispersion of the emulsified and colloidal components.

Data on the physical properties of milk are important since such parameters can influence the design and operation of dairy processing equipment (e.g., thermal conductivity or viscosity) or can be used to determine the concentration of specific components in milk (e.g., use of the elevation in freezing point to estimate added water or specific gravity to estimate solids-not-fat), or to assess the extent of biochemical changes in the milk during processing (e.g., acidification by starter or the development of a rennet coagulum). Some important physical properties of milk are summarized in Table 8.1 and were reviewed by McCarthy and Singh (2009).

8.1 Ionic Strength

The ionic strength, I, of a solution is defined as:

$$I = \frac{1}{2}\Sigma c_i z_i^2$$

(8.1)

where c_i is the molar concentration of the ion of type i and z_i is its charge.

The ionic strength of milk is ~0.08 M.

© Springer International Publishing Switzerland 2015
P.F. Fox et al., *Dairy Chemistry and Biochemistry*,
DOI 10.1007/978-3-319-14892-2_8

Table 8.1 Some physical properties of milk

Osmotic pressure	~700 kPa
a_w	~0.993
Boiling point elevation	~0.15 K
Freezing point depression	~0.522 K
E_h	+0.20 to +0.30 V
Refractive index,n_D^{20}	1.3440–1.3485
Specific refractive index	~0.2075
Density (20 °C)	~1,030 kg m^{-3}
Specific gravity (20 °C)	~1.0330
Specific conductance	0.0040–0.0050 Ω^{-1} cm^{-1}
Ionic strength	~0.08 M
Surface tension	~52 N m^{-1} at 20 °C

8.2 Density

The density (ρ) of a substance is its mass per unit volume, while its specific gravity (SG) or relative density is the ratio of the density of the substance to that of water (ρ_w) at a specified temperature:

$$\rho = m / V \tag{8.2}$$

$$SG = \rho / \rho_w \tag{8.3}$$

$$\rho = SG\rho_w \tag{8.4}$$

The thermal expansion coefficient governs the influence of temperature on density and therefore it is necessary to specify temperature when discussing density or specific gravity. The density of milk is of consequence since fluid milk is normally retailed by volume rather than by mass. Measurement of the density of milk using a hydrometer (lactometer) has also been used to estimate its total solids content.

The density of bulk milk (4 % fat and 8.95 % solids-not-fat) at 20 °C is approximately 1,030 kg m^{-3} and its specific gravity is 1.0321. Milk fat has a density of ~902 kg m^{-3} at 40 °C. The density of a given milk sample is influenced by its storage history since it is somewhat dependent on the liquid to solid fat ratio and the degree of hydration of proteins. To minimise effects of thermal history on its density, milk is usually pre-warmed to 40–45 °C to liquefy the milk fat and then cooled to the assay temperature (often 20 °C).

The density and specific gravity of milk vary somewhat with breed. Milk from Ayrshire cows has a mean specific gravity of 1.0317 while that of Jersey and Holstein milks is 1.0330. Density varies with the composition of the milk and its measurement has been used to estimate the total solids content of milk. The density of a multicomponent mixture (like milk) is related to the density of its components by:

$$1 / \rho = \Sigma \left(m_x / \rho_x \right) \tag{8.5}$$

where m_x is the mass fraction of component x, and ρ_x its apparent density in the mixture. This apparent density is not normally the same as the true density of the substance since a contraction usually occurs when two components are mixed.

Equations have been developed to estimate the total solids content of milk based on % fat and specific gravity (usually estimated using a lactometer). Such equations are empirical and suffer from a number of drawbacks; for further discussion see Jenness and Patton (1959). The principal problem is the fact that the coefficient of expansion of milk fat is high and it contracts slowly on cooling and therefore the density of milk fat (see Chap. 3) is not constant. Variations in the composition of milk fat and in the proportions of other milk constituents have less influence on these equations than the physical state of the fat.

In addition to lactometry (determination of the extent to which a hydrometer sinks), the density of milk can be measured by pycnometry (determination of the mass of a given volume of milk), by hydrostatic weighing of an immersed bulb (e.g., Westphal balance), by dialatometry (measurement of the volume of a known mass of milk) or by measuring the distance that a drop of milk falls through a density gradient column.

8.3   Redox Properties of Milk

Oxidation-reduction (redox) reactions involve the transfer of an electron from an electron donor (reducing agent) to an electron acceptor (oxidizing agent). The species that loses electrons is said to be oxidized while that which accepts electrons is reduced. Since there can be no net transfer of electrons to or from a system, redox reactions must be coupled and the oxidation reaction occurs simultaneously with a reduction reaction.

The tendency of a system to accept or donate electrons is measured using an inert electrode (typically platinum). Electrons can pass from the system into this electrode, which is thus a half cell. The Pt electrode is connected via a potentiometer to another half-cell of known potential (usually, a saturated calomel electrode). All potentials are referred to the hydrogen half-cell:

$$\tfrac{1}{2}H_2 \rightleftharpoons H^+ + e^- \tag{8.6}$$

which by convention is assigned a potential of zero when an inert electrode is placed in a solution of unit activity with respect to H^+ (i.e., pH = 0) in equilibrium with H_2 gas at a pressure of 1.013×10^5 Pa (1 atm). The redox potential of a solution, E_h, is the potential of the half-cell at the inert electrode and is expressed as volts. E_h depends not only on the substances present in the half-cell but also on the concentrations of their oxidized and reduced forms. The relationship between E_h and the concentrations of the oxidized and reduced forms of the compound is described by the Nernst equation:

$$E_h = E_o - RT / nF \ln a_{red} / a_{ox} \tag{8.7}$$

where E_o is the standard redox potential (i.e., potential when reactant and product are at unit activity), n is the number of electrons transferred per molecule, R is the Universal Gas Constant (8.314 J K^{-1} mol^{-1}), T is temperature (in Kelvin), F is the Faraday constant (96.5 kJ V^{-1} mol^{-1}) and a_{red} and a_{ox} are activities of the reduced and oxidized forms, respectively. For dilute solutions, it is normal to approximate activity by molar concentration. Equation 8.7 can be simplified, assuming a temperature of 25 °C, a transfer of one electron and that activity $\approx$ concentration:

$$E_h = E_o + 0.059 \log[Ox]/[Red] \tag{8.8}$$

Thus, E_h becomes more positive as the concentration of the oxidized form of the compound increases. E_h is influenced by pH since pH affects the standard potential of a number of half-cells. The above equation becomes:

$$E_h = E_o + 0.059 \log[Ox]/[Red] - 0.059 \, pH \tag{8.9}$$

The E_h of milk is usually in the range $+0.25$ to $+0.35$ V at 25 °C, at pH 6.6–6.7 and in equilibrium with air (Singh et al. 1997). The influence of pH on the redox potential of a number of systems is shown in Fig. 8.1.

The concentration of dissolved oxygen is the principal factor affecting the redox potential of milk. Milk is essentially free of O_2 when secreted but in equilibrium with air, its O_2 content is ~0.3 mM. The redox potential of anaerobically-drawn milk or milk which has been depleted of dissolved oxygen by microbial growth or by displacement of O_2 by other gases is more negative than that of milk containing dissolved O_2.

The concentration of ascorbic acid in milk (11.2–17.2 mg L^{-1}) is sufficient to influence its redox potential. In freshly drawn milk, all ascorbic acid is in the reduced form but can be oxidized reversibly to dehydroascorbate, which is present as a hydrated hemiketal in aqueous systems. Hydrolysis of the lactone ring of dehydroascorbate, which results in the formation of 2,3-diketogulonic acid, is irreversible (Fig. 8.2).

The oxidation of ascorbate to dehydroascorbate is influenced by O_2 partial pressure, pH and temperature and is catalyzed by metal ions (particularly Cu^{2+}, but also Fe^{3+}). The ascorbate/dehydroascorbate system in milk stabilizes the redox potential of oxygen-free milk at ~0.0 V and that of oxygen-containing milk at $+0.20$ to $+0.30$ V (Sherbon 1988). Riboflavin can also be oxidized reversibly but its concentration in milk (~4 μM) is thought to be too low to have a significant influence on redox potential. The lactate-pyruvate system (which is not reversible unless enzyme-catalyzed) is thought not to be significant in influencing the redox potential of milk since it, too, is present at very low concentrations. At the concentrations at which they occur in milk, low molecular mass thiols (e.g., free cysteine) have an insignificant influence on the redox potential of milk. Thiol-disulphide interactions between cysteine residues of proteins influence the redox properties of heated milks in which the proteins are denatured. The free aldehyde group of lactose can be oxidised to a carboxylic acid (lactobionic acid) at alkaline pH but this system contributes little to the redox properties of milk at pH 6.6.

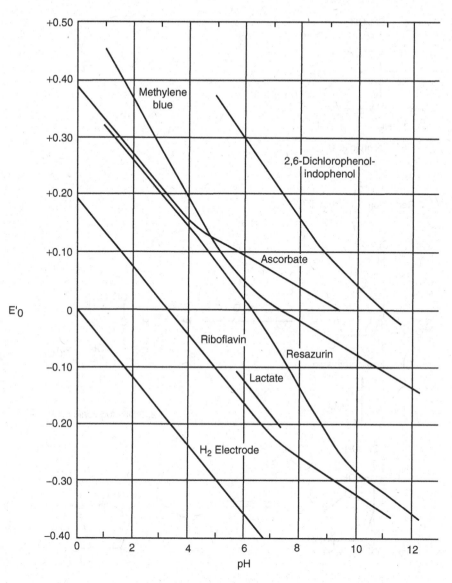

Fig. 8.1 Effect of pH on the oxidation–reduction potential of various systems (from Sherbon 1988)

The E_h of milk is influenced by exposure to light and by a number of processing operations, including those which cause changes in the concentration of O_2 in the milk. Addition of metal ions (particularly Cu^{2+}) also influences the redox potential. Heating of milk causes a decrease in its E_h, due mainly to the denaturation of β-lactoglobulin (and the consequent exposure of –SH groups) and loss of O_2. Compounds formed by the Maillard reaction between lactose and proteins can also influence the E_h of heated milk, particularly dried milk products.

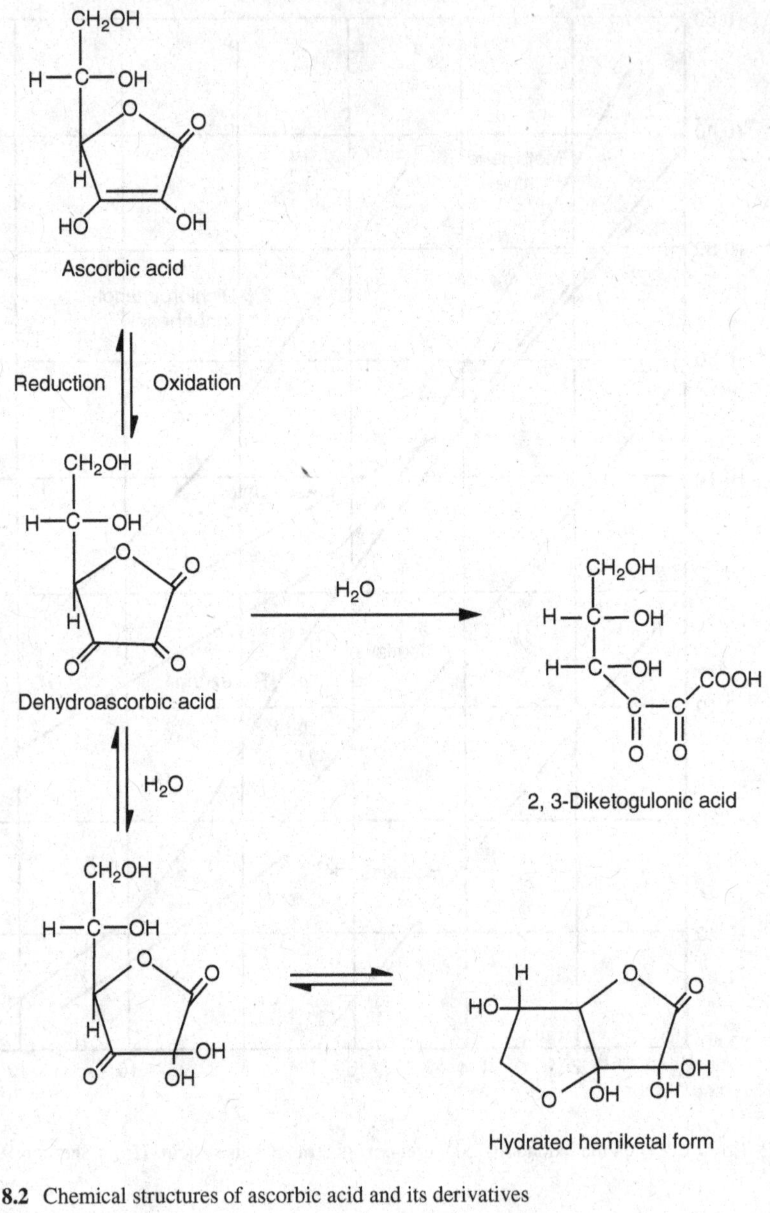

Fig. 8.2 Chemical structures of ascorbic acid and its derivatives

Fermentation of lactose during the growth of microorganisms in milk has a major effect on its redox potential. The decrease in the E_h of milk caused by the growth of lactic acid bacteria is shown in Fig. 8.3. A rapid decrease in E_h occurs after the available O_2 has been consumed by the bacteria. Therefore, the redox potential of cheese and fermented milk products is negative. Reduction of redox indicators (e.g., resazurin or methylene blue) can be used as an index of the bacterial quality of milk by measuring the "reduction time", at a suitable temperature, of milk containing the dye.

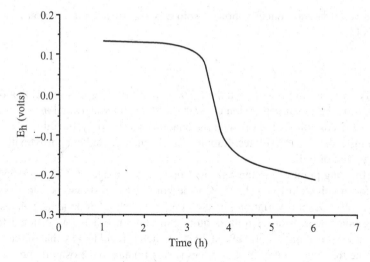

Fig. 8.3 Decrease in the redox potential of milk caused by the growth of *Lactococcus lactis* subsp. *lactis* at 25 °C

Riboflavin absorbs light maximally at about 450 nm and in doing so can be excited to a triplet state. This excited form of riboflavin can interact with triplet O_2 to form a superoxide anion O_2^- (or H_2O_2 at low pH). Excited riboflavin can also oxidize ascorbate, a number of amino acids and proteins and orotic acid. Riboflavin-catalyzed photooxidation results in the production of a number of compounds, most notably methional:

which is the principal compound responsible for the off-flavour in milk exposed to light.

8.4 Colligative Properties of Milk

Colligative properties are those physical properties which are governed by the number, rather than the kind, of particles present in solution. The important colligative properties of milk are its freezing and boiling points (*ca.* −0.522 and 100.15 °C, respectively) and its osmotic pressure (~700 kPa at 20 °C), all of which are interrelated. Since the osmotic pressure of milk remains essentially constant (because it is regulated by that of the cow's blood), the freezing point is also relatively constant.

The freezing point of an aqueous solution is governed by the concentration of solutes in the solution. The relationship between the freezing point of a simple aqueous solution and concentration of solute is described by a relation based on Raoult's Law:

$$T_f = K_f m \tag{8.10}$$

where T_f is the difference between the freezing point of the solution and that of the solvent, K_f is the molal depression constant (1.86 °C for water) and m is the molal concentration of solute. However, this equation is valid only for dilute solutions containing undissociated solutes. Raoult's Law is thus limited to approximating the freezing point of milk.

The freezing point of bovine milk is usually in the range −0.512 to −0.550 °C, with a mean value close to −0.522 °C (Sherbon 1988) or −0.540 °C (Jenness and Patton 1959). Despite variations in the concentrations of individual solutes, the freezing point depression of milk is quite constant since it is proportional to the osmotic pressure of milk (~700 kPa at 20 °C) which is regulated by that of the cow's blood. The freezing point of milk is more closely related to the osmotic pressure of mammary venous blood than to that of blood from the jugular vein.

Owing to their large particle or molecular mass, fat globules, casein micelles and whey proteins do not have a significant effect on the freezing point of milk to which lactose makes the greatest contribution. The freezing point depression in milk due to lactose alone has been calculated to be 0.296 °C. Assuming a mean freezing point depression of 0.522 °C, all other constituents in milk depress the freezing point by only 0.226 °C. Chloride is also an important contributor to the colligative properties of milk. Assuming a Cl$^-$ concentration of 0.032 M and that Cl$^-$ is accompanied by a monovalent cation (i.e., Na$^+$ or K$^+$), the freezing point depression caused by Cl$^-$ and its associated cation is 0.119 °C. Therefore, lactose, chloride and its accompanying cations together account for ~80 % of the freezing point depression in milk. Since the total osmotic pressure of milk is regulated by that of the cow's blood, there is a strong inverse correlation between lactose and chloride concentrations (see Chap. 5).

Natural variation in the osmotic pressure of milk (and hence freezing point) is limited by the physiology of the mammary gland. Variations in the freezing point of milk have been attributed to seasonality, feed, stage of lactation, water intake, breed of cow, heat stress and time of day. These factors are often interrelated but have relatively little influence on the freezing point of milk. Likewise, unit operations in dairy processing which do not influence the net number of osmotically active molecules/ions in solution do not influence the freezing point. Cooling or heating milk causes transfer of salts to or from the colloidal state. However, evidence for an effect of cooling or moderate heating (e.g., HTST pasteurization or minimum UHT processing) on the freezing point of milk is contradictory, perhaps since such changes are slowly reversible over time. Direct UHT-treatment involves the addition of water (through condensed steam). This additional water should be removed during flash cooling, which also removes gasses from the milk, e.g., CO_2, removal of which causes a small increase in freezing point. Vacuum treatment of milk, i.e., vacreation (to remove taints), has been shown to increase its freezing point, presumably by degassing. However, if vacuum treatment is severe enough to cause a significant

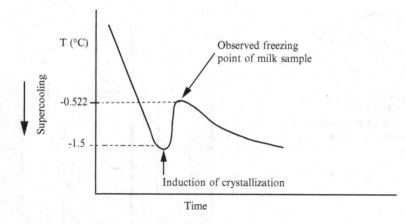

Fig. 8.4   Temperature–time curve for the freezing of milk

loss of water, the freezing point will be reduced, thus compensating fully or partially for the loss of CO_2. Fermentation of milk has a large effect on its freezing point since fermentation of 1 mol lactose results in the formation of 4 mol lactic acid. Likewise, fermentation of citrate influences the freezing point of milk.

Accurate measurement of the freezing point depression in milk requires great care. The principle used is to supercool the milk sample (by 1.0–1.2 °C), to induce crystallization of ice, after which the temperature increases rapidly to the freezing point of the sample (Fig. 8.4). For water, the temperature at the freezing point will remain constant until all the latent heat of fusion has been removed (i.e., until all the water is frozen). However, for milk the temperature is stable at this maximum only momentarily and falls rapidly because ice crystallization causes concentration of solutes which leads to a further depression of freezing point. The observed freezing point of milk (maximum temperature after initiation of crystallization) is not the same as its true freezing point since some ice crystallization will have occurred before the maximum temperature is reached. Correction factors have been suggested to account for this but, in practice, it is usual to report the observed freezing point when other factors (particularly the degree of supercooling) have been standardized. Therefore, the observed freezing point of milk is empirical and great care is necessary to standardize methodology.

The Hortvet technique (originally described in 1921) was used widely to estimate the freezing point of milk. The original apparatus consisted of a tube, containing the milk sample and a thermometer calibrated at 0.001 °C intervals, which was placed in ethanol in a Dewar flask which was cooled indirectly by evaporation of ether (caused by pulling or pumping air through the ether, Fig. 8.5). This apparatus has been modified to include mechanical refrigeration and various stirring or tapping devices to initiate crystallization. The early Hortvet cryoscopes used thermometers calibrated in degrees Hortvet (°H; values in °H are about 3.7 % lower than in °C). The difference between °H and °C originates from differences in the freezing points of sucrose solutions measured using the Hortvet cryoscope and procedure and their

Fig. 8.5 Schematic representation of a Hortvet cryoscope. 1,4, Inlet and outlet for air or vacuum supply; 2, thermometer calibrated at 0.001 °C intervals; 3, agitator; 5, milk sample; 6, glass tube; 7, alcohol; 8, ether cooled by evaporation; 9, insulated jacket

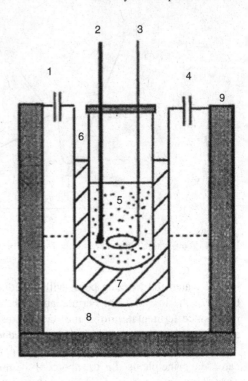

true freezing points. IDF (1983) suggested the following formulae to interconvert °H and °C:

$$°C = 0.96418°H + 0.00085$$

$$°H = 1.03711°C - 0.00085$$

However, it is now recommended that thermometers be calibrated in °C. More recently, thermistors have been used instead of mercury thermometers. Cryoscopes based on dew point depression have also been approved for use. These latter instruments also use thermistors and are based on changes in osmotic pressure. Thermistor cryoscopes are now used more widely than Hortvet instruments.

Measurement of the freezing point depression of milk is used to estimate the degree of adulteration of milk with added water. Assuming an average freezing point of −0.550 °C, the amount of added water can be calculated from:

$$\% \text{ added water} = \frac{0.550 - \Delta T}{0.550} \times (100 - TS) \tag{8.11}$$

where ΔT is the observed freezing point depression of the test sample and TS is the % total solids in the milk. Interpretation of freezing point values when assaying milk suspected of being adulterated with water requires care. Milk with a freezing point of $-0.525\,°C$ or below is usually presumed to be unadulterated. Due to greater variation in the freezing point of milks drawn from individual animals than of bulk milk, specifications for the freezing point of bulk milk are more stringent than those for milks from individual animals. Finally, it should be noted that estimation of the adulteration of milk with water depends on the constancy of the freezing point (as discussed above). Adulteration of milk with isotonic solutions, e.g., ultrafiltration permeate (which is being considered for standardization of the protein content of milk) will not be detected by this technique.

8.5 Interfacial Tension

A phase can be defined as a domain bounded by a closed surface in which parameters such as composition, temperature, pressure and refractive index are constant but change abruptly at the interface. The principal phases in milk are its serum and fat and the most important interfaces are air/serum and fat/serum. If present, air bubbles, and ice, fat or lactose crystals will also constitute phases. Forces acting on molecules or particles in the bulk of a phase differ from those at an interface since the former are attracted equally in all directions while those at an interface experience a net attraction towards the bulk phase (Fig. 8.6).

This inward attraction acts to minimise the interfacial area and the force which causes this decrease in area is known as the interfacial tension (γ). If one phase is air, the interfacial tension is referred to as surface tension. Interfacial tension can be expressed as force per unit length ($N\,m^{-1}$) or the energy needed to increase the interfacial area by a unit amount ($J\,m^{-2}$ or $N\,m^{-1}$).

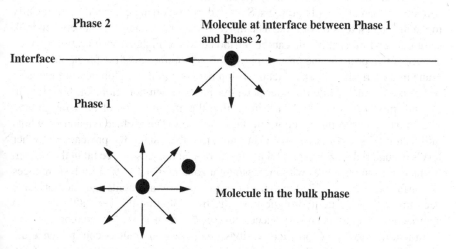

Fig. 8.6 Schematic representation of the forces acting on a molecule or particle in a bulk phase or at an interface

In addition to temperature (which decreases γ), the properties of interfaces are governed by the chemistry of the molecules present, their concentration and their orientation with respect to the interface. Solutes adsorbed at an interface which reduce interfacial tension are known as surface active agents or surfactants. Surfactants reduce interfacial tension by an amount given, under ideal conditions, by the Gibbs' equation:

$$d\gamma = - RT\Gamma d\ln a \qquad (8.12)$$

where Γ is the excess concentration of the solute at the interface over that in the bulk solution, a is the activity of the solute in the bulk phase and R and T are the Universal Gas Constant and temperature (in Kelvin), respectively. Therefore, the most effective surfactants are those which accumulate most readily at an interface.

Interfacial tension may be measured by a number of techniques, including determining how far a solution rises in a capillary, by measuring the weight, volume or shape of a drop of solution formed at a capillary tip, measuring the force required to pull a flat plate or ring from the surface or the maximum pressure required to form a bubble at a nozzle immersed in the solution. Ring or plate techniques are most commonly used to determine γ of milk.

Reported values for the interfacial tension between milk and air vary from 40 to 60 N m^{-1}, with an average of ~52 N m^{-1} at 20 °C (Singh et al. 1997). At 20–40 °C, the interfacial tension between milk serum and air is ~48 N m^{-1} while that between sweet cream buttermilk and air is ~40 N m^{-1} (Walstra and Jenness 1984). Surface tension values for rennet whey, skim milk, whole milk, 25 % fat cream and sweet-cream buttermilk are reported to be 51–52, 52–52.5, 46–47.5, 42–45 and 39–40 N m^{-1}, respectively (Jenness and Patton 1959).

The principal surfactants in milk are its proteins, phospholipids, mono- and diglycerides and salts of free fatty acids. The immunoglobulins are less effective surfactants than other milk proteins. Salts and lactose do not contribute significantly to the interfacial tension of milk. The difference in interfacial tension between milk serum/air and buttermilk/air can be attributed to the higher concentration of very surface active proteins and protein-phospholipid complexes of the fat globule membrane in buttermilk. The interfacial tension between milk fat globules and the milk serum is ~2 Nm^{-1}, while the interfacial tension between non-globular, liquid, milk fat and milk serum is ~15 Nm^{-1}, indicating the effectiveness of milk fat globule membrane material in reducing interfacial tension. The surface tension of whole milk is a little lower than that of skim milk, possibly due to the presence of higher levels of material from the fat globule membrane and traces of free fat in the former. Surface tension decreases with increasing fat content up to ~4 %. Lipolysis reduces the surface tension of milk due to the liberation of free fatty acids and attempts have been made to estimate hydrolytic rancidity by exploiting this fact, although such approaches have not been very successful (see Sherbon 1988, for references).

In addition to its composition, various processing parameters can influence the surface tension of milk. The surface tension of whole and skim milk decreases with increasing temperature. Surface tension also varies with the temperature history and age of the milk and with the time required for measurement. Homogenization of raw milk reduces surface tension because lipolysis by the indigenous milk lipase is stimulated and surface-active fatty acids released. Homogenization of pasteurized milk causes a slight increase in surface tension. Pasteurization of milk has little effect on its surface tension although heating milk to sterilization temperatures causes a slight increase in surface tension, resulting from denaturation and coagulation of proteins which are then less effective as surfactants.

8.6 Acid-Base Equilibria

The acidity of a solution is normally expressed as its pH, which may be defined as:

$$pH = -\log_{10} a_{H+} \tag{8.13}$$

or

$$pH = -\log_{10} f_H \left[H^+ \right] \tag{8.14}$$

where a_{H+} is the activity of the hydrogen ion, $[H^+]$ its concentration and f_H its activity coefficient. For many dilute solutions, $f_H \approx 1$ and pH can thus be closely approximated by the negative logarithm of the hydrogen ion concentration.

The pH of milk at 25 °C is usually in the range 6.5–6.7, with a mean value of 6.6. The pH of milk is influenced much more by temperature than is the pH of dilute buffers, principally due to the temperature dependence of the solubility of calcium phosphate (see Chap. 5). pH varies with stage of lactation; colostrum can have a pH as low as 6.0. Mastitis tends to increase the pH since increased permeability of the mammary gland membranes means that more blood constituents gain access to the milk; the pH of cow's blood is 7.4. The difference in pH between blood and milk results from the active transport of various ions into the milk, precipitation of colloidal calcium phosphate (CCP; which results in the release of H^+) during the synthesis of casein micelles, higher concentrations of acidic groups in milk and the relatively low buffering capacity of milk between pH 6 and 8 (see Singh et al. 1997).

An important characteristic of milk is its buffering capacity, i.e., resistance to changes in pH on addition of acid or base. A pH buffer resists changes in the $[H^+]$ (ΔpH) in the solution and normally consists of a weak acid (HA) and its corresponding anion (A$^-$, usually present as a fully dissociatable salt). An equilibrium thus exists:

$$HA \rightleftharpoons H^+ + A^- \tag{8.15}$$

The addition of H^+ to this solution favours the back reaction while the addition of base favours the forward reaction. The weak acid/salt pair thus acts to minimise ΔpH. An analogous situation exists for buffers consisting of a weak base and its salt. The pH of a buffer can be calculated from the concentration of its components by the Henderson-Hasselbach equation:

$$pH = pK_a + \log\frac{\left[A^-\right]}{\left[HA\right]} \tag{8.16}$$

where pK_a is the negative logarithm of the dissociation constant of the weak acid, HA. A weak acid/salt pair is most effective in buffering against changes in pH when the concentrations of acid and salt are equal, i.e., at $pH = pK_a$ of HA. The effectiveness of a buffer is expressed as its buffering index

$$\frac{dB}{dpH} \tag{8.17}$$

Milk contains a range of groups which are effective in buffering over a wide pH range. The principal buffering compounds in milk are its salts (particularly soluble calcium phosphate, citrate and bicarbonate) and acidic and basic amino acid side-chains on proteins (particularly the caseins). The contribution of these components to the buffering of milk was discussed in detail by Singh et al. (1997) and McCarthy and Singh (2009).

In theory, it should be possible to calculate the overall buffering properties of milk by combining the titration curves for all components but in practice this is not done since K_a values for many milk constituents are uncertain. Titration curves obtained for milk are very dependent on the technique used, and forward and back titrations may show a marked hysteresis in buffering index (Fig. 8.7a). The buffering curve for milk titrated from pH 6.6 to pH 11.0 (Fig. 8.7b) shows decreasing buffering from pH 6.6 to ~pH 9. Milk has good buffering capacity at high pH values (> pH 10), due principally to lysine residues and carbonate anions. When milk is back-titrated from pH 11.0 to pH 3.0, little hysteresis is apparent (Fig. 8.7b). Buffering capacity increases below pH 6.6 and reaches a maximum around pH 5.1. This increase, particularly below pH 5.6, is a consequence of the dissolution of CCP. The resulting phosphate anions buffer against a decrease in pH by combining with H^+ to form HPO_4^{2-} and $H_2PO_4^-$. If an acidified milk sample is back titrated with base (Fig. 8.7a), buffering capacity is low at ~pH 5.1 and the maximum in the buffering curve occurs at a higher pH value (~6.3) due to the formation of CCP from soluble calcium phosphate with the concomitant release of H^+. Ultrafiltration (UF) causes a steady increase in the buffering capacity of UF retentates due to increased concentrations of caseins, whey proteins and colloidal salts and makes it difficult to obtain an adequate decrease in pH during the manufacture of cheese from UF retentates.

Acid-base equilibria in milk are influenced by processing operations. Pasteurization causes some change in pH due to the loss of CO_2 and precipitation of calcium phosphate. Higher heat treatments (>100 °C) result in a decrease in pH due to the degradation of lactose to various organic acids, especially formic acid (see Chap. 9). Slow freezing of milk causes a decrease in pH since the formation of ice crystals during slow freezing concentrates the solutes in the aqueous phase of milk,

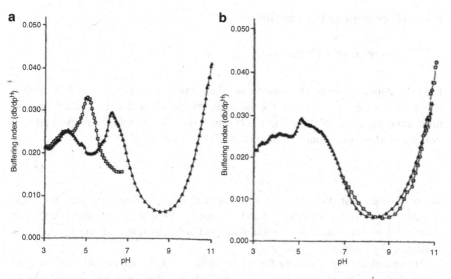

Fig. 8.7 (a) Buffering curves of milk titrated from its initial pH (6.6) to pH 3.0 with 0.5 N HCl [*open square*] and back-titrated to pH with 11.0 with 0.5 N NaOH [*filled triangle*]. (b) Buffering curves of milk titrated from its initial pH (6.6) to pH 11.0 with 0.5N NaOH [*open square*] and back-titrated to pH with 3.0 with 0.5 N HCl [*filled triangle*] (From Singh et al. 1997)

with the precipitation of calcium phosphate and a concomitant release of H⁺. Rapid freezing does not have this effect since there is insufficient time for the above changes to occur. Concentration of milk by evaporation of water causes a decrease in pH as the solubility of calcium phosphate is exceeded, resulting in the formation of more colloidal calcium phosphate. Conversely, dilution causes colloidal calcium phosphate to go into solution, with a corresponding decrease in [H⁺] (see Chap. 5).

The buffering capacity of milk is often estimated by determining its titratable acidity, which involves titrating a sample of milk, containing a suitable indicator (usually phenolphthalein), with NaOH and thus is a measure of the buffering capacity of the milk between its natural pH and the phenolphthalein end-point (i.e., between ~pH 6.6 and *ca.* 8.3). Titratable acidity is normally used to estimate the freshness of milk and to monitor the production of lactic acid during fermentation. Fresh milk typically requires 1.3 to 2.0 milliequivalents OH⁻ to titrate 100 ml from pH 6.6 to pH 8.3 (13 to 20 ml of 0.1 M NaOH), i.e., fresh milk has a titratable acidity of 0.14 to 0.16 %, expressed as lactic acid.

A high titratable acidity for fresh milk suggests high concentrations of proteins and/or other buffering constituents. Titratable acidity varies only slightly with the breed of cow, although the values for individual cows can vary more widely (0.08 to 0.25 % as lactic acid). The liberation of fatty acids on lipolysis can interfere with the estimation of titratable acidity in high-fat products. Precipitation of calcium phosphate (with a concomitant decrease in pH) and "fading of the phenolphthalein end-point" can occur during titration and thus the titratable acidity value obtained is influenced by the speed of titration.

8.7   Rheological Properties

8.7.1   Newtonian Behaviour

Under certain conditions (e.g., moderate shear rates, at fat contents < 40 % and at temperatures >40 °C, at which the fat is liquid and no cold agglutination occurs) milk, skim milk and cream are, in effect, fluids with Newtonian rheological properties. Newtonian behaviour can be described by the equation

$$\tau = \eta \dot{\gamma} \tag{8.18}$$

where τ is the shear stress (force per unit area, Pa), $\dot{\gamma}$ the shear rate (rate of change of velocity across the stream, s^{-1}) and η is the coefficient of viscosity (Pa s). The coefficient of viscosity for a Newtonian fluid is independent of shear rate but is influenced by temperature and pressure.

The coefficient of viscosity for whole milk at 20 °C but not affected by cold agglutination of fat globules is ~2.127 mPa s. Values for water and milk plasma at 20 °C are 1.002 and 1.68 mPa s, respectively. Casein, and to a lesser extent fat, are the principal contributors to the viscosity of milk; whey proteins and low molecular mass species have less influence.

The viscosity of milk and Newtonian milk products is influenced by composition, concentration, pH, temperature, thermal history and processing operations.

The Newtonian coefficient of viscosity at a given temperature for milk, creams and some concentrated milk products is related to the concentration of individual components by Eiler's equation

$$\eta = \eta_\circ \frac{1 + 1.25 \Sigma \left(\varphi_i\right)^2}{1 - \Sigma \left(\varphi_i\right)/\varphi_{max}} \tag{8.19}$$

where η_0 is the coefficient of viscosity of the portion of the fluid consisting of water and low molecular mass species other than lactose and φ is the volume fraction of all dispersed particles that are at least an order of magnitude larger than water. The volume fraction of any component is given by:

$$\varphi_i = V_i c_{v,i} \tag{8.20}$$

where V_i is the voluminosity of component i (in m^3 kg^{-1} dry component) and $c_{v,i}$ is the volume concentration of the component in the product (m^3 kg^{-1} product). The voluminosity of fat in fat globules is ~1.11×10^{-3} m^3 kg^{-1}, that of casein micelles is ~3.9×10^{-3} m^3 kg^{-1}, whey proteins ~1.5×10^{-3} m^3 kg^{-1} and lactose ~1×10^{-3} m^3 kg^{-1}. For milk

$$\varphi \approx \varphi_f + \varphi_c + \varphi_w + \varphi_l \tag{8.21}$$

where φ_f, φ_c, φ_w, φ_l are the volume fractions of fat, casein, whey proteins and lactose, respectively. φ_{max} is the assumed value of $\Sigma(\varphi_i)$ for maximum packing of all dispersed particles (0.9 for fluid milk products).

Increasing pH increases viscosity slightly (perhaps as a consequence of micellar swelling) while a small decrease in pH reduces viscosity although a large decrease in pH causes aggregation of casein micelles. Viscosity is inversely related to temperature. The viscosity of milk shows thermal hysteresis; it usually shows greater viscosity during heating than during subsequent cooling, probably due to the melting and crystallization behaviour of milk triglycerides.

The viscosity of milk and creams tends to increase slightly with age, due in part to changes in ionic equilibria. Heating skim milk to an extent that denatures most of the whey proteins increases its viscosity by about 10 %. Homogenization of whole milk has little effect on its viscosity. The increase in the volume fraction of fat on homogenization is compensated by a decrease in the volume fractions of casein and whey proteins because some skim milk proteins are adsorbed at the fat-oil interface. Pasteurization has no significant effect on the rheology of whole milk.

8.7.2 Non-Newtonian Behaviour

Raw milks and creams exhibit non-Newtonian rheological properties when they are held under conditions which favour cold agglutination of fat globules (<40 °C and low shear rates). Under such conditions, they show thixotropic (shear thinning) behaviour, i.e., their apparent viscosity (η_{app}) is inversely related to shear rate. Aggregates of fat globules and the milk serum trapped in their interstitial spaces have a large effective volume due to their irregular shapes. Increasing the shear rate causes increased shear stress to be applied to the aggregates which can disperse, yielding smaller or more rounded ones. Disaggregation reduces the interstitial space between fat globules, thereby reducing the total volume fraction of the fat phase and consequently reducing the η_{app} of the product. When the shearing force applied to the fluid increases in excess of the forces which hold the aggregates together, increases in shear rate cause increasingly smaller changes in apparent viscosity. Thus, at high shear rates the fluid will exhibit Newtonian behaviour.

Increasing the fat content and/or reducing the temperature favours non-Newtonian behaviour. Low temperatures promote cold agglutination of fat globules and thus increase both η_{app} and deviation from Newtonian behaviour. The temperature at cream separation also influences the rheological properties of the resulting cream. Cream prepared by separation above 40 °C shows less deviation from Newtonian behaviour since cryoglobulins are lost in the skim milk, resulting in less agglutination. Apparent viscosity is also influenced by the shear history of the product. The reformation of bonds between fat globules in aggregates requires time and thus the η_{app} versus shear rate ($\dot{\gamma}$) curves exhibit hysteresis. η_{app} increases after cessation of shearing (as aggregates are reformed) but usually does not return to its original value. Hysteresis is apparent in products containing aggregates caused by cold agglutination or homogenization.

Coalescence of fat globules does not change η_{app} since the volume fraction of the fat is not changed. However, partial coalescence can result in an increase in η_{app} due to entrapment of milk serum in aggregates. Indeed, high-fat creams can exhibit rheopectic (shear thickening) behaviour since shearing can cause partial coalescence of fat globules.

In addition to the general decrease in viscosity with increasing temperature, heating milk can also influence its rheology by heat-induced denaturation of cryoglobulins and/or other whey proteins. Concentration of milk, e.g., by ultrafiltration, prior to heating results in a greater increase in η_{app} than in milk heated before concentration.

The addition of hydrocolloids (e.g., carrageenans, pectins or carboxymethyl cellulose) as thickening agents will greatly increase the apparent viscosity of the product. The production of extracellular polysaccharides by certain bacteria will also increase the viscosity of milk products.

8.7.3   Rheology of Milk Gels

Gels are viscoelastic bodies, the rheological properties of which can be described by two parameters, the storage modulus (G', which is a measure of its elasticity) and the modulus (G'', which is a measure of its viscous nature). The combined viscoelastic modulus (G^*) is a measure of the overall resistance of a gel to deformation. These moduli are often highly dependent on the time-scale of deformation. Another important parameter of a food gel is its yield stress.

Although the gelation properties of whey proteins are of great importance in many foods (see Mulvihill 1992) and it is possible to form a weak gel in creams by the formation of a continuous network of fat globules, most important milk gels are those involving casein micelles which can be made to form a gel matrix either by isoelectric precipitation (acid-induced gel) or by the action of a proteolytic enzyme (rennet-induced gel). Both gel types are relatively similar but, over long deformation times, rennet-induced gels have more liquid character than acid gels, which means that the former can flow under their own weight while acid gels are more likely to retain their shape. Rennet-induced gels also have a greater tendency to synerese and have a higher yield stress than acid-induced gels.

The firmness of acid- and rennet-induced milk gels is increased by such factors as time elapsed after aggregation of the micelles, gelation at elevated temperature, increasing casein and calcium phosphate concentrations and reduced pH (see Walstra and Jenness 1984). Heat-induced denaturation of whey proteins decreases the firmness of rennet-induced gels but increases the firmness of acid-induced gels. Fat globules weaken casein gels by interrupting the gel matrix. Casein molecules on the surface of fat globules in homogenized milk can participate in formation of the gel network. However, in practice this is influenced by a number of other factors, including preheating, homogenization pressure and temperature, and type of gel (Walstra and Jenness 1984). Indeed, the yield stress of a rennet-induced milk gel may be reduced by homogenization.

8.7.4 Rheological Properties of Milk Fat

The rheological properties of milk fat are greatly influenced by the ratio of solid to liquid fat and by the crystal form of the solid fat. At room temperature (20 °C), milk fat is partially solid and has a plastic consistency, i.e., it exhibits viscoelastic properties; at small deformations (<1 %), it is almost completely elastic due to interactions between the fat crystals which form a weak network but it will begin to flow when subjected to greater deformations. The important parameters in determining the firmness of milk fat include the fraction of solid fat, the shape and size of fat crystals, heterogeneity throughout the fat and the extent to which fat crystals form a network throughout the mass of fat.

The structure of butter and other dairy spreads are further complicated by the presence of aqueous phase droplets and intact fat globules. Water droplets tend to weaken the structure and fat crystals inside intact fat globules cannot participate in the formation of a network throughout the product (see Chap. 3).

8.8 Electrical Conductivity

The specific resistance (ρ, ohm cm) of a substance is related to its dimensions by

$$\rho = \alpha R / l \tag{8.22}$$

where α is the cross-sectional area (cm^2), l is length (cm) and R the measured resistance (ohms). The specific conductance, K (ohm^{-1} cm^{-1}), is the reciprocal of specific resistance. The specific conductance of milk is usually in the range 0.0040 to 0.0055 Ω^{-1} cm^{-1}. Ions (particularly Na$^+$, K$^+$ and Cl$^-$) are responsible for most of the electrical conductivity of milk which is increased by the bacterial fermentation of lactose to lactic acid. Measurement of the specific conductance of milk has been used as a rapid method for detecting subclinical mastitis. The conductivity of solutions is altered by concentration and dilution. However, the usefulness of this in the context of milk (e.g., to detect adulteration with water) is reduced considerably by the influence of concentration or dilution on the precipitation or solubilization of colloidal calcium phosphate. Direct conductivity measurements are thus unsuitable for assessing the amount of water added to milk.

8.9 Thermal Properties of Milk

The specific heat of a substance is the amount of heat energy, in kJ, required to increase the temperature of 1 kg of the substance by 1 K. The specific heat of skim milk increases from 3.906 to 3.993 kJ kg^{-1} K^{-1} from 1 to 50 °C. Values of 4.052 and 3.931 kJ kg^{-1} K^{-1} have been reported for skim and whole milks, respectively, at 80 °C (see Sherbon 1988). The specific heat of milk is inversely related to its total solids content, although discontinuities have been observed around 70 to 80 °C. Skim milk powder usually has a specific heat in the range 1.172 to 1.340 kJ kg^{-1} K^{-1} at 18 to 30 °C.

The specific heat of milk fat (solid or liquid) is ~2.177 kJ kg^{-1} K^{-1}. The specific heat of milk and cream is therefore strongly influenced by their fat content. Over most commonly encountered temperature ranges, the specific heat of high-fat dairy products is complicated by the latent heat absorbed by melting fat (~84 J g^{-1}). The observed specific heat of these products at temperatures over which milk fat melts is thus the sum of the true specific heat and the energy absorbed to provide the latent heat of fusion of milk fat. Specific heat is thus influenced by factors such as the proportion of fat in the solid phase at the beginning of heating, and thus the composition of the fat and its thermal history. The apparent specific heat of high-fat dairy products (sum of "true" specific heat and the energy absorbed by melting of fat) is usually maximal at 15–20 °C and often has a second maximum or inflexion around 35 °C.

The rate of heat transfer through a substance by conduction is given by the Fourier equation for heat conduction

$$\frac{dQ}{dt} = -kA\frac{dT}{dx} \tag{8.23}$$

where dQ/dt is the quantity of heat energy (Q) transferred per unit time (t), A is the cross-sectional area of the path of heat flow, dT/dx is the temperature gradient and k is the thermal conductivity of the medium. The thermal conductivity of whole milk (2.9 % fat), cream and skim milk is ~0.559, ~0.384 and ~0.568 W m^{-1} K^{-1}, respectively. The thermal conductivity of skim milk, whole milk and cream increases slowly with increasing temperature but decreases with increasing levels of total solids or fat, particularly at higher temperatures. In addition to their composition, the thermal conductivity of dried milk products depends on bulk density (weight per unit volume) due to differences in the amount of air entrapped in the powder.

Thermal diffusivity is a measure of the ability of a material to dissipate temperature gradients within it. Thermal diffusivity (α, m^2 s^{-1}) is defined as the ratio of thermal conductivity (k) to volumetric specific heat (density times specific heat, ρc)

$$\alpha = k / \rho c \tag{8.24}$$

The thermal diffusivity of milk (at 15–20 °C) is ~1.25 × 10^{-7} m^2 s^{-1}.

8.10   Interaction of Light with Milk and Dairy Products

The refractive index (n) of a transparent substance is expressed by the relation

$$n = \frac{\sin i}{\sin r} \tag{8.25}$$

where i and r are the angles between the incident ray and the refracted ray of light, respectively, and a perpendicular to the surface of the substance. The refractive index of milk is difficult to estimate due to light scattering by casein micelles and fat globules. However, it is possible to make accurate measurements of the refractive index of milk using refractometers in which a thin layer of sample is used, e.g., the Abbé refractometer. The refractive index of milk at 20 °C using the D-line of the sodium spectrum (~589 nm), n_D^{20}, is normally in the range 1.3440 to 1.3485. The refractive index of milk fat is usually in the range 1.4537 to 1.4552 at 40 °C. Although there is a linear relationship between the solids content (weight per unit volume) and refractive index, determination of percent solids in milk by refractometry is difficult, since the contributions of various milk components differ and are additive. The relationship between the refractive index of milk and its total solids content varies with changes in the concentration and composition of the solutes in milk. However, attempts have been made to measure the total contribution of solids and casein in milk and milk products by estimating the refractive index (see Sherbon 1988; McCarthy and Singh 2009). The specific refractive index (refractive constant), K, is calculated from

$$K = \frac{n^2 - 1}{n^2 + 2} \cdot \frac{1}{\rho} \qquad (8.26)$$

where n is the refractive index and ρ is density. Milk has a specific refractive index of ~0.2075.

Milk contains not only numerous dissolved chemical components but it is also an emulsion with a colloidal continuous phase. Therefore, milk absorbs light of a wide range of wavelengths and also scatters ultraviolet (UV) and visible light due to the presence of particles. Milk absorbs light of wavelengths between 200 and ~380 nm due to the proteins present and between 400 and 520 nm due to fat-soluble pigments (carotenoids). A number of functional groups in milk constituents absorb in the infra-red (IR) region of the spectrum; the $-OH$ groups of lactose absorb at ~9.61 μm, the amide groups of proteins at 6.465 μm and the ester carbonyl groups of lipids at 5.723 μm (Singh et al. 1997; McCarthy and Singh 2009). Since light scattering is reduced at longer wavelengths in the IR region, the absorbance of IR light of specific wavelengths can be used to measure the concentrations of fat, protein and lactose in milk. Instruments using this principle are now widely used in the dairy industry. However, since milk contains ~87.5 % water (which absorbs IR light strongly), it is opaque to light throughout much of the IR region of the spectrum.

Milk contains ~1.62 mg kg^{-1} riboflavin which fluoresces strongly on excitation by light of wavelengths from 400 to 500 nm, emitting light with a $\lambda_{max} = 530$ nm. Milk proteins also fluoresce due to the presence of aromatic amino acid residues; part of the light absorbed at wavelengths around 280 nm is emitted at longer wavelengths.

8.11   Colour of Milk and Milk Products

The white colour of milk results from scattering of visible light by casein micelles and fat globules. Homogenization of milk results in a whiter product due to increased scattering of light by smaller, homogenized, fat globules. The serum phase of milk is greenish due to the presence of riboflavin which is responsible for the character- istic colour of whey.

The colour of dairy products such as butter and cheese is due to fat-soluble pig- ments, especially carotenoids which are not synthesized by the animal but are obtained from plant sources in the diet. Therefore, feed has a major effect on the colour of milkfat. Cows fed on grass produce a more yellow-coloured fat than ani- mals fed on hay or concentrates. The ability of cattle to metabolize carotenes to Vitamin A varies between breeds and between individuals (see Chap. 6).

The most widely used added colorant in dairy products is annatto (E160b) which is a yellow-orange preparation containing apocarotenoid pigments obtained form the pericarp of the seeds of the tropical shrub, *Bixa orellana*. The principal pigment in annatto is *cis*-bixin (methyl 9'-*cis*-6, 6'-diapocarotene-6, 6'-dioate) with smaller amounts of norbixin (*cis*-6, 6'-diapocarotene-6, 6'-dioic acid) (Fig. 8.8). The heat treatments used in extraction normally converts *cis*-bixin to *trans*-bixin which is red and soluble in oil. Annatto is used to give a yellow colour to margarine and to colour "red" Cheddar and other cheeses.

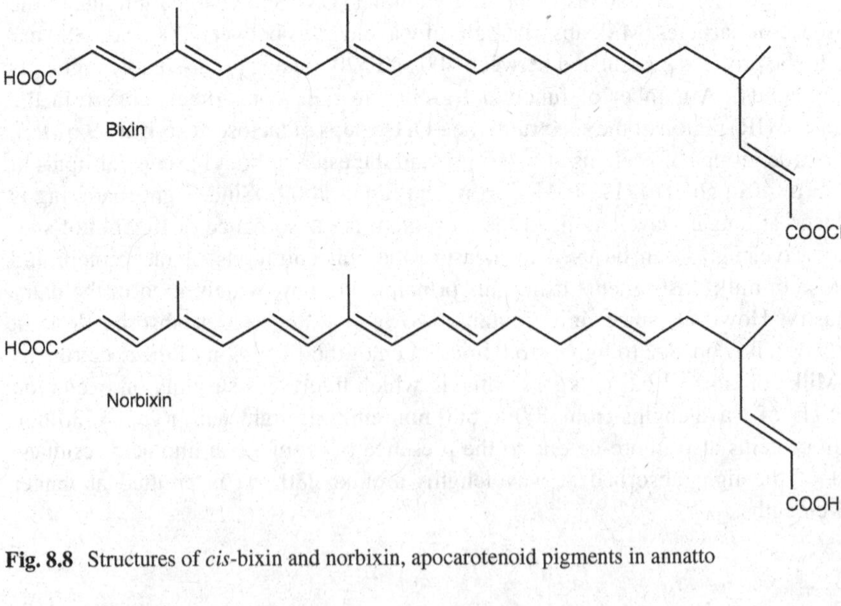

Fig. 8.8 Structures of *cis*-bixin and norbixin, apocarotenoid pigments in annatto

References

IDF. (1983). *Measurement of extraneous water by the freezing point test, Bulletin 154*. Brussels: International Dairy Federation.

Jenness, R., & Patton, S. (1959). *Principles of dairy chemistry*. New York, NY: John Wiley & Sons.

McCarthy, O. J., & Singh, H. (2009). Physico-chemical properties of milk. In P. L. H. McSweeney & P. F. Fox (Eds.), *Advanced dairy chemistry—3—lactose, water, salts and vitamins* (3rd ed., pp. 691–758). New York, NY: Springer.

Mulvihill, D. M. (1992). Production, functional properties and utilization of milk protein products. In P. F. Fox (Ed.), *Advanced dairy chemistry—1—proteins* (pp. 369–404). London, NY: Elsevier Applied Science.

Sherbon, J. W. (1988). Physical properties of milk. In N. P. Wong, R. Jenness, M. Keeney, & E. H. Marth (Eds.), *Fundamentals of dairy chemistry* (3rd ed., pp. 409–460). New York, NY: Van Nostrand Reinhold.

Singh, H., McCarthy, O. J., & Lucey, J. A. (1997). Physico-chemical properties of milk. In P. F. Fox (Ed.), *Advanced dairy chemistry—3—lactose, water, salts and vitamins* (2nd ed., pp. 469–518). London: Chapman & Hall.

Walstra, P., & Jenness, R. (1984). *Dairy chemistry and physics*. New York, NY: John Wiley & Sons.

Chapter 9
Heat-Induced Changes in Milk

9.1 Introduction

In modern dairy technology, milk is almost always subjected to a heat treatment; typical examples are:

Thermization	e.g., 65 °C × 15 s
Pasteurization	
LTLT (low temperature, long time)	63 °C × 30 min
HTST (high temperature, short time)	72 °C × 15 s
Forewarming for sterilization	e.g., 90 °C × 2–10 min, 120 °C × 2 min
Sterilization	
UHT (ultra-high temperature)	130–140 °C × 3–5 s
In-container	110–115 °C × 10–20 min

The objective of the heat treatment varies with the product being produced. Thermization is generally used to kill temperature-sensitive microorganisms, e.g., psychrotrophs, and thereby reduce the microflora of milk for low-temperature storage. The primary objective of pasteurization is to kill pathogens but it also reduces the number of non-pathogenic microoganisms which may cause spoilage, thereby standardizing the milk as a raw material for various products. Many indigenous enzymes, e.g., lipoprotein lipase, are also inactivated, thus contributing to milk stability. Fore-warming (pre-heating) increases the heat stability of milk for subsequent sterilization (as discussed in Sect. 9.7.1). Sterilization renders milk shelf-stable for very long periods, although gelation and flavour changes occur during storage, especially of UHT sterilized milks.

Although milk is a very complex biological fluid containing complex protein, lipid, carbohydrate, salts, vitamins and enzyme systems in soluble, colloidal or emulsified states, it is a very heat-stable system which allows it to be subjected to severe heat treatments without major changes in comparison to other foods if

© Springer International Publishing Switzerland 2015
P.F. Fox et al., *Dairy Chemistry and Biochemistry*,
DOI 10.1007/978-3-319-14892-2_9

subjected to similar treatments. However, numerous biological, chemical and physico-chemical changes occur in milk during thermal processing which affect its nutritional, organoleptic and/or technological properties. The temperature dependence of these changes varies widely, as depicted in general terms in Fig. 9.1 and Table 9.1. The most significant of these changes, with the exception of the killing of bacteria, will be discussed below. In general, the effect(s) of heat on the principal constituents of milk will be considered individually although there are interactions between constituents in many cases.

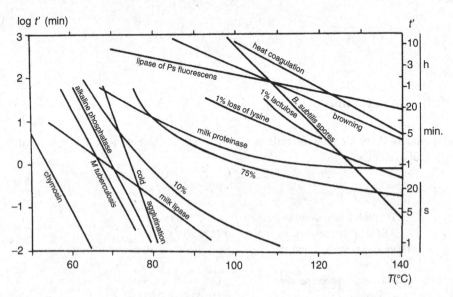

Fig. 9.1 The time needed (t') at various temperatures (T) to inactivate some enzymes and cryoglobulins; to kill some bacteria and spores; to cause a certain degree of browning; to convert 1 % of lactose to lactulose; to cause heat coagulation; to reduce available lysine by 1 %; and to make 10 and 75 % of the whey proteins insoluble at pH 4.6 (from Walstra and Jenness 1984)

Table 9.1 Approximate values for the temperature dependence of some reactions in heated milk (modified from Walstra and Jenness 1984)

Reaction	Activation energy (kJ mol⁻¹)	Q_{10} at 100 °C
Many chemical reactions	80–130	2.0–3.0
Many enzyme-catalyzed reactions	40–60	1.4–1.7
Autoxidation of lipids	40–100	1.4–2.5
Maillard reactions (browning)	100–180	2.4–5.0
Dephosphorylation of caseinate	110–120	2.6–2.8
Heat coagulation of milk	150	3.7
Degradation of ascorbic acid	60–120	1.7–2.8
Heat denaturation of protein	200–600	6.0–175.0
Typical enzyme inactivation	450	50.0
Inactivation of milk proteinase	75	1.9
Killing vegetative bacteria	200–600	6.0–175.0
Killing of spores	250–330	9.0–17.0

9.2 Lipids

Of the principal constituents, the lipids are the least affected by heat. However, significant changes do occur in milk lipids, especially in their physical properties, during heating.

9.2.1 Physico-Chemical Changes

9.2.1.1 Creaming

The principal effect of heat treatments on milk lipids is on creaming of the fat globules. As discussed in Chap. 3, the fat in milk exists as globules, 0.1–20 μm in diameter (mean, 3–4 μm). The globules are stabilized by a complex membrane acquired within the secretory cell and during excretion from the cell. Owing to differences in density between the fat and aqueous phases, the globules float to the surface to form a cream layer. In cows' milk, creaming is faster than predicted by Stokes' Law owing to aggregation of the globules which is promoted by cryoglobulins (a group of immunoglobulins). Buffalo, ovine or caprine milks do not undergo cryoglobulin-dependent agglutination of fat globules and cream very slowly with the formation of a compact cream layer.

When milk is heated to a moderate temperature (e.g., 70 °C×15 min), the cryoglobulins are irreversibly denatured and hence the creaming of milk is impaired or prevented; HTST pasteurization (72 °C×15 s) has little or no effect on creaming potential but slightly more severe conditions have an adverse effect (Fig. 9.2).

Homogenization, which reduces mean globule diameter to <1 μm, retards creaming due to the reduction in globule size but more importantly to the denaturation of cryoglobulins which prevents agglutination. In fact, there are probably two classes of cryoglobulin, one of which is denatured by heating, the other by homogenization.

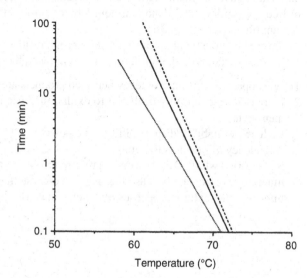

Fig. 9.2 Time-temperature curves for the destruction of *M. tuberculosis* (*dotted line*), inactivation of alkaline phosphatase (*solid line*) and creaming ability of milk (*dashed line*) (from Webb and Johnson 1965)

9.2.1.2   Changes in the Fat Globule Membrane

The milk fat globule membrane (MFGM) itself is altered during thermal processing. Milk is usually agitated during heating, perhaps with foam formation. Agitation, especially of warm milk in which the fat is liquid, may cause changes in globule size due to disruption or coalescence; significant disruption occurs during direct UHT processing. Foaming probably causes desorption of some membrane material and its replacement by adsorption of skim milk proteins. In these cases, it may not be possible to differentiate the effect of heating from the total effect of the process.

Heating to higher than 70 °C denatures membrane proteins with the exposure and activation of various amino acid residues, especially cysteine. This may cause the release of H_2S (which results in the development of an off-flavour) and disulphide interchange reactions with whey proteins, leading to the formation of a layer of denatured whey proteins on the fat globules at high temperatures (>100 °C). The membrane and/or whey proteins may participate in Maillard browning with lactose and the cysteine may undergo β-elimination to dehydroalanine which may react with lysine to form lysinoalanine or with cysteine residues to form lanthionine, leading to covalent cross linking of protein molecules (see Sect. 9.6.3). Membrane constituents, both proteins and phospholipids, are lost from the membrane to the aqueous phase at high temperatures. Much of the indigenous copper in milk is associated with the MFGM and some of it is transferred to the serum on heat processing. Thus, severe heat treatment of cream improves the oxidative stability of butter made from it as a result of the reduced concentration of pro-oxidant Cu in the fat phase and the antioxidant effect of exposed sulphydryl groups.

The consequences of these changes in the MFGM have been the subject of little study, possibly because severely heated milk products are usually homogenized and an artificial membrane, consisting mainly of casein and some whey proteins, is formed; consequently, changes in the natural membrane are not important. Damage to the membrane of unhomogenized products leads to the formation of free (non-globular) fat and consequently to "oiling off" and the formation of a "cream plug" (see Chap. 3).

Severe heat treatment, as during roller drying, results in at least some demulsification of milk fat, with the formation of free fat, which causes (see Chap. 3):

1. The appearance of fat droplets when such products are used in tea or coffee.
2. Increased susceptibility of the fat to oxidation, since it is not protected by a membrane.
3. Reduced wettability/dispersibility of the powder.
4. A tendency of powders to clump.
5. Roller-dried whole milk powder is preferred to spray-dried powder for the manufacture of chocolate because the free fat co-crystallizes with the cocoa fat and improves the textural properties of the chocolate (Liang and Hartel 2004).

9.2.2   Chemical Changes

Severe heat treatments, e.g., frying, may convert hydroxyacids to lactones which have strong, desirable flavours and contribute to the desirable attributes of milk fat in cooking.

Release of fatty acids and some inter-esterification may also occur but such changes are unlikely during the normal processing of milk.

Naturally occurring polyunsaturated fatty acids are methylene-interrupted (i.e., there is a –CH_2– group between each pair of double bonds) but may be converted to conjugated isomers at high temperatures. The four principal isomers of conjugated linoleic acid (CLA) are shown in Fig. 9.3. It is claimed that CLA has anti-carcinogenic properties (Bauman and Lock 2006; Parodi 2006). Rather high concentrations of CLA have been found in heated dairy products, especially processed cheese (Table 9.2). It has been suggested that whey proteins catalyse isomerization.

Conjugated PUFAs occur in the adipose tissue and milk fats of ruminants; they are formed by bacterial action in the rumen (see Bauman and Lock 2006).

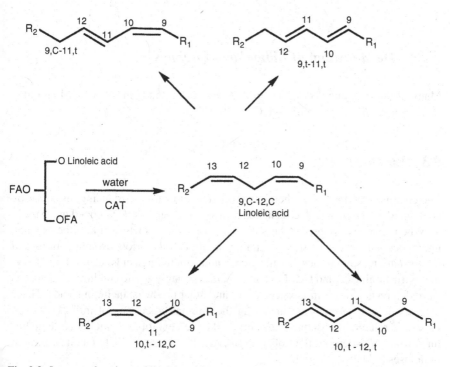

Fig. 9.3 Isomers of conjugated linoleic acid

Table 9.2 Concentration (mg/kg) of conjugated linoleic acid (CLA) isomers in selected foods (modified from Ha et al. 1989)

Sample	mg/CLA/kg food	Fat (%)	CLA in fat (mg/kg)
Parmesan cheese	622.3±15.0	32.3±0.9	1,926.7
Cheddar cheese	440.6±14.5	32.5±1.7	1,355.7
Romano cheese	356.9±6.3	32.1±0.8	1,111.9
Blue cheese	169.3±8.9	30.8±1.5	549.8
Processed cheese	574.1±24.8	31.8±1.1	1,805.3
Cream cheese	334.5±13.3	35.5±1.0	942.3
Blue spread	202.6±6.1	20.2±0.8	1,003.0
Cheese whiz	1815.0±90.3	20.6±1.1	8,810.7
Milk (whole)			
Pasteurized	28.3±1.9	4.0±0.3	707.5
Non-pasteurized	34.0±1.0	4.1±0.1	829.3
Ground beef			
Grilled	994.0±30.9	10.7±0.3	9,289.7
Uncooked	561.7±22.0	27.4±0.2	2,050.0

9.2.3   Denaturation of Indigenous Enzymes

Many of the indigenous enzymes in milk are concentrated in the MFGM and may be denatured on heating milk (see Chap. 10)

9.3   Lactose

The chemistry and physico-chemical properties of lactose, a reducing disaccharide containing galactose and glucose linked by a β-(1–4) bond, were described in Chap. 2.

When severely heated in the solid or molten state, lactose, like other sugars, undergoes numerous changes, including mutarotation, various isomerizations and the formation of numerous volatile compounds, including acids, furfural, hydroxy-methylfurfural, CO_2 and CO. In solution under strongly acidic conditions, lactose is degraded on heating to monosaccharides and other products, including acids. These changes do not normally occur during the thermal processing of milk. However, lactose is relatively unstable under mild alkaline conditions at moderate temperatures where it undergoes the Lobry de Bruyn-Alberda van Ekenstein rearrangement of aldoses to ketoses (Fig. 9.4).

Lactose undergoes at least three heat-induced changes during the processing and storage of milk and milk products.

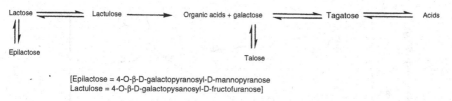

[Epilactose = 4-O-β-D-galactopyranosyl-D-mannopyranose
Lactulose = 4-O-β-D-galactopysanosyl-D-fructofuranose]

Fig. 9.4   Heat-induced changes in lactose under mild alkaline conditions

9.3.1   Formation of Lactulose

On heating at low temperatures under slightly alkaline conditions, the glucose moiety of lactose is epimerized to fructose with the formation of lactulose, which does not occur in nature. The significance of lactulose has been discussed in Chap. 2. Lactulose is not formed during HTST processing but is formed during UHT sterilization (more during indirect than direct heating) and especially during in-container sterilization; therefore, the concentration of lactulose in milk is a useful index of the severity of the heat treatment to which the milk has been subjected (see Fig. 2.21). The concentration of lactulose is probably the best index available at present for differentiating between UHT and in-container sterilized milks and a number of assay procedures have been developed, using HPLC or enzymatic/spectrophotometric principles.

9.3.2   Formation of Acids

Milk as secreted by the cow contains about 200 mg CO_2/L. Owing to its low concentration in air, CO_2 is rapidly and, in effect, irreversibly lost from milk on standing after milking; its loss is accelerated by heating, agitation and vacuum treatment. This loss of CO_2 causes an increase in pH of about 0.1 unit and a decrease in the titratable acidity of nearly 0.02 %, expressed as lactic acid. Under relatively mild heating conditions, this change in pH is more or less off-set by the release of H^+ on precipitation of $Ca_3(PO_4)_2$, as discussed in Sect. 9.4.

On heating at a temperature above 100 °C, lactose is degraded to acids with a concomitant increase in titratable acidity (Figs. 9.5 and 9.6). Formic acid is the principal acid formed; lactic acid represents only ~5 % of the acids formed. Acid production is significant in the heat stability of milk, e.g., when assayed at 130 °C, the pH falls to ~5.8 at the point of coagulation (after ~20 min) (Fig. 9.7). About half of this decrease is due to the formation of organic acids from lactose; the remainder is due to the precipitation of calcium phosphate and dephosphorylation of casein, as discussed in Sect. 9.4.

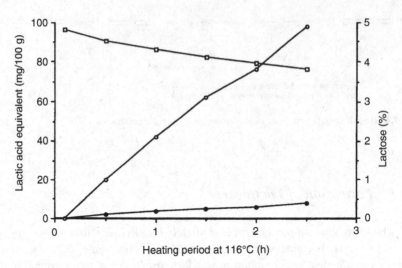

Fig. 9.5 Changes in titratable acidity (*open circle*), lactic acid (*filled circle*) and lactose (*open square*) on heating homogenized milk in sealed cans at 116 °C. Titratable acidity expressed as mg lactic acid/100 g milk (from Gould 1945)

Fig. 9.6 Effect of temperature on the rate of heat-induced production of acid in milk (from Jenness and Patton 1959)

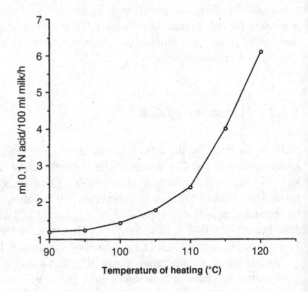

In-container sterilization of milk at 115 °C causes the pH to decrease to ~6 but much of this is due to the precipitation of calcium phosphate; the contribution of acids derived from lactose has not been quantified accurately. Other commercial heat treatments, including UHT sterilization, cause insignificant degradation of lactose to acids.

Fig. 9.7 The pH of samples of milk after heating for various periods at 130 °C with air (*open circle*). O_2 (*filled circle*) or N_2 (*open triangle*) in the headspace above the milk; ↑ coagulation time (from Sweetsur and White 1975)

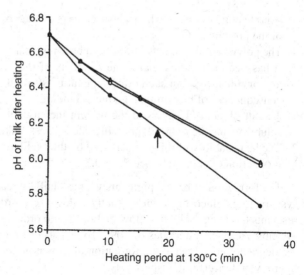

9.3.3 Maillard Browning

The mechanism and consequences of the Maillard reaction were discussed in Chap. 2. The reaction is most significant in severely heat-treated products, especially in-container sterilized milks. However, it may also occur to a significant extent in milk powders stored under conditions of high humidity and high ambient temperature, resulting in a decrease in the solubility of the powder. If cheese contains a high level of residual lactose or galactose (due to the use of a starter unable to utilize galactose; see Chap. 12), it is susceptible to Maillard browning, especially during cooking, e.g., Mozzarella cheese on pizza. Browning may also occur in grated cheese during storage if the cheese contains residual sugars; in this case, the water activity of the cheese ($a_w \sim 0.6$) is favourable for the Maillard reaction. Poorly washed casein and especially whey protein concentrates (which contain 30–60 % lactose) may undergo Maillard browning when used as ingredients in heat-treated foods.

Maillard browning in milk products is undesirable because:

1. The final polymerization products (melanoidins) are brown and hence dairy products which have undergone Maillard browning are discoloured and aesthetically unacceptable.
2. Some of the by-products of Maillard browning have strong flavours (e.g., furfural, hydroxymethylfurfural) which alter the typical flavour of milk.
3. The reactions up to, and including, the Schiff base are reversible and therefore the products are digestible but after the Amadori rearrangement, the products are not metabolically available. Since lysine is the amino acid most likely to be involved and is an essential amino acid, Maillard browning reduces the biological value of proteins. Interaction of lysine with lactose renders the adjacent

peptide bond resistant to hydrolysis by trypsin, thereby reducing the digestibility of the protein.
4. The polymerized products of Maillard browning can bind metals, especially Fe.
5. It has been suggested that some products of the Maillard reaction are toxic and/or mutagenic but such effects are, at most, weak and possibly due to other consequences of browning, e.g., metal binding.
6. The attachment of sugars to the protein increases its hydrophilicity; however, solubility may be reduced, probably due to cross-linking of protein molecules.
7. The heat stability of milk is increased by the Maillard reaction, probably *via* the production of carbonyls (see Sect. 9.7).

The formation of brown pigments via the Maillard reaction, especially in model systems (e.g., glucose-glycine), usually follows zero-order kinetics but the loss of reactants has been found to follow first or second order kinetics in foods and model systems. Activation energies of 109, 116 and 139 kJ mol^{-1} have been reported for the degradation of lysine, the formation of brown pigments and the production of hydroxymethylfurfural, respectively.

Browning can be monitored by measuring the intensity of brown colour, the formation of hydroxymethylfurfural (which may be measured spectrophotometrically, after reaction with thiobarbituric acid, or by HPLC), loss of available lysine (e.g., by reaction with 2,4-dinitrofluorobenzene) or by the formation of furosine. Furosine is formed on the acid hydrolysis of lactulosyl lysine (the principal Maillard product formed during the heating of milk). During acid hydrolysis, lactulosyl lysine is degraded to fructosylysine which is then converted to pyridosine, furosine and carboxymethyl lysine (Fig. 9.8). Furosine may be determined by ion-exchange chromatography, GC or HPLC and is considered to be a very good indicator of Maillard browning and the severity of heat treatment of milk (Erbersdobler and Dehn-Müller 1989). The effects of time and temperature on the formation of furosine are shown in Fig. 9.9. The concentration of furosine is highly correlated with the concentrations of HMF and carboxymethyl lysine. The concentration of furosine in commercial UHT milks is shown in Fig. 9.10.

Dicarbonyls, which are among the products of the Maillard reaction, can react with amines in the Strecker reaction, producing a variety of flavourful compounds (Fig. 2.32). The Maillard and especially the Strecker reactions can occur in cheese and may be significant contributors to flavour; in this case, the dicarbonyls are probably produced via biological, rather than thermal, reactions.

9.4 Milk Salts

Although the organic and inorganic salts of milk are relatively minor constituents in quantitative terms, they have major effects on many aspects of milk, as discussed in Chap. 5. Heating has little effect on milk salts with two exceptions, carbonates and calcium phosphates. Most of the potential carbonate occurs as CO_2 which is lost on heating, with a consequent increase in pH. Among the salts of milk, calcium

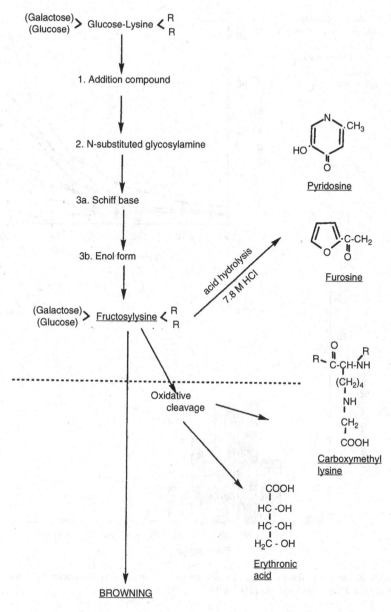

Fig. 9.8 Initial steps of the Maillard reaction with the formation of furosine (after hydrolysis with 7.8 M HCl) as well as of *N*-ε-carboxymethyl lysine and erythronic acid (from Erbersdobler and Dehn-Müller 1989)

phosphate is unique in that its solubility decreases with increasing temperature. On heating, the solubility of calcium phosphate decreases and is transferred to the colloidal phase, associated with the casein micelles, with a concomitant decrease in the concentration of calcium ions and pH (see Chap. 5). These changes are

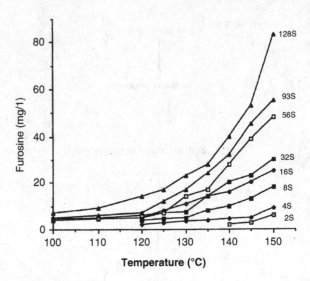

Fig. 9.9 Effect of heating temperature and time on the concentration of furosine in directly heated UHT milks (from Erbersdobler and Dehn-Müller 1989)

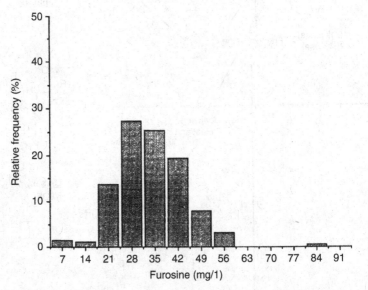

Fig. 9.10 Relative distribution of the furosine concentration in 190 commercial UHT milks in increments of 7 mg furosine (from Erbersdobler and Dehn-Müller 1989)

reversible on cooling if the heat treatment was not severe. Following severe heat treatment, the heat-induced colloidal calcium phosphate is probably insoluble but some indigenous colloidal calcium phosphate dissolves on cooling to partly restore the pH and salts equilibrium. The situation becomes rather complex in severely heated milk due to the decrease in pH caused by thermal degradation of lactose and dephosphorylation of casein.

The cooling and freezing of milk also cause shifts in the salts equilibria in milk, including changes in pH, as discussed in Chaps. 2 and 5.

9.5 Vitamins

Many of the vitamins in milk are relatively heat labile, as discussed in Chap. 6.

9.6 Proteins

The proteins of milk are probably the constituents most affected by heating. Some of the changes involve interaction with salts or sugars and although not always fully independent of changes in other constituents, the principal heat-induced changes in proteins are discussed in this section.

9.6.1 Enzymes

As discussed in Chap. 10, milk contains about 60 indigenous enzymes derived from the secretory cells or from blood. Stored milk may also contain enzymes produced by microorganisms. Both indigenous and bacterial enzymes can have undesirable effects in milk and dairy products. Although not the primary objective of thermal processing, some of the indigenous enzymes in milk are inactivated by the commercially used heat processes, although many are relatively heat stable (Fig. 9.11).

The thermal denaturation of indigenous milk enzymes is important from two main viewpoints:

1. To increase the stability of milk products. Lipoprotein lipase is probably the most important in this regard as its activity leads to hydrolytic rancidity. It is extensively inactivated by HTST pasteurization but heating at 78 °C × 10 s is

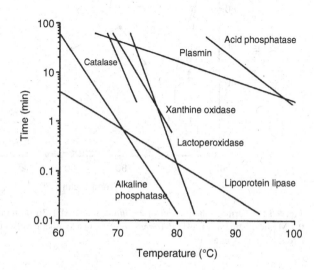

Fig. 9.11 Time-temperature combinations to which milk must be heated to inactivate some indigenous milk enzymes (from Walstra and Jenness 1984)

required to prevent lipolysis. Plasmin activity is actually increased by HTST pasteurization due to inactivation of inhibitors of plasmin and/or of plasminogen activators.

2. The activity of selected enzymes is used as indices of thermal treatments, e.g., alkaline phosphatase (HTST pasteurization), γ-glutamyl transferase (index of heating in the range 72–80 °C) or lactoperoxidase (80–90 °C).

9.6.1.1   Microbial Enzymes

The widespread use of refrigerated storage of milk at farm and factory for extended periods has led to psychrotrophs, especially *Pseudomonas fluorescens*, becoming the dominant microorganisms in raw milk supplies. Psychrotrophs are quite heat labile and are readily killed by HTST pasteurization and even by thermization. However, they secrete extracellular proteinases, lipases and phospholipases that are extremely heat stable—some are not completely inactivated by heating at 140 °C for 1 min and thus partially survive UHT processing. If the raw milk supply contains high numbers of psychrotrophs (>10⁶/mL), the amounts of proteinase and lipase that survive UHT processing may be sufficient to cause off-flavours, such as bitterness, unclean and rancid flavours, and perhaps gelation.

Surprisingly, the proteinases and lipases secreted by many psychrotrophs have relatively low stability in the temperature range 50–65 °C, Fig. 9.12 (the precise value depends on the enzyme). Thus, it is possible to reduce the activity of these enzymes

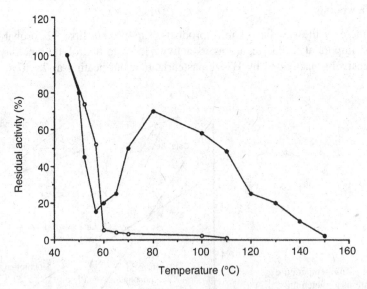

Fig. 9.12 Thermal inactivation of *Ps. fluorescens* AFT 36 proteinase on heating for 1 min in 0.1 M phosphate buffer, pH 6.6 (*open circle*) or in a synthetic milk salts buffer, pH 6 (*filled circle*) (from Stepaniak et al. 1982)

in milk by a low temperature inactivation (LTI) treatment (e.g., 60 °C × 5–10 min) before or after UHT processing. Inactivation of the proteinase by LTI appears to be due mainly to proteolysis; in the native state, the enzyme is tightly folded and resistant to proteolysis by other proteinase molecules in its neighbourhood but at ~60 °C, some molecules undergo conformational changes, rendering them susceptible to proteolysis by proteinase molecules which are still active. On increasing the temperature further, all proteinase molecules are denatured and inactive but they can renature on cooling. Since this mechanism does not apply to purified lipase, the mechanism of LTI of lipase is not clear [for reviews on enzymes from psychrotrophs see Driessen (1989) and McKellar (1989)].

9.6.2 Denaturation of Other Biologically-Active Proteins

Milk contains a range of biologically active proteins, e.g., vitamin-binding proteins, immunoglobulins, metal-binding proteins, antibacterial proteins (lactoferrin, lysozyme, lactoperoxidase), various growth factors and hormones (see Chaps. 4 and 11). These proteins play important nutritional and physiological functions in the neonate. All these proteins are relatively heat labile—some are inactivated by HTST pasteurization and probably all are inactivated by UHT and more severe heat treatments. Inactivation of these biologically active proteins may not be particularly important when milk is used in the diet of adults but may be highly significant in infant formulae; consequently, infant formulae may be supplemented with some of these proteins, e.g. lactoferrin.

9.6.3 Denaturation of Whey Proteins

The whey proteins, which represent about 20 % of the proteins of bovine milk, are typical globular proteins with high levels of secondary and tertiary structures, and are, therefore, susceptible to denaturation by various agents, including heat. The denaturation kinetics of whey proteins, as measured by loss of solubility in saturated NaCl at pH 4.6, are summarized in Fig. 9.13. Thermal denaturation is a traditional method for the recovery of proteins from whey as "lactalbumin" (coagulation is optimal at pH 6 and ~90 °C for 10 min, see Chap. 4) or Ricotta or Queso Blanco cheese (see Chap. 12).

The order of heat stability of the whey proteins, measured by loss of solubility, is: α-lactalbumin (α-la) > β-lactoglobulin (β-lg) > blood serum albumin (BSA) > immunoglobulins (Ig) (Fig. 9.14). However, when measured by differential scanning calorimetry, quite a different order is observed: Ig > β-lg > α-la > BSA. In the case of α-la, the discrepancy appears to be explained by the fact that it is a metallo (Ca)-protein which renatures following thermal denaturation. However, the Ca-free apoprotein is quite heat labile, a fact which is exploited in the isolation of α-la. The Ca^{2+} is bound

Fig. 9.13 Heat denaturation of whey proteins on heating skim milk at various temperatures (°C) as measured by precipitability with saturated NaCl (from Jenness and Patton 1959)

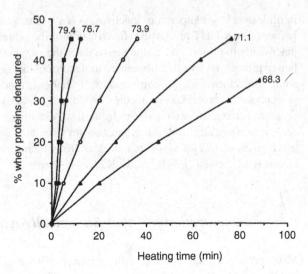

Fig. 9.14 The denaturation of the total (*open square*) and individual whey proteins in milk, heated at various temperatures for 30 min; β-lactoglobulin (*filled square*), α-lactalbumin (*open circle*), proteose peptone (*filled circle*), immunoglobulins (*open triangle*), and serum albumin (*filled triangle*) (from Webb and Johnson 1965)

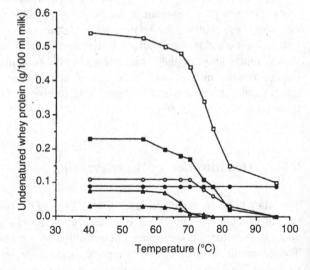

in a pocket to the carboxylic acid groups of three Asp residues and the carbonyls of an Asp and a Lys residue (see Chap. 4). The carboxylic acid groups become protonated $\lesssim$pH 5 and lose their ability to bind Ca^{2+}; the apo-protein can be aggregated by heating to ~55 °C, leaving mainly β-lg in solution. Apo-lactoferrin is also considerably less stable than the intact protein.

The denaturation of α-la and β-lg in milk follows first and second order kinetics, respectively (Fig. 9.15). Both proteins show a change in the temperature-dependence of denaturation at ~90 °C (Fig. 9.15).

Fig. 9.15 Arrhenius plot of
the rate constant for the heat
treatment of α-lactalbumin
(*open square*) and
β-lactoglobulin (*open circle*)
(from Lyster 1970)

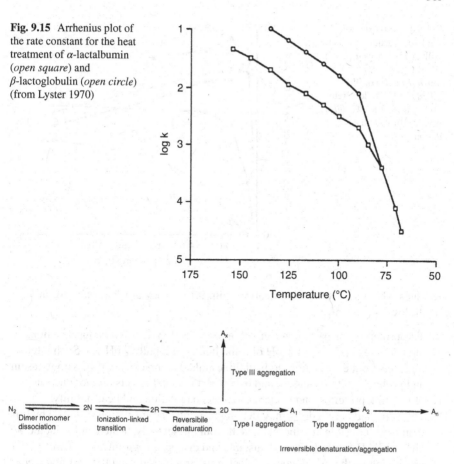

Fig. 9.16 Stages in the thermal denaturation of β-lactoglobulin (from Mulvihill and Donovan 1987)

The mechanism of the thermal denaturation of β-lg has been studied extensively; the sequence of events is shown schematically in Fig. 9.16. At ~20 °C in the pH range 5.5–7.0, β-lg exists as an equilibrium between its dimeric (N_2) and monomeric (2N) forms. Between pH 7 and 9, it undergoes a reversible conformational change, referred to as the N ↔ R transition. Both equilibria are pushed to the right as the temperature is increased, i.e., $N_2 \rightarrow 2N \rightarrow 2R$. Above about 65 °C, β-lg undergoes reversible denaturation (R ↔D) but at ~70 °C, denaturation becomes irreversible via a series of aggregation steps. The initial Type I aggregation involves the formation of intermolecular disulphide bonds while the later Type II aggregation involves non-specific interactions, including hydrophobic and electrostatic bonding. Type III aggregation involves non-specific interactions and occurs when the sulphydryl groups are blocked.

Some of the most important consequences of the heat denaturation of whey proteins are due to the fact that these proteins contain sulphydryl and/or disulphide

Fig. 9.17 Exposure of sulphydryl groups by heating milk at 75 °C (*open circle*), 80 °C (*filled circle*), 85 °C (*open triangle*) or 95 °C (*filled triangle*); de-aerated milk heated at 85 °C (*filled square*) (from Jenness and Patton 1959)

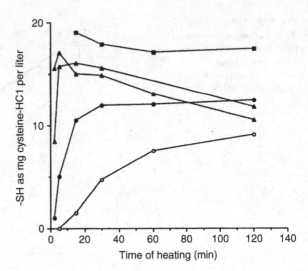

residues which are exposed on heating (Fig. 9.17). They are important for at least the following reasons:

1. The proteins can participate in sulphydryl-disulphide interchange reactions at temperatures $\gtrsim 75$ °C at the pH of milk, but more rapidly $\geq$pH 7.5. Such interactions lead to the formation of disulphide-linked complexes of β-lg with κ-casein and probably with α_{s2}-casein and α-la with profound effects on the functionality of the milk protein system, such as rennet coagulation and heat stability.

2. The activated sulphydryls may decompose with the formation of H_2S which is mainly responsible for the cooked flavour of severely heated milk, including UHT milk. H_2S is volatile and unstable and disappears within about 1 week after processing so that the flavour of UHT milk improves during the first few weeks after processing.

3. Serine, serine phosphate, glycosylated serine, cysteine and cystine residues can undergo β-elimination with the formation of dehydroalanine. Dehydroalanine is very reactive and can react with various amino acid residues, especially lysine, leading to the formation of lysinoalanine, and to a lesser extent with cysteine with the formation of lanthione (Fig. 9.18). These reactions lead to intra- or inter-molecular crosslinking which reduce protein solubility, digestibility and nutritive value (because the bonds formed are not hydrolyzed in the intestinal tract and lysine is an essential amino acid).

9.6.4   Effect of Heat on Caseins

As discussed in Chap. 4, the caseins are rather unique proteins. They are rather small (20–25 kDa), relatively hydrophobic molecules, with little higher structure, few disulphide bonds (present only in the two minor caseins, α_{s2} and κ) and no

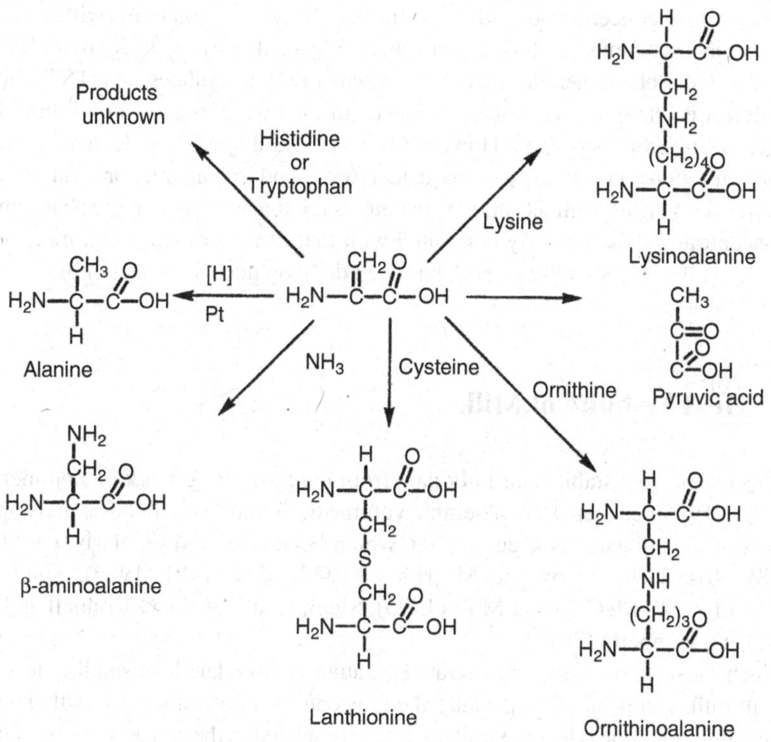

Fig. 9.18 Interaction of dehydroalanine with amino acids

sulphydryl groups. All the caseins are phosphorylated (8–9, 10–13, 4–5 and 1 mol P/mol protein for α_{s1}-, α_{s2}-, β- and κ-casein, respectively); due to their high levels of phosphorylation, α_{s1}-, α_{s2} and β-caseins bind calcium strongly, causing them to aggregate and precipitate and affecting their general stability, including heat stability.

Within the strict sense of the term, the caseins are not susceptible to thermal denaturation, e.g., sodium caseinate (pH 6.5–7.0) may be heated at 140 °C for >1 h without visible physico-chemical changes. However, severe heat treatments do cause substantial changes, e.g., dephosphorylation (~100 % in 1 h at 140 °C), aggregation (as indicated by changes in urea-PAGE or gel permeation chromatography), possibly due to the formation of inter-molecular disulphide and inter-molecular isopeptide bonds, cleavage of peptide bonds (formation of pH 4.6- or 12 % TCA-soluble peptides). The sensitivity of sodium caseinate and micellar casein to ethanol and calcium are reduced by severe heat treatment, probably due to modification of lysine and arginine residues (O'Connell and Fox 1999). β-Elimination of serine, serine phosphate and cysteine residues may also occur, especially at pH values >7. Such heat-induced changes are evident in commercial sodium caseinate.

The remarkably high heat stability of the caseins allows heat-sterilized dairy products to be produced without major changes in physical properties. The heat stability of unconcentrated milk is almost always adequate to withstand the temperature treatments to which it is normally subjected; only rarely is a defect known as the "Utrecht phenomenon" encountered, when milk coagulates on HTST heating. This defect is due to a very high Ca^{2+} concentration owing to a low concentration of citrate, arising from poor feed. However, the heat stability of milk decreases sharply on concentration and is usually inadequate to withstand in-container or UHT processing unless certain adjustments and/or treatments are made. Although the heat stability of concentrated milk is poorly correlated with that of the original milk, most of the research on the heat stability of milk has been done on unconcentrated milk.

9.7 Heat Stability of Milk

Studies on the heat stability of milk date from the pioneering work of Sommer and Hart in 1919. Since then, a considerable volume of literature on thr basic and applied aspects of the subject has accumulated, which been reviewed regularly, e.g., Pyne (1958), Rose (1963), Fox and Morrissey (1977), Fox (1981, 1982), Singh and Creamer (1992). McCrae and Muir (1995), Singh et al. (1995), O'Connell and Fox (2003) and Huppertz (2015).

Much of the early work concentrated on attempts to relate heat stability to variations in milk composition, especially the concentrations of milk salts. Although the heat coagulation time (HCT) of milk is inversely related to the concentrations of divalent cations (Ca^{2+} and Mg^{2+}) and positively with the concentrations of polyvalent anions (i.e., phosphate and citrate), the correlations are poor and unable to explain the natural variations in HCT. This failure was largely explained in 1961 by Dyson Rose who showed that the HCT of most milks is extremely sensitive to small changes in pH in the neighbourhood of 6.7. In effect, the influence of all other factors on the HCT of milk must be considered against the background of the effect of pH.

For the majority of individual-cow and all bulk milk samples, the HCT increases with increasing pH from 6.4 to ~6.7, then decreases abruptly to a minimum at pH ~6.9 but increases continuously with further increases in pH (Fig. 9.19). The HCT decreases sharply below pH 6.4. Milks which show a strong dependence of heat stability on pH are referred to as Type A milks. Occasionally, the HCT of milk from individual cows increases continuously with increasing pH, which is as would be expected due to increasing protein charge with increasing pH; these are referred to as Type B milks.

The maximum HCT and the shape of the HCT-pH profile are influenced by several compositional factors, of which the following are the most significant:

1. Ca^{2+} reduce HCT throughout the pH range 6.4–7.4.
2. Ca-chelators, e.g., citrate and polyphosphates, increase stability.
3. β-Lg, and probably α-la, increase the stability of casein micelles at pH 6.4–6.7 but reduce it at pH 6.7–7.0; in fact, the occurrence of a maximum-minimum in the HCT-pH profile depends on the presence of β-lg.

Fig. 9.19 Effect of pH on the heat stability of type A milk (*filled triangle*), type B milk (*filled circle*) and whey protein-free casein micelle dispersions (*open circle*) (from Fox 1982)

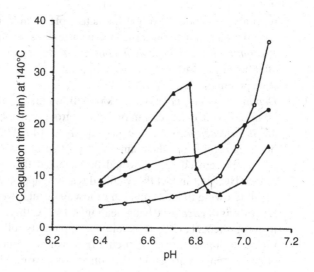

4. Addition of κ-casein to milk increases stability in the pH range of the HCT minimum, but not at pH values < the maximum.
5. Reducing the level of colloidal calcium phosphate increases stability in the region of the HCT maximum.
6. Natural variations in HCT are due mainly to variations in the concentration of indigenous urea due to changes in the animals' feed.

The cause of the maximum-minimum in the HCT-pH profile has attracted much attention, the current explanation is that on heating, κ-casein dissociates from the micelles; at pH values $\leqslant$6.7, β-lg reduces the dissociation of κ-casein, but at pH values >6.7, it accentuates dissociation (Singh and Fox 1987). In effect, coagulation in the pH range of minimum stability involves aggregation of κ-casein-depleted micelles, in a manner somewhat analogous to rennet coagulation although the mechanism by which the altered micelles are produced is very different.

The objective heat stability assay (the nitrogen depletion curve) shows that heat-induced coagulation at the pH of the minimum is a two-step process in which the large, κ-casein-deficient micelles coagulate prematurely [i.e., at the pH of the minimum] while the small, κ-casein-rich micelles coagulate later (O'Connell and Fox 2000).

As would be expected, heating milk at 140 °C for an extended period causes very significant chemical and physical changes in milk (see O'Connell and Fox 2003), of which the following are probably the most significant:

1. *Decrease in pH*; after heating at 140 °C for 20 min, the pH of milk has decreased to ~5.8 due to acid production from pyrolysis of lactose, precipitation of soluble calcium phosphate as $Ca_3(PO_4)_2$, with the release of H^+, and dephosphorylation of casein with subsequent precipitation of the liberated phosphate as $Ca_3(PO_4)_2$ with the release of H^+. The heat-induced precipitation of $Ca_3(PO_4)_2$ is partially

reversible on cooling so that the actual pH of milk at 140 °C at the point of coagulation is much lower than the measured value and is probably <5.0.

2. *Precipitation of soluble calcium phosphate as* $Ca_3(PO_4)_2$ *with the release of H+*; after heating at 140 °C for 5–10 min, most (>90 %) of the soluble phosphate has been precipitated.

3. *Dephosphorylation of casein*, which follows first order kinetics; after heating at 60 min, ~90 % of the casein phosphate groups have been hydrolyzed.

4. *Maillard reaction*, which occurs rapidly at 140 °C; since Maillard browning involves blocking of the ε-amino group of proteins with a concomitant reduction in protein charge, it would be expected that Maillard browning would reduce HCT but in fact the Maillard reaction appears to increase heat stability, possibly owing to the formation of low molecular weight carbonyls.

5. *Hydrolysis of caseins*; during heating at 140 °C there is a considerable increase in non-protein N (12 % TCA soluble), apparently following zero-order kinetics. κ-Casein appears to be particularly sensitive to heating and ~25 % of the N-acetyl neuraminic acid (a constituent of κ-casein) is soluble in 12 % TCA at the point of coagulation.

6. *Cross-linking of proteins*; covalent cross-linking of caseins is evident (by gel electrophoresis) after even the come-up time (~2 min) at 140 °C and it is not possible to resolve the heat-coagulated caseins by urea- or SDS-PAGE.

7. *Denaturation of whey proteins*. Whey proteins are denatured very rapidly at 140 °C; as discussed in Sect. 9.6.3, the denatured proteins associate with the casein micelles, via sulphydryl-disulphide interactions with κ-casein, and probably with $α_{s2}$-casein, at pH values <6.7. The whey proteins can be seen in electron photomicrographs as appendages on the casein micelles.

8. *Dissociation of micellar caseins*: Caseins, especially κ-casein, dissociates from the micelles on heating; the extent of dissociation increases with increasing pH, increasing temperature, decreasing micelle size and in the presence of β-lg at pH ⪞6.7.

9. *Aggregation and shattering of micelles*: electron microscopy shows that the casein micelles aggregate initially, then disintegrate and finally aggregate into a three-dimensional network.

10. *Changes in hydration*; as would be expected from many of the changes discussed above, the hydration of the casein micelles decreases with the duration of heating at 140 °C. The decrease appears to be due mainly to the fall in pH— if samples are adjusted to pH 6.7 after heating, there is an apparent increase in hydration on heating.

11. *Surface (zeta) potential*; it is not possible to measure the zeta potential of casein micelles at the assay temperature but measurements on heated micelles after cooling suggest no change in zeta potential, which is rather surprising since many of the changes discussed above would be expected to reduce surface charge.

All the heat-induced changes discussed would be expected to cause major alterations in the casein micelles, but the most significant change with respect to heat coagulation appears to be the decrease in pH—if the pH is readjusted occasionally to 6.7, milk can be heated for several hours at 140 °C without coagulation. The stabilizing effect of urea is, at least partially, due to the heat-induced formation of NH_3 which reduces or delays the fall in pH; however, other mechanisms for the stabilizing effect of urea have been proposed.

9.7.1 Effect of Processing Operations on Heat Stability

1. Concentration
   Concentration by thermal evaporation markedly reduces the heat stability of milk, e.g., concentrated skim milk containing ~18 % total solids coagulates in ~10 min at 130 °C. The stability of the concentrate is strongly affected by pH, with a maximum at ~pH 6.6, but stability remains low at all pH values ≳6.8 (Fig. 9.20). Concentration by ultrafiltration has a much smaller effect on HCT than thermal evaporation due to a lower concentration of soluble salts in the retentate.
2. Homogenization
   Homogenization of skim milk has no effect on HCT but it destabilizes whole milk, the extent of destabilization increasing with fat content and homogenization pressure (Fig. 9.21). Destabilization probably occurs because the fat globules formed on homogenization are stabilized by casein and consequently they

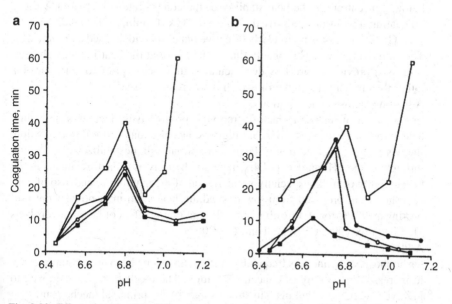

Fig. 9.20 Effect of total solids (TS) content on the heat stability at 130 °C of skim milk *open square*, 9.3 % TS; *filled circle*, 12.0 % TS; *open circle*, 15.0 % TS; *filled square*, 18.4 % TS. (**a**) Concentrated by ultrafiltration, (**b**) concentrated by evaporation (from Sweetsur and Muir 1980)

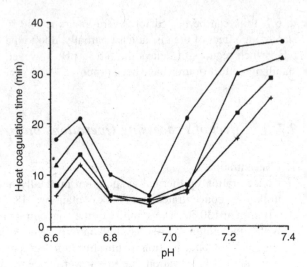

Fig. 9.21 Effect of pressure (Rannie homogenizer) on the heat coagulation time (at 140 °C) of milk, unhomogenized (*filled circle*) or homogenized at 3.5 MPa; (*filled triangle*); 10.4 MPa (*filled square*) or 20.7/3.5 MPa (*plus*) (from Sweetsur and Muir 1983)

behave like "casein micelles", in effect increasing the concentration of coagulable material.

3. Forewarming (preheating)

   Heating an unconcentrated milk, especially at 90 °C×10 min, before a heat stability assay reduces its heat stability, mainly by shifting its natural pH; maximum heat stability is affected only slightly or not at all. However, if milk is preheated before concentration, the heat stability of the concentrate is increased. Various preheating conditions are used, e.g., 90 °C×10 min, 120 °C×2 min or 140 °C×5 s; the last is particularly effective but is not widely used commercially. The stabilizing effect is probably due to the fact that the heat-induced changes discussed previously are less detrimental if they occur prior to concentration rather than in concentrated milk which is inherently less stable.

4. Adjusting lactose concentration

   The addition of lactose to lactose-free milk progressively increases maximum heat stability up to ~1 %, w/v, but higher concentrations cause destabilization. Increasing the concentration of lactose in normal milk destabilizes type-A milk throughout the pH range 6.4–7.4. Enzymatic hydrolysis of lactose enhances the heat stability of milk, especially in the region of the maximum, and also of concentrated milk prepared from low or medium heat skim milk powder but preheating ay 90 °C for 10 min eliminates the stabilizing effect of lactose hydrolysis (Tan and Fox 1996; O'Connell and Fox 2003)

5. Additives

   Orthophosphates, and less frequently citrates, have long been used commercially to increase the stability of concentrated milk. The mechanism was believed to involve Ca-chelation but pH adjustments may be the principal mechanism.

   Numerous compounds increase heat stability (e.g., various carbonyls, including diacetyl, and ionic detergents) but few are permitted additives. Although added urea has a major effect on the stability of unconcentrated milk, it does not stabilize concentrated milks, although it does increase the effectiveness of carbonyls.

Several other compounds increase the heat stability of milk, including, κ-carrageenan, sodium dodecyl sulphate, oxidizing agents (e.g., $KBrO_4$, KIO_3), polyphenols (O'Connell and Fox 2001) and lecithins (see O'Connell and Fox 2003).

6. Treatment with Transglutaminase
   Transglutaminase (TGase), which cross-links proteins by forming isopeptide bonds between lysine and glutamyl residues, prevents the dissociation of κ-casein and strongly affects the HCT-pH profile of unconcentrated, concentrated and whey protein-free milk (O'Sullivan et al. 2002; Mounsey et al. 2005; Huppertz 2015).

9.8 Effect of Heat Treatment on Rennet Coagulation of Milk and Related Properties

The primary step in the manufacture of most cheese varieties and rennet casein involves coagulation of the casein micelles to form a gel. Coagulation involves two steps (phases), the first of which involves enzymatically hydrolyzing the micelle-stabilizing protein, κ-casein, by selected proteinase preparations, called rennets. The second step of coagulation involves coagulation of rennet-altered micelles by $Ca^{2+} \gtrsim 20$ °C (see Chap. 12).

The rate of rennet-induced coagulation is affected by many compositional factors, including the concentrations of Ca^{2+}, casein and colloidal calcium phosphate and pH. Coagulation is adversely affected by heat treatment of the milk at a temperature $\gtrsim 70$ °C due to interaction of denatured β-lg (and α-la) with κ-casein. The primary and, especially, the secondary phases of rennet coagulation are adversely affected by the interaction and if the heat treatment is sufficiently severe (e.g., 80 °C × 5–10 min), the milk does not coagulate on renneting. The effect on the primary phase is presumably due to blockage of the rennet-susceptible bond of κ-casein following interaction with β-lg. The adverse effect of heating on the second phase arises because the whey protein-coated micelles are unable to interact properly because the aggregation sites, which are unknown, are blocked.

The adverse effects of heat treatment on the rennetability of milk can be off-set by acidifying or acidifying-reneutralizing the heated milk or supplementing it with Ca^{2+}. The mechanism by which acidification off-sets the adverse effects of heating is not known but may involve changes in Ca^{2+} concentration.

The strength of the rennet-induced gel is also adversely affected by heat treatment of the milk, again presumably because the whey protein-coated micelles are unable to participate properly in the gel network. Gels from severely heat-treated milk have poor syneresis properties, resulting in high moisture cheese which does not ripen properly. Syneresis is undesirable in fermented milks, e.g., yoghurt, the milk for which is severely heat-treated (e.g., 90 °C × 10 min) to reduce the risk of syneresis.

9.9   Age Gelation of Sterilized Milk

Two main problems limit the shelf-life of UHT sterilized milks: off-flavour development and gelation. Age gelation, which also occurs occasionally with in-container sterilized concentrated milks, is not related to the heat stability of the milk (provided that the product withstands the sterilization process) but the heat treatment applied does have a significant influence on gelation, e.g., indirectly heated UHT milk is more stable to age gelation than directly heated product (the former is the more severe heat treatment). Plasmin may be responsible for the gelation of unconcentrated UHT milk produced from good quality milk while proteinases from psychrotrophs are probably responsible if the raw milk was of poor quality. It is possible that physicochemical phenomena are also involved, e.g., interaction between whey proteins and casein micelles.

In the case of concentrated UHT milks, physicochemical effects appear to predominate, although proteolysis also occurs, e.g., the propensity of UHT concentrated milk reconstituted from high-heat milk powder to age gelation is less than those from medium- or low-heat powders although the formation of sediment is greatest in the concentrate prepared from the high-heat powder (see Harwalkar 1992).

9.10   Heat-Induced Changes in Flavour of Milk

Flavour is a very important attribute of all foods; heating/cooking makes a major contribution to flavour, both positively and negatively. Good quality fresh liquid milk products are expected to have a clean, sweetish taste and essentially no aroma; any departure therefrom can be considered as an off-flavour. Heat treatments have a major impact on the flavour/aroma of dairy foods, either positively or negatively.

On the positive side, thermization and minimum pasteurization should not cause the formation of undesirable flavours and aromas and should, in fact, result in improved flavour by reducing bacterial growth and enzymatic activity, e.g., lipolysis. If accompanied by vacuum treatment (vacreation), pasteurization removes indigenous off-flavours, i.e., those arising from the cow's metabolism or from feed, thereby improving the organoleptic qualities of milk.

Also on the positive side, severe heat treatment of cream improves the oxidative stability of butter produced therefrom due to the exposure of antioxidant sulphydryl groups. As discussed in Sect. 9.2.2, lactones formed from hydroxyacids are major contributors to the desirable cooking quality of milk fats but contribute to off-flavours in other heated products, e.g., milk powders.

UHT processing causes substantial deterioration in the organoleptic quality of milk. Freshly processed UHT milk is described as "cooked" and "cabbagy", but the intensity of these flavours decreases during storage, giving maximum flavour acceptability after a few days. These off-flavours are due to the formation of sulphur compounds from the denatured whey proteins, as discussed in Sect. 9.6.3. After this period of maximum acceptability, quality deteriorates, the milk being described as stale. At least 400 volatiles have been detected in UHT milk, about 50 of which (Table 9.3)

Table 9.3 Substances making a strong contribution to the flavour of indirectly heated UHT milk, those contributing to differences in flavour of milk heat-treated in different ways, and those used in a synthetic UHT flavour preparation (from Manning and Nursten 1985)

	UHT-i[a]	UHT-i-LP[b]	UHT-i-UHT-d[c]	Synthetic UHT flavour[d] (mg/kg LP)
Dimethyl sulphide	+	0	1	
3-Methylbutanal	+	1	1	
2-Methylbutanal	+	0	1	
2-Methyl-1-propanethiol	+	1	1	0.008
Pentanal	+	1	1	
3-Hexanone	+			
Hexanal	+	1	1	2
2-Heptanone	+	4	2	
Styrene	+			0.40
Z-4-Heptenal[e]	+	1	0	
Heptanal	+			
2-Acetylfuran	+			
Dimethyl trisulphide	+	2	0	
Cyanobenzene	+			
1-Heptanol	+			
1-Octen-3-one[e]	+			
Octanal	+	1	1	
p-Cymene	+			
Phenol	+			
Indene	+			
2-Ethyl-1-hexanol	+			
Benzyl alcohol	+			
Unknown	+			
Acetophenone	+	1	0	
1-Octanol	+			
2-Nonanone	+	4	2	0.21
Nonanal	+			
p-Cresol	+			
m-Cresol	+			
E-2,Z-6-Nonadienal	+			
E-2-Nonenal	+			
3-Methylindene	+			
Methylindene	+			
Ethyldimethylbenzene	+			
Decanal	+			
Tetraethylthiourea	+			
Benzothiazole	+	1	0	0.005
γ-Octalactone	+	1	0	0.025
2,3,5-Trimethylanisole	+			
δ-Octalactone	+	1	0	
1-Decanol	+	1	1	

(continued)

Table 9.3 (continued)

	UHT-i[a]	UHT-i-LP[b]	UHT-i-UHT-d[c]	Synthetic UHT flavour[d] (mg/kg LP)
2-Undecanone	+	2	1	0.18
2-Methylnaphthalene	+			
Indole	+			
δ-Decalactone	+	1	0	0.650
Hydrogen sulphide		2	1	0.03
Diacetyl		2	1	0.005
Dimethyl disulphide		2	1	0.002
2-Hexanone		2	1	
γ-Dodecalactone		2	1	0.025
δ-Dodecalactone		2	1	0.1
Methanethiol		1	1	0.002
2-Pentanone		1	1	0.29
Methyl isothiocyanate		1	1	0.01
Ethyl isothiocyanate		1	1	0.01
Furfural		1	1	
Benzaldehyde		1	0	
2-Octanone		1	0	
Naphthalene		1	0	
γ-Decalactone		1	0	
2-Tridecanone		1	0	
Acetaldehyde		−1	0	
1-Cyano-4-pentene		−1	0	
2-Methyl-1-butanol		−1	1	
Ethyl butyrate		−1	0	
3-Buten-1-yl isothiocyanate		−1	0	
E-2,E-4-nonadienal		−1	0	
2,4-Dithiapentane			1	
2,4-Dithiapentane				10.00

[a]Indirectly heated UHT milk; + indicates is a component that makes a strong contribution to the flavour. In addition to the components listed, a further 12 unknowns made strong contributions
[b]Components contributing to a difference in flavour between indirectly heated UHT milk and low temperature pasteurized (LP) milk. Scale for difference: 1 = slight; 2 = moderate; 3 = strong; 4 = very strong
[c]Components contributing to a difference in flavour between indirectly and directly heated UHT milks. Scale for difference as in C
[d]Composition of synthetic UHT flavour
[e]Tentative identification

are considered to make a significant contribution to flavour (see Manning and Nursten 1985; McSweeney et al. 1997; Cadwallader and Singh 2009). The shelf-life of UHT milk is usually limited by gelation and/or bitterness, both of which are due to proteolysis, as discussed in Sect. 9.6.1.

Since sulphur compounds are important in the off-flavour of UHT milk, attempts to improve its flavour have focussed on reducing the concentration of these, e.g., by adding thiosulphonates, thiosulphates or cystine (which react with mercaptans) or sulphydryl oxidase, an indigenous milk enzyme (which oxidizes sulphydryls to disulphides; see Chap. 10).

The products of Maillard browning have a significant negative impact on the flavour of heated milk products, especially in-container sterilized milks and milk powders.

References

Bauman, D. E., & Lock, A. L. (2006). Conjugated linoleic acid: Biosynthesis and nutritional significance. In P. F. Fox & P. L. H. McSweeney (Eds.), *Advanced dairy chemistry* (Lipids 2nd ed., Vol. 2, pp. 93–136). New York, NY: Springer.

Cadwallader, K. K., & Singh, T. K. (2009). Flavour and off-flavour in milk and dairy products. In P. L. H. McSweeney & P. F. Fox (Eds.), *Advanced dairy chemistry* (Lactose, water, salts and minor constitutes 3rd ed., Vol. 3, pp. 631–690). New York, NY: Springer.

Driessen, F. M. (1989). Inactivation of lipases and proteinases (indigenous and bacterial). In P. F. Fox (Ed.), *Heat-induced changes in milk* (Bulletin, Vol. 238, pp. 71–93). Brussels: International Dairy Federation.

Erbersdobler, H. F., & Dehn-Müller, B. (1989). Formation of early Maillard products during UHT treatment of milk. In P. F. Fox (Ed.), *Heat-induced changes in milk* (Bulletin, Vol. 238, pp. 62–67). Brussels: International Dairy Federation.

Fox, P. F. (1981). Heat-induced changes in milk preceding coagulation. *Journal of Dairy Science, 64*, 2127–2137.

Fox, P. F. (1982a). Heat-induced coagulation of milk. In P. F. Fox (Ed.), *Developments in dairy chemistry* (Proteins, Vol. 1, pp. 189–228). London, UK: Applied Science.

Fox, P. F., & Morrissey, P. A. (1977). Reviews on the progress of dairy science: The heat stability of milk. *Journal of Dairy Research, 44*, 627–646.

Gould, I. A. (1945). Lactic acid in dairy products. III. The effect of heat on total acid and lactic acid production and on lactose destruction. *Journal of Dairy Science, 28*, 367–377.

Ha, Y. L., Grimm, N. K., & Pariza, M. W. (1989). Newly recognized anticarcinogenic fatty acids. Identification and quantification in natural and processed cheeses. *Journal of Agricultural and Food Chemistry, 37*, 75–81.

Harwalkar, V. R. (1992). Age gelation of sterilized milks. In P. F. Fox (Ed.), *Advanced dairy chemistry* (Proteins 2nd ed., Vol. 1, pp. 691–734). London, UK: Elsevier Applied Science.

Huppertz, T. (2015). Heat stability of milk. In P. L. H. McSweeney & J. A. O'Mahony (Eds.), *Advanced dairy chemistry* (Proteins 4th ed., Vol. 1) in press. New York, NY: Springer.

Jenness, R., & Patton, S. (1959). *Principles of dairy chemistry*. New York, NY: John Wiley & Sons.

Liang, D., & Hartel, R. W. (2004). Effects of milk powders in chocolate. *Journal of Dairy Science, 87*, 20–31.

Lyster, R. L. J. (1970). The denaturation of α-lactalbumin and β-lactoglobulin in heated milk. *Journal of Dairy Research, 37*, 233–243.

Manning, D. J., & Nursten, H. E. (1985). Flavour of milk and milk products. In P. F. Fox (Ed.), *Developments in dairy chemistry* (Lactose and minor constituents, Vol. 3, pp. 217–238). London, UK: Elsevier Applied Science.

McCrae, C. H., & Muir, D. D. (1995). Heat stability of milk. In P. F. Fox (Ed.), *Heat-induced changes in milk* (Special Issue 2nd ed., 9501, pp. 206–230). Brussels: International Dairy Federation.

McKellar, R. C. (Ed.). (1989). *Enzymes of psychrotrophs in raw food*. Boca Raton, FL: CRC Press.

McSweeney, P. L. H., Nursten, H. E., & Urbach, G. (1997). Flavour and off-flavour in milk and dairy products. In P. F. Fox (Ed.), *Advanced dairy chemistry* (Lactose, water, salts and vitamins 3rd ed., Vol. 3, pp. 406–468). London, UK: Chapman & Hall.

Mounsey, J. S., O'Kennedy, B. T., & Kelly, P. M. (2005). Influence of transglutaminase treatment on properties of micellar casein and products made therefrom. *Le Lait, 85*, 405–418.

Mulvihill, D. M., & Donovan, M. (1987). Whey proteins and their thermal denaturation—A review. *Irish Journal of Food Science and Technology, 11*, 43–75.

O'Connell, J. E., & Fox, P. F. (1999). Heat-induced changes in the calcium sensitivity of casein. *International Dairy Journal, 9*, 839–847.

O'Connell, J. E., & Fox, P. F. (2000). The two-stage coagulation of milk proteins in the minimum of the heat coagulation time-pH profile of milk: Effect of casein micelle size. *Journal of Dairy Science, 83*, 378–386.

O'Connell, J. E., & Fox, P. F. (2001). Significance and applications of phenolic compounds in the production and quality of milk and dairy products. *International Dairy Journal, 11*, 103–120.

O'Connell, J. E., & Fox, P. F. (2003). Heat-induced coagulation of milk. In P. F. Fox & P. L. H. McSweeney (Eds.), *Advanced dairy chemistry* (Part B, Proteins 3rd ed., Vol. 1, pp. 879–945). New York, NY: Springer.

O'Sullivan, M. M., Kelly, A. L., & Fox, P. F. (2002). Effect of transglutaminase on the heat stability of milk: A possible mechanism. *Journal of Dairy Science, 85*, 1–7.

Parodi, P. W. (2006). Nutritional significance of milk lipids. In P. F. Fox & P. L. H. McSweeney (Eds.), *Advanced dairy chemistry* (Lipids 2nd ed., Vol. 2, pp. 136, 601–639). New York, NY: Springer.

Pyne, G. T. (1958). The heat coagulation of milk. II. Variations in the sensitivity of caseins to calcium ions. *Journal of Dairy Research, 25*, 467–474.

Rose, D. (1963). Heat stability of bovine milk: A review. *Dairy Science Abstracts, 25*, 45–52.

Singh, H., & Creamer, L. K. (1992). Heat stability of milk. In P. F. Fox (Ed.), *Advanced dairy chemistry* (Proteins 2nd ed., Vol. 1, pp. 621–656). London, UK: Elsevier Applied Science.

Singh, H., Creamer, L. K., & Newstead, D. F. (1995). Heat stability of concentrated milk. In P. F. Fox (Ed.), *Heat-induced changes in milk* (Special Issue 2nd ed., Vol. 9501, pp. 256–278). Brussels: International Dairy Federation.

Singh, H., & Fox, P. F. (1987). Heat stability of milk: Role of β-lactoglobulin in the pH-dependent dissociation of κ-casein. *Journal of Dairy Research, 54*, 509–521.

Stepaniak, L., Fox, P. F., & Daly, C. (1982). Isolation and general characterization of a heat-stable proteinase from *Pseudomonas fluorescens* AFT 36. *Biochimica et Biophysica Acta, 717*, 376–383.

Sweetsur, A. W. M., & Muir, D. D. (1980). Effect of concentration by ultrafiltration on the heat stability of skim milk. *Journal of Dairy Research, 47*, 327–335.

Sweetsur, A. W. M., & Muir, D. D. (1983). Effect of homogenization on the heat stability of milk. *Journal of Dairy Research, 50*, 291–300.

Sweetsur, A. W. M., & White, J. C. D. (1975). Studies on the heat stability of milk proteins. III. Effect of heat-induced acidity in milk. *Journal of Dairy Research, 42*, 73–88.

Tan-Kintia, R. H., & Fox, P. F. (1996). Effect of the enzymatic hydrolysis of lactose on the heat stability of milk or reconstituted milk. *Netherlands Milk and Dairy Journal, 50*, 267–277.

Walstra, P., & Jenness, R. (1984a). *Dairy chemistry and physics*. New York, NY: John Wiley & Sons.

Webb, B. H., & Johnson, A. H. (1965). *Fundamentals of dairy chemistry*. Westport, CT: AVI Publishing Company.

Suggested Reading

Fox, P. F. (Ed.). (1982b). *Developments in dairy chemistry* (Proteins, Vol. 1). London, UK: Applied Science.

Fox, P. F. (Ed.). (1989). *Heat-induced changes in milk* (Bulletin, Vol. 238). Brussels: International Dairy Federation.

Fox, P. F. (Ed.). (1995). *Heat-induced changes in milk* (Special Issue 2nd ed., Vol. 9501). Brussels: International Dairy Federation.

Walstra, P., & Jenness, R. (1984b). *Dairy chemistry and physics*. New York, NY: John Wiley & Sons.

Wong, N. P. (Ed.). (1980). *Fundamentals of dairy chemistry* (3rd ed.). Westport, CT: The AVI Publishing Company.

Chapter 10
Enzymology of Milk and Milk Products

10.1 Introduction

Like all other foods of plant or animal origin, milk contains several indigenous enzymes. The principal constituents of milk (lactose, lipids and proteins) can be modified by exogenous enzymes, added to induce specific changes; being a liquid, milk is more amenable to enzyme action than solid foods Exogenous enzymes may also be used to analyse for certain constituents in milk. In addition, milk and most dairy products contain viable microorganisms which secrete extracellular enzymes or release intracellular enzymes after the cells have lysed. Some of these microbial enzymes may cause undesirable changes, e.g., hydrolytic rancidity in milk and dairy products, bitterness and/or age gelation of UHT milk, bittiness in cream, malty flavours or bitterness in fluid milk, or they may cause desirable flavours, e.g., in ripened cheese.

This chapter is devoted mainly to the significance of indigenous enzymes in milk. The principal applications of exogenous enzymes in dairy products are dealt with in other chapters, i.e., rennets and lipases in cheese production (Chap. 12) and β-galactosidase to modify lactose (Chap. 2). Some minor or potential applications of exogenous enzymes are described here. Enzymes derived from contaminating bacteria, which may be significant in milk and some dairy products, will not be discussed. The significance of enzymes from microbial cultures in cheese ripening is discussed in Chap. 12.

© Springer International Publishing Switzerland 2015
P.F. Fox et al., *Dairy Chemistry and Biochemistry*,
DOI 10.1007/978-3-319-14892-2_10

10.2   Indigenous Enzymes of Bovine Milk

10.2.1   Introduction

At least 60 indigenous enzymes have been reported in normal bovine milk. They arise from three principal sources:

(a) The blood *via* defective mammary cell membranes.
(b) Secretory cell cytoplasm, some of which is occasionally entrapped within fat globules by the encircling fat globule membrane (MFGM) (see Chap. 3).
(c) The MFGM itself, the outer layers of which are derived from the apical membrane of the secretory cell, which, in turn, originates from the Golgi membranes; this is probably the principal source of the indigenous enzymes in milk.

Thus, most enzymes enter milk due to peculiarities of the mechanism by which milk constituents, especially the fat globules, are excreted from the secretory cells. Milk does not contain substrates for many of the enzymes present, while others are inactive in milk owing to unsuitable environmental conditions, e.g., pH.

Many indigenous milk enzymes are technologically significant from five viewpoints:

1. Deterioration [lipase (potentially, the most significant enzyme in milk), proteinase, acid phosphatase and xanthine oxidoreductase] or preservation (sulphydryl oxidase, superoxide dismutase) of milk quality.
2. As indices of the thermal history of milk: alkaline phosphatase, γ-glutamyl transferase, lactoperoxidase.
3. As indices of mastitic infection: the concentration of several enzymes increases on mastitic infection especially catalase, N-acetyl-β-D-glucosaminidase and acid phosphatase;
4. Antimicrobial activity: lysozyme, lactoperoxidase (which is exploited as a component of the lactoperoxidase-thiocyanate system for the cold pasteurization of milk).
5. As a potential commercial source of enzymes: ribonuclease, lactoperoxidase.

With a few exceptions (e.g., lysozyme and lactoperoxidase), the indigenous milk enzymes do not have a beneficial effect on the nutritional or organoleptic attributes of milk, and hence their inactivation by heat is one of the objectives of many dairy processes.

The principal indigenous enzymes in milk and their catalytic activity are listed in Table 10.1. Research on the indigenous enzymes in milk dates from 1881 and a very extensive literature has accumulated, which has been reviewed; a selection of reviews is included at the end of the chapter. The indigenous enzymes in milk were the subject of an International Dairy Federation Symposium in Cork, Ireland, in 2005.

In this chapter, the occurrence, distribution, isolation and characterization of the principal indigenous enzymes will be discussed, with an emphasis on their commercial significance in milk and dairy products. The review will focus on bovine milk, with reference to the principal activities in the milk of other important species. Studies on the enzymes in non-bovine milk have been concerned mainly

Table 10.1 Indigenous enzymes of significance to milk

Enzyme	Reaction	Importance
Lipase	Triglycerides + $H_2O \rightarrow$ fatty acids + partial glycerides + glycerol	Off flavours in milk; flavour development in Blue cheese
Proteinase (plasmin)	Hydrolysis of peptide bonds, particularly in β-casein	Reduced storage stability of UHT products; cheese ripening
Catalase	$2H_2O_2 \rightarrow O_2 + 2H_2O$	Index of mastitis; pro-oxidant
Lysozyme	Hydrolysis of mucopolysaccharides	Bacteriocidal agent
Xanthine oxidase	Aldehyde + $H_2O + O_2 \rightarrow$ Acid + H_2O_2	Pro-oxidant; cheese ripening
Sulphydryl oxidase	$2RSH + O_2 \rightarrow RSSR + H_2O_2$	Amelioration of cooked flavour
Superoxide dismutase	$2O_2^- + 2H^+ \rightarrow H_2O_2 + O_2$	Antioxidant
Lactoperoxidase	$H_2O_2 + AH_2 \rightarrow 2H_2O + A$	Index of pasteurization; bactericidal agent; index of mastitis; pro-oxidant
Alkaline phosphomonoesterase	Hydrolysis of phosphoric acid esters	Index of pasteurization
Acid phosphomonoesterase	Hydrolysis of phosphoric acid esters	Reduce heat stability of milk; cheese ripening

with quantifying the activity and the effects of various animal-related factors, storage and heat treatments thereon. Few of the enzymes in non-bovine milk have been isolated and characterized and it is assumed that these enzymes are similar to the corresponding enzymes in bovine milk. The milk of all species probably contains the same range of enzymes as bovine milk.

Not surprisingly, many of the enzymes in human milk have been studied extensively and their significance in human nutrition highlighted. Human milk contains a very high level of lysozyme (~4 % of total protein) and of bile salts-activated lipase (which the milk of other species lacks) and α-amylase. Reviews on the indigenous enzymes in human, buffalo, ovine, caprine, equine, and porcine milk are included at the end of the chapter.

10.2.2 Proteinases (EC 3.4.-.-)

In 1897, S.M. Babcock and H.L. Russell extracted a trypsin-like proteolytic enzyme from separator slime, which they called "galactase" [derived from *gala*, Greek for milk; genative, *galaktos*]. The presence of an indigenous proteinase in milk, mainly in separator slime, was confirmed by R.W. Tatcher and A.C. Dahlberg in 1917 but there appears to have been no further publications on milk proteinase until R.G. Warner and E. Polis reported in 1945 a low level of proteolytic activity in acid casein. In 1960, W.J. Harper, using aseptically-drawn, low-bacterial count milk with added antibiotics, showed that milk does contain an indigenous proteinase(s),

10.2.2.1   Plasmin (EC 3.4.21.7)

Milk is now known to contain several indigenous proteinases, the principal of which
is plasmin (fibrinolysin; EC 3.4.21.7). The physiological function of plasmin is to
dissolve blood clots. Milk contains the complete plasmin system: plasmin, plasminogen, plasminogen activators (PAs) and inhibitors of PAs and of plasmin. This
system enters milk from blood and plasmin activity increases during a mastitic
infection and in late lactation, when there is an increased influx of blood constituents into milk. In milk, there is about four times as much plasminogen as plasmin
and both, as well as plasminogen activators, are associated with the casein micelles,
from which they dissociate when the pH is reduced to ~4.6. The inhibitors of plasmin and of plasminogen activators are in the milk serum.

Owing to changes in practices in the dairy industry, e.g., improved bacterial quality, extended storage on farms and at factories and the introduction of high-
temperature processed milk (plasmin is very heat-stable), plasmin has become a very
significant enzyme in milk and, consequently, the subject of considerable research.

Plasmin is well-characterised, as are the other components of the plasmin system. Bovine plasminogen is a single-chain glycoprotein containing 786 amino acid
residues, with a calculated molecular mass of 88,092 Da, which exists as five
disulphide-linked loops ("kringles"). Plasminogen is converted to plasmin by cleavage of the Arg_{557}-Ile_{558} bond by specific proteinases, of which there are two types,
urokinase-type and tissue-type plasminogen activators. Plasmin is a serine proteinase (inhibited by diisopropylfluorophosphate, phenylmethyl sulphonyl fluoride and
trypsin inhibitor) with a high specificity for peptide bonds to which lysine or arginine supplies the carboxyl group. Its molecular weight is ~81 kDa.

Plasmin is usually extracted at pH 3.5 from rennet-coagulated casein and purified by precipitation with $(NH_4)_2SO_4$ and various forms of chromatography, including affinity chromatography. It is optimally active at ~pH 7.5 and ~35 °C; it exhibits
~20 % of maximum activity at 5 °C and is stable over the pH range 4–9. Plasmin is
quite heat stable: it is partially inactivated by heating at 72 °C × 15 s but its activity
in milk increases following HTST pasteurization, probably through inactivation of
the indigenous inhibitors of plasmin or of plasminogen activators. It partly survives
UHT sterilization but is inactivated by heating at 80 °C × 10 min at pH 6.8; its stability decreases with increasing pH in the range 3.5–9.2.

Plasmin is highly specific for peptide bonds containing Lys or Arg at the
N-terminal side. The specificity of plasmin on α_{s1}-, α_{s2}- and β-caseins in solution has
been determined; it has little or no activity on κ-casein (CN), β-lg or α-la (in fact,
denatured β-lg is an inhibitor). In milk, the principal substrate for plasmin is β-CN,
from which it produces γ^1- (β-CN f29–209), γ^2- (β-CN f106–209) and γ^3- (β-CN
f108–209) caseins and proteose peptone (PP)5 (β-CN f1-105/107), $PP8_{slow}$ (β-CN
f29–105/107) and $PP8_{fast}$ (β-CN f1–29).

α_{s2}-Casein in solution is also hydrolysed very rapidly by plasmin at bonds Lys_{21}-
$Gln22$, Lys_{24}-Asn_{25}, Arg_{114}-Asn_{115}, Lys_{149}-Lys_{150}, Lys_{150}-Thr_{151}, Lys_{181}-Thr_{182}, Lys_{187}-
$Thr188$ and Lys_{188}-Ala_{189} but it is not known if it is hydrolysed in milk. Although less
susceptible than α_{s2}- or β-caseins, α_{s1}-casein in solution is also readily hydrolysed by
plasmin and the λ-casein fraction in milk includes several α_{s1}-casein-derived

peptides which could have been produced by either plasmin or cathepsin D. Although κ-casein contains several Lys and Arg residues, it appears to be quite resistant to plasmin, presumably due to a relatively high level of secondary and tertiary structure. β-Lactoglobulin, especially when denatured, inhibits plasmin, *via* sulphydryl-disulphide interactions which rupture the structurally important krinkles.

Significance of Plasmin Activity in Milk

Plasmin and plasminogen accompany the casein micelles on the rennet-induced coagulation of milk and are concentrated in cheese in which plasmin contributes to primary proteolysis of the caseins, especially in high-cook cheeses, e.g., Swiss and some Italian varieties, in which the coagulant is totally or largely inactivated (see Chap. 12). Plasmin activity may contribute to age gelation in UHT milk. It has been suggested that plasmin activity contributes to the poor cheesemaking properties of late-lactation milk. Reduced yields of cheese and casein can be expected to result from plasmin action since the proteose peptones are, by definition, soluble at pH 4.6.

Human milk contains about the same level of plasmin as bovine milk but about four times more plasminogen. It also contains several other proteinases and peptidases, including amino- and carboxy-peptidases.

10.2.2.2   Cathepsin D (EC 3.4.23.5)

The second proteinase identified in milk was cathepsin D, which is a lysosomal enzyme but is present in acid whey. As with plasmin, cathepsin D is part of a complex system, including an inactive precursor. The principal form of cathepsin D in milk is the zymogen, procathepsin D. The level of cathepsin D in milk is correlated significantly with somatic cell count (SCC), although it is not clear whether this reflects increased production of cathepsin D and/or increased activation of precursors.

The principal peptide produced from α_{s1}-casein by cathepsin D is α_{s1}-CN (f24–199), which is the primary peptide produced by chymosin. The proteolytic specificity of cathepsin D on β-casein is also similar to that of chymosin. Cathepsin D can cleave κ-casein but has very poor milk clotting properties. Two cleavage sites of cathepsin D on α-lactalbumin have been identified but native β-lactoglobulin is resistant to cathepsin D. At least some cathepsin D is incorporated into cheese curd and may contribute to proteolysis in cheese but its activity is normally overshadowed by chymosin, which is present at a much higher level.

Other Proteinases

Somatic cells contain several other proteinases, including cathepsins B, L and G, and elastase, which have received limited attention. Lysosomes contain several enzymes in addition to cathepsins B, L and G, including cathepsins S, K, T, N and O, dipeptidyl peptidase I, fructose-1,6-bisphosphatase (aldolase)-converting enzyme and legumain (in legumes). Presumably, most of these enzymes are present in milk but may be inactive owing to the high redox potential of milk, under which conditions the active site sulphydryl group would be oxidised; we do not know if attempts have been made to assay these enzymes under reducing conditions.

10.2.3   Lipases and Esterases (EC 3.1.1.-)

Milk contains carboxylester hydrolases (EC 3.1.1.1.1), glycerolester hydrolases, lipases (EC 3.1.1.3), some of which are lipoprotein lipases (EC 3.1.1.34), arylesterases (EC 3.1.1.2) and cholinesterases (EC 3.1.1.7, EC 3.1.1.8); lipoprotein lipase is by far the most technologically important. Classically, lipases hydrolyse ester bonds in emulsified esters, i.e., at a water/oil interface, although some may have activity on soluble esters; they are usually activated by blood serum albumin and Ca^{2+} which bind inhibitory free fatty acids.

The presence of lipolytic activity in milk was reported by E. Moro in 1902 and this was confirmed by several workers during the next few years, but there was uncertainty because some investigators used soluble esters as substrate, In 1922, using a new assay method, F.E. Rice and A.L. Markley presented strong evidence indicating the presence of lipase in milk. A lipase was purified from separator slime in 1963 by R.C. Chandan and K.M. Shahani with an 88×-fold increase in specific activity. This enzyme, which had a MW of only ~7 kDa, and was inhibited strongly by the principal milk proteins, probably originated from somatic cells and was considered to be a minor lipase in milk.

A lipase was isolated from skim milk by Fox and Tarassuk (1968); this enzyme had serine at the active site, a MW of ~210 kDa and was optimally active at pH 9.2 and 37 °C. In 1958, T.N. Quigley reported that milk contains a lipoprotein lipase (LPL; i.e., a lipase which is activated by plasma apolipoprotein C-II). LPL was isolated from milk by T. Egelrud and T. Olivecrona in 1972; it is a homodimer of monomers containing 450 amino acid residues, i.e., a MW of ~90 kDa. The molecule has been characterized at the molecular, genetic, enzymatic and physiological levels The lipase isolated by Fox and Tarassuk has been shown to be a LPL. Reflecting its importance in the biosynthesis of milk fat and its role in hydrolytic rancidity, mammary/milk lipase/LPL has been the subject of extensive research.

The pH and temperature optima of LPL are ~9 and 37 °C, respectively. Under optimum conditions the k_{cat} of LPL is ~3,000 s^{-1} and milk contains sufficient lipase (1–2 mg L^{-1}; 10–20 nM) to cause hydrolytic rancidity in 10 s. Little lipolysis normally occurs in milk because most (>90 %) of the lipase is associated with the casein micelles while the triglyceride substrates are in fat globules surrounded, and protected, by the fat globule membrane (MFGM). When the MFGM is damaged, e.g., by agitation, foaming, cooling/warming, freezing or homogenization, lipolysis occurs rapidly, causing hydrolytic rancidity The milk of some cows undergoes spontaneous lipolysis, i.e., without the need for an activation step. Initially, it was proposed that such milk contains a second (membrane) lipase, but it now appears that they contain either a high level of apolipoprotein CII, which activates LPL, or that normal milk has a higher level of proteose peptone 8, which inhibits LPL. Dilution of 'spontaneous milk' with normal milk prevents spontaneous rancidity, which consequently is not normally a problem with bulk herd milks because dilution with normal milk reduces the lipoprotein content of the mixture to below the threshold necessary for lipase adsorption. Natural variations in the levels of free fatty acids in normal milk and the susceptibility of normal milks to lipolysis may be due to variations in the level of blood serum in milk.

Although caprine milk contains only ~4 % as much lipolytic activity as bovine milk, it is prone to spontaneous rancidity, which enhances its "goaty" flavour. LPL in caprine milk is concentrated in the cream phase (unlike bovine milk, in which LPL is mainly on the casein micelles) and LPL activity is strongly correlated with a particular genetic variant of α_{s1}-casein. Ovine milk has 10 % of the LPL activity in bovine milk.

10.2.3.1 Significance of Lipase

Technologically, lipase is arguably the most significant indigenous enzyme in milk. Although it may play a positive role in cheese ripening, undoubtedly the most industrially important aspect of milk lipase is its role in hydrolytic rancidity which renders liquid milk and dairy products unpalatable and eventually unsaleable. As discussed in Chap. 3, all milk contains an adequate level of lipase for rapid lipolysis, but become rancid only after the fat globule membrane has been damaged.

10.2.3.2 Bile Salts-Stimulated Lipase

It has been known since early in the twentieth Century that human milk has considerably higher lipolytic activity than bovine milk. In fact, human milk, and that of some other species, contains a second lipase in addition to LPL, i.e., bile salts-stimulated lipase (BSSL) which is similar to the broad-specificity pancreatic carboxylic ester hydrolase (CEH; also called cholesterol ester hydrolase). BSSL is considered to be very important for the digestion of lipids by human babies who secrete low levels of both pancreatic lipase and bile salts. The pre-duodenal lipases (lingual lipase, pre-gastric esterase and gastric lipase) are important for fat digestion by human infants.

BSSL is synthesised in the mammary gland and represents ~1 % of the total protein in human milk. It is inactivated by HTST pasteurisation, as a result of which the absorption of lipids by pre-term infants is reduced by ~30 %. The sequence of human BSSL consists of 722 amino acid residues with a total molecular mass of ~105 kDa, including 15–20 % carbohydrate. BSSL shows high homology with lysophospholipase from rat pancreas and acetylcholine esterase, as well as to CEH.

10.2.3.3   Phospholipase and Esterases

Some authors have reported that milk possesses significant phospholipase-D activity but others failed to detect this enzyme in milk.

Milk contains several esterases the most significant of which are arylesterases (EC 3.1.1.7), cholinesterase (EC 3.1.1.8) and carboxylesterase (3.1.1.1). Arylesterase (also called solalase) was among the first enzymes reported in milk;. It has been isolated from milk and characterised. Arylesterase activity is high in colostrum and during mastitis but it probably has no technological significance in normal milk.

10.2.4   Phosphatases (EC 3.1.3 -)

Milk contains several phosphatases, the principal being alkaline and acid phosphomonoesterases, which are of technological significance, and ribonuclease, which has no known function or significance in milk, although it may be significant in the mammary gland. The alkaline and acid phosphomonoesterases in milk have been studied extensively. Alkaline phosphatase (AlP; EC 3.1.31) has been isolated from several sources and characterised; it is a membrane-bound glycoprotein which is widely distributed in animal tissues and in micro-organisms; there are four principal types of mammalian AlP: intestinal, placental, germ-cell and tissue-non-specific. The intestine and placenta are particularly rich sources. AlP is a very important enzyme in clinical chemistry, its activity in various tissues being an indicator of diseased states. AlPs are very active subjects for research, mainly with a clinical focus; there has been relatively little recent research on milk AlP, with the exception of assay methodology.

10.2.4.1   Milk Alkaline Phosphatase (EC 3.1.3.1)

The occurrence of a phosphatase in milk was first recognised by F. Demuth in 1925. Subsequently characterised as an alkaline phosphatase, it became significant when it was shown that the time-temperature combinations required for the thermal inactivation of alkaline phosphatase were slightly more severe than those required to kill *Mycobacterium tuberculosis*, then the target micro-organism for pasteurisation. The enzyme is readily assayed and a test procedure based on its inactivation was developed as a routine quality control test for the HTST pasteurisation of milk.

The AlP activity of bovine milk varies considerably between individuals and herds, and throughout lactation; activity varies inversely with milk yield but is independent of fat content, breed and feed; AlP activity in human milk also varies.

Isolation and Characterisation

AlP is concentrated in cream and released into buttermilk, where it occurs in lipoprotein particles, on churning (about 50 % of AlP is in the skimmed milk but the specific activity is higher in cream). AlP is released from the lipoprotein particles by treatment with *n*-butanol, which, combined with salting-out and ion-exchange or

gel permeation chromatography, formed the basis of all early methods for the isolation of AlP from milk; chromatography of n-butanol extracts of MFGM on
Concanavalin-A Agarose/Sepharose 4B/Sephacryl S-200 has been used in recent
methods. AlP is optimally active at pH 10.5 when assayed on *p-nitrophenylphosphate*
but at ~6.8 on caseinate; its optimum temperature is ~37 °C. It is a homodimer of
85 kDa monomers; it contains four atoms of Zn which are essential for activity and
is also activated by Mg^{2+}, AlP is inhibited by metal chelators; the apo-enzyme may
be reactivated by Zn and a number of other metals, which is used as the principle of
methods for determination of very low concentrations of zinc in biological systems.
It is also inhibited by inorganic phosphate. The amino acid composition of milk AlP
is known but its amino acid sequence has not been reported.

The indigenous AlP in milk is similar to the AlP in mammary tissue. The AlP in
human milk is similar, but not identical, to human liver AlP (i.e., tissue non-specific
type). It has been suggest that there are two AlPs in milk, one of which is from
sloughed-off mycoepithelial cells, the other originating from lipid microdroplets
and acquired intra-cellularly, which is probably the AlP found in the MFGM. Most
or all studies on milk AlP have been on that isolated from cream/MFGM, i.e., the
minor form of AlP in milk.

Assay Methods

In 1935, H.D. Kay and W.R. Graham developed a method based on the inactivation
of AlP for determination of the adequate pasteurisation of milk. The principle of
this method is still used throughout the world and several modifications have been
published. The usual substrates are phenylphosphate, *p*-nitrophenyl phosphate, phenolphthalein phosphate or fluorophos, which are hydrolysed to inorganic phosphate
and phenol, *p*-nitrophenol, phenolphthalein or Fluoroyellow, respectively:

$$X-O-\overset{\overset{\displaystyle O}{\|}}{\underset{\underset{\displaystyle O^-}{|}}{P}}-OH \quad \xrightarrow{\text{H}_2\text{O}} \quad H_2PO_4^- + XOH$$

where XOH is phenol, *p*-nitrophenol, phenolphthalein or Fluoroyellow. The liberated phosphate could be measured but the increase is small against a high background of inorganic phosphate in milk; therefore, the liberated alcohol is quantified.
Reflecting the widespread assay of AlP in routine dairy laboratories, many assay
methods have been developed.

Reactivation of Alkaline Phosphatase

Much work has focussed on 'phosphatase reactivation', first recognised in 1953 by
R.C. Wright and J. Tramer, who observed that UHT-treated milk was phosphatase-
negative immediately after processing but became positive on storage; microbial
phosphatase was shown not to be responsible. HTST-pasteurised bulk milk does
not undergo reactivation, although some samples from individual cows may.

HTST pasteurisation after UHT treatment usually prevents reactivation, which is never observed in in-container sterilized milk. Reactivation can occur following heating at a temperature as low as 84 °C for milk or 74 °C for cream. The optimum storage temperature for reactivation is 30 °C, at which reactivation is detectable after 6 h and may continue for up to 7 days. The greater reactivation in cream than in milk may be due to protection of the enzyme by fat.

A number of attempts have been made to explain the mechanism of reactivation of AlP. The enzyme which becomes reactivated is membrane-bound and several factors which influence reactivation have been established. Mg^{2+} and Zn^{2+} strongly promote reactivation but Sn^{2+}, Cu^{2+}, Co^{2+} and EDTA are inhibitory, while Fe^{2+} has no effect. Sulphydryl (SH) groups are essential for reactivation; perhaps this is why phosphatase becomes reactivated in UHT milk but not in HTST milk. The role of -SH groups, supplied by denatured whey proteins, is considered to be chelation of heavy metals, which would otherwise react with SH groups of the enzyme (also activated on denaturation), thus preventing renaturation. It has been proposed that Mg^{2+} or Zn^{2+} cause a conformational change in the denatured enzyme, which is necessary for renaturation. Maximum reactivation occurs in products heated at ~104 °C, adjusted to pH 6.5, containing 64 mM Mg^{2+} and incubated at 30 °C; homogenisation of products before heat treatment reduces the extent of reactivation.

Reactivation of alkaline phosphatase is of considerable practical significance since regulations for HTST pasteurisation specify the absence of phosphatase activity. Methods for distinguishing between renatured and residual native alkaline phosphatase are based on the increase in phosphatase activity resulting from addition of Mg^{2+} to the reaction mixture.

Significance of AlP

Alkaline phosphatase in milk is significant mainly because it is used universally as an index of HTST pasteurisation. However, the enzyme may not be the most appropriate for this purpose because:

• Reactivation of alkaline phosphatase under certain conditions complicates interpretation of the test.
• The enzyme appears to be fully inactivated by temperature × time combinations (e.g., 70 °C × 16 s) less severe than full HTST conditions.
• The relationship between $\log_{10}$ % initial activity and pasteurisation equivalent is less linear than the relationship for lactoperoxidase or γ-glutamyl transpeptidase.

Although AlP can dephosphorylate casein under suitable conditions, as far as is known, it has no direct technological significance in milk. Perhaps its pH optimum is too far removed from that of milk, especially in fermented milk products, although the pH optimum on casein is reported to be ~7. Moreover, it is inhibited by inorganic phosphate.

Proteolysis is a major contributor to the flavour and texture of cheese. Most of the small water-soluble peptides in cheese are derived from the N-terminal half of α_{s1}- or β-casein; many are phosphorylated and show evidence of phosphatase activity (i.e., they are partially dephosphorylated) possibly by indigenous acid phosphatase and bacterial phosphatases are probably responsible. Further work on the significance of indigenous alkaline and acid phosphatases in the dephosphorylation of phosphopeptides in cheese is warranted.

10.2.4.2   Acid Phosphatase (EC 3.1.3.2)

The occurrence of an acid phosphomonoesterase (AcP) in milk was first reported by C. Huggins and P. Talaly in 1948 and confirmed by J. E. C. Mullen, who reported that AcP is optimally active at pH 4.0 and is very heat-stable (heating at 88 °C for 10 min is required for complete inactivation). The enzyme is not activated by Mg^{2+} (as is AlP), but it is activated slightly by Mn^{2+} and is strongly inhibited by fluoride. The level of AcP activity in milk is only ~2 % that of AlP; activity reaches a maximum 5–6 days *post partum*.

Isolation and Characterization

About 80 % of the AcP in milk is in the skimmed milk but the specific activity is higher in cream; it is strongly attached to the MFGM and is not released by non-ionic detergents. Acid phosphatase in milk has been purified to homogeneity by various forms of chromatography. Adsorption onto Amberlite IRC50 resin is a very effective first step in purification but only ~50 % of the total AcP in skim milk is adsorbed by Amberlite IRC50, even after re-extracting the skim milk with fresh batches of Amberlite, suggesting that skim milk contains at least two AcP isozymes. About 40 % of the AcP in skim milk partitions into the whey on rennet coagulation; this enzyme, which does not adsorb on Amberlite IRC50, has been partly purified. The AcP isolated from skim milk by adsorption on Amberlite IRC50 has been well characterised. It is a glycoprotein with a molecular weight of ~42 kDa and a pI of 7.9. It is inhibited by many heavy metals, F^-, oxidising agents, orthophosphates and polyphosphates and activated by thiol-reducing agents and ascorbic acid; it is not affected by metal chelators. It contains a high level of basic amino acids and no methionine

Since milk AcP is quite active on phosphoproteins, including caseins, it has been suggested that it is a phosphoprotein phosphatase. Although casein is a substrate for milk AcP, the major caseins, in the order α_s $(\alpha_{s1} + \alpha_{s2}) > \beta > \kappa$, are competitive inhibitors when assayed on *p*-nitrophenylphosphate, probably due to binding of the enzyme to the casein phosphate groups (the effectiveness of the caseins as inhibitors is related to their phosphate content).

Assay Methods

Acid phosphatase may be assayed at pH *ca.* 5, on the same substrates as used for AlP. If *p*-nitrophenol phosphate or phenolphthalein phosphate is used, the pH must be adjusted to >8 after incubation to induce the colour of *p*-nitrophenol or phenolphthalein.

Significance of AcP

Although AcP is present in milk at a much lower level than AlP, its greater heat stability and lower pH optimum may make it technologically significant. Dephosphorylation of casein markedly changes its physico-chemical properties. As discussed under AlP, several small partially-dephosphorylated peptides have been isolated from cheese and AcP may be involved. Dephosphorylation may be rate-limiting for proteolysis in ripening cheese since most proteinases and peptidases are inactive on phosphoproteins or phosphopeptides.

The suitability of AcP as an indicator enzyme for super-pasteurization of milk has been assessed but it is not as suitable for this purpose as γ-glutamyl transferase or lactoperoxidase.

10.2.5 Ribonuclease

RNase occurs in various tissues and secretions, including milk. Bovine pancreatic RNase A has been studied in detail; it was the first enzyme to have its complete amino acid sequence determined. It contains 124 amino acid residues, with a calculated MW of 13,683 Da, and has a pH optimum of 7.0–7.5.

The first study on the indigenous RNase in milk is that of E.W. Bingham and C. A. Zittle in 1962, who reported that bovine milk contains a much higher level of RNase than the blood serum or urine of human, rat or guinea pig, and that most or all of the activity is in the serum phase; bovine milk could potentially serve as a commercial source of RNase. Like pancreatic RNase, the RNase in milk is optimally active at pH 7.5 and is more heat-stable at acid pH values than at pH 7; in acid whey, adjusted to pH 7, 50 % of RNase activity is lost on heating at 90 °C for 5 min and 100 % after 20 min, but it is completely stable in whey at pH 3.5 when heated at 90 °C for 20 min. The enzyme was purified 300-fold by adsorption on Amberlite IRC-50 resin and desorption by 1 M NaCl, followed by precipitation with cold (4 °C) acetone (46–66 % fraction).

Elution from Amberlite IRC-50 using a NaCl gradient resolves two isoenzymes, A and B, of RNase in milk, at a ratio of about 4:1, as for pancreatic RNase. Amino acid analysis, electrophoresis and immunological studies showed that milk RNase is identical to pancreatic RNase. It is presumed that the RNase in milk originates in the pancreas and is absorbed through the intestinal wall into the blood, from which it enters milk; however, the level of RNase activity in milk is considerably higher than in blood serum, suggesting active transport.

Bovine and buffalo milks contain about equal amounts of RNase and about three times as much as human, ovine or caprine milk and porcine milk contains a very low level of RNase. A high molecular weight (80 kDa) RNase (hmRNase) has been purified from human milk and characterised as a single-chain glycoprotein, with a pH optimum in the range 7.5–8.0. It has been suggested that hmRNase is synthesised in the mammary gland and passes into milk, rather than being transferred from blood, as are RNase A and B.

Little or no RNase activity survives UHT sterilisation (121 °C for 10 s) but about 60 % survives heating at 72 °C for 2 min or at 80 °C for 15 s. RNase activity in raw or heat-treated milk is stable to repeated freezing and thawing and to frozen storage for at least a year. Although RNase has no technological significance in milk, which contains very little RNA, it may have significant biological functions.

10.2.6 Lysozyme (EC 3.1.2.17)

The presence of natural antibacterial factor(s) in fresh raw bovine milk was reported by Kitasoto in 1889 and by Fokker in 1890; these inhibitors are now called lactenins, one of which is lactoperoxidase. In 1922, A. Fleming identified an antibacterial agent, shown to be an enzyme, in nasal mucus, tears, sputum, saliva and other body fluids which caused lysis of many types of bacteria (*Micrococcus lysodeikiticus* was used for assays), which he called lysozyme. Chicken egg white is the richest source of lysozyme which constitutes ~3.5 % of egg white protein which is the principal commercial source of lysozyme. Milk was not among the several fluids in which Fleming found lysozyme but shortly afterwards the milk of several species was shown to contain lysozyme; human milk is a comparatively rich source but bovine milk is a poor source.

Lysozyme (muramidase, mucopeptide N-acetyl-muramyl hydrolase) hydrolyses the $\beta(1 \rightarrow 4)$-linkage between muramic acid and N-acetylglucosamine of mucopolysaccharides in the cell wall of certain bacteria, resulting in cell lysis. Egg white lysozyme (EWL) has been studied extensively.

Lysozyme was isolated from human milk in 1961 by P. Jolles and J. Jolles, who believed that bovine milk was devoid of lysozyme. Lysozyme was soon found in the milk of several species; the milk of cattle, goats, sheep, pigs and guinea pigs contain a low and variable level of lysozyme. Human and equine milk contain ~400 mg L^{-1} and ~800 mg L^{-1} lysozyme, respectively (3,000 and 6,000 times the level in bovine milk); these levels represent ~4 and ~3 % of the total protein in human and equine milk, respectively Although lysozyme is a lysosomal enzyme, it is found in soluble form in many body fluids (milk, tears, mucus, egg white); the lysozyme in milk is usually isolated from whey.

In addition to the lysozyme in human, equine and bovine milk, lysozyme has been isolated and partially characterised from the milk of baboon, camel, buffalo and dog. The properties of these lysozymes are generally similar to those of HML, but there are substantial differences, even between the lysozymes of closely-related species, e.g., cow and buffalo. Lysozyme has been isolated from the milk of a wider range of species than any other milk enzyme, which may reflect the perceived importance of lysozymes as a protective agent in milk or perhaps because it can be isolated from milk relatively easily.

The pH optimum of HML, bovine milk lysozyme (BML) and EWL is 7.9, 6.35 and 6.2, respectively. The MW these lysozymes is ~15 kDa. The amino acid sequence of HML and EWL are highly homologous, but there are several differences; HML consists of 130 amino acid residues, compared with 129 in EWL, the extra residue in the former being Val_{100}. The amino acid sequence of equine milk lysozyme also consists of 129 amino acid residues, but shows only 51 % homology with HML and 50 % homology with EWL.

The amino acid sequence of lysozyme is highly homologous with that of α-lactalbumin (α-la), a whey protein which is an enzyme modifier in the biosynthesis of lactose. The gene sequence and three-D structure of α-la and c-type lysozymes are similar. α-La binds a Ca^{2+} in an Asp-rich loop but most c-type lysozymes do not bind a Ca^{2+}, those of equine and canine milk being exceptions. All lysozymes are relatively stable to heat at acid pH values (3–4) but are relatively labile at pH > 7. More than 75 % of the lysozyme activity in bovine milk survives heating at 75 °C × 15 min or 80 °C × 15 s and is affected little by HTST pasteurization

Presumably, the physiological role of lysozyme is to act as a bactericidal agent. In the case of milk, it may simply be a 'spill-over' enzyme or it may have a definite protective role. If the latter is true, then the exceptionally high level of lysozyme in human, equine and assinine milk may be significant. Breast-fed babies generally suffer less enteric problems than bottle-fed babies. While there are many major compositional and physico-chemical differences between bovine and human milk that may be responsible for the observed nutritional characteristics, the disparity in lysozyme content may be significant. Fortification of bovine milk-based infant formulae with EWL, especially for premature babies, has been recommended but feeding studies are equivocal on the benefits of this practice; it appears that EWL is inactivated in the human GIT.

10.2.7   N-Acetyl-β-D-Glucosaminidase (EC 3.2.1.30)

N-Acetyl-β-D-glucosaminidase (NAGase) hydrolyses terminal, non-reducing N-acetyl-β-D-glucosamine residues from glycoproteins. It is a lysosomal enzyme and originates mainly from somatic cells and mammary gland epithelial cells. Consequently, NAGase activity increases markedly and correlates highly with the intensity of mastitis. A field test for mastitis based on NAGase activity has been developed, using chromogenic N-acetyl-β-D-glucosamine-p-nitrophenol as substrate; hydrolysis yields yellow p-nitrophenol. NAGase is optimally active at 50 °C and pH 4.2 and is inactivated by HTST pasteurization.

10.2.8   γ-Glutamyl Transpeptidase (Transferase) (EC 2.3.2.2)

γ-Glutamyl transferase (GGT) catalyses the transfer of γ-glutamyl residues from γ-glutamyl-containing peptides:

$$\gamma\text{-glutamyl-peptide} + X \rightarrow \text{peptide} + \gamma\text{-glutamyl-X},$$

where X is an amino acid.

GGT, which has been isolated from the fat globule membrane, has a molecular mass of about 80 kDa and consists of two subunits of 57 and 26 kDa. It is optimally active at pH 8–9, has a pI of 3.85 and is inhibited by iodoacetate, diisopropylfluorophosphate and metal ions, e.g., Cu^{2+} and Fe^{3+}.

It plays a role in amino acid transport in the mammary gland. γ-Glutamyl peptides have been isolated from cheese but since γ-glutamyl bonds do not occur in milk proteins, their synthesis may be catalysed by GGT. The enzyme is relatively heat stable and has been proposed as a marker enzyme for milks pasteurized in the range 72–80 °C$\times$15 s. GGT is absorbed from the gastrointestinal tract, resulting in high levels of GGT activity in the blood serum of newborn animals fed colostrum or early breast milk.

10.2.9   Amylases (EC 3.2.1.-)

Amylase (diastase) was identified in milk in 1883. During the next 40 years, several workers reported that the milk of several species contains an amylase. The principal amylase in milk is α-amylase (EC 3.2.1.1), with a lesser amount of β-amylase (EC 3.2.1.2); the enzymes partition mainly into skimmed milk and whey. A highly-concentrated preparation of α-amylase was obtained from whey by E.J. Guy and R. Jenness in 1958 but there appears to have been no further work on the isolation of amylase from bovine milk. Amylase is quite heat-labile and loss of amylase activity was proposed in the 1930s as an index of the intensity of heat treatment applied to milk.

Human milk and colostrum contain 25–40 times more α-amylase than bovine milk; α-amylase has been purified from human milk by gel permeation chromatography and its stability to pH and pepsin determined. The level of α-amylase in human milk is 15–140 times higher than in blood plasma, suggesting that it is not transferred from blood but is synthesised in the mammary gland. Milk α-amylase is similar to salivary amylase.

Since milk contains no starch, the function of amylase in milk is unclear. Human milk contains up to 130 oligosaccharides, at a total concentration up to 15 mg mL^{-1} but α-amylase can not hydrolyse the oligosaccharides in milk. Since human babies secrete low levels of salivary and pancreatic amylases (0.2–0.5 % of the adult level), the high level of amylase activity in human milk may enable them to digest starch. By hydrolysing the polysaccharides in the cell wall of bacteria, it has been suggested that milk amylase may have anti-bacterial activity. The amylase activity of human milk is an active area of research at present but there appears to be little or no recent research on the amylase in bovine milk or that of other species.

10.2.10   Catalase (EC 1.11.1.6)

Catalase (H_2O_2:H_2O_2 oxidoreductase; EC 1.11.1.6) catalyses the decomposition of H_2O_2, as follows:

$$2H_2O_2 \rightarrow 2H_2O + O_2$$

Catalase activity may be determined by quantifying the evolution of O_2 manometrically or by titrimetrically measuring the reduction of H_2O_2. Catalases are haem-containing enzymes that are distributed widely in plant, microbial and animal tissues and secretions; liver, erythrocytes and kidney are particularly rich sources.

A catalase was among the first enzymes demonstrated in milk by Babcock and Russell in 1897, who reported that an extract of separator slime (somatic cells and other debris) could decompose H_2O_2, The catalase activity in milk varies with feed, stage of lactation and especially during mastitis when the level of activity increases markedly, making catalase a useful indicator of mastitis but it is now rarely used for this purpose, determination of somatic cell count, N-acetylglucosaminidase activity or electrical conductivity being usually used.

Most, ~70 %, of the catalase in milk is in the skimmed milk but the specific activity in the cream is 12-fold higher than in skimmed milk and the MFGM is usually used as the starting material for the isolation of catalase from milk. Although the level of catalase in milk is relatively high and the enzyme is easily assayed, catalase was not isolated from milk until 1983, when O. Ito and R. Akuzawa purified catalase from milk and crystallised the enzyme, which had a MW of 225 kDa (by gel permeation). Bovine liver catalase is a homotetramer of 60–65 kDa subunits (total MW ~250 kDa); it seems likely that the structure of catalase in milk is similar. Milk catalase is a heme protein with a MW of 200 kDa, and an isoelectric pH of 5.5; it is stable between pH 5 and 10 but rapidly loses activity outside this range. Catalase is relatively heat-labile; heating at 70 °C for 1 h causes complete inactivation. Like other catalases, it is strongly inhibited by Hg^{2+}, Fe^{2+}, Cu^{2+}, Sn^{2+}, CN^- and NO_3^-. It may act as a lipid pro-oxidant *via* its heme iron.

Catalase was among the first indicators of pasteurisation investigated and recently, it has been considered as an indicator of cheese made from sub-pasteurized milk. Although the inactivation of catalase is a useful index of thermisation of milk (it was almost completely inactivated by heating at 65 °C for 16 s), it is not suitable as an index of cheese made from thermised milk owing to the production of catalase in the cheese during ripening, especially by coryneform bacteria and yeasts, if present.

10.2.11 Lactoperoxidase (EC 1.11.1.7)

Peroxidases, which are widely distributed in plant, animal and microbial tissues and secretions, catalyse the following reaction:

$$2HA + H_2O_2 \rightarrow A + 2H_2O$$

where HA is an oxidisable substrate or a hydrogen donor.

Lactoperoxidase (LPO) was first demonstrated in milk by C. Arnold in 1881, using "*guajaktinctur*" as reducing agent; he reported that the activity of LPO is lost on heating milk at 80 °C. Legislation was introduced in Denmark in 1898 requiring that all skim milk returned by creameries to farmers should be flash (i.e., no holding period) pasteurized at 80 °C. Various tests were proposed to ensure that the milk was adequately pasteurised, but the most widely used method was that developed in 1898 by V. Storch, who assayed LPO activity using *p*-phenylenediamine as reducing agent; this test is still used to identify super-pasteurised milk, i.e., milk heated ≥76 °C for 15 s.

Work on the isolation of LPO was commenced by S. Thurlow in 1925 and LPO was isolated and crystallised by H. Theorell and A. Aokeson in 1943. It is a heam protein containing protoporphyrin IX with 0.069 % Fe, a Soret band at 412 nm, an $A_{412}:A_{280}$ ratio of 0.9, has a mass of 82 kDa and occurs as two isozymes, A and B. Since then, several improved methods for the isolation of LPO have been published and its characteristics refined. Since LPO is cationic at the pH of milk, as are lactoferrin and some minor proteins, it can be easily isolated from milk or whey using a cationic exchange resin (e.g., Amberlite $CG-50-NH_4$). There are ten isozymes of LPO, arising from differences in the level of glycosylation and deamination of Gln or Asn. LPO consists of 612 amino acids and shows 55, 54 and 45 % identity with human myeloperoxidase, eosinophil peroxidase and thyroperoxidase, respectively. LPO binds a Ca^{2+}, which has a major effect on its stability, including its heat stability; at a pH below ~5.0, the Ca^{2+} is lost, with a consequent loss of stability.

LPO is synthesised in the mammary gland and is the second most abundant enzyme in milk (next to xanthine oxidoreductase), constituting ~0.5 % of the total whey proteins (~0.1 % of total protein; 30 mg L^{-1}). Human milk contains mainly myeloperoxidase, with a low level of LPO. Apparently, human colostrum contains a high level of myeloperoxidase, derived from leucocytes, and a lower level of LPO. The level of myeloperoxidase decreases rapidly *post partum* and LPO is the principal peroxidase in mature human milk.

10.2.11.1 Significance of LPO

Apart from its exploitation as an index of flash or super-HTST pasteurization, LPO is technologically significant for a number of other reasons also.

  i. It is a possible index of mastitic infection but is not well correlated with somatic cell count.
 ii. LPO causes non-enzymic oxidation of unsaturated lipids, acting through its heme group; the heat-denatured enzyme is more active than the native enzyme.
iii. Milk contains bacteriostatic or bactericidal substances referred to as lactenins, one of which is LPO, which requires H_2O_2 and thiocyanate (SCN^-) to cause inhibition. The nature, mode of action and specificity of the $LPO-SCN^--H_2O_2$ system has been widely studied. LPO and thiocyanate, which is produced in the rumen by enzymic hydrolysis of thioglycosides from *Brassica* plants, occur naturally in milk, but H_2O_2 does not occur naturally in milk. However, H_2O_2 can be generated metabolically by catalase-negative bacteria, produced *in situ* through the action of exogenous glucose oxidase on glucose, or it may be added directly.

In the presence of low levels of H_2O_2 and SCN^-, LPO exhibits very potent bactericidal activity; this system is 50–100 times more effective than H_2O_2 alone. The LPO system has been found to have good bactericidal efficiency for the cold pasteurization of fluids or sanitization of immobilized enzyme columns. A self-contained $LPO-H_2O_2-SCN^-$ system using coupled β-galactosidase and glucose oxidase, immobilized on porous glass beads, to generate H_2O_2 *in situ* from lactose in milk containing 0.25 mM thiocyanate has been developed. Indigenous xanthine oxidoreductase, acting on added hypoxanthine, may also be exploited to produce H_2O_2. The bactericidal effects of the $LPO-H_2O_2-SCN^-$ system may be used to cold pasteurize milk in situations where refrigeration and/or thermal pasteurization is lacking. Addition of isolated LPO to milk replacers for calves or piglets reduces the incidence of enteritis.

iv. It has been proposed that the LPO system may be exploited for bleaching coloured whey.

10.2.12   Xanthine Oxidoreductase (XOR) [EC, 1.13.22; 1.1.1.204]

In 1902, F. Schardinger showed that milk contains an enzyme capable of oxidising aldehydes to acids, accompanied by the reduction of methylene blue; this enzyme was commonly called the "Schardinger enzyme". In 1922, it was shown that milk contains an enzyme capable of oxidising xanthine and hypoxanthine, with the concomitant reduction of O_2 to H_2O_2, this enzyme was called xanthine oxidase (XO). In 1938, V. H. Booth showed that the Schardinger enzyme is, in fact, xanthine oxidase and partially purified it. XO requires FAD for catalytic activity. Under certain circumstances XO can dehydrogenate xanthine and is now called xanthine oxidoreductase (XOR). XOR exists as two forms, xanthine oxidase (XO; EC 1.1.3.22) and xanthine dehydrogenase (XDH; 1.1.1.204) which can be inter-converted by sulphydryl reagents and XDH can be converted irreversibly to XO by specific proteolysis.

XOR is concentrated in the MFGM, in which it is the second most abundant protein, after butyrophilin; it represents ~20 % of the protein of the MFGM (~0.2 % of total milk protein; ~120 mg L^{-1}). Therefore, all isolation methods use cream as the starting material; the cream is washed and churned to yield a crude MFGM preparation; dissociating and reducing agents are used to liberate XOR from membrane lipoproteins and some form of chromatography is used for purification. Milk XOR is a homodimer of 146 kDa sub-units, each containing ~1,332 amino acid residues; each monomer contains 1 atom of Mo, 1 molecule of FAD and 2 Fe_2S_2 redox centres. NADH acts as a reducing agent and the oxidation products are H_2O_2 and O_2^-. Cows deficient in Mo have low XOR activity. The quaternary structure of XDH and XO has been described. Milk is a very good source of XOR, at least part of which is transported to the mammary gland via the blood stream.

10.2.12.1   Activity in Milk and Effect of Processing

Various processing treatments affect the XOR activity of milk. Activity is increased by ~100 % on storage at 4 °C for 24 h, by 50–100 % on heating at 70 °C for 5 min and by 60–90 % on homogenization. These treatments cause the transfer to XO from the fat phase to the aqueous phase, rendering the enzyme more active. The heat stability of XOR is very dependent on whether it is a component of the fat globules or is in the aqueous phase;. XOR is most heat stable in cream and least in skim milk. Homogenization of concentrated milk prepared from heated milk (90.5 °C for 15 s) partially reactivates XOR, which persists on drying the concentrate, but no reactivation occurs following more severe heating (93 °C for 15 s).

The XOR activity in human milk is low because 95–98 % of the enzyme molecules lack Mo. The level of XOR activity in goat and sheep milk is also low and be increased by supplementing the diet with Mo.

10.2.12.2   Assay

Xanthine oxidase activity can be assayed manometrically, potentiometrically, polarographically or spectrophotometrically. The latter may involve the reduction of colourless triphenyltetratetrazolium chloride to a red product or the conversion of xanthine to uric acid which is quantified by measuring absorbance at 290 nm.

10.2.12.3   Significance of Xanthine Oxidase

*   *As an index of heat treatment*: XOR has been considered as a suitable indicator of milk heated in the temperature range 80–90 °C but the natural variability in the level of XOR activity in milk is too high for its use as a reliable index of heat treatment.
*   *Lipid oxidation*: XOR can excite stable triplet oxygen ($^3O_2$) to singlet oxygen ($^1O_2$), a potent pro-oxidant. Some individual-cow milk samples, which undergo spontaneous oxidative rancidity, contain about ten times the normal level of XOR, and spontaneous oxidation can be induced in normal milk by the addition of XOR to ~4× the normal level. Heat-denatured or FAD-free enzyme is not a pro-oxidant
*   *Atherosclerosis*: It has been suggested that XOR enters the vascular system from homogenised milk and may be involved in atherosclerosis *via* oxidation of plasmalogens in cell membranes; this aspect of XOR attracted considerable attention in the early 1970s but the hypothesis has been discounted.
*   *Reduction of nitrate in cheese*: Sodium nitrate is added to milk for many cheese varieties to prevent the growth of *Clostridium tyrobutyricum*, which causes flavour defects and late gas blowing; XOR reduces nitrate to nitrite, which is bactericidal, and then to NO.
*   *Production of H_2O_2*: The H_2O_2 produced by the action of XOR can serve as a substrate for lactoperoxidase in its action as a bactericidal agent.
*   *Purine catabolism* XOR catalyses the catabolism of purines and may be involved in the regulation of blood pressure.
*   *Bactericidal activity*: XOR has strong antibacterial activity in the human intestine, probably *via* the production of peroxynitrite ($ONOO^-$).
*   *Secretion of milk fat*: Probably the most important role of XOR in milk is now considered to be in the secretion of milk fat globules from the mammary secretory cells. The triglycerides in milk are synthesised in the endoplasmic reticulum (ER), which is located toward the basal membrane of the cell. In the ER, the TGs are formed into micro lipid droplets and released through the involvement of the protein, acidophilin (ADPH), which surrounds the globules. The ADPH-covered globules move toward the apical membrane of the cell, probably through

a microtubular/microfilament system, and acquire additional coat material, cytoplasmic proteins and phospholipids. At the apical membrane, ADPH forms a disulphide-linked complex with butyrophlin (BTN) and dimeric XOR. Somehow, XOR causes blebbing of the fat globule through the membrane and it is eventually pinched off and released into the alveolar lumen. In the secretion of milk fat globules, XOR does not function as an enzyme (see Chap. 3).

- The production of reactive oxygen and nitrogen species by XOR and the involvement of these in various pathophysiological conditions including cardiovascular disease has attracted much research attention.

10.2.13   Sulphydryl Oxidase (EC 1.8.3.-)

Milk contains sulphydryl oxidase (SO), capable of oxidizing sulphydryl groups of cysteine, glutathione and proteins to the corresponding disulphide. The enzyme is an aerobic oxidase which catalyses the following reaction:

$$2RSH + O_2 \rightarrow RSSR + H_2O_2$$

It undergoes marked self-association and can be purified readily by chromatography on porous glass. The enzyme has a molecular mass of ~89 kDa, a pH optimum of 6.8–7.0, and a temperature optimum of 35 °C. Its amino acid composition, its requirement for iron but not for molybdenum and FAD, and the catalytic properties of the enzyme, indicate that sulphydryl oxidase is distinct from xanthine oxidoreductase and thiol oxidase (EC 1.8.3.2).

SO is capable of oxidizing reduced ribonuclease and restoring enzymatic activity, suggesting that its physiological role may be the non-random formation of protein disulphide bonds during protein biosynthesis. SO immobilized on glass beads has the potential to ameliorate the cooked flavour arising from sulphydryl groups exposed upon protein denaturation.

The production of sulphur compounds is believed to be very important in the development of Cheddar cheese flavour. Residual sulphydryl oxidase activity may play a role in reoxidizing sulphydryl groups exposed upon heating cheesemilk; the sulphydryl groups thus protected may be reformed during ripening.

10.2.14   Superoxide Dismutase (EC 1.15.1.1)

Superoxide dismutase (SOD) scavenges superoxide radicals, O_2^- according to the reaction:

$$2O_2^- + 2H^+ \rightarrow H_2O_2 + O_2$$

The H_2O_2 formed may be reduced by catalase, peroxidase or a suitable reducing agents. SOD occurs in many animal and bacterial cells; its biological function is to protect tissue against oxygen free radicals in anaerobic systems.

SOD, isolated from bovine erythrocytes, is a blue-green protein due to the presence of copper, removal of which by treatment with EDTA results in loss of activity which is restored by adding Cu^{2+}; it also contains Zn^{2+}, which appears not to be at the active site. The enzyme consists of two identical 16 kDa subunits held together by one or more disulphide bonds. The amino acid sequence has been established.

Milk contains trace amounts of SOD which has been isolated and characterized; it appears to be identical to the bovine erythrocyte enzyme. SOD inhibits lipid oxidation in model systems. The level of SOD in milk parallels that of XO (but at a lower level), suggesting that SOD may be excreted in milk in an attempt to offset the pro-oxidant effect of XO. However, the level of SOD in milk is probably insufficient to explain observed differences in the oxidative stability of milk. The possibility of using exogenous SOD to retard or inhibit lipid oxidation in dairy products has been considered.

In milk, SOD is stable at 71 °C for 30 min but loses activity rapidly at even slightly higher temperatures. Slight variations in pasteurization temperature are therefore critical to the survival of SOD in heated milk products and may contribute to variations in the stability of milk to oxidative rancidity.

10.2.15 Other Enzymes

In addition to the enzymes describe above, a number of other indigenous enzymes (Table 10.2) have been isolated and partially characterized. Although fairly high levels of some of these enzymes occur in milk, they have no apparent function in milk and will not be discussed further. Many other enzymatic activities have been detected in milk but have not been isolated and limited information on their molecular and biochemical properties in milk is available; some of these are listed in Table 10.3.

Table 10.2 Other enzymes that have been isolated from milk and partially characterized but which are of no known significance (see Farkye 2003)

Enzyme		Reaction catalysed	Comment
Glutathione peroxidase	EC 1.11.1.9	$2\,GSH + H_2O \rightleftharpoons GSSH$	Contains Se
Ribonuclease	EC 3.1.27.5	Hydrolysis of RNA	Milk a very rich source; similar to pancreatic RNase
α-Amylase	EC 3.2.1.1	Starch	
β-Amylase	EC 3.2.1.2	Starch	
α-Mannosidase	EC 3.2.1.24		Contains Zn^{2+}
β-Glucuronidase	EC 3.2.1.31		
5′-Nucleotidase	EC 3.1.3.5	5′ nucleotides + $H_2O \rightleftharpoons$ Ribonucleosides + Pi	Diagnostic test for mastitis
Adenosine triphosphatase	EC 3.6.1.3	$ATP + H_2O \rightleftharpoons ADP + Pi$	
Aldolase	EC 4.1.2.13	Fructose 1,6 diP $\rightleftharpoons$ glyceraldehyde-3-P dihydroxyacetone-P	

Table 10.3 Partial list of minor enzymes in milk (modified from Farkye 2003)

Enzyme	Reaction catalyzed	Source	Distribution in milk
EC 1.1.1.1 Alcohol dehydrogenase	Ethanol+NAD$^+$ ⇌ acetaldehyde+NADH+H$^+$	–	SM
EC 1.1.1.14 L-Iditol dehydrogenase	L-Iditol+NAD$^+$ ⇌ L-sorbose+NADH		SM
EC 1.1.1.27 Lactate dehydrogenase	Lactic acid+NAD$^+$ ⇌ pyruvic acid+NADH+H$^+$	Mammary gland	SM
EC 1.1.1.37 Malate dehydrogenase	Malate+NAD$^+$ ⇌ oxaloacetate+NADH	Mammary gland	SM
EC 1.1.1.40 Malic enzyme	Malate+NADP$^+$ ⇌ pyruvate+CO$_2$+NADH	Mammary gland	SM
EC 1.1.1.42 Isocitrate dehydrogenase	Isocitrate+NADP$^+$ ⇌ 2-oxogluterate+CO$_2$+NADH	Mammary gland	SM
EC 1.1.1.44 Phosphoglucuronate dehydrogenase (decarboxylating)	6-Phospho-D-gluconate+NADP$^+$ ⇌ D-ribose-5 phosphate+CO$_2$+NADPH	Mammary gland	SM
EC 1.1.1.49 Glucose-6-phosphate dehydrogenase	D-Glucose-D-gluconate+NADP$^+$ ⇌ D-glucono-1,5-lactone-6-phosphate+NADPH	Mammary gland	SM
EC 1.4.3.6 Amine oxidase (Cu-containing)	RCH$_2$NH$_2$+H$_2$O+O$_2$ ⇌ RCHO+NH$_3$+H$_2$O$_2$	–	SM
– Polyamine oxidase	Spermine → spermidine → putrescine	–	SM
– Fucosyltransferase	Catalyses the transfer of fucose form GDP L-fucose to specific oligosaccharides and glycoproteins	–	SM
EC 1.6.99.3 NADH dehydrogenase	NADH+acceptor ⇌ NAD$^+$+reduced acceptor	–	FGM
EC 1.8.1.4 Dihydrolipoamide dehydrogenase (Diaphorase)	Dihydrolipoamide+NAD$^+$ ⇌ lipoamide+NADH	–	SM/FGM
EC 2.4.1.22 Lactose synthetase A protein: UDP-galactose: D-glucose, 1-galactosyltransferase; B protein: α-lactalbumin	UDP galactose+D-glucose ⇌ UDP+lactose	Golgi apparatus	SM
EC 2.4.1.38 Glycoprotein 4-β-galactosyltransferase	UDP galactose+N-acetyl D-glucosaminyl-glycopeptide ⇌ UDP+4,β-D-galactosyl-N-acetyl-D-glucosaminyl glycopeptide	–	FGM
EC 2.4.1.90 N-Acetyllactosamine synthase	UDP galactose+N-acetyl-D-glucosamine ⇌ UDP N-acetyllactosamine	Golgi apparatus	–
EC 2.4.99.6 CMP-N-acetyl-N-acetyllactosaminide α-2,3-sialyltransferase	CMP-N-acetylneuraminate+β-D-galactosyl 1,4-N-acetyl D-glucosaminyl glycoprotein ⇌ CMP+α-N-acetylneuraminyl 1-2,3-β-D-galactosyl-1,4-N-acetyl-D-glucosaminyl-glycoprotein	–	SM

EC number	Enzyme	Reaction	Source	Location
EC 2.5.1.3	Thiamine-phosphate pyrophosphorylase	2-Methyl-4-amino-5-hydroxymethyl/pyrimidine diphosphate + 4-methyl-5-(2-phosphonooxyethyl)-thiazole $\rightleftharpoons$ pyrophosphate + thiamine monophosphate	–	FGM
EC 2.6.1.1	Aspartate aminotransferase	L-Aspartate + 2-oxogluterate $\rightleftharpoons$ oxaloacetate + L-glutamate	Blood	SM
EC 2.6.1.2	Alanine aminotransferase	L-Alanine + 2-oxogluterate $\rightleftharpoons$ pyruvate + L-glutamate	Blood	SM
EC 2.7.5.1	Phosphoglucomutase			
EC 2.7.7.49	RNA-directed DNA polymerase	n Deoxynucleoside triphosphate $\rightleftharpoons$ n pyrophosphate + DNA$_n$	–	SM
EC 2.8.1.1	Thiosulphate sulphur transferase	Thiosulphate + cyanide $\rightleftharpoons$ sulphite + thiocyanate	–	SM
EC 3.1.1.8	Cholinesterase	An acylcholine + H$_2$O $\rightleftharpoons$ choline + a carboxylic acid anion	Blood	FGM
EC 3.1.3.9	Glucose-6-phosphatase	D-Glucose 6-phosphate + H$_2$O $\rightleftharpoons$ D-glucose + inorganic phosphate		FGM
EC 3.1.4.1	Phosphodiesterase			
EC 3.1.6.1	Arylsulphatase	Phenol sulphate + H$_2$O $\rightleftharpoons$ phenol + sulphate	–	–
EC 3.2.1.21	β-Glucosidase	Hydrolysis of terminal non-reducing β-D-glucose residues	Lysosomes	FGM
EC 3.2.1.23	β-Galactosidase	Hydrolysis of terminal non-reducing β-D-galactose residues in β-D-galactosides	Lysosomes	FGM
EC 3.2.1.51	α-Fucosidase	An α-L-fucoside + H$_2$O $\rightleftharpoons$ an alcohol + L-fucose	Lysosomes	–
EC 3.4.11.1	Cytosol aminopeptidase (Leucine aminopeptidase)	Aminoacyl-peptide + H$_2$O $\rightleftharpoons$ amino acid + peptide	–	SM
EC 3.4.11.3	Cystyl-aminopeptidase (Oxytocinase)	Cystyl-peptides + H$_2$O $\rightleftharpoons$ amino acid + peptide	–	SM
EC 3.4.21.4	Trypsin	Hydrolyses peptide bonds, preferentially Lys-X, Arg-X	–	SM
EC 3.6.1.1	Inorganic pyrophosphatase	Pyrophosphate + H$_2$O $\rightleftharpoons$ 2 orthophosphate	–	SM/FGM
EC 3.6.1.1	Pyrophosphate phosphorylase			
EC 3.6.1.9	Nucleotide pyrophosphate	A dinucleotide + H$_2$O $\rightleftharpoons$ 2 mononucleotides	–	SM/FGM
EC 4.2.1.1	Carbonate dehydratase	H$_2$CO$_3$ $\rightleftharpoons$ CO$_2$ + H$_2$O	–	SM
EC 5.3.1.9	Glucose-6-phosphate isomerase	D-glucose-6-phosphate $\rightleftharpoons$ fructose-6-phosphate	–	SM
EC 6.4.1.2	Acetyl-CoA carboxylase	ATP + acetyl-CoA + HCO$_3$ $\rightleftharpoons$ ADP + orthophosphate + malonyl-CoA	–	FGM

SM skim milk, FGM fat globule membrane

10.3 Exogenous Enzymes in Dairy Technology

10.3.1 Introduction

Crude enzyme preparations have been used in food processing since pre-historic times; classical examples are rennets in cheesemaking, malt in brewing and papaya leaves to tenderize meat. Added (exogenous) enzymes are attractive in food processing because they can induce specific changes in contrast to chemical or physical methods which may cause non-specific undesirable changes. For some applications, there is no alternative to enzymes, e.g., rennet-coagulated cheeses, whereas in some cases, enzymes are preferred over chemical methods because they cause fewer side reactions and consequently give superior products, e.g., hydrolysis of starch.

Although relatively few enzymes are used in the dairy industry on a significant scale, the use of rennets in cheesemaking is one of the principal of all industrial applications of enzymes.

The applications of exogenous enzymes in dairy technology can be divided into two groups:

1. Technological, in which an enzyme is used to modify a milk constituent or to improve its microbiological, chemical or physical stability.
2. Enzymes as analytical reagents. Although the technological applications are quantitatively the more important, many of the analytical applications of enzymes are unique and are becoming increasingly important.

Since the principal constituents of milk are proteins, lipids and lactose, proteinases, lipases and β-galactosidase (lactase) are the principal exogenous enzymes used in dairy technology. Apart from these, there are, at present, only minor applications for glucose oxidase, catalase, superoxide dismutase and lysozyme. Lactoperoxidase, xanthine oxidase and sulphydryl oxidase might also be included, although at present the indigenous form of these enzymes is exploited.

10.3.2 Proteinases

There is one major (rennet) and several minor applications of proteinases in dairy technology.

10.3.2.1 Rennets

The use of rennets in cheesemaking is the principal application of proteinases in food processing. The sources of rennets and their role in milk coagulation and cheese ripening were discussed in Chap. 12 and will not be considered here.

10.3.2.2   Accelerated Cheese Ripening

Cheese ripening is a slow, expensive and not fully controlled process; consequently, there is increasing interest, at both the research and industrial levels, in accelerating ripening. Various approaches have been investigated to accelerate ripening, including a higher ripening temperature (especially for Cheddar-type cheese which is usually ripened at 6–8 °C), exogenous proteinases and peptidases, modified starters (e.g., heat-shocked or lactose-negative) and genetically engineered starters or starter adjuncts The possible use of exogenous proteinases and peptidases attracted considerable attention for a period but uniform distribution of the enzymes in the cheese curd is a problem, microencapsulation of enzymes offers a possible solution. Exogenous proteinases/peptidases are not used commercially in natural cheeses but are being used to produce "Enzyme Modified Cheese" for use in processed cheese, cheese dips and sauces. Selected genetically modified and adjunct cultures appear to be more promising.

10.3.2.3   Protein Hydrolyzates

Protein hydrolyzates are used as flavourings in soups and gravies and in dietetic foods. They are generally prepared from soy, gluten, milk, meat or fish proteins by acid hydrolysis. Neutralization results in a high salt content which is acceptable for certain applications but may be unsuitable for dietetic foods and food supplements. Furthermore, acid hydrolysis causes total or partial destruction of some amino acids. Partial enzymatic hydrolysis is a viable alternative for some applications but bitterness due to hydrophobic peptides is frequently encountered. Bitterness may be eliminated or at least reduced to an acceptable level by treatment with activated carbon, carboxypeptidase, aminopeptidase, ultrafiltration or hydrophobic chromatography. Caseins yield very bitter hydrolyzates but the problem may be minimized by the judicious selection of the proteinase(s) and by using exopeptidases (especially aminopeptidases) together with the proteinase.

A novel, potentially very significant application of proteinases in milk protein technology is the production of biologically active peptides (see Chap. 11). Carefully selected proteinases of known specificity are required for such applications but the resulting products have high added value.

The functional properties of milk proteins may be improved by limited proteolysis. Acid-soluble casein, free of off-flavour and suitable for incorporation into beverages and other acid foods (in which casein is insoluble) can be produced by limited proteolysis. The antigenicity of casein is destroyed by proteolysis and the hydrolysate is suitable for use in milk protein-based foods for infants allergic to cows' milk formulations. Controlled proteolysis improves the meltability of directly-acidified cheese but excessive proteolysis causes bitterness. Partial proteolysis of lactalbumin (heat-coagulated whey proteins), which is insoluble and has very poor functional properties, yields a product that is almost completely soluble above pH 6; although the product is slightly bitter, it appears promising as a food ingredient. Limited proteolysis of whey protein concentrate reduces its emulsifying capacity, increases its specific foam volume but reduces foam stability and increases heat stability.

10.3.3   β-Galactosidase

β-Galactosidase (lactase; EC 3.1.2.23), which hydrolyses lactose to glucose and galactose, is probably the second most significant enzyme in dairy technology. In the 1970s, β-galactosidase was considered to have very considerable potential but this has not materialized although there are a number of significant technological and nutritional applications. The various aspects of lactose and applications of β-galactosidase are considered in Chap. 2.

β-Galactosidase has transferrase as well as hydrolase activity and under certain conditions produces several oligosaccharides which have interesting nutritional and physic-chemical properties.

10.3.4   Lipases

The principal application of lipases in dairy technology is in cheese manufacture, particularly certain hard Italian varieties. The characteristic "piccante" flavour of these cheeses is due primarily to short-chain fatty acids resulting from the action of lipase(s) in the rennet paste traditionally used in their manufacture. Rennet paste is prepared from the stomachs of calves, kids or lambs slaughtered after suckling; the stomachs and contents are aged and then macerated. The lipase in rennet paste, pre-gastric esterase (PGE), is secreted by a gland at the base of the tongue, which is stimulated by suckling; the secreted lipase is washed into the stomach with the ingested milk. The physiological significance of PGE, which is secreted by several species, is to assist in lipid digestion in the neonate which has limited pancreatic function. PGE has a high specificity for short chain fatty acids, especially butanoic acid, esterified on the sn-3 position of glycerol, although some inter-species differences in specificity have been reported. Semi-purified preparations of PGE from calf, kid and lamb are commercially available and give satisfactory results; slight differences in specificity renders one or other more suitable for particular applications. Connoisseurs of Italian cheese claim that rennet paste gives superior results to semi-purified PGE, and it is cheaper.

Rhizomucor miehei secretes a lipase that is reported to give satisfactory results in Italian cheese manufacture. This enzyme has been characterized and is commercially available as "Piccantase". Lipases secreted by selected strains of *Penicillium roqueforti* and *P. candidum* are considered to be potentially useful for the manufacture of Italian and other cheese varieties.

Extensive lipolysis occurs in Blue cheese varieties in which the principal lipase is secreted by *P. roqueforti* (see Chap. 12). It is claimed that treatment of Blue cheese curd with PGE improves and intensifies its flavour but this practice is not widespread. Several techniques have been developed for the production of fast-ripened Blue cheese-type products suitable for use in salad dressings, cheese dips, etc. Lipases, usually of fungal origin, are used in the manufacture of these products or to pre-hydrolyse fats/oils used as ingredients in their production.

Although Cheddar cheese undergoes relatively little lipolysis during ripening, it is claimed that addition of PGE, gastric lipase or selected microbial lipases improves the flavour of Cheddar, especially that made from pasteurized milk, and accelerates ripening. It is also claimed that the flavour and texture of Feta and Egyptian Ras cheese can be improved by adding kid or lamb PGE or a low level of selected microbial lipases to the cheese milk, especially if milk concentrated by ultrafiltration is used.

Lipases are used to hydrolyze milk fat for a variety of uses in the confectionary, candy, chocolate, sauce and snack food industries and there is interest in using immobilized lipases to modify fat flavours for such applications. Enzymatic interesterification of milk lipids to modify rheological properties is also feasible.

10.3.5   Lysozyme

As discussed in Sect. 10.2.5, lysozyme has been isolated from the milk of several species; human and equine milks are especially rich sources. In view of its antibacterial activity, the large difference in the lysozyme content between human and bovine milks may have significance in infant nutrition. It is claimed that supplementation of baby food formulae based on cows' milk with egg white lysozyme is beneficial, especially with premature babies but views on this not unanimous, and it is not used commercially.

Nitrate is added to many cheese varieties to prevent the growth of *Clostridium tyrobutyricum* which causes off-flavours and late gas blowing. However, the use of nitrate in foods is considered to be undesirable because of its involvement in nitrosamine formation and many countries have reduced the permitted level or prohibited its use. Lysozyme, which inhibits the growth of vegetative cells of *Cl. tyrobutyricum* and hinders the germination of its spores, is an alternative to nitrate for the control of late gas blowing in cheese. Lysozyme also kills *Listeria* spp. Co-immobilized lysozyme has been proposed for self-sanitizing immobilized enzyme columns.

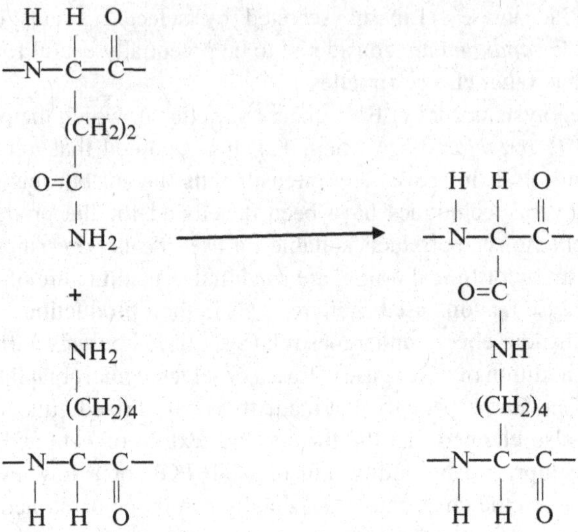

$$
\begin{array}{c}
\overset{\displaystyle H}{|}\ \ \overset{\displaystyle H}{|}\ \ \overset{\displaystyle O}{\|} \\
-N-C-C- \\
| \\
(CH_2)_2 \\
| \\
O=C \\
| \\
NH_2 \\
+ \\
NH_2 \\
| \\
(CH_2)_4 \\
| \\
-N-C-C- \\
|\ \ \ |\ \ \ \| \\
H\ \ H\ \ O
\end{array}
\qquad\longrightarrow\qquad
\begin{array}{c}
\overset{\displaystyle H}{|}\ \ \overset{\displaystyle H}{|}\ \ \overset{\displaystyle O}{\|} \\
-N-C-C- \\
| \\
O=C \\
| \\
NH \\
| \\
(CH_2)_4 \\
| \\
-N-C-C- \\
|\ \ \ |\ \ \ \| \\
H\ \ H\ \ O
\end{array}
$$

Fig. 10.1 Formation of an ε(γ-glutamyl) lysine isopeptide bond between proteins or peptides

10.3.6 Transglutaminase

Transglutaminase (TGase; EC 2.3.1.13; γ-glutamylpeptidase, amine-γ-glutamyl transferase) catalyses the acyl transfer between the γ-carboxyl amine group of a peptide-bound glutamine residue and the primary amine group of various substrates, lysine being of special interest (Fig. 10.1).

Because of their open structure the caseins are very good substrates for TGase but native whey proteins are not good. Benefits of using TGase in the dairy industry include:

1. The emulsifying capacity of sodium caseinate is scarcely affected by TGase treatment but the stability of the emulsion is improved,
2. The firmness of fermented milk products is improved by TGase treatment of the milk,
3. The rennet coagulability of milk is adversely affected because the accessibility of κ-casein is hindered
4. The heat stability of milk is greatly increased and the minimum in the HCT-pH profile is removed.

10.3.7 Catalase (EC 1.1.1.6)

Hydrogen peroxide is a very effective chemical sterilant and although it causes some damage to the physico-chemical properties and nutritional value of milk protein, principally by oxidizing methionine, it is used as a milk preservative, especially in warm regions lacking refrigeration. Excess H_2O_2 may be reduced following treatment by soluble exogenous catalase (from beef liver, *Aspergillus niger* or *Micrococcus lysodeiktieus*). Immobilized catalase has been investigated for this purpose but the immobilized enzyme is rather unstable.

As discussed in Sect. 10.3.8, catalase is frequently used together with glucose oxidase in many of the food applications of the latter; however, the principal potential application of glucose oxidase in dairy technology is for the *in situ* production of H_2O_2 for which the presence of catalase is obviously undesirable.

10.3.8   Glucose Oxidase (EC 1.1.3.4)

Glucose oxidase (GO) catalyses the oxidation of glucose to gluconic acid (*via* gluconic acid-δ-lactone) according to the following reaction:

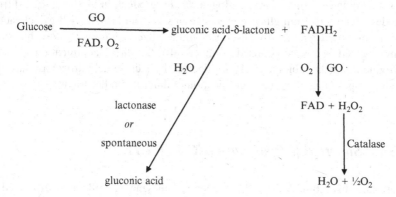

The H_2O_2 formed is normally reduced by catalase present as a contaminant in commercial GO preparations (from *P. notatum, P. glaucum* or *A. niger*) or added separately. Glucose oxidase, which has a pH optimum of ~5.5, is highly specific for D-glucose and may be used to assay specifically for D-glucose in the presence of other sugars.

In the food industry, glucose oxidase has four principal applications:

1. *Removal of residual trace levels of glucose*: This application, which is particularly useful for the treatment of egg white prior to dehydration (although alternative procedures using yeast fermentation are used more commonly), is of little, if any, significance in dairy technology.
2. *Removal of trace levels of oxygen*: Traces of oxygen in wines and fruit juices cause discolouration and/or oxidation of ascorbic acid. Chemical reducing agents may be used to scavenge oxygen but enzymatic treatment with GO may be preferred. Glucose oxidase has been proposed as an antioxidant system for high-fat products such as mayonnaise, butter and whole milk powder but it does not appear to be widely used for this purpose, probably because of cost *vis-à-vis*

chemical antioxidants (if permitted) and the relative effectiveness of inert gas
flushing in preventing lipid oxidation in canned milk powder.

3. *Generation of H_2O_2 in situ*: The H_2O_2 generated by glucose oxidase has a direct
bactericidal effect (which appears to be a useful side-effect of GO applied to egg
products) but its bactericidal properties can be much more effectively exploited
as a component of the lactoperoxidase/H_2O_2/SCN⁻ system. Glucose required for
GO activity may be added or produced by the action of β-galactosidase on lactose (both β-galactosidase and glucose oxidase have been immobilized on porous
glass beads). H_2O_2 may also be generated *in situ* by the action of xanthine
oxidoreductase on added hypoxanthine. It is likely that exogenous H_2O_2 will
be used in such applications rather than H_2O_2 generated by glucose oxidase or
xanthine oxidase.

4. *Production of acid in situ*: Direct acidification of dairy products, particularly
Cottage and Mozzarella cheeses, is fairly common. Acidification is normally
performed by addition of acid or acidogen (usually gluconic acid-δ-lactone) or
by a combination of acid and acidogen. *In situ* production of gluconic acid from
added glucose or from glucose produced *in situ* from lactose by β-galactosidase
or from added sucrose by invertase; immobilized glucose oxidase has been
investigated but is not is used commercially for direct acidification of milk.
Production of lactobionic acid from lactose by lactose dehydrogenase has also
been proposed for the direct acidification of dairy and other foods.

10.3.9   Superoxide Dismutase (EC 1.15.1.1)

Superoxide dismutase (SOD), an indigenous enzyme in milk, was discussed in
Sect. 10.2.10. A low level of exogenous SOD, coupled with catalase, is a very effective inhibitor of lipid oxidation in dairy products, particularly for preserving the
flavour of long-life UHT milk which is prone to lipid oxidation. Obviously, the
commercial feasibility of using SOD as an antioxidant depends on cost, particularly
vis-à-vis chemical methods, if permitted.

10.3.10   Glucose Isomerase (EC 5.3.1.5)

Glucose isomerase, which converts glucose to fructose, is widely use to produce
high-fructose syrups from glucose produced from starch by amylase and glucoamy-
lase. It has potential in the dairy industry in conjunction with β-galastosidase for the
production of galactose-glucose-fructose syrup (which is sweeter than galactose-
glucose syrup) from lactose.

10.3.11 Exogenous Enzymes in Food Analysis

Exogenous enzymes have several applications in food analysis. One of the principal attractions of enzymes as analytical reagents is their specificity which eliminates the need for extensive clean-up of the sample and makes it possible to quantify separately closely related molecules, e.g., D- and L-glucose, D- and L-lactic acid. Enzymatic assays can be very sensitive; some can detect concentrations at the picomole level. Enzymes can be immobilized as enzyme electrodes and as such can be used continuously to monitor changes in the concentration of a substrate in a product stream. Disadvantages of enzymes as analytical reagents are their relatively high cost, especially when few samples are to be analysed, relatively poor stability (due to denaturation or inhibition) and the need to use highly purified enzymes.

Enzymes are rarely used by industrial food laboratories but find regular application in more specialized analytical or research laboratories. Important applications are summarized in Table 10.4 (information on products and methods may be obtained from R-Biopharm AG at www.r-biopharm.com). There are alternative chemical and/ or physical methods, especially some form of chromatography, for all these applications, but extensive clean-up and perhaps concentration may be required.

The use of luciferase to quantify ATP in milk is the principle of modern rapid methods for assessing the bacteriological quality of milk based on the production of ATP by bacteria. Such methods have been automated and mechanized.

Table 10.4 Some examples of compounds in milk that can be analysed by enzymatic assays

Substrate	Enzyme
D-Glucose	Glucose oxidase; Glucokinase; Hexokinase
Galactose	Galactose dehydrogenase
Fructose	Fructose dehydrogenase
Lactose	β-Galactosidase, then analyse for glucose or galactose
Lactulose	β-Galactosidase, then analyse for fucose or galactose
D- and L-lactic acid	D- and L-lactic dehydrogenase
Citric acid	Citrate dehydrogenase
Acetic acid	Acetate kinase + pyruvate kinase + lactic dehydrogenase
Ethanol	Alcohol dehydrogenase
Glycerol	Glycerol kinase
Fatty acids	Acyl-CoA synthetase + Acyl-CoA oxidase
Amino acids	Decarboxylases; Deaminases
Metal ions (inhibitors or activators)	Choline esterase; Luciferase; Invertase
ATP	Luciferase
Pesticides (inhibitors)	Hexokinase; Choline esterase
Inorganic phosphate	Phosphorylase a
Nitrate	Nitrate reductase

10.3.11.1 Enzyme-Linked Immunosorbent Assays

An indirect application of enzymes in analysis is as a marker or label in enzyme-linked immunosorbent assays (ELISA). In ELISA, the enzyme does not react with the analyte; instead, an antibody is raised against the analyte (antigen or hapten) and labelled with an easily-assayed enzyme, usually a phosphatase or a peroxidase. The enzyme activity is proportional to the amount of antibody in the system, which in turn is proportional, directly or indirectly depending on the arrangement used, to the amount of antigen present.

Either of two approaches may be used: competitive and non-competitive, each of which may be used in either of two modes.

1. Competitive ELISA
    On the basis of enzyme-labelled antigen
    The antibody (Ab) is adsorbed to a fixed phase, e.g., the wells of a microtiter plate. An unknown amount of antigen (Ag, analyte) in the sample to be assayed together with a constant amount of enzyme-labelled antigen (Ag-E) are then added to the well (Fig. 10.2). The Ag and Ag-E compete for the fixed amount of Ab and amount of Ag-E bound is inversely proportional to the amount of Ag present in the sample. After washing away the excess of unbound antigen (and other materials), a chromogenic substrate is added and the intensity of the colour determined after incubation for a fixed period. The intensity of the colour is inversely proportional to the concentration of antigen in the sample (Fig. 10.2a).

    On the basis of enzyme-labelled antibody
    In this mode, a fixed amount of unlabelled antigen (Ag) is bound to microtiter plates. A food sample containing antigen is added, followed by a fixed amount of enzyme-labelled antibody (Ab-E) (Fig. 10.2b). There is competition between the fixed and free antigen for the limited amount of Ab-E. After an appropriate reaction time, unbound Ag (and other materials) are washed from the plate and the amount of bound enzyme activity assayed. As above, the amount of enzyme activity is inversely proportional to the concentration of antigen in the food sample.

2. Non-competitive ELISA
    The usual principle here is the sandwich technique, which requires the antigen to have at least two antibody binding sites (epitopes). Unlabelled antibody is first fixed to microtiter plates; a food sample containing antigen (analyte) is then added and allowed to react with the fixed unlabelled antibody (Fig. 10.3). Unadsorbed material is washed out and enzyme-labelled antibody then added which reacts with a second site on the bound antigen. Unadsorbed Ab-E is washed off and enzyme activity assayed; activity is directly related to the concentration of antigen.

Examples of ELISA in dairy analyses include:

Quantifying denaturation of β-lactoglobulin in milk products
Detection and quantitation of adulteration of milk from one species with that from other species, e.g., sheeps' milk by bovine milk.

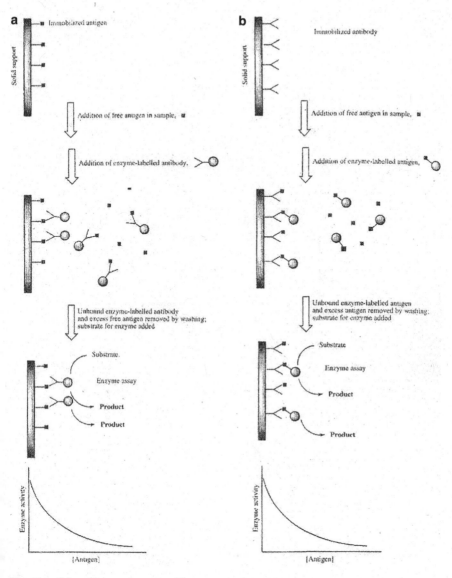

Fig. 10.2 Schematic representation of a competitive enzyme-linked immunosorbent assay using (**a**) immobilized antigen or (**b**) immobilized antibody

Authentication of cheese, e.g., sheeps' milk cheese.

Detection and quantitation of bacterial enzymes in milk, e.g., from psychrotrophs.

Quantitation of antibiotics.

Potential application of ELISA include monitoring proteolysis in the production of protein hydrolyzates or in cheese

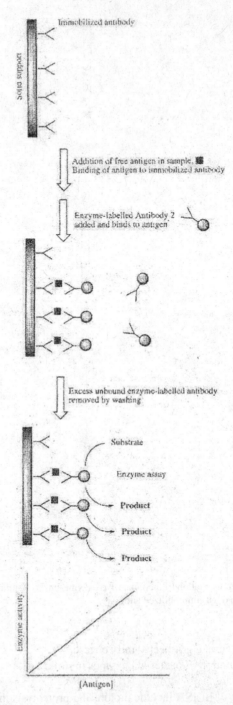

Fig. 10.3 Schematic representation of a non-competitive enzyme-linked immunosorbent assay using the "sandwich" technique

References and Suggested Reading

Indigenous Enzymes in Milk

Abd El-Salam, M. H., & El-Shibiny, S. (2011). A comprehensive review on the composition and properties of buffalo milk. *Dairy Science and Technology, 91*, 663–699.

Anderson, M., & Cawston, T. E. (1975). Reviews in the progress of dairy science. The milk fat globule membrane. *Journal of Dairy Research, 42*, 459–483.

Andrews, A. T., Olivecrona, T., Bengtsson-Olivecrona, G., Fox, P. F., Björck, L., & Farkye, N. Y. (1991). Indigenous enzymes in milk. In P. F. Fox (Ed.), *Food enzymology* (pp. 53–129). London, UK: Elsevier Applied Science.

Andrews, A. T., Olivecrona, T., Vilaro, S., Bengtsson-Olivecrona, G., Fox, P. F., Björck, L., et al. (1992). Indigenous enzymes in milk. In P. F. Fox (Ed.), *Advanced dairy chemistry* (Proteins, Vol. 1, pp. 285–367). London, UK: Elsevier Applied Science.

Bastian, E. D., & Brown, R. J. (1996). Plasmin in milk and dairy products, an update. *International Dairy Journal, 6*, 435–457.

Blanc, B. (1982). Les proteines du lait, à activete enzymatique et harmonal. *Le Lait, 62*, 352–395.

Booth, V. H. (1938). The specificity of xanthine oxidase. *Biochemical Journal, 32*, 494–502.

Brockerhoff, H., & Jensen, R. G. (1974). *Lipolytic enzymes*. New York, NY: Academic.

Chandan, R. C., & Shahani, K. M. (1964). Milk lipase: A review. *Journal of Dairy Science, 47*, 471–480.

Corry, A. M. (2004). *Purification of bile salts-stimulated lipase from breast milk and ligand affinity purification of a potential receptor*. M.Sc. Thesis, National University of Ireland, Cork.

Deeth, H. C. (2006). Lipoprotein lipase and lipolysis in milk. *International Dairy Journal, 16*, 555–562.

Deeth, H. C., & Fitz-Gerald, C. H. (1995). Lipolytic enzymes and hydrolytic rancidity in milk and milk products. In P. F. Fox (Ed.), *Advanced dairy chemistry* (Lipids, Vol. 2, pp. 247–308). London: Chapman & Hall.

Deeth, H. C., & Fitz-Gerald, C. H. (2006). Lipolytic enzymes and hydrolytic activity. In P. F. Fox & P. L. H. McSweeney (Eds.), *Advanced dairy chemistry* (3rd ed., Vol. 2, pp. 481–555). New York, NY: Kluwer Academic-Plenum.

Dwivedi, B. K. (1973). The role of enzymes in food flavors. Part I. Dairy products. *CRC Critical Reviews in Food Technology, 3*, 457–478.

Enroth, C., Eger, B. T., Okamoto, K., Nishino, T., Nishino, T., & Pai, E. (2000). Crystal structures of bovine xanthine dehydrogenase and xanthine oxidase: Structure-based mechanism of conversion. *Proceedings of the National Academy of Sciences of the United States of America, 97*, 10723–10728.

Everse, J., Everse, K. E., & Grisham, M. B. (Eds.). (1991). *Peroxidases in chemistry and biology* (Vol. I & II). Boca Raton, FL: CRC Press.

Farkye, N. Y. (2003). Indigenous enzymes in milk; other enzymes. In P. F. Fox & P. L. H. McSweeney (Eds.), *Advanced dairy chemistry* (Proteins 3rd ed., Vol. 1, pp. 571–603). New York, NY: Kluwer Academic-Plenum.

Flynn, N. K. R. (1999). *Isolation and characterization of bovine milk acid phosphatase*. M.Sc. Thesis, National University of Ireland, Cork.

Fox, P. F. (2003a). Significance of indigenous enzymes in milk and dairy products. In J. R. Whitaker, A. G. J. Voragen, & D. W. S. Wong (Eds.), *Handbook of food enzymology* (pp. 255–277). New York, NY: Marcel Dekker.

Fox, P. F., & Kelly, A. L. (2006a). Indigenous enzymes in milk: Overview and historical aspects—Part 1. *International Dairy Journal, 16*, 500–516.

Fox, P. F., & Kelly, A. L. (2006b). Indigenous enzymes in milk: Overview and historical aspects—Part 2. *International Dairy Journal, 16*, 517–532.

Fox, P. F., & Morrissey, P. A. (1981). Indigenous enzymes of bovine milk. In G. G. Birch, N. Blakeborough, & K. J. Parker (Eds.), *Enzymes and food processing* (pp. 213–238). London, UK: Applied Science.

Fox, P. F., Olivecrona, T., Vilaro, S., Olivecrona, G., Kelly, A. L., Shakeel-ur-Rheman, et al. (2003). Indigenous enzymes in milk. In P. F. Fox & P. L. H. McSweeney (Eds.), *Advanced dairy chemistry* Proteins (Vol. 1, Part A, 3rd ed., pp. 467–603). New York, NY: Kluwer Academic-Plenum.

Fox, P. F., and Tarassuk N. P. (1968). Bovine milk lipase. 1. Isolation from skim milk. *Journal of Dairy Science., 51,* 826–833.

Gallagher, D. P., Cotter, P. F., & Mulvihill, D. M. (1997). Porcine milk proteins: A review. *International Dairy Journal, 7,* 99–118.

Garattini, E., Mendel, R., Romao, M. J., Wright, R., & Terao, M. (2003). Mammalian molybdo-flavoenzymes, an expanding family of proteins: Structure, genetics, regulation, function and pathophysiology. *Biochemical Journal, 372,* 15–32.

Got, R. (1971). Les enzymes des laits. *Annal Nutr l'Aliment, 25,* A291–A311.

Groves, M. L. (1971). Minor milk proteins and enzymes. In H. A. McKenzie (Ed.), *Milk proteins, chemistry and molecular biology* (Vol. 1, pp. 367–418). New York, NY: Academic.

Grufferty, M. B., & Fox, P. F. (1988). Milk alkaline proteinase. *Journal of Dairy Research, 55,* 609–630.

Hamosh, M. (1995). Enzymes in human milk. In R. G. Jensen (Ed.), *Handbook of milk composition* (pp. 388–427). San Diego, CA: Academic.

Harrison, R. (2000). Milk xanthine oxidoreductase: Hazard or benefit? *Journal of Nutritional and Environmental Medicine, 12,* 231–238.

Harrison, R. (2002). Structure and function of xanthine oxidoreductase: Where are we now? *Free Radical Biology and Medicine, 33,* 774–797.

Harrison, R. (2004). Physiological roles of xanthine oxidoreductase. *Drug Metabolism Reviews, 36,* 363–375.

Harrison, R. (2006). Milk xanthine oxidase: Properties and physiological roles. *International Dairy Journal, 16,* 546–554.

Hernell, O., & Lonnerdal, B. (1989). Enzymes in human milk. In S. A. Atkinson & B. Lonnerdal (Eds.), *Proteins and non-protein nitrogen in human milk* (pp. 67–75). Boca Raton, FL: CRC Press.

Herrington, B. L. (1954). Lipase: A review. *Journal of Dairy Science, 37,* 775–789.

Humbert, G., & Alais, C. (1979). Review of the progress of dairy science. The milk proteinase system. *Journal of Dairy Research, 46,* 559–571.

Hurley, M. J., Larsen, L. B., Kelly, A. L., & McSweeney, P. L. H. (2000). The milk acid proteinase, cathepsin D: A review. *International Dairy Journal, 10,* 673–681.

IDF. (2006). Proceedings of the first IDF symposium on indigenous enzymes in milk. *International Dairy Journal, 16,* 499–715.

Jensen, R. G., & Pitas, R. E. (1976). Milk lipoprotein lipase: A review. *Journal of Dairy Science, 59,* 1203–1214.

Johnson, H. A. (1974). The composition of milk. In B. H. Webb, A. H. Johnson, & J. A. Alford (Eds.), *Fundamentals of dairy chemistry* (2nd ed., pp. 1–57). Westport, CT: AVI Publishing Co.

Kato, A. (2003). Lysozyme. In J. R. Whitaker, A. G. J. Voragen, & D. W. S. Wong (Eds.), *Handbook of food enzymology* (pp. 971–978). New York, NY: Marcel Dekker.

Kelly, A. L., & McSweeney, P. L. H. (2003). Indigenous proteinases in milk. In P. F. Fox & P. L. H. McSweeney (Eds.), *Advanced dairy chemistry* (Proteins 2nd ed., Vol. 1, pp. 495–521). New York, NY: Kluwer Academic-Plenum.

Kitchen, B. J. (1985). Indigenous milk enzymes. In P. F. Fox (Ed.), *Developments in dairy chemistry* (Lactose and minor constituents, Vol. 3, pp. 239–279). London, UK: Elsevier Applied Science.

Kitchen, B. J., Taylor, G. C., & White, I. C. (1970). Milk enzymes—Their distribution and activity. *Journal of Dairy Research, 37,* 279–288.

Linden, G., & Alais, C. (1976). Phosphatase alkaline du lait de vache. II. Structure sous-unitaire, nature metalloproteique et parameters cinetiques. *Biochimica et Biophysica Acta, 429*, 205–213.

Linden, G., & Alais, C. (1978). Alkaline phosphatase in human, cow and sheep milk: Molecular and catalytic properties and metal ion action. *Annales De Biologie Animale, Biochimie, Biophysique, 18*, 749–758.

Lonnerdal, B. (1985). Biochemistry and physiological function of human milk. *American Journal of Clinical Nutrition, 42*, 1299–1317.

Massey, V., & Harris, C. M. (1997). Milk xanthine oxidoreductase: The first one hundred years. *Biochemical Society Transactions, 25*, 750–755.

Moatsou, G. (2010). Indigenous enzymatic activities in caprine and ovine milks. *International Journal of Dairy Technology, 63*, 16–31.

O'Keefe, R. B., & Kinsella, J. E. (1979). Alkaline phosphatase from bovine mammary tissue: Purification and some molecular and catalytic properties. *International Journal of Biochemistry, 10*, 125–134.

O'Mahony, J. A., Fox, P. F., & Kelly, A. L. (2013). Indigenous enzymes in milk. In P. L. H. McSweeney & P. F. Fox (Eds.), *Advanced dairy chemistry* (Proteins 4th ed., Vol. 1A, pp. 337–385). New York, NY: Springer.

Olivecrona, T., & Bengtsson-Olivecrona, G. (1991). Indigenous enzymes in milk, Lipases. In P. F. Fox (Ed.), *Food enzymology* (Vol. 1, pp. 62–78). London, UK: Elsevier Applied Science.

Olivecrona, T., Vilaro, S., & Bengtsson-Olivecrona, G. (1992). Indigenous enzymes in milk, Lipases. In P. F. Fox (Ed.), *Advanced dairy chemistry* (Proteins, Vol. 1, pp. 292–310). London, UK: Elsevier Applied Science.

Olivecrona, T., Vilaro, S., & Olivecrona, G. (2003). Indigenous enzymes in milk, Lipases. In P. F. Fox & P. L. H. McSweeney (Eds.), *Advanced dairy chemistry* (Proteins 3nd ed., Vol. 1, pp. 473–494). New York, NY: Kluwer Academic-Plenum.

Palmquist, D. L. (2006). Milk fat: Origin of fatty acids and influence of nutritional factors thereon. In P. F. Fox & P. L. H. McSweeney (Eds.), *Advanced dairy chemistry* (3rd ed., Vol. 2, pp. 43–92, 555). New York, NY: Kluwer Academic-Plenum.

Robert, A. M., & Robert, L. (2014). Xanthine oxido-reductase, free radicals and cardiovascular disease. A critical review. *Pathology and Oncology Research, 20*, 1–10.

Seifu, E., Buys, E. M., & Donkin, E. F. (2005). Significance of the lactoperoxidase system in the dairy industry and its potential applications: A review. *Trends in Food Science and Technology, 16*, 137–154.

Shahani, K. M. (1966). Milk enzymes: Their role and significance. *Journal of Dairy Science, 49*, 907–920.

Shahani, K. M., Harper, W. J., Jensen, R. G., Parry, R. M., Jr., & Zittle, C. A. (1973). Enzymes of bovine milk: A review. *Journal of Dairy Science, 56*, 531–543.

Shahani, K. M., Kwan, A. J., & Friend, B. A. (1980). Role and significance of enzymes in human milk. *American Journal of Clinical Nutrition, 33*, 1861–1868.

Shakeel-ur-Rehman, Fleming, C. M., Farkye, N. Y., & Fox, P. F. (2003). Indigenous phosphatases in milk. In P. F. Fox & P. H. L. McSweeney (Eds.), *Advanced dairy chemistry* (Proteins, Vol. 1, pp. 523–543). New York, NY: Kluwer Academic-Plenum.

Sindhu, J. S., & Arora, S. (2011). Buffalo milk. In J. W. Fuquay, P. F. Fox, & P. L. H. McSweeney (Eds.), *Encyclopedia of dairy sciences* (2nd ed., Vol. 3, pp. 503–511). Oxford, UK: Academic.

Tkadlecova, M., & Hanus, J. (1973). [Enzymes in cows' milk]. *Die Nahrung, 17*, 565–577.

Uniacke-Lowe, T., & Fox, P. F. (2011). Equid milk. In J. W. Fuquay, P. F. Fox, & P. L. H. McSweeney (Eds.), *Encyclopedia of dairy sciences* (2nd ed., Vol. 3, pp. 518–529). Oxford, UK: Academic.

Whitaker, J. R., Voragen, A. G. J., & Wong, D. W. S. (2003). *Handbook of food enzymology*. New York, NY: Marcel Dekker.

Whitney, R. M. L. (1958). The minor proteins of milk. *Journal of Dairy Science, 41*, 1303–1323.

Yuan, Z. Y., & Jiang, T. J. (2003). Horseradish peroxidase. In J. R. Whitaker, A. G. J. Voragen, & D. W. S. Wong (Eds.), *Handbook of food enzymology* (pp. 403–411). New York, NY: Marcel Dekker.

Exogenous Enzymes in Dairy Technology and Analysis

Brown, R. J. (1993). Dairy products. In T. Nagodawithana & J. Reed (Eds.), *Enzymes in food processing* (3rd ed., pp. 347–361). San Diego, CA: Academic.

Dekker, P. J. T., & Daamen, C. B. G. (2011). β-D-Galactosidase. In J. W. Fuquay, P. F. Fox, & P. L. H. McSweeney (Eds.), *Encyclopedia of dairy sciences* (Vol. 2, pp. 276–283). Oxford, UK: Academic.

El-Soda, M., & Awad, S. (2011). Accelerated cheese ripening. In J. W. Fuquay, P. F. Fox, & P. L. H. McSweeney (Eds.), *Encyclopedia of dairy sciences* (Vol. 1, pp. 795–798). Oxford, UK: Academic.

Fox, P. F. (1991). *Food enzymology* (Vol. 1 & 2). London, UK: Elsevier Applied Science.

Fox, P. F. (1993). Exogenous enzymes in dairy technology—A review. *Journal of Food Biochemistry, 17,* 173–199.

Fox, P. F. (1998/99). Acceleration of cheese ripening. *Food Biotechnology 2,* 133–185.

Fox, P. F. (2003b). Exogenous enzymes in dairy technology. In J. R. Whitaker, A. G. J. Voragen, & D. W. S. Wong (Eds.), *Handbook of food enzymology* (pp. 279–301). New York, NY: Marcel Dekker.

Fox, P. F., & Grufferty, M. B. (1991). Exogenous enzymes in dairy technology. In P. F. Fox (Ed.), *Food enzymology* (Vol. 1 & 2, pp. 219–269). London, UK: Elsevier Applied Science.

Fox, P. F., & Stepaniak, L. (1993). Enzymes in cheese technology. *International Dairy Journal, 3,* 509–530.

Guilbault, G. G. (1970). *Enzymatic methods of analysis.* Oxford, UK: Pergamon Press.

IDF. (1998). *The use of enzymes in dairying* (Bulletin, Vol. 332, pp. 8–53). Brussels: International Dairy Federation.

Jaros, D., & Rohm, H. (2011). Transglutaminase. In J. W. Fuquay, P. F. Fox, & P. L. H. McSweeney (Eds.), *Encyclopedia of dairy sciences* (Vol. 2, pp. 297–300). Oxford, UK: Academic.

Kilara, A. (2011). Lipases. In J. W. Fuquay, P. F. Fox, & P. L. H. McSweeney (Eds.), *Encyclopedia of dairy sciences* (Vol. 2, pp. 284–288). Oxford, UK: Academic.

Kilara, A. (1985). Enzyme-modified lipid food ingredients. *Process Biochemistry, 20*(2), 35–45.

McSweeney, P. L. H. (2011). Catalase, glucose oxidase, glucose isomerase, and hexose oxidase. In J. W. Fuquay, P. F. Fox, & P. L. H. McSweeney (Eds.), *Encyclopedia of dairy sciences* (Vol. 2, pp. 301–303). Oxford, UK: Academic.

Morris, B. A., & Clifford, M. N. (1984). *Immunoassays in food analysis.* London, UK: Elsevier Applied Science.

Mottola, N. A. (1987). Enzymes as analytical reagents: Substrate determinations with soluble and immobilized enzyme preparations. *Analyst, 112,* 719–727.

Nagodawithana, T., & Reed, J. (Eds.). (1993). *Enzymes in food processing* (3rd ed.). San Diego, CA: Academic.

Nelson, J. H., Jensen, R. G., & Pitas, R. E. (1977). Pregastric esterase and other oral lipases: A review. *Journal of Dairy Science, 60,* 327–362.

Nongonierma, A. B., & FitzGerald, R. J. (2011). Proteinases. In J. W. Fuquay, P. F. Fox, & P. L. H. McSweeney (Eds.), *Encyclopedia of dairy sciences* (Vol. 2, pp. 289–296). Oxford, UK: Academic.

O'Sullivan, M. M., Kelly, A. L., & Fox, P. F. (2001). Effect of transglutaminase on the heat stability of milk. *Journal of Dairy Science, 85,* 1–7.

Whitaker, J. R. (1991). Enzymes in analytical chemistry. In P. F. Fox (Ed.), *Food enzymology* (Vol. 2, pp. 287–308). London, UK: Elsevier Applied Science.

Wilkinson, M. G., Doolan, I. A., & Kilcawley, K. N. (1911). Enzyme-modified cheese. In J. W. Fuquay, P. F. Fox, & P. L. H. McSweeney (Eds.), *Encyclopedia of dairy sciences* (Vol. 1, pp. 799–804). Oxford, UK: Academic.

Chapter 11
Biologically Active Compounds in Milk

11.1 Introduction

Despite a significant amount of research in many areas, the definition of a bioactive compound remains ambiguous and unclear (see Guaadaoui et al. 2014). Bioactive compounds in foods are generally regarded as components that affect biological processes or substrates and, hence, have an impact on body function or condition and ultimately health (Schrezenmeir et al. 2000). However, this definition has been refined by two caveats:

(1) To be considered bioactive, a dietary component should impart a measurable biological effect at a physiologically realistic level, and
(2) The bioactivity measured must have the potential to affect health in a beneficial manner, thus excluding potential damaging effects from the definition such as toxicity, allergenicity and mutagenicity (Schrezenmeir et al. 2000; Möller et al. 2008)

Milk is secreted by all mammals for nutrition of the young; milk secretion is one of the distinguishing characteristic of mammals (~4,500 species). Milk is required to meet the nutritional and physiological requirements of neonates which are born at different, species-specific, states of development and have different growth rates; therefore, the milk of different species differs in composition and, in effect, is species-specific. The nutritional value of milk is due to the presence of lactose, proteins, lipids and inorganic elements (metals). Bioactive compounds in milk perform many functions other than nutritional, e.g., immune system, hormones and related compounds, antibacterial agents, enzymes (~60), enzyme inhibitors and cryptic peptides (various functions). Biologically active milk compounds in the form of immunoglobulins (Igs), antibacterial peptides, antimicrobial proteins, oligosaccharides (OSs) and lipids protect neonates and adults against

© Springer International Publishing Switzerland 2015
P.F. Fox et al., *Dairy Chemistry and Biochemistry*,
DOI 10.1007/978-3-319-14892-2_11

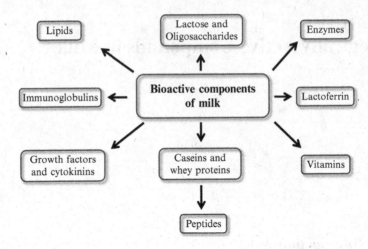

Fig. 11.1 The principal groups of bioactive compounds in milk (from Park 2009)

pathogens and illnesses. Gobbetti et al. (2007) categorised the bioactivity of milk compounds into 4 major groups:

1. Gastrointestinal development, activity and function
2. Infant development
3. Immunological development and function
4. Microbial activity, including antibiotic and probiotic action

A schematic representation of the major biologically active compounds in milk is shown in Fig. 11.1. In this chapter, the great diversity of biologically active compounds in milk will be described. While most scientific work has been carried out on the bioactive compounds in bovine milk, reference is made throughout to bioactive components in human milk and their significance in infant growth and development.

For neonates, colostrum produced during the first 48–72 h *post-partum* is the only source of essential nutrients and protection against microbial infection *via* Igs. For several further months, breast-fed infants benefit from bacteria, such as bifidobacteria, in the gut which reduce enteric disorders (Zinn 1997). Other compounds also provide the neonate with immunological protection and facilitate the development of neonatal immune competence; the two major categories are cytokines, which are not produced efficiently by the neonate, and peptides (Politis and Chronopoulou 2008).

Individually or collectively, milk proteins, together with other bioactive components, influence the health of the neonate and may influence the health and milk production capacity of the lactating female (Zinn 1997). Women who lactate have a significantly reduced risk of breast cancer compared to those who do not (Freudenheim et al. 1994). Some milk proteins [e.g., the feedback inhibitor of lactation (FIL)] may influence total milk production (Peaker and Wilde 1996). The advantages of mother's milk may extend well beyond the first few hours *post-partum* and may have significant consequences later in life. It has been reported that human infants who are breast-fed for 6 months have fewer health-related problems later in life than formula-fed infants, including a reduced risk of allergies (Saarinen and Kajosaari 1995), respiratory and gastrointestinal disorders (Koletzko et al. 1989), childhood lymphomas (Davies 1988; Schwartzbaum et al. 1991), type-I diabetes (Borch-Johnsen, et al. 1984) and pre- and post-menopausal cancer (Byers et al. 1985). The act of suckling may influence the growth rate of neonates (Zinn 1997). It has been reported that piglets nursed by the sow grow more slowly than those raised on milk replacer and it is postulated that the sow's milk limits the availability of nutrients to the piglet although it is not known if the slower growth rate positively correlates with long-term health benefits (Boyd et al. 1995).

11.2 Bioactive Milk Lipids

The principal function of milk lipids is as an energy source but components of the lipid fraction serve some specific biological functions; the significance of the essential fatty acids, linoleic and linolenic acids, and the fat-soluble vitamins is well known. Bioactive milk lipids include, triacylglycerols (triglycerides), fatty acids, sterols and phospholipids. Anticarcinogenic activity has been attributed to conjugated linoleic acid (CLA), sphingomyelin, butanoic acid, ether lipids (plasmalogens), β-carotene, vitamin A (retinol) and vitamin D; the subject is reviewed by Parodi (1999).

11.2.1 Medium Chain Fatty Acids

Medium chain fatty acids (MCFAs), containing six to ten carbons, are very different from long chain fatty acids in terms of their chemical and physical properties, e.g., MCFAs do not require binding to proteins for transport (Marten et al. 2006). In bovine milk, MCF's make up 4–12 % of all fatty acids (Jensen 2002). MCFAs are hydrolysed rapidly and completely in the body after absorption across the epithelial barrier (Bach and Babayan 1992). Research has focussed recently on the ability of MCFA's to reduce body weight and particularly body fat and their usefulness as supplements in functional foods is the subject of many studies (for review see Marten et al. 2006).

11.2.2 Conjugated Linoleic Acid

More recently, the importance of CLA [9, 11- or 10,12-octadecadienoic acid (8 cis, trans isomers)] has been highlighted. Ruminant milk and body lipids contain a relatively high level of CLA (0.24–2.8 % of total fatty acids in milk fat), being produced in the rumen by incomplete biohydrogenation of n-6 octadecadienoic acid

(LA) (see Whigham et al. 2000; Bauman and Lock 2006; Collomb et al. 2006); CLA can be increased by feeding cows with polyunsaturated fatty acid (PUFA) -rich oils (Stanton et al. 2003). CLA is also formed in mammary tissue from *trans* vaccenic acid (Aminot-Gilchrist and Anderson 2004). CLA has several desirable effects in the diet, some of the positive health effects attributed to it include suppression of carcinogenesis, anti-obesity agent, modulator of the immune system and control of atherogenesis and diabetes. Current research on CLA is focussed on its potential in the treatment or prevention of type-2 diabetes and prevention of heart disease and other health problems (see Aminot-Gilchrist and Anderson 2004).

11.2.3   Polar Milk Lipids

Milk contains several biologically important polar lipids including phospholipids and sphingolipids (glycosylceramides), which are located primarily in the milk fat globule membrane (MFGM). Sphingolipids are a class of lipids with a backbone of sphingoid bases and a set of aliphatic amino alcohols, including sphingosine, which are involved in signal transmission and cell recognition. They protect cell surfaces against harmful environmental factors by forming mechanically stable and chemically resistant outer leaflets of the plasma membrane 'lipid bilayer' (Rombaut and Dewettinck 2006). Both phospholipids and sphingolipids are regarded as being highly bioactive, with anti-cancer, bacteriostatic and cholesterol-lowering properties. The nutritional and technological properties of phospholipids and sphingolipids were reviewed by Rombaut and Dewettinck (2006).

11.2.4   Fatty Acids with Significant Bioactivity

The most abundant fatty acids in human milk are oleic ($c_{18:1}$), palmitic ($C_{16:0}$), linoleic ($C_{18:2}$, ω-6) and α-linolenic ($C_{18:3}$, ω-3). Some unsaturated fatty acids in human milk may provide protection against microorganisms, probably by disrupting viral envelopes, while others defend against enteric parasites, e.g., *Giardia lamblia,* a flagellated protozoan parasite that colonizes and reproduces in the small intestine, causing giardiasis (Thormar and Hilmarsson 2007).

Both *n*-6 and *n*-3 fatty acids are essential in human metabolism as components of membrane phospholipids, precursors of eicosanoids, ligands for membrane receptors and transcription factors that regulate gene expression. The importance of LA (*n*-6 $C_{18:2}$) has been known for many years but the significance of α-linolenic acid (ALA, *n*-3 $C_{18:3}$) was not recognized until the late 1980s and has since been identified as a key component of the diet for the prevention of atopic dermatitis (Horrobin 2000). LA and ALA are not interconvertible but are the parent acids of the *n*-6 and *n*-3 series of long-chain polyunsaturated fatty acids, respectively [e.g., *n*-6 $C_{20:4}$, arachidonic acid (AA); *n*-3 $C_{20:5}$, eicosapentaenoic acid (EPA) and *n*-3 $C_{22:6}$, docosahexaenoic acid (DHA)] which are components of cellular membranes and precursors of other essential metabolites such as prostaglandins and prostacyclins (Cuthbertson 1999; Innis 2007). DHA and AA are now recognised as being crucial for normal neurological development (Carlson 2001). Both DHA and EPA are critical for the normal growth and development of the central nervous system and the retina (Uauy et al. 1990). Humans have evolved on a diet with a ratio of n-6 to n-3 fatty acids of ~ 1:1 but current Western diets have a ratio of 15:1 to 16.7:1. As a species, humans are generally deficient in n-3 fatty acids and have excessive levels of n-6 which is associated with the pathogenesis of cardiovascular, cancerous, inflammatory and autoimmune diseases (Simopoulos 2002).

Butyric acid (C4:0) represents ~10 % of all fatty acids in bovine milk and is generated from carbohydrates by bacteria in the rumen, then transported via the blood to the mammary gland where it is reduced to butanoic acid (Jensen 1999). Butyrate is reported to exert many effects on intestinal function (Hamer, et al. 2008), especially colonic mucosa. Its anti-proliferate, anti-inflammatory and apoptotic properties were reviewed by Mills et al. (2009). Ingested milk butyrate does not reach the large intestine in humans but it can undergo lipase-mediated hydrolysis in the stomach which enables absorption by the proximal small intestine from where it is transported to the liver (Parodi 1997a).

11.2.5 Gangliosides

The name ganglioside was first applied by the German scientist Ernst Klenk in 1942 to lipids isolated from ganglion cells of the brain. A ganglioside is a molecule composed of a glycosphingolipid (ceramide and OS) with one or more residues (e.g., N-acetylneuraminic acid, NANA) linked to the OS chain. More than 60 gangliosides are known, which differ from each other mainly in the position and number of NANA residues. Gangliosides are components of the cell plasma membrane that modulates cell signal transduction events and appear to be concentrated in lipid rafts. Gangliosides are very important molecules in immunology. The OSs on gangliosides extend beyond the surface of cell membranes and act as distinguishing surface markers that can serve as specific determinants in cellular recognition and cell-to-cell communication. These carbohydrate head groups also act as specific receptors for certain pituitary glycoprotein hormones and certain bacterial protein toxins, such as cholera toxin. The functions of gangliosides as specific determinants suggest their important role in the growth and differentiation of tissues as well as in carcinogenesis. As well as bovine and human milk, buffalo and goat milk also contain gangliosides (Guo 2012; Park 2012).

11.2.6 Milk Fat Globule Membrane

The MFGM contains many bioactive glycoproteins and glycolipids. Many of the indigenous enzymes in milk are concentrated in the MFGM. The glycoproteins in the MFGM of human, rhesus monkey, chimpanzee, dog, sheep, goat, cow, grey seal, camel, horse and alpaca have been studied; large intra- and inter-species differences have been found (see Keenan and Mather 2006). Very highly glycosylated proteins occur in the MFGM of primates, horse, donkey, camel and dog. Long (0.5–1 μm) filamentous structures, comprised of mucins (highly glycosylated proteins), extend from the surface of the fat globules in equine and human milk (Welsch et al. 1988). These filaments dissociate from the surface into the milk serum on cooling and are lost on heating. For unknown reasons, the filaments on bovine milk fat globules are lost much more easily than those in equine or human milk. The filaments facilitate the adherence of fat globules to the intestinal epithelium and probably improve the digestion of fat (Welsch et al. 1988). The mucins prevent bacterial adhesion and may protect mammary tissue against tumours (Patton 1999). Mucin (MUC1), lactadherin and butyrophilin are the principal bioactive components of the human MFGM glycoprotein fraction. MUC1 in human milk is reported to bind rotavirus; the glycoprotein, lactadherin, also binds rotavirus but in a different manner (Yolken et al. 1992). Lactadherin is especially resistant to degradation in the neonatal stomach and is abundant in the gastrointestinal tract of (GIT) breast-fed infants. Another specific protein isolated from the MFGM called fatty-acid binding protein (FABP) has been shown to inhibit some breast cancer cell lines (Spitsberg et al. 1995).

Butyrophilin, acidophilin and xanthine oxidoreductase have been identified in bovine, human and equine MFGMs. All three proteins in equine milk appear to be similar to the corresponding proteins of the human MFGM, as does lactadherin which shares 74 % identity with that of the human lactadherin (Barello et al. 2008). Both xanthine oxidoreductase and acidophilin are involved in fat globule secretion with butyrophilin while lactadherin is thought to have a protective function against rotavirus in the intestinal tract (Barello et al. 2008). The bioactivity and associated health benefits of the MFGM were reviewed by Spitsberg (2005).

11.2.7 Phospholipids

The three principal phospholipids found in milk are sphingomyelin, phosphatidyl choline and phosphatidyl ethanolamine; all three are involved in many cellular processes including, growth and development and myelination of the central nervous system (Oshida et al. 2003 a, b). Sphingomyelin has shown some anticancer effects (Parodi 2001), inhibition of the absorption of cholesterol in the intestine of rats (Noh and Koo 2004) and activation and regulation of the immune system (Cinque et al. 2003).

11.3 Bioactive Milk Carbohydrates

Bioactive carbohydrates in milk include monosaccharides (glucose and galactose), disaccharides (lactose) and oligosaccharides (OSs).

11.3.1 *Lactose*

Lactose is the principal carbohydrate in the milk of most mammals but all contain other carbohydrates, e.g., galactosamine, glycoproteins and especially OSs. As well as being a major energy source for the neonate, lactose affects bone mineralization during the first few months *post-partum* as it stimulates the intestinal absorption of calcium (Schaafsma 2003). Heat treatment of lactose produces lactulose, an osmotic laxative and bifidus factor, which is added to infant formulae and lacto-oligosaccharides are also produced from lactose through enzymatic processes.

11.3.2 *Oligosaccharides*

OSs are polymers containing three to nine simple sugars which are present in all mammalian milk (Chap. 2). The OSs in milk are important protective factors and inhibit the binding of enteropathogenic *E. coli*, *Campylobacter jejuni* and *Streptococcus pneumonia* to target cells (Shah 2000). Fucosylated OSs, glycoproteins and glycolipids are reported to protect the human infant against enterotoxigenic *E. coli* (Newburg et al. 1990; Bode 2006).

There is a very high concentration of OSs in human (>15 g L^{-1}; >130 OSs) and in elephant and bear milk. The milk of monotremes and marsupials contains very little free lactose but mainly OSs. Colostrum of all species is especially rich in OSs (see Chap. 2).

OSs are relatively resistant to hydrolysis by β-galactosidase and are indigestible in the infant GIT, i.e., they serve as soluble fibre and promote the growth of bifidobacter. They may be absorbed from lower intestine by pinocytosis and are hydrolysed by lysosomal enzymes and the monosaccharides catabolised for energy. OSs are a major source of energy for monotremes, marsupials and bears, i.e., species with immature (altrical) young. In human milk, the concentration of OSs, especially fucosyloligosaccharides, varies significantly over the course of lactation and between individual mothers, suggesting that the protective effects of milk against intestinal pathogens vary among individuals (Chaturvedi et al. 2001).

Galactose and especially sialic acids are required for the biosynthesis of glycoproteins and glycolipids, which are essential for brain development (see Urashima et al. 2009, 2011, 2014). In humans, the highest concentration of sialic acid (as N-acetylneuraminic acid) occurs in the brain where it forms an integral part of ganglioside structure in synaptogenesis and neural transmission. Human milk has an exceptionally high level of sialic acids attached to the terminals of free OSs and, while the metabolic fate and biological role are largely unknown, it has been postulated that it confers a developmental advantage to breast-fed infants over bottle-fed infants (Wang and Brand-Miller 2003).

11.3.3   Bifidus Factors

Bifidobacterium is a genus of Gram-positive microorganisms which are ubiquitous endosymbiotic inhabitants of the GIT, vagina and mouth of mammals, including humans. It has been recognized for many years that breast-fed babies are more resistant to gastroenteritis than bottle-fed babies. This is undoubtedly a multifactorial phenomenon, including better hygiene, more appropriate milk composition, several antibacterial systems [especially Igs, lysozyme (Lyz), lactoferrin (Lf), vitamin-binding proteins and lactoperoxidase (LPO)], and a lower intestinal pH.

The mean pH of the faeces of breast-fed infants is ~5.1 while that of bottle-fed infants is ~6.4 due, in part, to the difference in composition between human and bovine milk—the former contains much less protein and phosphate and therefore has a lower buffering capacity but the intestinal microflora of breast-fed and bottle-fed differ widely; the microflora in the faeces of breast-fed infants is mainly *B. bifidum,* while that of bottle-fed infants is mainly *B. longum,* with lower numbers of *B. bifidum.*

Bifidobacteria are acid producers and their growth in breast-fed infants is promoted by the bifidus factors in human milk. The growth of bifidobacteria is stimulated by several factors, but the principal one in this case is *N*-acetylglucosamine-containing saccharides (Bifidus Factor 1) which is present at high levels in human milk and colostrum and bovine colostrum but not in bovine, goat or sheep milk. Human milk also contains several non-dialysable bifidus-promoting factors which are glycoproteins, referred to as bifidus factor 11. Many of the glycoproteins have been isolated and characterized (see Fox and Flynn 1992).

11.3.4   Fucose

L-Fucose (6-deoxy-L-galactose) is a monosaccharide that is a common component of many *N*-and *O*-linked glycans and glycolipids produced by mammalian cells. Two structural features distinguish fucose from other six-carbon sugars present in mammals, the lack of a hydroxyl group on the carbon at the 6-position (C-6) and the L-configuration (Becker and Lowe 2003). Human milk OSs are highly fucosylated, whereas fucosylated OSs are either absent, or present at very low concentrations, in bovine milk (Finke et al. 2000; Tao et al. 2008; Nwosu et al. 2012). Saito et al. (1987) reported the presence of fucosylated OSs in bovine colostrum. By contrast, bovine milk is highly sialylated, whereas human milk is not (Tao et al. 2008; Nwosu et al. 2012). Free fucose has been reported not to occur in human milk (Barfoot et al. 1988) although later studies reported its presence, albeit at a low concentration, along with free N-acetylneuraminic acid and N-aceetylhexosamine (Newburg and Weiderschain 1997; Wiederschain and Newburg 2001). α-L-fucosidase activity has been reported in human milk and activity increases over the course of lactation (Newburg and Weiderschain 1997); however, even after 16 h storage at normal body temperature, free fucose levels in human milk represent only ~ 5 % of available bound fucose, although it may be significant in preventing intestinal infection in the neonate (Wiederschain and Newburg 2001).

11.4 Vitamins

Vitamins are vital bioactive compounds in milk as they are essential, in minute amounts, for normal physiological functions and are not synthesized by the host in adequate amounts to meet such needs (Combs 2012). Ruminants obtain vitamins from feed and can also absorb some synthesized by intestinal microorganisms; this does not occur in humans and only a small amount of vitamin K is reabsorbed after synthesis by bacteria in the colon (Nohr 2011). Milk is an important source of vitamins A and C, thiamine, biotin (B_7), riboflavin (B_2), pyridoxine and cobalamin (B_{12}). Table 11.1 shows

Table 11.1 Average quantity of vitamins in bovine and human milk with recommended daily allowances (RDA) and approximate % of RDA supplied by bovine milk

Vitamin	Bovine (L^{-1})	Human (L^{-1})	Physiological function	RDA[a]	~ RDA % in 1 L bovine milk
Fat-soluble vitamins					
Retinol (A), mg	0.31	0.6	Visual pigments; epithelial cell differentiation	1	38
Cholecalciferol (D_3), µg	0.2	0.3	Calcium homostasis; bone mineralization; insulin release	5–10	10
α-Tocopherol (E), mg	0.9	3.5	Membrane antioxidant	12	10
Phylloquinone (K), mg	0.6	0.15	Blood clotting; calcium metabolism	90–120 µg	44
Water-soluble vitamins					
Ascorbic acid (C), mg	20	38	Formation of collagen and carnitine	60–75	25
Thiamine (B_1), mg	0.4	0.16	Co-enzyme for decarboxylation of 2-keto acids	1–1.2	33
Riboflavin (B_2), mg	1.9	0.3	Co-enzyme in redox reactions of fatty acids and TCA cycle	1.2–1.4	139
Niacin (B_3), mg[b]	0.8	2.3	Co-enzyme for several dehydrogenases	13–17	53
Pantothenic (B_5), mg	0.36	0.26	Co-enzyme in fatty acid metabolism	6	70
Pyridoxine (B_6), mg	0.4	0.06	Co-enzyme in amino acid metabolism	1.2–1.5	39
Biotin (B_7), µg	20	7.6	Co-enzyme for carboxylations	30–60	100
Folate (B_9), mg	0.05	0.05	Co-enzyme in single carbon metabolism	400 µg[c]	13
Cobalamin (B_{12}), µg	4	1.0	Co-enzyme in metabolism of propionate, the amino acids and single-carbon units	3	167

Data compiled from Schaafma (2003), Combs (2012), Morrissey and Hill (2009) and Nohr and Biesalski (2009)
[a]Values depend on age and sex
[b]Niacin (mg equivalents per day); 1 mg niacin equivalent = 60 mg tryptophan
[c]Calculated for sum of folates in normal nutrition

comparative values for the principal vitamins of bovine and human milk and includes the percentage of the recommended daily amount present in bovine milk. In comparison to bovine milk, human milk contains more vitamins A, E and C but less K, thiamine, riboflavin and pyridoxine. Vitamins in milk are discussed in detail in Chap. 6.

11.5 Bioactive Milk Proteins

Casein and whey proteins have been found to be increasingly important for physiological and biochemical functions and play a crucial role in human metabolism and health (Korhonen and Pihlanto-Leppälä 2004; Gobbetti et al. 2007). The principal function of the caseins is in nutrition as a source of amino acids, Ca^{2+} and Pi for the neonate; however, their amino acid sequences contain cryptic peptides which are biologically active when released by proteolysis. The major whey proteins α-lactalbumin (α-La), β-lactoglobulin (β-Lg) and immunoglobulins (Igs), have important biological roles. Many bioactive peptides, especially those from caseins, act as antioxidants in reducing cholesterol and blood pressure, others have anticarcinogenic, anti-inflammatory, immunomodulatory, antimicrobial and wound-healing properties and provide protection for tooth enamel. The concentrations of casein, whey proteins and other biologically active proteins in bovine and human milk are shown in Table 11.2.

11.5.1 Caseins

About 80 % of the protein of bovine milk is casein (see Chap. 4) which is primarily a source of essential amino acids and bioactive peptides, as well as a carrier of calcium and phosphate, for the neonate. In human milk, casein constitutes 20–30 % of total protein (Hambræus 1984). The casein fraction of most species consists of four gene products: α_{s1}-, α_{s2}-, β- and κ-caseins, of which the first three are calcium sensitive; the approximate proportions of each casein in bovine milk are 38, 11, 38 and 13 %, respectively. In other species, not all of these types of casein are present and the relative and absolute concentrations of caseins differ (Dalgleish 2011). In human milk, >85 % of the casein is β-casein (β-CN) and there is little (Rasmussen et al. 1995) or no α_s-casein (α-CN, see Hambræus and Lönnerdal 2003). The amino acid sequences of the individual caseins are not well conserved between species (Martin et al. 2013). The biological function of the caseins lies in their ability to form macromolecular structures, casein micelles, which transfer large amounts of calcium to the neonate with a minimal risk of pathological calcification of the mammary gland. The high level of calcium in milk is important for the development, strength and density of bones in children and in the prevention of osteoporosis in adults. Calcium also reduces cholesterol absorption and controls body weight and blood pressure. A possible role of calcium in the prevention of colon cancer has been investigated and

Table 11.2 The concentration of the major milk proteins of bovine and human milk and their physiological functions

Protein	Bovine milk (g L^{-1})	Human milk (g L^{-1})	Function
Total protein	34	9	
Total casein	26	2.4	Ion carriers (Ca, PO_4, Fe, Zn, Cu); precursors of bioactive peptides
α_{s1}-Casein	10.7	0.77[a]	Precursor of bioactive peptides
α_{s2}-Casein	2.8	–	Precursor of bioactive peptides
β-Casein	8.6	3.87	Precursor of bioactive peptides
κ-Casein	3.3	0.14	Precursor of CMP: antiviral; bifidogenic
Total whey protein	6.3	6.2	
β-Lactoglobulin	3.2	–	Retinol carrier; binding fatty acids; antioxidant
α-Lactalbumin	1.2	2.5	Synthesis of lactose; Ca carrier; immunomodulation; anticarcinogenic
Serum albumin	0.4	0.48	
Proteose peptone	1.2	0.8	Not characterised?
Total immunoglobulins	0.8	0.96	Immune protection
$IgG_{1,2}$	0.65	0.03	Immune protection
IgA	0.14	0.96	Immune protection
IgM	0.05	0.02	Immune protection
Lactoferrin	0.1	1.65	Antimicrobial; antioxidative; immunomodulation; anticarcinogenic
Lysozyme	126×10^{-6}	0.34	Antimicrobial; synergistic effect with lactoferrin and immunoglobulins
Lactoperoxidase	0.03		Antimicrobial
Miscellaneous	0.8	1.1	

Adapted from Shah (2000) and Uniacke-Lowe and Fox (2012)
[a]See Uniacke-Lowe et al. 2010

it is hypothesized that, as bile salts are one of the main promoters of colon cancer, ingestion of milk, which provides calcium phosphate to bind bile salts, may prevent their toxic effect (van der Meer et al. 1991).

11.5.2 Whey Proteins

The major whey proteins in bovine milk are β-Lg, α-La, Igs, blood serum albumin (BSA), lactoferrin (Lf) and lysozyme (Lyz), which are discussed in Chap. 4. Except for β-Lg, all these proteins are also present in human milk; however, the relative amounts of the whey proteins differ considerably between these milks. Many whey proteins have physiological properties including, metal-binding, immunomodulatory, growth factor activity and hormonal activity. The principal anti-microbial agent in bovine milk is Lyz and to a lesser extent Lf (the latter predominates in human milk (Table 11.2)). Both Lf and Lyz are present at low levels in bovine milk, in which Igs are the main defense against microbes (Malacarne et al. 2002). Together, IgA, IgG, IgM, Lf and Lyz provide the neonate with immune and non-immune protection against infection (Baldi et al. 2005).

11.5.2.1 Bioactivity of β-Lactoglobulin

β-Lg is the major whey protein in the milk of ruminants and is also present in milk of monogastrics and marsupials, but is absent from the milk of humans, camels, lagomorphs and rodents (see Chap. 4). Although several biological roles for β-Lg have been proposed, e.g., facilitator of retinol uptake and an inhibitor, modifier or promoter of enzyme activity, conclusive evidence for a specific biological function of β-Lg is not available (Sawyer 2003; Creamer et al. 2011). β-Lg binds retinol and that of many species, but not equine or porcine, binds fatty acids also (Pérez et al. 1993). During digestion, milk lipids are hydrolysed by pre-duodenal lipases, greatly increasing the amount of free fatty acids which could potentially bind to β-Lg, displacing any bound retinol, and implying that fatty acid metabolism, rather than retinol transport, is the more important function of β-Lg (Pérez and Calvo 1995). Bovine β-Lg is very resistant to peptic digestion and can cause allergenic reactions on consumption. Resistance to digestion may not be uniform among species; ovine β-Lg is reported to be far more digestible than bovine β-Lg (El-Zahar et al. 2005).

11.5.2.2 Bioactivity of α-Lactalbumin

α-La is a modifier of UDP-galactosyl transferase and regulates lactose biosynthesis (see Chap. 2). There is a substantial concentration in milk, ~ 4 % of total protein in bovine milk and ~ 25 % of total protein in human milk. α-La, a unique milk protein, is homologous with c-type lysozymes. It is a calcium metalloprotein, in which the Ca^{2+} plays a crucial role in folding and structure and has a regulatory function in the synthesis of lactose (Larson 1979; Brew 2003, 2013; Neville 2009). A high molecular weight form of α-La isolated from acid-precipitated human casein causes apoptosis of tumour cells, the native protein has no such effect but can be converted to an active anti-tumour form, HAMLET [human α–La made lethal to tumour cells] by reaction with oleic acid and conditions in an infant's stomach can cause this change (Svensson et al. 2003). Bovine α-La can also be transformed to an anti-tumour agent, BAMLET [bovine α–La made lethal to tumour cells]. The α-La/oleic acid complex and its cytotoxic activity was reviewed by Jøhnke and Petersen (2012).

11.5.2.3 Immunoglobulins

The concentration of Igs is significantly elevated in the colostrum of all ruminants (10 % of total N *vs* 3 % in mature milk, Fox and Kelly 2003) and equids as maternal Igs are passed from mother to neonate after birth when the small intestine is capable of absorbing intact proteins. After a few days, the gut 'closes' and further significant passage of proteins is prevented and within 2–3 days, the serum level of IgG in the neonate is similar to adult levels (Widdowson 1984). In contrast, *in utero* transfer of Igs occurs in humans and in some carnivores Igs are passed to the newborn both before and after birth. The milk of species that provide pre-natal passive immunization tends to have relatively small differences in protein content between colostrum and mature milk compared to species that depend on post-natal passage of maternal Igs. In the latter cases, of which all ungulates are typical, colostrum is rich in Igs and there are large quantitative differences in protein content between colostrum and mature milk (Langer 2009). Chapter 4 provides a detailed description of Igs.

11.5.2.4 Lactoferrin

Lf is an iron-binding glycoprotein, comprised of a single polypeptide chain of MW ~78 kDa (Conneely 2001). Lf is structurally similar to transferrin (Tf), a plasma iron transport protein, but has a much higher (~300-fold) affinity for iron (Brock 1997). Lf is not unique to milk but is especially abundant in colostrum, with small amounts in tears, saliva and mucus secretions and in the secondary granules of neutrophils. The expression of Lf in the bovine mammary gland is dependent on prolactin (Green and Pastewka 1978); its concentration is very high during early pregnancy and involution and is expressed predominantly in the ductal epithelium close to the teat (Molenaar et al. 1996). Human and bovine milk contain 1.65 g and 0.1 g Lf per L, respectively (Table 11.1).

Shimazaki et al. (1994) purified Lf from equine milk (~0.6 g per L) and compared its iron-binding ability with that of human and bovine Lfs and with bovine Tf. The iron-binding capacity of equine Lf is similar to that of human Lf and higher than that of bovine Lf and Tf. Various biological functions have been attributed to Lf but its exact role in iron-binding in milk is unknown and there is no relationship between the concentrations of Lf and Tf and the concentration of iron in milk (human milk is very rich in Lf but low in iron) (Masson and Heremans 1971).

Lf is a bioactive protein with nutritional and health-promoting properties (Baldi et al. 2005). Bacterial growth is inhibited by its ability to sequester iron (chelated Fe is unavailable to intestinal microorganisms) and also to permeabilize bacterial cell walls by binding to lipopolysaccharides through its N-terminus. Lf can inhibit viral infection by binding tightly to the envelope proteins of viruses and is also thought to stimulate the establishment of a beneficial microflora in the GIT (Baldi et al. 2005). Ellison and Giehl (1991) suggested that Lf and Lyz work synergistically to effectively eliminate Gram-negative bacteria; Lf binds OSs in the outer bacterial membrane, thereby opening 'pores' for Lyz to hydrolyse glycosidic linkages in the interior of the peptidoglycan matrix. This synergistic process leads to inactivation of both Gram-negative bacteria, e.g., *E. coli* (Rainhard 1986) and Gram-positive bacteria, e.g., *Staph. epidermis* (Leitch and Willcox 1999). A proteolytic digestion product of bovine and human Lf, lactoferricin, has bactericidal activity (see below and Bellamy et al. 1992). Bovine and human Lf are reported to have antiviral activity and act as growth factors (Lönnerdal 2003, 2013). Lf in human milk is reported to increase the production and release of cytokines such as, IL-1, IL-8, tumour necrosis factor α, nitric oxide and granulocytic-macrophage colony-stimulating factor which may have a positive effect on the immune system (Hernell and Lönnerdal 2002). Nowadays, most infant formulae are fortified with Lf (O'Regan et al. 2009; Lönnerdal and Suzuki 2013).

11.5.2.5 Serum Albumin

Bovine serum albumin is the most abundant protein in the circulatory system and is multifunctional as it transports a variety of ligands, including long-chain fatty acids, steroid hormones, bilirubin and various metal ions. It is believed to enter milk via leakage through para-cellular means or via uptake with other molecules (Fox and Kelly 2003). Its physiological significance in milk is relatively insignificant as it is present at very low concentrations compared to blood plasma.

11.5.3 Vitamin-Binding Proteins

Milk contains specific binding proteins for retinol (vitamin A), vitamin D, riboflavin (vitamin B_2), folate and cyanocobalamin (vitamin B_{12}). Such proteins improve the absorption of these vitamins by protecting and transferring them to receptor proteins in the intestine, or they may have antibacterial activity by rendering vitamins required by intestinal bacteria unavailable. The activity of these proteins is reduced or destroyed by heat treatment (see Wynn and Sheehy 2013).

11.5.3.1 Retinol-Binding Protein

β-Lg binds retinol in a hydrophobic pocket and protects it against oxidation (see Chap. 4). It improves the absorption of retinol which it exchanges with a retinol-binding protein in the gut. It also binds fatty acids and thereby activates lipases, the significance of which is not known. Human and rodent milk lacks β–Lg; it is not known if this is significant.

11.5.3.2 Vitamin D-Binding Protein

Vitamin D-binding protein (DBP), also called Gc-globulin (group-specific component), is a member of a gene family that includes serum albumin and α-fetoprotein, occurs in the plasma, ascetic fluid, cerebrospinal fluid and the surface of many cells types in most vertebrates. DBP is a 51–58 kDa multifunctional serum glycoprotein synthesized in large quantities by hepatic parenchymal cells and is secreted into the circulatory system as a monomeric peptide of 458 residues. Two binding regions are well characterized—a vitamin D/fatty acid binding domain located between residues 35 and 49, and an actin binding domain between residues 350 and 403 (Malik et al. 2013). DBP circulates in amounts far in excess of normal vitamin D metabolite concentrations in blood (Haddad 1995). DBP binds vitamin D and its plasma metabolites and transports them to target tissues, it prevents polymerization of actin (G-actin) by binding its monomers and it may have significant anti-inflammatory and immunoregulatory functions (Malik et al. 2013). DBP has been detected at a low level in the milk of several species and occurs at a higher concentration in colostrum than in milk. DBP variants have attracted attention in recent years as genetic factors with major roles in several chronic disease outcomes, e.g., pancreatic, prostate and bladder cancers (for reviews see Chun 2012 and Malik et al. 2013).

11.5.3.3 Riboflavin-Binding Protein

Riboflavin-binding protein (RfBP) has been partially purified from bovine milk; it has a MW of ~38 kDa (Kanno et al. 1991). The RfBP—riboflavin complex has good antioxidant properties, similar to riboflavin bound to egg white RfBP (Toyosaki and Mineshita 1988). The RfBP in milk is probably derived from blood serum.

11.5.3.4 Folate-Binding Protein

The folate-binding properties of milk have been known since the late 1960s and the involvement of a specific protein was confirmed by Ghitis et al. (1969). Later, a minor whey protein in milk was identified as having specific folate-binding properties (Salter et al. 1981). A folate-binding protein (FBP) isolated from bovine and human milk is a glycoprotein with a MW of ~35 kDa (Salter et al. 1981). Its concentration is higher (~5×) in colostrum than in milk (Nygren-Babol et al. 2004). FBP is crucial for the assimilation, distribution and retention of folic acid (Davis and Nichol 1988) and is antibacterial through reducing the availability of folate to microorganisms. The effectiveness of FBP is reduced by heat treatment of milk (Gregory 1982; Achanta et al. 2007). FBP is found in both soluble and particulate forms in human milk; ~ 22 % of the soluble form is glycosylated, providing protection against digestive enzymes. FBP in human milk is thought to slow the release and uptake of folate in the neonatal small intestine to allow gradual release and absorption of folate which may improve tissue utilization (Pickering et al. 2004). FBP and its role in folate nutrition has been reviewed by Parodi (1997b), while its biochemistry and physiology were reviewed by Nygren-Babol and Jägerstad (2012).

11.5.3.5 Haptocorrin

Haptocorrin (formerly called vitamin B_{12}-binding protein) is involved in the protection of acid-sensitive vitamin B_{12} as it passes through the GIT and is produced by the salivary glands in response to the ingestion of food. In all, three proteins are involved in the uptake of vitamin B_{12} (cobalamin). Gastric intrinsic factor (GIF) binds free B_{12} released from foods on digestion and transports it to the intestine where it is transferred to another protein, transcobalamin (TC). The B_{12}-TC complex and free TC are released into portal blood. Haptocorrin binds vitamin B_{12} to form a halohaptocorrin complex which can attach to human intestinal brush border membranes where the associated vitamin is absorbed by intestinal cells (Adkins and Lönnerdal 2001). Thus, absorption of vitamin B_{12} is improved by haptocorrin in the neonate and is antibacterial as protein-bound B_{12} is unavailable to gut microflora. Binding and inhibitory effects are reduced by heating.

11.5.4 Hormone-Binding Proteins

11.5.4.1 Corticosteroid-Binding Protein

Human milk and colostrum contain two corticosteroid-binding proteins (Rosner et al. 1976); similar proteins occur in blood. The function and significance of these proteins in milk are unknown.

11.5.4.2 Thyroxine-Binding Protein

The whey fraction of human milk contains a thyroxine-binding protein analogous to serum thyroxine-binding globulin at a concentration of ~0.3 mg mL^{-1} (Oberkotter and Farber 1984). The function of this protein in milk is unknown.

11.5.5 Metal-Binding Proteins

Milk contains many metal-binding proteins (or peptides therefrom), some of which have a nutritional function while others are enzymes, for the activity of which, the metal is essential. The most important inorganic elements in milk from a bone-health point of view are calcium, phosphorus, magnesium, sodium, potassium and zinc (for review see Cashman 2006). The significance of milk metals on the risk factors for heart disease, diabetes, stroke and other illnesses was reviewed extensively by Scholtz-Ahrens and Schrezenmeir (2006). Table 11.3 shows some of the principal metal-binding peptides in milk. While casein phosphopeptides (CPPs, see 11.7.2.4) are the main carriers of metals in milk, some metal binding peptides are found in whey protein hydrolysates, α-La, β-Lg and Lf. These are not phosphorylated peptides but metals bind to them *via* other sites which may be influenced by the protein conformation. α-La- and β-Lg-derived peptides have a greater affinity for iron than their parent protein (Vegarud et al. 2000).

Table 11.3 Metal-binding peptides derived from milk proteins

Protein	Enzyme	Phosphoresidues	Net charge	Metals bound
Casein-derived phosphopeptides				Fe, Mn, Cu, Se, Ca, Zn
α_{s1}-CN, f 43–58	Trypsin	2	–7	Ca, Fe
α_{s1}-CN, f 43–79	Trypsin	7		
α_{s1}-CN, f 59–64	Trypsin	4		Ca
α_{s1}-CN, f 59–79	Trypsin	5	–9	Ca, Fe
α_{s2}-CN, f 1–21/–32	Trypsin	4		
α_{s2}-CN, f 46–70	Trypsin	4	–11	
α_{s2}-CN, f55–64	Trypsin	4		
α_{s2}-CN, f 66–74	Trypsin	3		Ca
β-CN, f 1–25/28	Trypsin	4	–9/–8	Ca, Fe
β-CN, f 33–48		1	–6	
Whey protein-derived peptides				
α-La	Pepsin, trypsin/chymotrypsin			Cu, Ca, Zn, Fe
β-Lg	Thermolysin			Fe
β-Lg	Pepsin, trypsin/chymotrypsin			Fe
Lf (30 kDa)	Pepsin/trypsin			Fe
Lf (40 kDa)	Trypsin			Fe
Lf (50 kDa)	Trypsin			Fe

CN casein, La lactalbumin, Lg lactoglobulin, Lf lactoferrin

11.6   Minor Biologically-Active Proteins in Milk

Together with growth factors, minor milk proteins elicit significant effects on the growth and development of the neonatal calf as well as on maternal physiology (Wynn and Sheehy 2013). Milk contains about 100 proteins at trace levels (Table 11.4). Many of these have a biological activity. Angiogenins play several roles, especially in the vascular and immune systems. β_2-Microglobulin, osteopontin, proteose peptone 3, lactoperoxidase (LPO), lysozyme (Lyz) and transforming growth factors (TGFs β_1 and β_2) all have significant biological roles in the immune system while insulin-like growth factors (IGFs 1 and 2), epidermal growth factors (EGFs) and TGF α, play important roles in facilitating maturation of the gastrointestinal epithelium. Several minor proteins bind vitamins while others play roles in mammary gland and maternal physiological regulatory functions (e.g., leptin, feedback inhibitor of lactation, parathyroid hormone-related peptide and relaxin). Minor proteins, including growth factors have been reviewed by many authors (see Wynn and Sheehy 2013).

Table 11.4 Minor biologically-active proteins in milk

Protein	Molecular mass (Da)	Concentration in mature bovine milk (mg L^{-1})	Source
β_2-Microglobulin	11,636	9.5	Monocytes
Osteopontin	60,000	3–10	Mammary
Proteose peptone 3	28,000	300	Mammary
Folate-binding Protein (FBP)	30,000	6–10	
Vitamin D-binding Protein	52,000	16	Blood
Vitamin B$_{12}$-binding Protein	43,000	0.1–0.2	
Angiogenin-1	14,577	4–8	Mammary
Angiogenin-2	14,522		
Kininogen	68,000/17,000		Blood
Serotransferrin	77,000		Blood
α_1-Acid glycoprotein	40,000	<20	Blood
Ceruloplasmin	132,000		Mammary
Prosaposin	66,000	6.0	Mammary
Enzymes (~ 60)	Various	trace	Blood, Mammary

Modified from Fox and Kelly (2003)

11.6.1   Heparin Affinity Regulatory Peptide

Heparin Affinity Regulatory Peptide (HARP) is a 136-amino acid growth factor with an MW of ~18 kDa which has a high affinity for the anticoagulant glycosaminoglycan heparin and is secreted in human milk and colostrum with a threefold higher concentration in colostrum (Wynn and Sheehy 2013). Several physiological functions have been ascribed to HARP including stimulation of cell replication and chemotaxis and promoting angiogenesis both *in vivo* and *in vitro* (Papadimitriou et al. 2001).

11.6.2   Colostrinin

Colostrinin is a complex of proline-rich phosphopeptides first isolated from the IgG$_2$ fraction of ovine colostrum and containing mainly β-CN f121–138; it is also present in the colostrum of other species. It has beneficial effects on Alzheimer's disease but it is not known how it is produced (see Bilikieweiz and Gaus 2004; Kurzel et al. 2004).

11.6.3 β_2-Microglobulin

β_2-Microglobulin (β_2-MG) is homologous to the "constant domain" of Ig and histo-compatibility antigen (Groves and Greenberg 1982). It is probably produced in milk by intra-mammary proteolysis of somatic cells. Its MW is 11,636 Da and it contains 98 amino acid residues. It was isolated initially as a tetramer called lactollin. β_2-MG occurs free in body fluids where it may help T-lymphocytes with antigen recognition but its significance in milk is not known (Fox and Kelly 2003).

11.6.4 Osteopontin

Osteopontin (OPN) is a highly phosphorylated glycoprotein (MW, 29,283 Da; 261 amino acid residues of which 27 are phosphoserines and 1 is phosphothreonine) with 50 Ca-binding sites (Fox and Kelly 2003). It is present in bone and many other tissues and fluids, including milk. It has many functions, including mineralization and resorption of bone and biological signalling. The significance of OPN in milk is unknown but it may be important for calcium binding or anti-infectious activity (Fox and Kelly 2003).

11.6.5 Proteose Peptone-3

Proteose peptone-3 (PP3) is a heat-stable, acid-soluble glycophosphoprotein. Unlike the other proteose peptones, PP3 is an indigenous protein in milk, occurring mainly in whey. It has a MW of 28 kDa but two proteolytic fragments (18 and 11 kDa) also occur in milk. PP3 forms an amphiphilic helix and behaves hydrophobically. It prevents contact between lipoprotein lipase and its substrate and thereby prevents spontaneous lipolysis. It has been proposed that PP3 should be called lactophorin or lactoglycoporin (Girardet and Linden 1996). The biological function of PP3 in milk is unknown but it may stimulate the growth of bifidobacteria or have some involvement in calcium ion-binding *via* its phosphorylated *N*-terminus (Fox and Kelly 2003).

11.6.6 Angiogenins

Angiogenins induce the growth of blood vessels (angiogenesis). They have sequence homology with RNase and have RNase activity which is important for angiogenesis. Two angiogenins (1 and 2) occur in bovine milk and blood serum; their MW is ~15 kDa. Both have strong ability to promote the growth of new blood vessels in a chicken membrane assay (Fox and Kelly 2003). Their function in milk is unknown but it has been suggested that they may have a protective effect in the mammary gland or the neonatal intestine (Strydom 1998).

11.6.7 Kininogens

There are two forms of kininogen in bovine milk, a high MW (>68 kDa, 626 amino acid residues, produced in the liver) and low a MW (16–17 kDa, produced in various tissues). Bradykinin, a biologically active peptide, is released from high MW kininogen by the action of the enzyme, killikrenin:

$$\text{High MW kininogen} \xrightarrow{\text{Kallikrenin}} \text{bradykinin} \left(a\, 9\, \text{AA peptide} \right) + \text{kellidin}$$

Bradykinin is secreted into milk from the mammary gland (Fox and Kelly 2003). It is believed that kininogens in milk are different from those in blood plasma.

Plasma kininogen is an inhibitor of thiol proteinases and has a role in the initiation of blood coagulation. Bradykinin has several functions: it affects smooth muscle contraction and causes vasodilation and hypotension. The function(s) of kininogen and its derivatives in milk are unknown.

11.6.8 Glycoproteins

Milk contains several minor glycoproteins, one of which is M-1 glycoprotein (MW ~10 kDa). Its sugars are galactose, galactosamine and NANA and it stimulates bifidobacteria, probably via its amino sugars (Fox and Kelly 2003).

Another glycoprotein, orosamucoid (α_1-acid glycoprotein), has been detected in bovine colostrum but not in mature milk. It is a member of the lipocalin family with a MW of 40 kDa. It can modulate the immune system and its concentration in milk increases during inflammatory diseases, malignancy and pregnancy (Fox and Kelly 2003).

11.6.8.1 Prosaposin

Prosaposin (or PSAP) is a highly conserved glycoprotein of ~66 kDa and a precursor for four cleavage products, saposins A, B, C and D (each contains ~80 amino acids and is glycosylated), which are required for the hydrolysis of some sphingolipids by specific lysosomal hydrolases. Prosapsin occurs in milk but the saposins do not. Prosaposin has been isolated from human milk (Kondoh et al. 1991; Hiraiwa et al. 1993) and bovine milk (Patton et al. 1997) as well as the milk of other species (chimpanzee, rhesus monkey, goat and rat). Prosaposin is located exclusively in milk serum and its exact function in milk is unclear. Prosaposin plays a broad role in the development, maintenance and repair of the nervous system and only a small portion of its saposin C segment is required for neurotrophic activity (Patton et al. 1997). Human milk contains a significant amount of prosaposin (5–10 mg L^{-1}) and it may have direct effects on the neonatal gut especially it could be directly absorbed (Patton et al. 1997).

11.7 Indigenous Milk Enzymes

Milk contains many indigenous enzymes (>60) which originate from the mammal's blood plasma, leucocytes (somatic cells), or cytoplasm of the secretory cells and the MFGM (see Fox and Kelly 2006). The principal enzymes found in milk include digestive enzymes (proteinases, lipases, amylases and phosphatases) and enzymes with antioxidant and antimicrobial characteristics (Lyz, catalase, superoxide dismutase, LPO, myeloperoxidase, xanthine oxidoreductase, ribonuclease) all of which are important for milk stability and protection of mammals against pathogens (Korhonen and Pihlanto 2006). The indigenous enzymes in bovine and human milk have been studied extensively but the enzymes in the milk of other species have been studied only sporadically. Indigenous milk enzymes are discussed in detail in Chap. 10.

11.7.1 Lysozyme

Lyz (EC 3.1.2.17) occurs at a very high level in equine, asinine and human milk, >6,000, 6,000 and 3,000 times more, respectively, than bovine milk (Salimei et al. 2004; Guo et al. 2007). It has been suggested, but research is scarce, that while the composition of breast milk varies widely between well-nourished and poorly-nourished mothers, the amount of Lyz is conserved. Similar to Lf, the concentration of Lyz in human milk increases strongly after the second month of lactation, and it has been suggested that Lyz and Lf play major roles in fighting infection in breast-fed infants during late lactation, and in protecting the mammary gland (Montagne et al. 1998). Interesently, the Lyz content of equid milks is one of the main attractions for use of these milks in cosmetology as it is reputed to have a smoothing effect on the skin and may reduce scalp inflammation when incorporated into shampoo. Equid milk has very good antibacterial activity, presumably due to its high level of Lyz. Lyz is discussed in detail in Chap.10.

11.7.2 Lactoperoxidase

LPO is a broad-specificity peroxidase present at a high concentration in bovine milk but at a low level in human milk. LPO, which has been isolated and well characterized (Chap. 10), has attracted considerable interest owing to its antibacterial activity in the presence of H_2O_2 and thiocyanate (SCN⁻); the active species is hypothiocyanate (OSCN⁻) or other higher oxidation species. Milk normally contains no indigenous H_2O_2, which must be added or produced in situ, e.g., by the action of glucose oxidase or xanthine oxidoreductases; it is usually necessary to supplement the indigenous SCN⁻ Commercial interest in LPO is focused on:

1. Activation of the indigenous enzyme for cold pasteurization of milk or protection of the mammary gland against mastitis; and
2. Addition of isolated LPO to calf or piglet milk replacers to protect against enteritis, especially when the use of antibiotics in animal feed is not permitted.

LPO, which is positively charged at neutral pH, can be isolated from milk or whey by ion-exchange chromatography which has been scaled up for industrial application. These methods isolate LPO together with Lf which is also cationic at neutral pH. LPO and Lf can be resolved by chromatography on CM-Toyopearl or by hydrophobic interaction chromatography on Butyl Toyopearl 650 M (see Mulvihill 1992).

11.8 Bioactive Milk Peptides

Milk proteins are susceptible to proteolysis during GIT processing and later via exposure to indigenous or intestinal bacteria-derived enzymes in the gut (Politis and Chronopoulou 2008). Fermentation of milk by cultures of proteolytic bacteria used in the production of dairy products also produces bioactive peptides (Michalidou 2008). There is growing interest in physiologically active peptides derived from milk proteins and their use as potential ingredients of health-promoting functional foods targeted at diet-related diseases such as cardiovascular disease, diabetes and obesity (Korhonen 2009).

All the principal milk proteins contain sequences which, when released on enzymatic digestion, exhibit biological activity (Clare and Swaisgood 2000; Gobbetti et al. 2002). These bioactive milk peptides are defined as specific protein fragments (3–20 amino acid residues) which have a positive impact on the physiological functions of the body, ultimately affecting the health of the living organism (Kitts and Weiler 2003; Möller et al. 2008). Research over the last 15 years or so has shown that these peptides possess antibacterial, antiviral, antithrombotic, antihypertensive, antioxidative, anticytotoxic, immunomodulatory, opioid, opioid antagonist, metal-binding or smooth muscle contraction activities. These peptides may also play an important role in reducing the risk of obesity and development of type-II diabetes

Table 11.5 Bioactivity of peptides derived from milk proteins

Protein precursor	Bioactive peptide	Bioactivity
α- and β-CNs	Casomorphins	Opioid agonist
α-La	α-Lactorphin	Opioid agonist
β-Lg	β-Lactorphin	Opioid agonist
Lf	Lactoferroxins	Opioid antagonist
κ-CN	Casoxins	Opioid antagonist
α$_{s2}$-CN	Casocidin	Opioid antagonist
α- β-CN	Casokinins	ACE-inhibitory
α-La, β-Lg	Lactokinins	ACE-inhibitory
α-CN, β-CN, β-Lg	Immunopeptides	Immunomodulatory
Lf	Lactoferricin B	Antimicrobial
α$_{s1}$-CN	Isracidin	Antimicrobial
κ-CN, transferrin	Casoplatelins	Antithrombotic
κ-CN	Caseinomacropeptide	Antimicrobial; antimicrobial
α- and β-CN	Caseinophosphopeptides	Mineral binding

Adapted from Shah (2000) and Gobbetti et al. (2002)
CN casein, *La* lactalbumin, *Lg* lactoglobulin, *Lf* lactoferrin

(Erdman et al. 2008; Haque and Chand 2008; Möller et al. 2008). In addition, such peptides have much lower allergenicity than their parent proteins, believed to be related to their lower molecular weights (Høst and Halken 2004). A summary of bioactive peptides, their precursors and reported bioactive roles is shown in Table 11.5.

To preserve their physiological activity, bioactive peptides must maintain their integral state during transport through the body and must be absorbed from the intestine in active form; however, there is currently little evidence that this is in fact the case and many proposed properties remain to be proven. Di- and tri-peptides can be absorbed relatively efficiently in the intestine but it is not clear whether peptides with more than three amino acid residues are absorbed and are capable of reaching a target organ (Shah 2000).

11.8.1 Production of Bioactive Peptides

A general schematic representing the mechanisms by which bioactive peptides are obtained from bovine milk proteins is shown in Fig. 11.2. Proteolysis is the most common process resulting in the formation of bioactive peptides (Korhonen and Pihlanto 2006). Milk proteins are susceptible to proteolysis during gastric processing and later *via* exposure to indigenous or intestinal bacteria-derived enzymes in the gut (Politis and Chronopoulou 2008). Digestive enzymes, such as pepsin and pancreatic enzymes (trypsin, chymotrypsin, carboxy- and amino-peptidases), release bioactive peptides from milk proteins in the GIT (Szwajkowska et al. 2011).

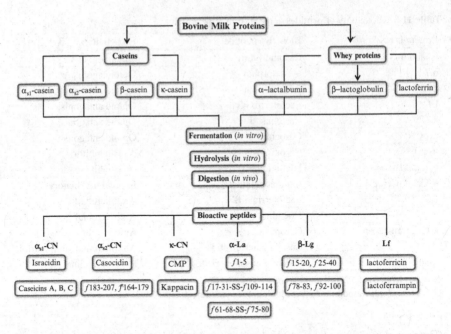

Fig. 11.2 Schematic representation of the mechanisms for the production of bioactive peptides from milk (adapted from Szwajkowska et al. 2011)

In vitro generation of bioactive peptides has been reported using pepsin and chymotrypsin. As well as digestive enzymes and proteinases such as chymotrypsin, pepsin and thermolysin, bacterial (e.g., alcalase from *Bacillus subtilis*) and fungal proteinases have also been used to produce bioactive peptides (Szwajkowska et al. 2011). Fermentation of milk by cultures of proteolytic bacteria used in the production of dairy products also produces bioactive peptides (Michalidou 2008).

11.8.2   Physiological Functionality of Bioactive Peptides

Milk-derived peptides may be multifunctional, i.e., peptide sequences may have two or more different bioactivities (Meisel 2004). Bioactive peptides are released in the stomach during protein digestion and the number and size of the peptides decreases between the stomach and the distal end of the duodenum but it is claimed that several long peptides, including casinomacropeptide (CMP) and an antihypertensive peptide sequence (residues $f24$–35) of α_{s1}-CN have been detected in blood plasma (Chabance et al. 1998). CMP is released intact from the stomach and is only partially hydrolysed by pancreatic enzymes (Fosset et al. 2002). Some peptides are capable of modulating specific physiological functions: anti-hypertensive, opioid, metal-binding, anti-bacterial and immunomodulatory activities have been reported

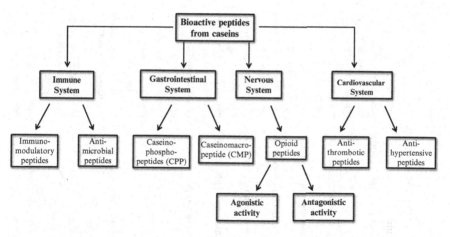

Fig. 11.3 Schematic of the principal physiological roles of casein-derived bioactive peptides (modified from Silva and Malcata 2005)

for casein-derived peptides (Abd El-Salam et al. 1996; Dziuba and Minkiewicz 1996; Brody 2000; Malkoski et al. 2001; Baldi et al. 2005; Silva and Malçata 2005; Thomä-Worringer et al. 2006; Michaelidou 2008) and whey protein-derived peptides (Nagaoka et al. 1991; Mullally et al. 1996; Pellegrini et al. 2001; Hernández-Ledesma et al. 2005; Chatterton et al. 2006; Yamauchi et al. 2006; Hernández-Ledesma et al. 2008). A schematic of the principal physiological roles of casein-derived bioactive peptides is shown in Fig. 11.3.

Bioactive peptides derived from casein, and to a lesser extent from whey proteins, have been shown to have effects on the cardiovascular, nervous, immune and nutritional systems.

Many milk protein-derived peptides have specific roles while others have multifunctional properties and specific peptide sequences may possess two or more distinct physiological roles (Gobbetti et al. 2002). Most β-casomorphins and cytokinins are both immunostimulatory and ACE-inhibitory while α- and β-lactorphin exhibit both opioid and ACE-inhibitory activity. Within the primary structure of caseins, overlapping peptide sequences exert different activities and many of these fragments are resistant to further proteolytic breakdown in the GIT (see Park 2009). Peptides with the same or different amino acid sequences may have the same or different bioactive functionalities (Park 2009). Table 11.6 shows many of the physiologically active peptides derived from milk proteins, their sequences and bioactivities.

11.8.2.1 Cardiovascular System

Many peptides derived from milk proteins have antithrombotic and antihypertensive activity (see Table 11.6).

Table 11.6 Multifunctional bioactive peptides encrypted in milk proteins (protein precursors are bovine unless otherwise indicated)

Substrate	Preparation	Fragment	Sequence	Name	Bioactivity
β-Casein	trypsin	β-CN(f1–25)4P	RELEELNVPGEIVE S*LS*S*S*EESITR[a]	Caseinophosphopeptide	mineral binding, immunomodulatory, cytomodulatory
β-Casein	In vitro and/or in vivo digestion	β-CN(f1–28)4P	RELEELNVPGEIVE S*LS*S*S*EESETRINK[a]	Caseinophosphopeptide	gastrointestinal effects, mineral binding
β-Casein	In vitro and/or in vivo digestion	β-CN(f2–28)4P	ELEELNVPGEIVE S*LS*S*S*EESETRINK[a]	Caseinophosphopeptide	gastrointestinal effects, mineral binding
β-Casein (human)	In vivo	β-CN(f1–18)	RETIESLS*S*S*EESITEYK[a]	Caseinophosphopeptide	gastrointestinal effects, mineral binding
β-Casein (human)	In vivo	β-CN(f1–23)	RETIESLS*S*S*EESITEYKQKVEK[a]	Caseinophosphopeptide	gastrointestinal effects, mineral binding
β-Casein (human)	In vivo	β-CN(f1–23)	RETIES*LS*S*S*EESITEYKQKVEK[a]	Caseinophosphopeptide	gastrointestinal effects, mineral binding
β-Casein (human)	In vivo	β-CN(f41–44)	YPSFQ	β–casomorphin	opioid agonistic
β-Casein (human)	In vivo	β-CN(f51–54)	YPFV		opioid agonist
β-Casein (human)	In vivo	β-CN(f54–59)	VEPIPY		immunostimulating
β-Casein (human)	In vivo	β-CN(f60–62)	GFL		immunostimulating
β-Casein (human)	In vivo	β-CN(f51–55/57)	YPFVE/PI		opioid agonist
β-Casein	Jejunum, fermentation+ pepsin+trypsin	β-CN(f60–70)	YPFPGPIPNSL[a]	β-Casomorphin-11	opioid, ACE-inhibitory, immunomodulatory
β-Casein	Trypsin	β-CN(f60–66)	YPFPGPI[a]	β-Casomorphin-7	opioid, ACE-inhibitory, immunomodulatory, cytomodulatory
β-Casein	Trypsin	β-CN(f60–64)	YPFPG	β-Casomorphin-5	opioid agonist
β-Casein	Trypsin	β-CN(f60–63)	YPFP-NH₂	Morphiceptin	opioid agonist

Protein	Production	Fragment	Sequence	Name	Bioactivity
β-Casein	Trypsin	β-CN(f63–68)	GPIPNS		immunomodulatory
β-Casein	Fermentation	β-CN(f74–76)	IPP		ACE-inhibitory
β-Casein	Fermentation	β-CN(f84–86)	VPP		ACE-inhibitory
β-Casein	Fermentation+pepsin+trypsin	β-CN(f108–113)	EMPFPK		anti-hypertensive
β-Casein	Lactobacillus CP790 protease	β-CN(f169–174)	KVLPVPQ	Antihypertensive peptide	anti-hypertensive
β-Casein (human)	Pepsin + pancreatin	β-CN(f125–129)	HLPLP		ACE-inhibitory
β-Casein (human)	Pepsin + pancreatin	β-CN(f154–160)	WSVPQPK		antioxidant
β-Casein (human)	Pepsin + pancreatin	β-CN(f169–173)	VPYPQ		antioxidant
β-Casein	Trypsin, fermentation	β-CN(f177–183)	AVPYPQR[a]	β-Casokinin-7	ACE-inhibitory, immunomodulatory, cytomodulatory
β-Casein	Trypsin, synthetic	β-CN(f191–193)	LLY	Immunopeptide	Immunomodulatory (stimulatory)
β-Casein (human)	Trypsin	β-CN(f54–59)	VEPIPY		immunomodulatory
β-Casein	Chymosin; synthesis	β-CN(f193–202)	YQEPVLGPVR	β-Casokinin-10	ACE-inhibitory, immunomodulatory (+/−)
β-Casein		β-CN(f193–209)	YQEPVLGPVRGPFPIIV		Immunomodulatory; antimicrobial
β-Casein	Fermentation+pepsin+trypsin	β-CN(f193–198)	YQQPVL	β-Immunocasokinin	ACE-inhibitory
β-Casein		β-CN(f199–204)	GPVRGP		ACE-inhibitory

(continued)

Table 11.6 (continued)

Substrate	Preparation	Fragment	Sequence	Name	Bioactivity
α$_{s1}$-Casein	Chymosin	α$_{s1-CN}$(f1–23)	RPKHPIKHQGLPQE VLNENLLRP	Isracidin	immunomodulatory; antimicrobial
α$_{s1}$-Casein (human)		α$_{s1-CN}$(f8–11)	YPER		ACE-inhibitory
α$_{s1}$-Casein	Trypsin	α$_{s1-CN}$(f23–24)	FF		anti-hypertensive
α$_{s1}$-Casein	Trypsin	α$_{s1-CN}$(f23–27)	FFVAP	α$_{s1}$-Casokinin-5	ACE inhibitor; anti-hypertensive
α$_{s1}$-Casein	Trypsin	α$_{s1-CN}$(f23–34)	FFVAPFPQVFGK	Casokinin	ACE-inhibitory
α$_{s1}$-Casein	In vitro and/or in vivo digestion	α$_{s1-CN}$(f43–58)	DIES*ES*TEDQAMEDIK	Caseinophosphopeptide	gastrointestinal effects; mineral-binding
α$_{s1}$-Casein	In vitro and/or in vivo digestion	α$_{s1-CN}$(f45–55)	GSESTEDQAME	Caseinophosphopeptide	gastrointestinal effects
α$_{s1}$-Casein	Trypsin and in vitro and/or in vivo digestion	α$_{s1-CN}$(f59–79)	QMEAES*IS*S*S*EEIVP NSVEQK	Caseinophosphopeptide	calcium binding and transport
α$_{s1}$-Casein	In vitro and/or in vivo digestion	α$_{s1-CN}$(f66–74)	SSSEEIVPN	Caseinophosphopeptide	gastrointestinal effects
α$_{s1}$-Casein	Pepsin	α$_{s1-CN}$(f90–95)	RYLGYL	α$_{s1}$-Exorphin	agonistic,opioid
α$_{s1}$-Casein	Pepsin	α$_{s1-CN}$(f90–96)	RYLGYLE	α$_{s1}$-Exorphin	agonistic,opioid
α$_{s1}$-Casein	Pepsin	α$_{s1-CN}$(f91–95)	YLGYL		agonistic,opioid
α$_{s1}$-Casein	Trypsin	α$_{s1-CN}$(f102–109)	KKYKVPQL		anti-hypertensive
α$_{s1}$-Casein	In vitro and/or in vivo digestion	α$_{s1-CN}$(f106–119)	VPQLEIVPNSAEER	Caseinophosphopeptide	gastrointestinal effects
α$_{s1}$-Casein (human)		α$_{s1-CN}$(f136–143)	YYPQIMQY		ACE inhibitory
α$_{s1}$-Casein (human)		α$_{s1-CN}$(f143–147)	YVPFP		opioid agonist

Protein	Treatment	Fragment	Sequence	Name	Activity
α_s1-Casein (human)		α_s1-CN(f143–149)	YVPFPPF		opioid antagonist
α_s1-Casein	Fermentation+pepsin+trypsin	α_s1-CN(f142–147)	LAYFYP		anti-hypertensive
α_s1-Casein	Fermentation+pepsin+trypsin	α_s1-CN(f157–164)	DAYPSGAW		anti-hypertensive
α_s1-Casein	Pepsin-chymotrypsin	α_s1-CN(f158–164)	YVPF PPF	Casoxin-D	opioid antagonist
α_s1-Casein	Trypsin	α_s1-CN(f194–199)	TTMPLW[a]	α-Immunocasokinin	immunomodulatory, ACE-inhibitory (hypotensive *in vivo*)
α_s2-Casein	Trypsin	α_s2-CN(f1–32)4P	KNTMEHVS*S*EE SIIS*QETYKQEKN	Caseinophosphopeptide	mineral binding, immunomodulatory
α_s2-Casein	*In vitro* and/or *in vivo* digestion	α_s2-CN(f2–21)	NTMEHVSSSEESIIS QETYK	Caseinophosphopeptide	gastrointestinal effects
α_s2-Casein	*In vitro* and/or *in vivo* digestion	α_s2-CN(f46–70)	NANEEEYSIGSSSEES AEVATEEVK	Caseinophosphopeptide	gastrointestinal effects
α_s2-Casein	*In vitro* and/or *in vivo* digestion	α_s2-CN(f55–75)	GSSSEES AEVATEEVKITVDD	Caseinophosphopeptide	gastrointestinal effects
α_s2-Casein	*In vitro* and/or *in vivo* digestion	α_s2-CN(f126–136)	EQLSTSEENSK	Caseinophosphopeptide	gastrointestinal effects
α_s2-Casein	*In vitro* and/or *in vivo* digestion	α_s2-CN(f138–149)	TVDMESTEVFTK	Caseinophosphopeptide	gastrointestinal effects
α_s2-Casein	Pepsin	α_s2-CN(f164–179)	LKKISQRYQKFALPQY		antimicrobial
α_s2-Casein	Synthetic peptide	α_s2-CN(f165–203)	LKKISQRYQKFALPQY LKTVYQHQKAMKP WIQPKTKVIPY	Casocidin-I	antimicrobial
α_s2-Casein	Trypsin	α_s2-CN(f174–179)	FALPQY		strong ACE-inhibitory
α_s2-Casein	Trypsin	α_s2-CN(f174–181)	FALPQYLK		strong ACE-inhibitory

(continued)

Table 11.6 (continued)

Substrate	Preparation	Fragment	Sequence	Name	Bioactivity
α_{s2}-Casein	Pepsin	$\alpha_{s2\text{-}CN}$(f183–207)	VYQHQKAMKP WIQPKTKVIPYVRYL		antimicrobial
α_{s2}-Casein		$\alpha_{s2\text{-}CN}$(f189–193)	AMKPW		weak anti-hypertensive
α_{s2}-Casein		$\alpha_{s2\text{-}CN}$(f189–197)	AMKPWIQPK		weak anti-hypertensive
α_{s2}-Casein		$\alpha_{s2\text{-}CN}$(f190–197)	MKPWIQPK		anti-hypertensive
α_{s2}-Casein		$\alpha_{s2\text{-}CN}$(f198–202)	TKVIP		weak anti-hypertensive
κ-Casein	Trypsin	κ-CN(f25–34)	YIPIQYVLSR	Casoxin-C	opioid (antagonist), ACE-inhibitory, smooth muscle (ileum, artery) contraction
κ-Casein	Pepsin	κ-CN(f33–38)	SRYPSY	Casoxin-6	opioid
κ-Casein(human)	Pepsin + pancreatin	κ-CN(f31–36)	YPNSYP		antioxidant
κ-Casein(human)	Pepsin + pancreatin	κ-CN(f53–58)	NPYVPR		antioxidant
κ-Casein	Pepsin+trypsin	κ CN(f35–41)	YPSYGLN	Casoxin-A	opioid (antagonist), ACE-inhibitory
κ-Casein	Synthesis	κ CN(f35–41)	YPSYGLN	κ-Immunocasokinin	immunomodulatory, ACE-inhibitory
κ-Casein	Trypsin	κ-CN(f58–61)	YPYY		antagonistic
κ-Casein	Trypsin	κ-CN(f103–111)	LSFMAIPPK		weak antithrombotic
κ-Casein	Trypsin	κ-CN(f106–110)	MAIPP	Casopiastrin	antithrombotic
κ-Casein	Trypsin	κ-CN(f106–116)	MAIPPKKNQDK	Casoplatelin	antithrombotic
κ-Casein	Trypsin	κ-CN(f106–112)	MAIPPKK	Casoplatelin	antithrombotic

κ-Casein	Rennet/chymosin	κ-CN(f106–169)	MAIPPKKNQDKTEIPTINTIASGE PTSTPTTEAVESTVATLEDSPEVIE SPPEINTVQVTSTAV	Caseinomacropeptide	nutrition system,i.e., probiotic (growth of bifidobacteria in GIT), antimicrobial
κ-Casein	Fermentation	κ-CN(f108–111)	IPP		ACE-inhibitory
κ-Casein (human)		κ-CN(f114–124)	IAIPPKKIQDK		antithrombotic
κ-Casein	Trypsin	κ-CN(f112–116)	KDQDK	Thrombin inhibitory peptide	antithrombotic
κ-Casein	Trypsin	κ CN(f113–116)	NQDK	Casoplatelin	antithrombotic
κ-Casein		κ-CN(f158–164)	EINTVQV		opioid antagonistic
γ-Casein	Trypsin	γ–CN(f108–113)	EMPFPK		bradykinin-potentiating activity
γ-Casein	Trypsin	γ–CN(f114–121)	YPVEPFTE		bradykinin-potentiating activity, ACE-inhibitory,opioid
α-Lactalbumin		α-La(f1–5)	EQLTK		bactericidal
α-Lactalbumin (human)		α-La(f1–5)[a]	KQFTK		bactericidal
α-Lactalbumin (bovine or human)		α-La(50–52)	YGL		ACE-inhibitory
α-Lactalbumin	Pepsin	α-La(50–53)	YGLF-NH₂[a]	α-Lactorphin	opioid (antagonist), ACE-inhibitory (hypotensive in vivo)
α-Lactalbumin (human)	Pepsin	α-La(50–53)	YGLF		opioid agonist
α-Lactalbumin (human)		α-La(51–53)	GLF		immunostimulating

(continued)

Table 11.6 (continued)

Substrate	Preparation	Fragment	Sequence	Name	Bioactivity
α-Lactalbumin		α-La(f50–51) (f18–19)		α-Immunolactokinin	immunostimulating
α-Lactalbumin		α-La(f17–31) (f109–114)	GYGGVSLPEWVC*TIF ALC*SEK		bactericidal
α-Lactalbumin (human)		α-La(f17–31) (f109–114)[a]	GYGGIALPELIC*TMF ALC*TEK		bactericidal
α-Lactalbumin		α-La(f61–68) (f75–80)	C*KDDQNPH ISC*DKF		bactericidal
α-Lactalbumin (human)		α-La(f61–68) (f75–80)[a]	C*KSSQVPQ ISC*DKF		bactericidal
α-Lactalbumin (bovine or human)	Trypsin	α-La(f104–108)	WLAHK	Lactokinin	ACE-inhibitory
β-Lactoglobulin		β-Lg(f15–20)	VAGTWY		antimicrobial; antihypertensive
β-Lactoglobulin		β-Lg(f40–42)	RVY		antihypertensive
β-Lactoglobulin		β-Lg(f46–48)	LKP		antihypertensive
β-Lactoglobulin	Synthetic or trypsin	β-Lg(f102–105)	YLLF-NH$_2$[a]	β-Lactorphin amide	opioid, ACE-inhibitory, smooth muscle contraction
β-Lactoglobulin		β-Lg(f122–124)	LVR		antihypertensive
β-Lactoglobulin	Trypsin	β-Lg(f142–148)	ALPMHIR	β-Lactorphin	ACE-inhibitory
β-Lactoglobulin	Trypsin	β-Lg(f146–149)	HIRL	β-Lactotensin	ileum contraction
Lactoferrin	Pepsin	Lf(f17–41)	FKCRRWQWRMK KLGAPSITCVRRAF[a]	Lactoferricin-B	antimicrobial, immunomodulatory (+), probiotic
Lactoferrin	Pepsin	(f39–42)	KRDS	Thrombin inhibitory peptide	antithrombotic

Lactoferrin	Pepsin	Lf(β318–323)	YLGSGY-OCH$_3$	Lactoferroxin A	opioid antagonist
Bovine serum albumin	Trypsin	BSA(β208–216)	ALKAWSVAR	Albutensin A	ACE-inhibitory, smooth muscle (ileum) contraction, artery relaxation
Bovine serum albumin	Pepsin	BSA(β399–404)	YGFQNA	Serorphin	weak opioid agonist

S* = Phosphoserine

Data from various sources including Clare and Swaisgood (2000), Meisel (2004), Silva and Malcata (2005), Park 2009, Wada and Lönnerdal (2014) and Raikos and Dassios (2014)

CN casein, *La* lactalbumin, *Lg* lactoglobulin

[a]Sequence also contains smaller bioactive peptides

Antithrombotic Peptides

The mechanisms involved in milk clotting and blood clotting are comparable.
Several peptide sequences in κ-casein (κ-CN) are similar to those of the γ-chain of
human fibrinogen, and peptides derived from both proteins have antithrombotic
properties. κ-CN f106–116 is produced from the (glyco)-macropeptide, κ-CN f106–
169, formed by the action of chymosin on κ- CN. The undecapeptide, f106–116 of
bovine κ- CN (a platelet-modifying peptide or casoplatelin, see Table 11.5) inhibits
the aggregation of ADP-treated blood platelets; its behaviour is similar to that of the
structurally related C-terminal dodecapeptide (f400–411) of human fibrinogen
γ-chain (Jollès et al. 1978, 1986; Maubois et al. 1991; Caen et al. 1992). When
bovine κ- CN f106–116 is reduced to the pentapeptide, KNQDK, the antithrombotic
activity is maintained (Caen et al. 1992). Casoplatelins, casein-derived peptides
(f106–116, f106–112, f113–116—see Table 11.5) are inhibitors of both the aggregation of ADP-activated platelets and the binding of human fibrinogen γ-chain to a
specific receptor on the platelet surface (Jollès et al. 1986; Silva and Malcata 2005).
κ-CN, f106–110, called casopiastrin, obtained by tryptic hydrolysis, exhibits antithrombotic properties by inhibiting fibrinogen binding (Jollès and Henschen 1982).
A second tryptic fragment, f103–111, inhibits platelet aggregation but does not
inhibit fibrinogen binding (Jollès et al. 1986). κ-CN f106–112 and f113–116, have
similar but weaker effects on platelet aggregation. Inhibition of platelet aggregation
is greatly enhanced if lysine is included in the peptide sequence; κ-CN f112–116 has
a lysine residue and is 222 times more active than κ-CN f113–116 (Maubois et al.
1991). A peptide with similar properties has been isolated from a hydrolysate of Lf
(Mazoyer et al. 1990). Bovine κ-CN f103–111 can prevent blood-clotting through
inhibition of platelet aggregation but it does not affect fibrinogen-binding to ADP-
treated platelets (Fiat et al. 1993). Chabance et al. (1995) reported antithrombotic
activity by human κ-CN f114–124. Table 11.7 compares IC50 (μM) values for
bovine κ-CN derived-peptides to peptides f400–411 of human fibrinogen γ-chain,
f39–42 human Lf and f572–575 of human fibrinogen α-chain.

Antihypertensive Peptides

The renin-angiotensin system is implicated in blood pressure regulation and hypertension. Renin acts on angiotensin and releases inactive angiotensin I, which is then
converted to the active peptide hormone, angiotensin II, by angiotensin I-converting
enzyme (ACE), a peptidyl dipeptidase (Fiat et al. 1993; Nakamura et al. 1995 a, b).
Angiotensin II is a vasoconstrictor, which inhibits bradykinin, a vasodilator
(Fig. 11.4). ACE inhibitors derived from casein, called casokinins, have been identified within the sequences of human β-CN (Kohmura et al. 1989; Bouzerzour et al.
2012) and k-CN (Kohmura et al. 1990) and bovine $α_{s1}$- and β-CNs (Maruyama and
Suzuki 1982; Meisel and Schlimme 1994). The dodecapeptide, $α_{s1}$-CN f23–34, from
tryptic hydrolysates of casein, inhibits ACE; the C-terminal sequence, $α_{s1}$-CN f194–
199, also has ACE inhibitory activity and both probably increase bradykinin

Table 11.7 Anti-aggregating (antithrombotic) milk-derived peptides compared to fibrinogen peptides

	Inhibition	
	ADP-induced human platelet aggregation	ADP-induced fibrinogen binding to human platelets
Peptides	IC50 (μM)[a]	
Bovine κ-casein		
f106–116 (MAIPPKKNQDK)	60	120
f106–112 (MAIPPKK)	>1,600	
f113–116 (NQDK)	400	
Human fibrinogen γ-chain		
f400–411 (H HLGGAKQAGD)	150	400
Human lactoferrin		
f39–42 (K RDS)	350	360
Human fibrinogen α-chain		
f572–575 (R GDS)	75	20

[a]Mean inhibitory concentration (concentration necessary to reduce the induced platelet aggregation by 50 % with respect to a control). Adapted from Fiat et al. (1993)

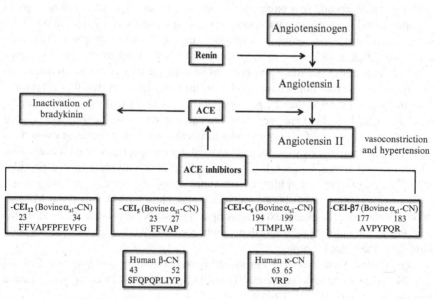

(angiotensin I-converting enzyme (ACE); converting enzyme inhibitor (CEI); CEI$_5$= N-terminal pentapeptide of CEI$_{12}$)

Fig. 11.4 Milk protein-derived antihypertensive peptides which act as angiotensin I-converting enzyme inhibitors (CEI); adapted from Fiat et al. (1993)

(Maruyama et al. 1987; Fiat et al. 1993). Peptides from the sequence *f*39–52 of human β-CN, especially β-CN *f*43–52 and human κ-CN *f*63–65, also have very potent ACE inhibitory activity *in vitro* while, the sequence, β-CN *f*43–52, exhibiting potent activity *in vivo* (Fiat et al. 1993). Several tryptic hydrolysates of caseins can inhibit activity of ACE *in vitro* (see Table 11.6), particularly *f*23–24, *f*23–27 and *f*194–199 of bovine α_{s1}-CN and *f*177–183 and *f*193–202 of β-CN. α_{s2}-CN peptides *f*189–193, *f* 189–197, *f* 190–197 and *f* 198–202 have been reported to have weak ACE-inhibitory activity. Some whey protein- derived peptides exhibit ACE-inhibitory activity, details of which are summarized in Table 11.6. It has been reported that ACE inhibitory peptides may also have immunomodulatory effects; because ACE cataly-ses the inactivation of angiotensin II, as a result of which bradykinin is inactivated, any peptide that inhibits ACE would favour bradykinin activity which is reported to include the stimulation of macrophages to enhance lymphocyte migration and increase the secretion of lymphokines (Maruyama et al. 1985; Paegelow and Werner 1986). Other biological activities of bradykinin are included under kininogens below.

11.8.2.2   Nervous System

Recent studies have shown that there is scientific merit to the belief that a glass of milk can aid sleep and that babies are calmed by breast or bottle-feeding. Opioid peptides found in milk can have agonistic or antagonistic activity (see Table 11.2). In pharmacology terms, agonist–antagonist is used to refer to a drug which exhibits some properties of an agonist (a substance that fully activates the neuronal receptor to which it attaches) and some properties of an antagonist (a substance that attaches to a receptor but does not activate it or if it displaces an agonist at that receptor it seemingly deactivates it thereby reversing the effect of the agonist). Opioid recep-tors (μ-, δ-, and κ-types) are present in the endocrine system, the immune system and the intestinal tract of mammals and will interact with endogenous or exogenous ligands. Endogenous ligands include typical opioid peptides, e.g., encephalin, endorphin and dynorphin, while exogenous ligands are atypical opioid peptides with either agnostic [exorphins or formones (food hormones)] or antagonistic (casoxins) activities (Shah 2000; Fitzgerald and Meisel 2000; Silva and Malcata 2005). Both endogenous and exogenous opioid peptides have Tyr at the N-terminus with Phe or Tyr at the third or fourth position, a structural feature which allows good binding to the opioid receptor.

   Opioid antagonists have been produced from both bovine and human κ-CN and α_{s2}-CN (Table 11.5). The sequence *f*41–44 from human β-CN is reported to have agnostic activity (Fiat et al. 1993).

β-Casomorphins

Endogenous opioid peptides (endorphins) are produced in many tissues. Opioid pep-tides are present in the hydrolyzates of many proteins, including milk proteins; these are called exorphins. Exorphins exhibit pharmaceutical properties similar to opium (morphine) and may induce apnea and irregular breathing, stimulate food intake and increase insulin output among many other properties (see Xu 1998). The first, and most effective, opioid peptides discovered from milk were the β-casomorphins (BCMs; Brantl et al. 1979). BCMs have been found in the intestinal contents and in blood plasma of very young babies but not of children or adults. They have a variety of physiological effects but whether they reach the brain is unclear.

BCMs are biologically active peptides with opioid activity, specifically μ-opioid agonists and antagonists (Clare and Swaisgood 2000; Teschemacher 2003). BCMs originate from β-casein and have a chain length of 4–11 amino acids, all starting with tyrosine 60 of β-CN (Kostyra et al. 2004). BCMs are found at analogous positions in sheep, water buffalo and human β-CN (Fiat and Jollès 1989; Teschemacher et al. 1990; Meisel and Schlimme 1996; Meisel 1997). BCMs have been detected in bovine milk (Cieślińska et al. 2007), milk products (Jarmolowska et al. 1999), human milk (Jarmolowska et al. 2007a, b) and infant formulae (Sturner and Chang 1988). The mechanisms involved in the intestinal transport of BCMs *in vivo* by the human body have been poorly researched. However, Iwan et al. (2008) demonstrated the transport of opioid peptides across human intestinal mucosa, specifically the transport of μ-opioid receptor agonists, human BCMs 5 and 7 (BCM5, BCM7) and the antagonist lactoferroxin A (LCF A). The physiological effects of BCMs are believed to be restricted to the gastrointestinal tract where they modulate general function, intestinal transit, amino acid uptake and water balance. However, β-CN-derived peptides may pass through the intestinal mucosa in neonates via active transport, thus producing a calming effect (Chang et al. 1985; Sturner and Chang 1988). Furthermore, the mammary tissue of pregnant or lactating women is reported to be permeable to BCMs (Clare and Swaisgood 2000). The best known bovine BCMs are β-CN *f*60–63/6 and β-CN *f*60–70); they are 300–4,000 times less active than morphine. The corresponding peptides from human β-CN are human β-CM 4, 5, 6, 7 (hβ-CN 51–54/57). Other caseinomorphines identified are human β-CN *f*59–63 and *f*41–44, α-CN *f*90–95/6 and fragments of α-La and β-Lg (see Table 11.6).

BCMs have unique structural features that impart a high and physiologically significant affinity for the binding sites of endogenous opioid receptors (Brantl et al. 1981; Meisel and FitzGerald 2000). Once formed, BCMs are resistant to proteolysis because of their proline-rich sequences and can reach significant levels in the stomach (Sun et al. 2003). BCMs are absorbed from the GIT and can cross the blood-brain barrier of newborns and young infants due to an immature central nervous system (Sun et al. 1999; Sun and Cade 1999; Sun et al. 2003). Indirect evidence suggests that adults who consume bovine casein produce BCMs in the GIT but are reported not to have circulating BCMs (Svedberg et al. 1985; Teschemacher et al. 1986).

In bovine β-CN, residue 67 is proline in variant A^2 but is histidine in variant A^1 and B (Groves 1969; Jinsmaa and Yoshikawa 1999) (Fig. 11.5). Structural differences between β-CNs A^1 and A^2 variants result in each releasing its own set of bioactive peptides on digestion by gastrointestinal enzymes. The one amino acid difference at position 67 allows cleavage by digestive enzymes of the peptide chain next to His_{67} but not next to Pro_{67} and, in the former case, BCM-7 is formed. BCM-7, *f*60 to 66 of β-CN (Kamiński et al. 2007) is believed to prevent the release of

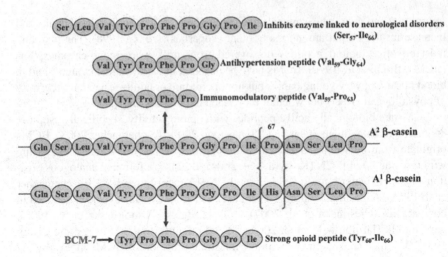

Fig. 11.5 Bioactive peptides from bovine b-casein variants A^1 and A^2, including β-casomorphin-7 (Tyr_{60}-Ile_{66}) released from the A^1 variant

many peptides with important bioactive properties (Fig. 11.5). BCM-7 has been isolated and identified in fresh bovine and human milk as well as in dried infant formulae (Sun et al. 2003; De Noni 2008) and in other dairy products (De Noni and Cattaneo 2010). BCM-7 is one of the first examples of a bioactive peptide derived from a food protein (Brantl et al. 1979) and it can be converted to BCM-5 by proteolysis in the GIT (Meisel et al. 1989). BCM-7 is reported to play a significant role in the aetiology of certain human diseases and can potentially affect numerous opioid receptors in the nervous, endocrine and immune systems (Bell et al. 2006). Epidemiological evidence suggests that the consumption of BCM-7 is associated with increased risk of ischaemic heart disease (Chin-Dusting et al. 2006), atherosclerosis (Tailford et al. 2003; Venn et al. 2006), Type 1 diabetes mellitus (Thorsdottir et al. 2000; Elliott et al. 1999), sudden infant death syndrome (Sun et al. 2003), autism and schizophrenia (Cade et al. 2000) although the European Food Safety Authority (EFSA 2009) have reported no cause and effect relationship between the oral intake of BCM-7 or related peptides and the aetiology or course of any suggested noncommunicable disease.

Neuropeptides

Galanin

Galanin is a neuropeptide consisting of 29 amino acids produced from cleavage of a 123-amino acid protein called preprogalanin. Galanin is widely distributed in the nervous and endocrine systems and in the intestine and has been identified in human milk (Hernández-Ledesma et al. 2007). Galanin facilitates the growth and repair of sensory neurons in the peripheral nervous system and the gut. Other important neuropeptides identified in human milk include neuropeptide Y, neurotensin, substance

P, somatostatin and vasoactive peptide. Some of these neuropeptides potentiate an immune response, while substance P induces the production of interleukin (IL-12) by macrophages; many cells in the neonatal immune system have receptors for these neuropeptides (Hendricks and Guo 2014).

Delta sleep–inducing peptide
Delta sleep–inducing peptide (DSIP) is a neuropeptide consisting of nine amino acids, including exyteenin, an amino acid normally produced by the hypothalamus. DSIP-like peptides are found in various tissues and fluids, including milk. DSIP promotes a particular type of sleep that is characterized by an increase in the delta rhythm of the electroencephalogram (EEG, Graf et al. 1984). A high level of DSIP is present in human milk with lower levels in the blood plasma of neonates, in which it induces sleep (Graf et al. 1984; Graf and Kastin 1986). DSIP is unusual in that it crosses the blood-brain barrier freely and is absorbed from the gut without degradation by proteolytic enzymes (Hendricks and Guo 2014).

Colostrinin
Colostrinin is a complex of proline-rich phosphopeptides, first isolated from the IgG_2 fraction of ovine colostrum and containing mainly β-CN *f*121–138; it is also present in the colostrum of other species. It has beneficial effects on Alzheimer's disease although it is not known how is it produced (see Bilikieweiz and Gaus 2004; Kurzel et al. 2001, 2004).

Anti-convulsant (anti-epileptic, calming) peptide
An anti-convulsive peptide present in the tryptic hydrolysate of casein has been identified as $α_{S1}$-CN *f*91–100; it has been called α-casozepine. Its significance *in vivo* is not known (Miclo et al. 2001).

11.8.2.3   Immune System

The defence systems of the human body are very complex and it is only relatively recently that the role played by diet, and specifically bioactive peptides, has been recognised. The two main activities of bioactive peptides in this area are stimulation of the immune system and inhibition of pathogenic bacteria.

Immunomodulation

During the digestion of human and bovine milk, peptides with immunomodulating properties are released (see Table 11.6). The immunomodulatory action of bioactive peptides is related to stimulation of the proliferation of human lymphocytes and the phagocytic activity of macrophages (Clare et al. 2003). Some cytochemical studies have shown that bioactive peptides can induce apoptosis of cancer cells (López-Expósito and Recio 2008).

The casein fraction of milk is a rich source of peptides which stimulate and aid the immune system. Isracidin, derived from $α_{S1}$-CN (*f*1–23) by the action of chymosin, is reported to have antibiotic properties against *Staphylococcus aureus* and *Candida albicans in vivo* (Shah 2000). Intermammary injection of isracidin protects cows and sheep against mastitic infection (Hayes et al. 2005; Haque and Chand 2008).

Anti-Microbial Whey Protein-Derived Peptides

The main biologically significant properties associated with whey proteins (e.g., Igs, Lf, LPO and Lyz) and peptides therefrom are bacteriostatic and bactericidal properties. Antibacterial peptides kill bacteria by destroying the cell membrane or mitochondrial membrane. The mechanism of action depends on the binding affinity (electrostatic interaction) of the peptide for the cell membrane surface (Brodgen 2005). Some of the biologically active peptides from whey proteins include α-lactorphin, β-lactorphin, albutensin A, serorphins and β-lactotensin (see Table 11.6). Both α- and β-lactorphin are believed to induce contraction of smooth muscle similar to morphine (Shah 2000).

Biologically active peptides are released from β-Lg during digestion with trypsin and have been reported to be active against several food pathogens, e.g., *Staphylococcus aureus*, some *Salmonella spp.* and *Escherichia coli* (Pellegrini et al. 2003). Some peptides derived from B-Lg ($f15$–20, $f25$–40, $f78$–83 and $f92$–100— see Fig. 11.2 and Table 11.6) have a negative electrostatic charge and their activity is restricted to Gram-positive bacteria (Pellegrini et al. 2001). Several peptides from β-Lg have been reported to have antihypertensive activity including, $f40$–42, $f122$–124, (Hernández-Ledesma et al. 2004) and $f15$–20 and $f46$–48 (Català-Clariana et al. 2010).

α-La has immunomodulatory properties but no antimicrobial activity whereas peptides produced as a result of its hydrolysis by trypsin or chymotrypsin ($f1$–5, $f17$–31, $f109$–114, $f61$–68, $f75$–80; see Fig. 11.4 and Table 11.6) show both immunomodulatory and antimicrobial properties against bacteria, viruses and fungi (Pellegrini et al. 1999; Kamau et al. 2010).

Lf exhibits strong antibacterial activity through its ability to bind iron. Lactoferricin (Lfcin) is a 25 amino acid peptide from the N-terminal region of Lf released by pepsin under acidic conditions (Haug et al. 2007). Lfcin has significantly greater antimicrobial activity than LF and is more heat resistant and active over a considerably broader pH range. Lfcin binds to the surface of Gram-negative bacteria resulting in the release of lipopolysaccharide from the bacterial cell wall which damages the cell wall and causes other morphological changes (Bellamy et al. 1992; Appelmelk et al. 1994; Tomita et al. 2001). Gifford *et al* (2005) reported the effective treatment of some cancers, such as leukemia and neuroblastoma, with Lfcin.

Lactoferrampin (Lfampin), another peptide from Lf, has strong antifungal and antibacterial properties. Its activity against *Candida* is reported to be much greater than that of Lf and it is also very active against *Bacillus subtillus*, *Escherichia coli* and *Pseudomonas aeruginosa* (van der Kraan et al. 2004, 2005). *In vitro*, Lf and its derived peptides show antibacterial activity against many pathogens, e.g.,

Clostridium perfringens, *Helicobacter pylori*, *Vibrio cholera* and many viruses including, hepatitis C, G and B, poliovirus, rotavirus and herpes simplex virus (Pan et al. 2007).

Antimicrobial Peptides from Casein

Caseicidin from chymosin-mediated hydrolysis of casein was one of the first antimicrobial peptides identified and purified. Later other antimicrobial peptides, caseicins A, B and C produced from α_{s1}-CN during fermentation of milk (See Fig. 11.4 and Table 11.6) by *Lactobacillus acidophilus* and exhibit antimicrobial activity were identified. Caseicins A and B have very high activity against *Escherichia coli* · O157:H7 and *Enterobacter sakazakii* (Hayes et al. 2005). The latter has been found in powdered infant formulae (FAO/WHO 2008; Oonaka et al. 2010) and could lead to severe neurological complications in infants with a mortality rate of 40–80 % (Korpysa-Dzirba et al. 2007). Caseicins have also been reported to be very active against some Gram-negative pathogens such as *Cronobacter sakazakii* and *Pseudomonas fluorescens* as well as Gram-positive *Staphylococcus aureus* (Norberg et al. 2011). Hydrolysis of α_{s2}-CN (by chymosin at neutral pH) results in the formation of casocidin (Fig. 11.2 and Table 11.6) which has antibacterial activity against, e.g., *Staphylococcus spp.* and *Bacillus subtillus* (Clare and Swaisgood 2000; Silva and Malcata 2005). Other peptides derived from α_{s2}-CN (*f*183–207 and *f*164–179) inhibit the growth of both Gram-positive and Gram-negative bacteria at low concentrations (Recio and Visser 1999).

11.8.2.4   Bioactive Peptides and Nutrition

Peptides can sequester calcium and other metal ions and are called caseinophosphopeptides (CPPs). Caseinomacropeptide (CMP) also has several effects on the nutritional system which are discussed below.

Casein Phosphopeptides

The caseins are phosphoproteins insoluble at pH 4.6 and are considered in detail in Chap. 4. Bovine α_{s1}-, α_{s2}-, β- and κ-CNs contain 8–9, 10–13, 4–5 and 1–2 mol P/mol, respectively. α_{s1}-, α_{s2}- and β-CNs bind Ca^{2+} (and other ions) strongly, causing charge neutralization, aggregation and micelle formation. Human β-CN contains 0 to 5P, κ-casein 1P and human milk contains little or no α_s-CN (see Uniacke-Lowe et al. 2010). CPPs have been identified in tryptic hydrolytes of α_{s1}-, α_{s2}- and β-CN (Kitts 1994) and are resistant to proteolysis unless dephosphorylated. Residues 1–25 of β-CN and, to a lesser extent, residues 59–79 of α_{s1}-CN have been investigated extensively (for review see Wada and Lönnerdal 2014). They stimulate Ca absorption and have been proposed as sources of bioavailable Fe but their effectiveness in this regard has been queried (see Fitzgerald 1998, for review). Research on CPPs has mostly been conducted on bovine caseins, but Ferranti et al. (2004) found β-CN peptides in human milk and concluded that plasmin-like activity in human milk played an important role in initiating the release of CPP, although biological roles have yet to be elucidated.

CPPs have been found in the stomach and duodenum following ingestion of milk or from *in vivo* and/or *in vitro* digestion of α_{s1}-, α_{s2}- or β-CN (Kitts 1994). As well as metal-binding, CPPs are reported to exhibit cytomodulatory effects (Meisel and Fitzgerald 2003). Most CPPs have a sequence of three phosphoseryl residues followed by two glutamic acid residues (Meisel 1997). The high concentration of negative charges on such phosphate peptides makes them resistant to further hydrolysis which is discussed by several authors, including, Meisel and Schlimme (1990), Clare and Swaisgood (2000) and Fitzgerald (1998). The negatively charged phosphate groups are the binding sites for metals such as Ca, Mg, Fe, Zn and Cu as well as the trace elements, Ba, Cr, Ni, Co and Se. Absorption of Zn and Fe by the body can be enhanced when these elements are bound to CPPs (Silva and Malcata 2005).

Ca-binding phosphopeptides have been reported to exhibit anticarcinogenic activity by inhibiting dental caries through their ability to recalcify dental enamel (Clare and Swaisgood 2000). The addition of CPPs to toothpaste has been suggested as a means of preventing enamel demineralization (Reynolds 1997).

In addition to its Ca-binding properties, β-CN $f1$–28 (proteose peptone 8 fast) which is produced from β–CN by plasmin, disrupts tight-junction integrity; it accelerates mammary involution and drying off and has been commercialized as an agent to accelerate drying off in goats (Shamay et al. 2002) and cows (Shamay et al. 2003).

Dietary Ca has many bioactive functions, some of which have been mentioned earlier, and its possible role in weight management is receiving attention over the last decade. Calcitrophic hormones, parathyroid hormone and 1,25 $(OH)_2D$ (calcitriol) all respond to low Ca in the diet and exert coordinated regulatory effects on human adipocyte, lipogenic and lipolytic systems (Zemel 2004). High Ca diets have been shown to increase faecal fat excretion in rats (Jacobsen et al. 2003). Dietary Ca and its effects on weight management was reviewed by Zemel (2004, 2005).

Caseinomacropeptide

Whole κ-CN is thought to play a major role in preventing the adhesion of *Helicobacter pylori* to human gastric mucosa (Strömqvist et al. 1995). It is likely that heavily glycosylated κ-CN provides protection due to its carbohydrate content and in breast-feeding infants is thought to be important, especially as *H. pylori* infection occurs at an increasingly younger age (Lönnerdal 2003).

Chymosin hydrolyses the Phe_{105}-Met_{106} bond of κ-CN, leading to the formation of two fragments, the hydrophobic N-terminal fragment, $f1$–105, which remains attached to the casein micelles and is referred to as para-κ-casein, and the hydrophilic phosphorylated and glycosylated C-terminal fragment, $f106$–169, which is released into the milk serum and is referred to as the CMP (see Chap. 12). At least six genetic variants of κ-CN have been identified in bovine milk; A and B are the most common. Cleavage of CMP [by the hydrolytic action of endoproteinase GluC (*Staphylococcus aureus* Protease 8)] results in the formation of kappacin, the non-glycosylated form of CMP, which has bactericidal properties (Malkoski et al. 2001). CMP and its derivatives also have immunomodulatory properties (Meisel 1997), can inhibit the binding of *Cholera* toxin (Kawasaki et al. 1992), may depress platelet aggregation (Boman 1991) and can inhibit hemagglutination caused by the influenza virus (Kawasaki et al. 1993). Furthermore, these peptides inhibit the adherence of *Streptococcus mutans* (Neeser et al. 1994; Strub et al. 1996; Vacca-Smith et al. 1994; Malkoski et al. 2001), which causes the development of dental caries and growth of Gram-negative bacteria such as *Porphyromonas gingivals* and *Escherichia coli* (Brody 2000; Malkoski et al. 2001; Rhoades et al. 2005; Haque and Chand 2008).

The bioactive potency of human and bovine CMP is different and may be explained by the fact that the human peptide is more highly glycosylated; human CMP contains ~55 % carbohydrate while that of bovine is only ~10 % (Fiat and Jollès 1989). The formation and absorption of CMP in the GIT has been shown in human studies and CMP has been detected in the plasma of breast-fed and formula-fed infants (Chabance et al. 1995).

Physiologically, the release of CMP in the mammalian stomach is important. CMP inhibits acid gastric secretions and gastrin activity and has been found in blood plasma (Yvon et al. 1994; Chabance et al. 1995, 1998; Fosset et al. 2002). CMP is the only peptide released during the first hour after ingestion of milk by the calf, with fragments $f165$–199 of α_{s1}- and $f193$–209 of β-CN being released within ~90 min (Yvon and Pelissier 1987). The release of CMP in the stomach increases the efficiency of the digestive process, promotes the growth of bifidobacteria and controls acid secretion (Stan and Chernikov 1982) while preventing neonatal hyper-sensitivity to ingested proteins and inhibiting gastric pathogens (Rhoades et al. 2005). Of the many physiological functions attributed to CMP, protection against toxins, bacteria and viruses and immunomodulation have been reported to be the most promising applications (Brody 2000). Supplementation of infant formula with CMP has been suggested as a means of promoting the growth of host-friendly colonic bacteria, especially bifidobacteria or lactobacilli which may help overcome or prevent some enteric infections (Brück et al. 2003).

11.9 Free Amino Acids

The free amino acid content of bovine and human milk is 578 and 3,019 μmol.L^{-1}, respectively (Rassin et al. 1978; Agostini et al. 2000). Glutamine, glutamate, gly-cine, alanine and serine are the most abundant free amino acids in bovine and human milk; taurine also is exceptionally high in human milk (Rassin et al. 1978; Sarwar et al. 1998; Carratù et al. 2003). Taurine is an essential metabolite for the human infant and may be involved in the structure and function of retinal photoreceptors (Agostini et al. 2000). In contrast to total amino acid composition, which is essentially similar in bovine and human milks, free amino acids show a pattern characteristic of each species which may be important for early post-natal development in different animals. Free amino acids are more easily absorbed than protein-derived amino acids and glutamic acid and glutamine, which comprise >50 % of the total free amino acids of human milk, are a source of α-ketoglutaric acid for the citric acid cycle and also act as neurotransmitters in the brain (Levy 1998; Agostini et al. 2000). Free amino acids are discussed in Chap. 4.

11.10   Hormones, Growth Factors and Cytokines

Milk contains many protein hormones, including several from the anterior pituitary gland (e.g., prolactin and somatotropin), from the hypothalamus (e.g., somatotropin-releasing hormone and somatostatin) and the GIT (e.g., vasoactive intestinal peptide, gastrin and substance P). In addition, milk contains many growth factors and a variety of other bioactive peptides, including IGFs I and II, IGF-binding proteins, epidermal growth factor (EGF), insulin and TGF-β, prostaglandin $F_2\alpha$ and E and Lf (Campana and Baumrucker 1995; Xu et al. 2000). Colostrum has much higher concentrations of hormones and growth factors than mature milk.

Steroidic hormones of gonadal and adrenal origin were first reported in bovine milk in the mid-1950s (Jouan et al. 2006). In the late 1970s and early 1980s, milk hormones were thought to have originated from endocrine hormones circulating in the body but, later research indicated that many bioactive compounds in milk originated from mammary tissue and are secreted into milk and are, in turn, able to regulate mammary cell proliferation and differentiation through autocrine or paracrine action. Hormones and growth factors from the maternal circulatory system may be transferred into milk *via* active transport or 'leaky' junctions of mammalian epithelial cells.

In bovine milk, the concentration of many hormones and growth factors is much greater than levels found in maternal plasma; oestrogen, gonadotropin-releasing hormone (GnRH), somatostatin, parathyroid hormone-related peptide, prolactin, insulin and insulin-like growth factor are all found in much higher concentrations in milk than in plasma. Evidence now suggests that these bioactive compounds play very significant roles in the growth of the neonate, development and maturation of its GIT and immune systems and play a significant role in endocrine and metabolic functions (Jouan et al. 2006). Table 11.8 summarizes the hormones and growth factors identified in bovine and human milk.

The protective role of breast milk may be due not only to its nutritional composition (e.g., low protein concentration) but also to the presence of several bioactive substances called adipokines which are involved in the development of many important physiological functions (Lönnerdal 2003). These hormones are involved in food intake regulation, energy balance and glucose homeostasis and act through neuroendocrine circuits between the hypothalamus and peripheral tissues (Bouret 2009).

Table 11.8 Hormones and growth factors identified in bovine and human milk

Mature bovine milk

Adrenal Gland	Gonadal	Hypothalmus	Thyroid and parathyroid	Pituitary	Gastro-intestinal	Growth factors
Corticosteroids (glucocorticoids and metal corticoids)	5-α Androstane-3, 17-dione	Lutenizing hormone-releasing hormone (LHRH)	Parathyroid hormone-related peptide (PTHrP)	Growth hormone (GH) or bovine somatotropin (bST)	Bombesin	Insulin-like growth factors (IGFs) (e.g., insulin, IGF-I, IGF-II and relaxin
	Testosterone	Gonadotropin hormone-releasing hormone (GnRH)	Thyroxin (T3 and T4)	Prolactin	Gastrin	IGF-binding proteins
	Estradiol 17-α	Somatostatin (SS)			Gastrin-releasing hormone	Mammary-derived growth inhibitor (MDGI)
	Estradiol 17-β (E2)	Thyrotropin-releasing hormone (TRH)			Neurotensin	Tissue plasminogen activator (tPA)
	Estriol (E3)	Gonadotropin-releasing hormone-associated peptide (GAP)				Transforming growth factor (TGF-α and TGF-β1,2)
	Estrone (E1)					Epithelial growth factor (EGF)
	Estrogen					Betacellulin
	Progesterone					Fibroblast growth factor (FGF)
						Platelet-derived growth factor

(continued)

Table 11.8 (continued)

Adrenal Gland	Gonadal	Hypothalmus	Thyroid and parathyroid	Pituitary	Gastro-intestinal	Growth factors
Mature human milk						
Cortisol	Progesterone	Gonadotropin hormone-releasing hormone (GnRH)	Parathyroid hormone-related peptide (PTHrP)	Growth hormone	Gastrin	Epithelial growth factor (EGF)
	Pregnane-3(α)20(β)-diol	Growth hormone-releasing factor (GRF)	Thyroxine	Prolactin	Gastric inhibitory polypeptide (GIP)	Insulin
	Estrogens	Somatostatin (SS)	Parathyroid hormone	Thyroid-stimulating hormone (TSH)	Gastrin-releasing peptide (GRP)	IGF-I
	Contraceptives	Vasoactive intestinal peptide (VIP)	Calcitonin-like		Neurotensin	Nerve growth factor (NGF)
		Thyrotropin-releasing hormone (TRH)	Triiodothyronine			Transforming growth factor (TGF-α)
					Peptide histidine methionine (PHM)	
			Reverse Triiodothyronine		Peptide YY (PYY) or Peptide tyrosine tyrosine	Other growth factors
					Vasoactive intestinal peptide (VIP)	

Adapted from Koldovský and Štrbák (1995), Campana and Baumrucker (1995) and Pouliot and Gauthier (2006)

Breast milk contains several hormones such as leptin, ghrelin and adiponectin as well as growth factors that are absent from infant formulae (Savino et al. 2009a).

11.10.1   Gonadal Hormones

The principal gonadal hormones in milk are estrogens, progesterone and androgens (Table 11.8). About 65 % of estradiol 17-β and 80 % of estrone are found in the milk fat fraction (Wolford and Argoudelis 1979). The estrone is the predominant estrogen in milk (Jouan et al. 2006) and, in general, the estrogen level is higher in milk than in blood, suggesting uptake of these hormones by the mammary gland. Progesterone is absent from colostrum but has been found in milk about 15 days *post partum* (Darling et al. 1974) with a much higher level in cream than in skim milk and higher concentrations in milk than in blood plasma (Heap et al. 1973).

The androgen, 5-α androstane-3, 17-dione, has been isolated from milk (Darling et al. 1974) and may be involved in the development of milk secretion although it is not generally present in colostrum, except immediately *post-partum*.

11.10.2   Adrenal Hormones

Human milk and colostrum contain two corticosteroid-binding proteins (see Table 11.8); similar proteins occur in blood. The function and significance of these proteins in milk are unknown. Cortisol and corticosterone are the main glucocorticoids found in the blood plasma of cows and during lactation their concentrations in milk are only ~4 % those in blood plasma (Tucker and Schwalm 1977). Glucocorticoid receptors have been demonstrated in the mammary gland, implying that these hormones may act in conjunction with other hormones to maintain lactation (Jouan et al. 2006). In bovine milk, corticosterone levels are greater than those of cortisol whereas in plasma the reverse is true, suggesting enhanced mammary gland generation of androgens with 19 carbons atoms.

11.10.3   Brain-Gut Hormones

11.10.3.1   Hypothalmic Hormones

Gonadotropin-releasing hormone (GnRH) was first detected in bovine milk by Baram et al. (1977) and has similar biological activity as the hypothalamic hormone. The concentration of GnRH in bovine milk exceeds that in plasma by at least fivefold. Gonadotropin-releasing hormone-associated peptide (GAP) has been detected in bovine colostrum (Zhang et al. 1990) and is believed to be a precursor

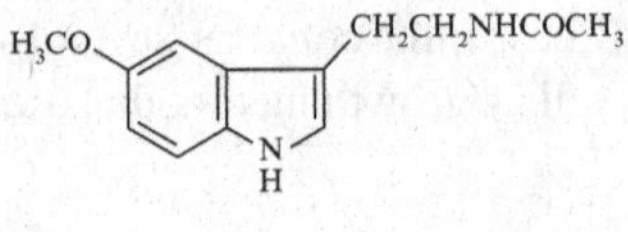

Fig. 11.6 Structure of
melatonin

of GnRH. Evidence suggests that GAP may be synthesised by the mammary gland (Zhang et al. 1990).

The concentration of TRH (thyrotropin-releasing) and LHRH (luteinizing hormone-releasing hormone) in bovine milk far exceeds the level found in maternal serum. However, evidence suggests that the TRH gene is not expressed in mammary tissue and, as yet, the origin of both these hormones in milk is unknown. LHRH is biologically active in milk and is absorbed intact and in an active form by the neonatal intestine from where it may be absorbed and used to stimulate secretion of pituitary gonadotropins. TRH hormone has been detected in both bovine milk and colostrum (Amarant et al. 1982) and is also absorbed from the neonate's intestine. Immunoreactive somatostatin (also called growth hormone-inhibiting hormone, GHIH), a hormone which regulates the endocrine system, has been found in the milk of many species.

Melatonin

Melatonin (N-acetyl-5-methoxytryptamine) is a sleep-inducing hormone which is present in milk in which its concentration follows diurnal rhythm, being highest at night. It is a biogenic amine (Fig. 11.6) that is found in animals and plants. In mammals, melatonin is produced by the pineal gland; melatonin is implicated in the regulation of sleep, mood, and reproduction. Melatonin is also an effective antioxidant.

Melatonin has been found in human, bovine and caprine milk. Some farmers produce melatonin-rich milk from cows milked in the middle of the night; it sells for a premium price (Eriksson et al. 1998; Valtonen et al. 2003); for review see Singh et al. (2011).

11.10.3.2   Pituitary Hormones

The principal pituitary hormones found in milk are growth hormone (GH), prolactin (PL) and prostaglandins (PG). GH or bovine somatotrophin, bST, has attracted much interest and has been approved by the USFDA for use to increase milk volume from cows. GH was first detected in milk in the late 1980s and is believed to act in the mammary gland through specific receptors (Jouan et al. 2006).

The level of PL in milk fluctuates seasonally and a significant proportion is associated with the milk fat globule. The level of PL in colostrum is much higher than that of milk and it probably originates from blood plasma (Jouan et al. 2006). PL exhibits several different patterns of immunoreactivity and biological activity.

Prostaglandins (E and F series; PG, PG_2, $PG\alpha$ and $PGF\alpha$) have been identified in bovine milk although the source is uncertain. The mammary gland synthesizes PG_2

and macrophages in milk may also synthesize and secrete prostaglandins (Campana and Baumrucker 1995).

11.10.3.3   Gut Hormones

The gut hormones, bombesin (gastrin-releasing peptide) and neurotensin have been found in bovine milk and occur at higher concentrations than in blood serum. Satiety, blood sugar level, gut acidity and the gastrointestinal concentrations of several hormones are influenced by bombesin. The levels of both gastrin and bombesin in bovine milk are greatest in *pre partum* secretions and decline significantly within a week *post partum*. Bombesin has been found in human milk, milk powder, whey and also in porcine milk (Koldovský 1989).

11.10.3.4   Thyroid and Parathyroid Hormones

Milk contains a thyroxine-binding protein (~0.3 mg mL^{-1}), the function of which is unknown. Triodothyroxine has also been reported in bovine milk. The level of thyroxine in bovine milk is very low. Many studies have reported the presence of parathyroid hormone-related protein (PTHrP) in milk. Similar to many other hormones, PTHrP is at a higher level in bovine milk than in maternal blood, although the level in milk varies considerably, depending on the breed of cow. Milk PTHrP level correlates positively with total milk calcium, which suggests that this hormone plays a role in mammary calcium transport from the blood to milk although its exact physiological function has not been established. PTHrP is relatively heat stable and is not affected by pasteurization (Rathcliffe et al. 1990).

Calcitonin has been detected in human milk (Koldovský 1989) and is believed to inhibit the liberation of prolactin (Jouan et al. 2006).

11.10.4   Growth Factors

Mammalian milk contains many growth factors, hormones and cytokines and the exact distinction between them is unclear as they are all involved in cell proliferation and differentiation (Pouliot and Gauthier 2006). The term 'growth factor' is applied to a group of potent hormone-like polypeptides which play a critical role in the regulation and differentiation of a variety of cells acting through cell membrane receptors. Growth factors may be transferred from mammary gland tissue directly into milk in their active or modified (glycosylated or phosphorylated form) or may occur complexed to other factors (Pouliot and Gauthier 2006). The milk and, especially, colostrum of several species contain several growth factors, including insulin-like growth factors (IGF1, IFG2), transforming growth factors (TGF$_{a1}$, TGF$_{a2}$, TGFβ), mammary-derived growth factors (MDGF I, MDGF 11), some epithelial growth factors (EGF), for example betacellulin (BTC), basic fibroblast growth factors (bFGF), fibroblast growth factor (FGF)-2, platelet-derived growth factor (PDGF), betacellulin and bombasin. Quantitatively, the relative concentrations of growth factors in milk are IGF-I > TGF-β2 > EGF $\approx$ IGF-II > bFGF. The concentrations of growth factors are highest in colostrum and gradually decrease through lactation, except for BTC, the concentration of which is equivalent in both colostrum and mature milk. The principal growth factors in milk were reviewed in detail by Gauthier et al. (2006). EGF and BTC stimulate proliferation of epidermal, epithelial and embryonic cells, inhibit the secretion of gastric acid and promote wound healing and bone reabsorption while TGF-βs are important in embryogenesis, tissue repair and control of immunity (Pouliot and Gauthier 2006). IGF-I and II are important for cell proliferation in general, while the former is also important for glucose uptake and glycogen synthesis. The role of EGF, IGFs-I and II, insulin and TGF-β in stimulating GI tissue growth and repair in the suckling neonate were discussed by Xu et. al. (2000). PDGF growth factors are involved in embryonic development and proliferation of many cell types. FGF is important for the proliferation and differentiation of epithelial, endothelial and fibroblast cell, promote collagen synthesis and are involved in angiogenesis and wound healing (Pouliot and Gauthier 2006).

The source of these polypeptides may be blood plasma, mammary gland or both. The biological significance of these growth-promoting activities in colostrum and mature milk is not clear. In terms of possible physiological significance, two potential targets may be considered, i.e., the mammary gland or the neonate. In general, most attention has focussed on the latter. It is not known whether the factors in milk that have the capacity to promote cell proliferation (1) influence growth of mammary tissue, (2) promote the growth of cells within the intestine of the recipient neonate, or (3) are absorbed in a biologically active form and exert an effect on enteric or other target organs.

Bovine mammary secretions contain many compounds that stimulate the growth of cells in cultures. These include:

1. IGFs, part of the insulin family which includes insulin, IGF-I, IGF-II and relaxin. IGFs act as mediators of growth, development and differentiation. They are heat and acid-stable and hence have potential bioactivity in the GIT of consumers.
2. Insulin occurs in bovine milk and colostrum at much higher concentrations than in blood and higher in turn in *pre partum* secretions than in those *post partum* (Malven 1977).
3. Transforming growth factors (TGF α and β), are present in bovine milk. TGF-β is important for cell proliferation and differentiation.

11.10.4.1   Epidermal Growth Factors

Epidermal growth factor (EGF) and heparin-binding EGF-like growth factor (HB-EGF) are members of the family of EGF-related peptides found in human milk and the milk of many other species but are not found in significant amounts in bovine milk (Playford et al. 2000). A common feature of EGF growth factors is that they are synthesized as larger trans-membrane precursor molecules that can be cleaved proteolytically to release the soluble form of the growth factor or they can function as membrane-anchored growth factors in juxtacrine signalling. All EGFs are characterized by extensive sequence similarity, including a six-cysteine consensus motif that forms three intra-molecular disulphide bonds and a core arginine residue that stabilizes protein orientation (Dunbar et al. 1999)

Human milk-borne EGF (also called urogastrone) is a 53-amino acid peptide produced by the salivary glands and Brunner's glands of the adult duodenum and is present in both human colostrum (200 μg L^{-1}) and milk (30–50 μg L^{-1}) (Playford et al. 2000). EGF conveys important regulatory signals to developing infants such as timing of eyelid opening, tooth eruption and development of intestinal, hepatic, pancreatic and lung systems (Donovan and Odle 1994). EGFs are heat stable and resistant to degradation in the GIT and retain their bioactivity in the neonatal intestine. In the early post-natal period EGFs from breast milk are crucial for the development and maturation of intestinal mucosa, possibly by interacting with EGF receptors in the neonatal small intestine (Lönnerdal 2003). EGF, administered as an oral physiological dose, has been shown to reduce the incidence and severity of necrotizing enterocolitis, a disease affecting premature infants (Dvorak 2010). In adults, EGF may be beneficial during recovery from gastrointestinal trauma (Donovan and Odle 1994; Dvorak 2010).

HB-EGF is found in amniotic fluid and breast milk. The concentration of HB-EGF is 1,000–10,000 times lower than EGF but it is also reported to be an effective treatment for necrotizing enterocolitis, although pharmacological doses are required (Dvorak 2010). HB-EGF may provide protection against injury in the small intestine of adult mammals (Pillai et al. 1998)

Betacellulin (BTC), a member of the epidermal growth factor (EGF) family of peptide growth factors, has been identified in human milk and it has been suggested that it has a major role in the growth and development of the neonatal GIT (Dunbar et al. 1999). The subject of growth factor polypeptides in colostrum and milk was reviewed by Playford et al. (2000) while, EGF and EGF-related peptides were reviewed by Barnard et al. (1995).

11.10.4.2   Growth Inhibitors

There is considerable interest at present in growth inhibitors of mammary tissue proliferation due to their potential in the treatment of breast cancer (Nevo et al. 2010). A 13 kDa polypeptide, called mammary-derived growth inhibitor, MDGI (also called FABP-3 or H-FABP), has been purified from bovine mammary tissue and from the MFGM (Skelwagen *et al* 1994).

11.10.4.3 Minor Growth Factors

Tissue plasminogen activator (tPA) which has been identified in bovine milk associated with casein micelles (Heegaard et al. 1994) and may be a potential mammary trophic factor involved in tissue remodelling. tPA catalyzes the conversion of plasminogen to plasmin. Both plasminogen and plasmin are present in bovine milk and it is believed that they originate primarily from white blood cells.

11.10.5 Cytokines

Cytokines are a broad and loose category of small proteins (~5 to 20 kDa) that are important in cell signalling. Cytokines include chemokines, interferons, interleukins, lymphokines, tumour necrosis factor but generally not hormones or growth factors (despite some overlap of terminology). They are different from hormones, which are also important cell-signalling molecules, in that hormones circulate at much lower concentrations and tend to be produced by specific kinds of cells. Cytokines are important in health and disease, specifically in host responses to infection, immune responses, inflammation, trauma, sepsis, cancer and reproduction.

In human milk, the principal cytokines that have been identified thus far are tumour necrosis factor α, transforming growth factor β, colony-stimulating factors and interleukins IL β, IL-6, IL-8 and IL-10, all of which are immunomodulatory and some are anti-inflammatory. Most cytokines are found in free form in milk and some can be released from specific cells

11.10.5.1 Colony-Stimulating Factors

Colony-stimulating factors (CSFs) are secreted cytokines which survive digestion and are known to augment neonatal defenses against microorganisms. CSFs also function by binding to receptor proteins on the surfaces of hemopoietic stem cells, activating intracellular signalling pathways that can cause the cells to proliferate and differentiate into a specific kind of blood cell (usually white blood cells). Thus, CSFs regulate cell proliferation and the various paths to cell differentiation during hemopoiesis. Three CSFs are known;

1. CSF1, macrophage colony-stimulating factor
2. CSF2, granulocyte macrophage colony-stimulating factor (GM-CSF or sargramostim)
3. CSF3, granulocyte colony-stimulating factor (G-CSF or Filgrastim) and all three have been identified in human milk, with the level of G-CSF being particularly high in the first 2 days *post partum* (Calhoun et al. 2000).

11.10.5.2 Erythropoietin

The glycoprotein hormone, erythropoietin (EPO), is a cytokine, produced by the kidney, which controls erythropoiesis (red blood cell production) and has been found in human milk (Grosvenor et al. 1993); its presence in bovine milk has not been established.

11.10.6 Adipokins

Adipokins are cytokines (immunomodulating agents) secreted by adipose tissue; however there is currently no clear distinction between cytokines and hormones. Adiponectin, leptin ghrelin and resistin, all of which have been identified in milk, are generally not considered to be cytokines as they do not act directly on the immune system. They are often referred to as adipokins but should be more accurately put into the larger and ever-growing list of adipose-derived hormones. The protective role of breast milk may be due not only to its nutritional composition (e.g., low protein concentration) but also to the presence of adipokins which are involved in the development of many important physiological functions (Lönnerdal 2003). Adipokins are involved in regulating food intake, energy balance and glucose homeostasis and act through neuroendocrine circuits between the hypothalamus and peripheral tissues (Bouret 2009).

11.10.6.1 Leptin

Leptin is a protein hormone of ~16 kDa and 167 amino acids which was discovered in human milk (Casabiell et al. 1997; Houseknecht et al. 1997; Smith-Kirwin et al. 1998; Uçar et al. 2000), it plays a key role in regulating energy intake and energy expenditure, including regulation of appetite and metabolism, as well as functioning in mammary cell proliferation, differentiation and apoptosis (Marchbank and Playford 2014). Leptin is an anorexigenic hormone and acts on receptors in the hypothalamus, inhibiting appetite by counteracting the feeding stimulators, neuropeptide Y and anandamide, while also stimulating the synthesis of α-melanocyte-stimulating hormones which suppress appetite (Marchbank and Playford 2014). The concentration of leptin in breast milk correlates well with maternal circulating leptin levels (Casabiell et al. 1997; Houseknecht et al. 1997), maternal body mass index and adiposity (Houseknecht et al. 1997; Uysal et al. 2002). Breast-fed infants have a higher serum leptin level than formula-fed infants (Savino et al. 2002) and the serum leptin level in breast-fed infants is positively correlated with the leptin level in maternal milk (Uçar et al. 2000; Ilcol et al. 2006; Schuster et al. 2011), suggesting that leptin in breast milk may be important for normal growth and development with both short- and long-term effects (Locke 2002; Agostoni 2005). Milk leptin may influence an infant's weight gain during the early stages of lactation and sufficient maternal leptin provides some protection to infants against excessive weight gain (Miralles et al. 2006). More recent studies have suggested that breast-fed infants, with lower weight gain in the first few months of life, have a significantly lower risk of obesity in childhood and adulthood than formula-fed infants (Gillman 2010; Taveras et al. 2011).

Mature bovine milk contains ~6.14 µg L^{-1} of leptin, 56 % less than the concentration, 13.90 µg L^{-1}, in colostrum and in both cases the level of leptin is positively correlated with fat and choline phospholipid concentrations (Pinotti and Rosi 2006). Mature human milk contains a wide range of leptin concentrations, 0.11–4.97 µg L^{-1}, while colostrum contains 0.16–7.0 µg L^{-1} (Ilcol et al. 2006). Human-like leptin has been isolated from equine milk at a level of 3.2–5.4 µg L^{-1}, which is similar to the levels reported for other mammals and showed little variation throughout lactation (Salimei et al. 2002).

11.10.6.2 Ghrelin

Ghrelin is a 28 amino acid peptide produced mainly in the stomach and its main function is thought to be the stimulation of growth hormone (GH) secretion. Ghrelin is also produced and secreted by the mammary gland and the level in breast milk is higher than that in blood plasma (Savino et al. 2009a). Ghrelin acts on the hypothalamus and stimulates food intake in rats and humans while also exerting adipogenic activity and is involved in long-term regulation of body weight (Cummings 2006). As ghrelin is involved in the short-term regulation of food intake by stimulating appetite and in long-term regulation of body weight by inducing adiposity its presence in breast milk may be an important factor through which breast-feeding influences infant feeding behaviour and body composition in later life (Savino et al. 2009b).

11.10.6.3 Resistin

A recently discovered adipokine is the hormone resistin, first identified in rodents (Steppan et al. 2001) which may link obesity and diabetes and contribute to insulin resistance *in vivo* although this hypothesis is supported only by *in vitro* studies (Bouret 2009). It is also speculated that resistin is a feedback regulator of adipogenesis and a signal to restrict adipose tissue formation (Stocker and Cawthorne 2008). Resistin has been identified and studied in human milk (Ilcol et al. 2008; Savino et al. 2012). Several studies showed a positive correlation between serum resistin and leptin levels in infants (Ng et al. 2004; Marinoni et al. 2010; Savino et al. 2012). It has been postulated that both hormones may play a role in modulating energy homeostasis and growth *in utero* (Ng et al. 2004; Marinoni et al. 2010).

11.10.6.4   Adiponectin

Adiponectin is the most abundant adipose-specific protein hormone and its presence was first detected in human milk in 2006 by Martin et al. (2006). The adiponectin level is high in human serum and its level is inversely related to the degree of adiposity and positively associated with insulin sensitivity; plasma level of adiponectin is reduced in individuals with obesity and type-II diabetes (Savino et al. 2009b). Currently, research is focussed on establishing whether exposure to adipokine in infancy determines the weight status of individuals in later life.

11.10.6.5   Obestatin

Obestatin is a relatively new bioactive compound discovered in human milk. It is a 23 amino acid peptide derived from pre-proghrelin, the ghrelin peptide precursor and is produced by the stomach, small intestine and salivary glands (Ozbay et al. 2008). Obestatin levels in colostrum and mature milk have been reported to be twice those in blood plasma (Aydin et al. 2008). It is unclear where obestatin originates nor is it clear what its exact function is; it has been reported that it reduces food intake and body weight gain while assisting gastric emptying and suppressing intestinal motility (Tang et al. 2008).

11.11   Minor Bioactive Compounds

11.11.1   Polyamines

Polyamines are small, organic aliphatic polycationic molecules found ubiquitously in all organisms and function in a wide variety of biological processes. Polyamines have variable hydrocarbon chains and two or more primary amino groups. Putrescine, a diamine, spermidine, a triamine and spermine, a tetraamine, are all associated with cellular growth and differentiation and are found at relatively high levels in human milk and that of other mammals. Polyamines are involved in various growth-related processes, including carcinogenesis, regulation and stimulation of DNA, RNA and protein synthesis, modulation of membrane function, stimulation of cell differentiation, modulation of intracellular messengers and acceleration of intestinal proliferation and maturation of biogenic amines (Kalač and Krausová 2005). The potential role of polyamines in the neonatal digestive system and maintenance of normal growth and general properties of the adult GIT was reviewed by Deloyer et al. (2001). Insufficient polyamine uptake may play a role in the induction of sensitization to dietary allergens (Kalač and Krausová 2005). The role of polyamines in human cell growth and proliferation has implicated them in the growth of many types of tumours and high levels of polyamines have been found in rapidly

Table 11.9 Concentration (nmol/dl) of polyamines in human and bovine milk (nmol/dl) on selected days *post partum*

	Putrescine	Spermidine	Spermine
Human milk			
(Day 7)	33.8	224.4	276.2
(Day 7)	129±21	711±109	663 ± 136
(Day 7)	24±3.5	220±20	313 ± 16
(Day 16)	77	454	376
(Day 5)		11.6±3	6.8 ± 1.7
Bovine milk			
(Day 30)		470±280	~ 400
(Day 28)		19.8 ± 0.7	8.4 ± 3
Full cream milk	100	100–300	100–300

Adapted from Löser (2000)

dividing cells and tissues (Thomas and Thomas 2003). One proposed beneficial effect of polyamines is their use in wound healing of post-operative patients. As well as intracellular polyamine *de novo* synthesis, uptake of extracellular polyamines from the gut lumen is very important for regulation of polyamine metabolism in the body (Löser 2000). Data on the concentration of polyamines in human and bovine milk are shown in Table 11.9. Human milk contains high levels of spermine and spermadine and a lower level of putrescine. Concentrations of all polyamines in human milk increases steadily for the first 2 weeks *post partum* and declines thereafter (Löser 2000).

Bovine milk has lower levels of polyamines than human milk (Table 11.9) due to the high rate of polyamine degradation by diamine oxidase and polyamine oxidase which are present at much higher levels than those in human milk (Löser 2000).

11.11.2   Amyloid A

Amyloid A3 (AA3) is a protein produced in the mammary gland and is encoded by a separate gene from that for serum amyloid A (serum AA) (Duggan et al. 2008). AA3 is believed to prevent attachment of pathogenic bacteria to the intestinal cell wall (Mack et al. 2003) and may prevent necrotizing enterocolitis in human infants (Larson et al. 2003). McDonald et al. (2001) demonstrated the presence of AA3 in the colostrum of cows, ewes, sows and horses. Bovine colostrum has a high concentration of AA3 but by ~3 days *post partum* the level declines. The presence of serum AA in bovine milk is an indicator of mastitic infection (Winter et al. 2006). In equine colostrum, the concentration of AA3 is considerably lower than in milk and consequently it may play a crucial role in protection of intestinal cells in the foal, especially after gut closure (Duggan et al. 2008).

11.11.3 Nucleotides

Nucleotides are organic molecules that serve as subunits of nucleic acids, e.g., DNA and RNA. Nucleotides serve to carry packets of energy within cells in the form of the nucleoside triphosphates (ATP, GTP, CTP and UTP) which have a central role in metabolism. Human milk has a significant concentration of nucleotides and their metabolic products which are naturally present and essential for rapidly dividing tissues such as intestinal epithelium and lymphoid cells. Nucleotides are also important for the immune system. Nucleotides may be obtained from the diet or synthesized *de novo* from amino acid precursors, both processes requiring significant amounts of energy; however, free nucleotides in human milk contribute as much as 25 % of the infants daily needs (Hendricks and Guo 2014).

11.11.4 Calmodulin-Inhibiting Peptide

Calmodulin (CaM) is an abbreviation for calcium-modulated protein and is a calcium-binding messenger expressed in all eukaryotic cells. CaM mediates processes such as inflammation, metabolism, short- and long-term memory and smooth muscle contraction. Many proteins that bind CaM are unable to bind calcium themselves. Peptides that inhibit CaM-dependent cyclic nucleotide phosphodiesterase have been isolated from peptic digests of α_{s1}-CN (α_{s1} plus α_{s2}) and identified as α_{s2}-CN f164–179, α_{s2}-CN fl83–206 and α_{s2}-CN fl83–207. The affinity of these peptides for CaM is comparable to the affinities of some endogenous neurohormones and proteins with CaM (Kizavva et al. 1995). The physiological significance of these peptides in milk is unknown (see Aluko 2010).

11.11.5 Cluster of Differentiation 14 (CD14)

CD14 is a glycosyl-phosphatidyl-inositol anchored membrane protein expressed in mature monocytes and functions as a co-receptor for bacterial liposaccharide (LPS) and triggers induction of inflammatory responses (Filipp et al. 2001). CD14 plays a pivotal role in the recognition of, and cell activation induced by, cell wall components of Gram-negative and Gram-positive bacteria as well as mycobacteria (Labéta et al. 2000) and is one of the best characterized bacterial pattern-recognition receptors (LPS receptor in this case). CD14 occurs in two forms, a membrane anchored form (mCD14) and in a soluble form (sCD14). sCD14 is found in amniotic fluid as well as in breast milk and exposure to reduced levels of sCD14 in the foetal or neonatal GIT is reported to be associated with the development of atopy (genetically mediated predisposition to excessive IgE reaction), eczema or both (Jones et al. 2002).

The concentration of sCD14 is 20-fold higher in breast milk than in blood serum (Labéta et al. 2000). It has been postulated that sCD14 in human milk plays a sentinel role during bacterial colonization of the neonatal gut and thus contributes to the innate immune mechanisms controlling gut homeostasis in the neonate (Labéta et al. 2000; Vidal et al. 2001; Oriquat et al. 2011).

11.11.6 Cysteine Protease Inhibitors

A 12 kDa cysteine protease inhibitor (CPI) has been purified from bovine milk protein and identified as bovine cystatin C (Matsuoka et al. 2002). CPIs are associated with bactericidal activity and protection from bone resorption, a process whereby osteoclasts secrete proteases to digest bone matrix proteins, such as collagen (Drake et al. 1996).

Both Lf and β-CN have some cysteine protease inhibiting ability (Ohashi et al. 2003).

11.11.7 Antioxidants and Prooxidants

Lipid oxidation causes major problems in the dairy industry: off-flavour, toxic effects and loss of PUFAs (nutritional). Milk contains several antioxidants, especially metal-binding proteins, tocopherols (vitamin E), carotenoids, ascorbic acid (vitamin C at low concentrations), sulphydryl groups in heated proteins, superoxidase dismutase, glutathione peroxidase, Lf and serotransferrin. Prooxidants include xanthine oxidoreductase, sulphydryl oxidase, polyvalent metals (Fe, Cu), vitamin C-metal complexes and denatured LPO and catalase. Some casein-derived peptides have been reported to have antioxidant activity, e.g., human β-CN ƒ154–160 (Hernández-Ledesma et al. 2007).

11.12   Effect of Processing Conditions on Bioactive Components in Milk

Food processing, storage conditions and physiological events dramatically affect food composition and bioactivity. To ensure safety and prolong shelf-life, raw milk is exposed to various processes including, heat treatments and homogenization which profoundly affect its physico-chemical properties and may, in turn, affect the bioactivity of many of the compounds discussed earlier. Scientific knowledge concerning the influence of processing and isolation procedures on the bioactive components in milk is very limited.

Heat treatments, centrifugation, churning and homogenization affect the nutritional and functional properties of the MFGM and alter its composition with loss of phospholipids and adsorption of caseins and whey proteins onto the membrane surface (Michalski and Januel 2006; Michalski 2007; Gallier et al. 2010). Homogenisation reduces fat globule size (Walstra 2003) although the effect of this structural change on the bioactivity of MFGM components is unknown.

Thermal treatment of Igs affects their unfolding and biological activity (Lindstrom et al. 1994; Li-Chan et al. 1995; Mainer et al. 1999). Heating at 72 °C for 15 s results in 10–30 % loss of Ig activity while UHT treatment (138 °C, 4 s) and evaporation destroy most Igs in milk (Lindstrom et al. 1994).

The effects of heat treatments and freeze-drying on the concentrations of Igs, TGF-β_2, IGF-I and GH in bovine colostrum were analysed by Elfstrand et al. (2002). The concentration of Igs decreased by 75 % while IGF-I and TGF-β_2 were unaffected by processing conditions; IgM was the most sensitive to thermal processing and freeze-drying and when filtration steps were included, IGF-I and TGF--β_2 concentrations were reduced by 25 % (Elfstrand et al. 2002).

Standard pasteurization temperature (72 °C, 15 s) has little effect on Lf structure, antibacterial activity or bacterial interaction. Preheating milk (70 °C, 3 min) followed by UHT treatment (130 °C, 2 s) causes only a 3 % reduction in residual iron-binding capacity; however, UHT processing prevents iron-saturated Lf from binding to bacteria and inhibits the bacteriostatic activity of iron-depleted Lf (apo-Lf, Pihlanto and Korhonen 2003).

Aspects of the stability of caseins and whey proteins to heat treatment are discussed in Chaps. 4 and 9. The effects of processing on β-Lg and α-La in bovine milk and milk products were reviewed by Chatterton et al. (2006).

Storage of human milk is a critically important issue as the milk is often banked for use in hospital neonatal units when breast-feeding is not an option. The effects of storage of human milk on its constituents and in turn, on its immunological properties, were investigated by Lawrence (1999). In his study, up to 72 h at 4 °C had little effect on immunological properties but freezing destroyed cellular activity and reduced the levels of vitamins B_6 and C. Boiling destroyed lipase activity and reduced the effectiveness of IgA and secretory IgA.

Nowadays, a minimal processing concept is recommended to optimize and maintain the beneficial properties of milk-based foods while ensuring delivery of the bioactive component(s) to target sites in the body (Korhonen 2002). Several novel processing techniques are in use currently, e.g., membrane separation, supercritical fluid extraction and high hydrostatic pressure.

High hydrostatic pressure processing of milk products as a means of preservation seems to be a promising technique to retain the bioactivity of components while retaining organoleptic characteristics and preventing the Maillard reaction (Heremans et al. 1997). The subject was reviewed Naik et al. (2013).

11.13   Commercial Production and Uses of Bioactive Compounds from Milk

The biological activity of milk proteins is exploited in the production of functional and nutritional products. Industrial or semi-industrial scale processing techniques are available for the fractionation and isolation of many bioactive components from bovine milk and colostrum. Proteins may be used in their intact state or peptides may be generated from them using proteolytic enzymes, microbial proteolytic enzymes and other food processing treatments such as heating or acidification (O'Regan et al. 2009; Mills et al. 2011). Products derived from milk proteins are used extensively in special dietary preparations for the ill or those convalescing from illness, for malnourished children and for people on therapeutic or weight-reducing diets.

Whey powder preparations with low mineral content are used to produce infant formulae that closely resemble human milk. Infant formulae may be supplemented with α-La as a rich source of tryptophan and its metabolites such as serotonin. βCN-, α-La- and Lf -enriched protein fractions have been used as ingredients in 'humanized' infant formulae (O'Regan et al. 2009).

Whey protein hydrolysates have been used in hypoallergenic, peptide-based infant formulae.

Whey protein-enriched diets are reported to reduce the growth of cancer tumours in the GIT and to inhibit cancer cell growth in the head and neck (Parodi 1999).

Biozate™, produced by Davisco Foods International, is a whey protein-based product which contains ACE-inhibitory peptides and has been shown to reduce systolic and diastolic blood pressure as well as aiding the body's immune defence systems.

Prolibra™, is a specialized whey protein preparation which is high in leucine, bioactive peptides and calcium and has been shown, in a randomized human clinical trial over 12 weeks, to promote the loss of body fat mass while increasing the preservation of lean muscle mass (Frestedt et al. 2008). Calpis®, a sour milk product from Japan and Evolus, a Ca-enriched fermented milk from Finland, both contain antihypertensive peptides which reduce blood pressure. Bioactive peptides are also produced on an industrial scale as ingredients for toothpaste, chewing gum (MI Paste, Trident Xtra Care) and food supplements (Capolac, Recaldent, Ameal peptide). Dziuba and Dziuba (2014) review the current products on the market which use milk protein-derived bioactive peptides and discuss the molecular, biological and technological aspects of production in detail.

Classically, Ig is prepared by salting-out, usually with ammonium sulphate [$(NH_4)_2SO_4$]. This method is effective but expensive and current commercial products are usually prepared by ultrafiltration of colostrum or milk from hyperimmunized cows. Some recently developed methods for the isolation of Ig, sometimes with Lf, use monoclonal antibodies, metal chelate or gel filtration chromatography (O'Regan et al. 2009; Mills et al. 2011).

Ig-rich preparations are commercially available for the nutrition of calves and other neonatal animals. While breast feeding is best for healthy full-term infants, it is frequently impossible to breast-feed pre-term or very-low-birth-weight infants, who may be fed on banked human milk. Such infants have high protein and energy requirements which may not be met by human milk and consequently special formulae have been developed. A milk immunological concentrate, prepared by diafiltration of acid whey from colostrum and early lactation milk from immunized cows, for use in such formulae has been described; the product contains approximately 75 % protein, 50 % of which is Ig, mainly IgG_1 and not IgA, which is predominant in human milk. The development of Ig in cows against human pathogens, e.g., rotavirus, an important cause of illness in children, is considered to be an attractive approach in human medicine. The Ig could be administered in milk or as a concentrate prepared from milk. Casein hydrolysates are used in specialised formulae for premature infants and those with intestinal disorders. These formulae are low in phenylalanine and are suitable for infants with phenylketonuria (O'Regan et al. 2009).

CMP (non-glycosylated) is used in weight control diet supplements where it stimulates the release of cholecystokinin which results in the production of insulin causing inhibition of gastric secretions and control of food intake and digestion (Yvon et al. 1994).

Opioid peptides have pharmacological properties similar to morphine and, as such, have potential for use in certain medical or dietary preparations; they may also stimulate pancreatic insulin and modify GIT function after eating which prolongs GIT transit time and prevents diarrhoea (O'Regan et al. 2009).

The challenge in research now is to find cost-effective large-scale methods to produce milk bioactive compounds. Foods containing bioactive peptides from protein hydrolysis have been approved for mass consumption but their availability on the market is quite limited (Dziuba and Dziuba 2014), although as consumer demands increase, the production of functional foods including nutraceuticals will also increase. Modern cloning techniques have allowed the incorporation of milk-derived bioactive peptides into food proteins of non-dairy origin (Mills et al. 2011).

Methods using ultrafiltration and chromatography have been developed for the concentration of growth factors from whey. In addition to possible food (nutritceutical) applications for such growth factors, a major potential application is in tissue cultures, for which foetal bovine serum is used as a source of growth factors. However, the supply of foetal bovine serum is limited, unreliable, expensive and of variable quality. Whey-derived growth factors have the potential to have a major impact on the biotechnological and pharmaceutical industries for the production of vaccines, hormones, drugs, monoclonal antibodies, and the production of tissue, especially skin for treatment of burns, ulcers and lacerations. A number of new technological methods have been developed to extract growth factors from milk for use as health products (see Pouliot and Gauthier 2006).

Several methods have been described for the production, characterization and evaluation of milk protein hydrolysates tailored for specific applications in the health-care, pharmaceutical, baby food and consumer product areas (O'Regan et al.

2009; Mills et al. 2011). Several peptides with specific properties may be prepared from milk proteins, either *in vivo* or *in vitro;* some may have commercial potential. Protein hydrolysates may be prepared with a low degree of hydrolysis or may be extensively hydrolysed where the latter are used as nutritional supplements. The most common method of preparation is by batch or continuous hydrolysis using proteolytic enzymes (e.g., pepsin or trypsin) followed by fractionation and enrichment of the peptides produced (O'Regan et al. 2009). Gel permeation, ion exchange, hydrophobic interaction and reverse phase chromatography have all been used to fractionate and purify biologically active peptides from milk protein hydrolysates. Several chromatographic techniques may be employed sequentially and purification may include an ultrafiltration step. For example, CPPs are produced on an industrial scale by enzymatic hydrolysis followed by ion-exchange chromatography or by acid precipitation, diafiltration and anion-exchange chromatography.

The high glutamine level in caseins is exploited in the preparation of formulae used by athletes as it is beneficial for maintenance of muscle protein mass (O'Regan et al. 2009).

Casein-based preparations containing high levels of TGF-β are reported to be beneficial to children with Crohn's disease (Fell et al. 2000).

Starter and non-starter lactic acid bacteria are used during fermentation of milk-based products to generate bioactive peptides (Gobbetti et al. 2007) which is dealt with in detail in Chap. 13.

11.14   Bioactive Components in Other Milks

To date, the majority of research has investigated the bioactive components of bovine and human milk and there is limited research on the milk of other species used in human nutrition. Buffalo milk is reported to contain most of the bioactive components found in bovine milk but has higher levels of protein, medium-chain fatty acids, CLA, retinol and tocopherols and gangliosides (Guo 2012). Caprine milk is reported to have therapeutic, hypoallergenic and nutritional advantages over bovine milk due to the presence of specific bioactive compounds including its content of short- and medium-chain fatty acids which may play a role in digestion, metabolism and some lipid malabsorption syndromes (Michaelidou 2008; Park 2012). Ovine milk is an excellent source of high quality protein, calcium and lipids especially rumenic acid, an isomer of CLA, which may be responsible for the anticarcinogenic and antiatherogenic properties of CLA (Michaelidou 2008; de la Fuente and Juarez 2012). The milk-borne factors present in porcine milk including, Igs, Lf, Lyz, LPO, leukocytes, epidermal growth factor, insulin-like growth factors (IGF I, II) and transforming growth factors (β1, β 2) and their possible effects on intestinal function and maturation in neonatal pigs have been reviewed comprehensively by Xu et al. (2002). Equine milk has been suggested as a substitute for human milk in infant nutrition. To be successful, it must be capable of performing many biological functions associated with human milk. The presence of high concentrations of Lf, Lyz, n-3 and n-6 fatty acids in equine milk are good indicators of its potential role (Uniacke-Lowe et al. 2010). Other characteristics of both equine and asinine milk of interest in human nutrition include an exceptionally high concentration of polyunsaturated fatty acids, low cholesterol content, high lactose and low protein levels (Solaroli et al. 1993; Salimei et al. 2004), as well as high levels of vitamins A, B and C. The low fat content and unique fatty acid profile of both equine and asinine milk result in low atherogenic and thrombogenic indices. Research has shown that human health is considerably improved when dietary fat intake is reduced and, more importantly, when the ratio of saturated to unsaturated fatty acids is reduced. The high lactose content of equid milk gives good palatability and improves intestinal absorption of calcium which is important for bone mineralization in children. The renal load of equine milk, based on levels of protein and inorganic substances, is equal to that of human milk, a further indication of its suitability as an infant food. Equine and asinine milk can be used for their prebiotic and probiotic activity and as alternatives for infants and children with cow's milk protein allergy (CMPA) and multiple food intolerances (Iacono et al. 1992; Carroccio et al. 2000). Levels of the bioactive peptides, ghrelin and insulin growth factor I, which play a direct role in metabolism, body composition and food intake, have also been reported for asinine milk at 4.5 pg mL^{-1} and 11.5 ng mL^{-1}, respectively, similar to levels in human milk (Salimei 2011).

The invigorating effect of equine milk may be, at least partially, due to its immuno-stimulating ability. Lyz, Lf and n-3 fatty acids have long been associated with the regulation of phagocytosis of human neutrophils *in vitro* (Ellinger et al. 2002). The concentration of these compounds is exceptionally high in equine milk and the consumption of frozen equine milk significantly inhibits chemotaxis and respiratory burst, two important phases of the phagocytic process (Ellinger et al. 2002). This result suggests a potential anti-inflammatory effect by equine milk.

11.15 Conclusion

Milk is a very complex system and contains many biologically active compounds, not fully appreciated, in addition to its gross composition. Many of these compounds may be spill-over constituents but some probably play valuable roles. Much discussion on the bioactive compounds in milk is speculative at best and whether these compounds actually survive in the lower GIT and are absorbed intact has not been proven to date for many compounds although some, e.g., cytokines are relatively resistant to digestion. Many of the physiological effects observed have been proven only *in vitro* or in animal models and have yet to be proven in humans. The fate of bioactive compounds in fermented milk products and cheese is not fully understood and requires much research in the future. The dairy industry now faces new technological challenges to exploit and maintain the bioactive properties of dairy components during the processing of milk.

References

Abd El-Salam, M. H., El-Shibinyand, S., & Buchheim, W. (1996). Characteristics and potential uses of the casein macropeptide. *International Dairy Journal, 6*, 327–341.

Achanta, K., Boeneke, C. A., & Aryana, K. J. (2007). Characteristics of reduced fat milks as influenced by the incorporation of folic acid. *Journal of Dairy Science, 90*, 90–98.

Adkins, Y., & Lönnerdal, B. (2001). High affinity binding of the transcobalamin II-cobalamin complex and mRNA expression of haptocorrin by human mammary epithelial cells. *Biochimica et Biophysica Acta, 1528*, 43–48.

Agostini, C., Carratù, B., Boniglia, C., Riva, E., & Sanzini, E. (2000). Free amino acid content in standard infant formulas: Comparison with human milk. *Journal of the American College of Nutrition, 19*, 434–438.

Agostoni, C. (2005). Ghrelin, leptin and the neurometabolic axis of breastfed and formula-fed infants. *Acta Paediatrica, 94*, 523–525.

Aluko, R. E. (2010). Food protein-derived peptides as calmodulin inhibitors. In Y. Mine, E. Li-Chan, & B. Jiang (Eds.), *Bioactive proteins and peptides as functional foods and nutraceuticals* (pp. 55–65). Ames, IA: Wiley-Blackwell.

Amarant, T., Fridkin, M., & Koch, Y. (1982). Luteinizing hormone-heleasing hormone and thyrotropin-releasing hormone in human and bovine milk. *European Journal of Biochemistry, 127*, 647–650.

Aminot-Gilchrist, D. V., & Anderson, H. D. I. (2004). Insulin resistance-associated cardiovascular disease: potential benefits of conjugated linoleic acid. *The American Journal of Clinical Nutrition, 79*, S1159–S1163.

Appelmelk, B. J., An, Y.-Q., Geerst, M., Thijs, B. G., Deboer, H. A., McClaren, D. M., et al. (1994). Lactoferrin is a lipid A-binding protein. *Infection and Immunity, 62*, 2628–2632.

Aydin, S., Ozkan, Y., Erman, F., Gurates, B., Kilic, N., Colak, R., et al. (2008). Presence of obestatin in breast milk: Relationship among obestatin, ghrelin, and leptin in lactating women. *Nutrition, 24*, 689–693.

Bach, A. C., & Babayan, V. K. (1992). Medium-chain triglycerides: An update. *The American Journal of Clinical Nutrition, 36*, 950–962.

Baldi, A., Politis, I., Pecorini, C., Fusi, E., Roubini, C., & Dell'Orto, V. (2005). Biological effects of milk proteins and their peptides with emphasis on those related to the gastrointestinal ecosystem. *Journal of Dairy Research, 72*, 66–72.

Baram, T., Koch, Y., Hazum, E., & Fridkin, M. (1977). Gonadotropin-releasing hormone in milk. *Science, 198*, 300–302.

Barello, C., Perono Garoffo, L., Montorfano, G., Zava, S., Berra, B., Conti, A., et al. (2008). Analysis of major proteins and fat fractions associated with mare's milk fat globules. *Molecular Nutrition & Food Research, 58*, 1448–1456.

Barfoot, R. A., McEnery, G., Ersser, R. S., & Seakins, J. W. (1988). Diarrhoea due to breast milk: Case of fucose intolerance? *Archives of Disease in Childhood, 63*, 311.

Barnard, J. A., Beauchamp, R. D., Russell, W. E., Dubois, R. N., & Coffey, R. J. (1995). Epidermal growth factor-related peptides and their relevance to gastrointestinal pathophysiology. *Gastroenterology, 108*, 564–580.

Bauman, D. E., & Lock, A. L. (2006). Conjugated linoleic acid: Biosynthesis and nutritional significance. In P. F. Fox & P. L. H. McSweeney (Eds.), *Advanced dairy chemistry, vol. 2, lipids* (3rd ed., pp. 93–136). New York, NY: Springer.

Becker, D. J., & Lowe, J. B. (2003). Fucose: Biosynthesis and biological function in mammals. Review. *Glycobiology, 13*, 41–53.

Bell, S. J., Grochoski, G. T., & Clarke, A. J. (2006). Health implications of milk containing β-casein with the A^2 genetic variant. *Critical Reviews in Food Science and Nutrition, 46*, 93–100.

Bellamy, W., Takase, M., Yamauchi, K., Wakabayasha, H., Kawase, K., & Tomita, M. (1992). Identification of the bactericidal domain of lactoferrin. *Biochimica et Biophysica Acta, 1121*, 130–136.

Bilikieweiz, A., & Gaus, W. (2004). Colostrinin (a naturally occurring, proline-rich, polypeptide mixture) in the treatment of Alzheimer's disease. *Journal of Alzheimer's Disease, 6*, 17–26.

Bode, L. (2006). Recent advances on structure, metabolism, and function of human milk oligosaccharides. *Journal of Nutrition, 136*, 2127–2130.

Boman, H. G. (1991). Antibacterial peptides: Key components needed in immunity. *Cell, 65*, 205–207.

Borch-Johnsen, K., Mandrup-Poulsen, T., Zachau-Christiansen, B., Joner, G., Christy, M., Kastrup, K., et al. (1984). Relation between breast-feeding and incidence rates of insulin-dependent diabetes mellitus. A hypothesis. *Lancet, 2*, 1083–1086.

Bouret, S. G. (2009). Early life origins of obesity: Role of hypothalamic programming. *Journal of Pediatric Gastroenterology and Nutrition, 48*, 531–538.

Bouzerzour, K., Morgan, F., Cuinet, I., Bonhomme, C., Jardin, J., Le Huëou-Luron, I., et al. (2012). In vivo digestion of infant formula in piglets: Protein digestion kinetics and release of bioactive peptides. *British Journal of Nutrition, 108*, 2105–2114.

Boyd, R. D., Kensinger, R. S., Harrell, R. J., & Bauman, D. E. (1995). Nutrient uptake and endocrine regulation of milk synthesis by mammary tissue of lactating sows. *Journal of Animal Science, 73*, 36–56.

Brantl, V., Teschemacher, H., Bläsig, J., Henschen, A., & Lottspeich, F. (1981). Opioid activities of β-casomorphins. *Life Sciences, 28*, 1903–1909.

Brantl, V., Teschemacher, H., Henschen, A., & Lottspeich, F. (1979). Novel opioid peptides derived from casein (β-casomorphins). I. Isolation from bovine casein peptone. *Hoppe-Seyler's Zeitschrift für physiologische Chemie, 360*, 1211–1216.

Brew, K. (2003). α-Lactalbumin. In P. F. Fox & P. L. H. McSweeney (Eds.), *Advanced dairy chemistry, vol. 1, proteins* (3rd ed., pp. 387–419). New York, NY: Kluwer Academic/Plenum Publishers.

Brew, K. (2013). α-Lactalbumin. In P. L. H. McSweeney & P. F. Fox (Eds.), *Advanced dairy chemistry, vol. IA, protein: Basic aspects* (4th ed., pp. 261–273). New York, NY: Springer.

Brock, J. H. (1997). Lactoferrin structure-function relationships. In T. W. Hutchens & B. Lonnerdal (Eds.), *Lactoferrin: Interactions and biological functions* (pp. 3–23). Totowa, NJ: Humana.

Brodgen, K. A. (2005). Antimicrobial peptides: Pore formers or metabolic inhibitors in bacteria. *Nature Reviews. Microbiology, 3*, 238–250.

Brody, E. P. (2000). Biological activities of bovine glycomacropeptide. *British Journal of Nutrition, 84*, S39–S46.

Brück, W. M., Graverholt, G., & Gibson, G. R. (2003). A two-stage continuous culture system to study the effect of supplemental a-La and glycomacropeptide on mixed cultures of human gut bacteria challenged with enteropathogenic *Escherichia coli* and *Salmonella* serotype Typhimurium. *Journal of Applied Microbiology, 95*, 44–53.

Byers, T., Graham, S., Rzepka, T., & Rzepka, T. (1985). Lactation and breast cancer: Evidence for a negative association in premenopausal women. *American Journal of Epidemiology, 121*, 664–674.

Cade, R., Privette, M., Fregly, M., Rowland, N., Sun, Z., Zele, V., et al. (2000). Autism and schizophrenia: Intestinal disorders. *Nutritional Neuroscience, 3*, 57–72.

Caen, J. P., Bal Dit Sollier, C., Fiat, A. M., Mazoyer, E., & Drovet, L. (1992). Activité antithrombotique de séquences peptidiques de protéins de lait. *Cahiers de Nutrition et de Dietetique, 27*, 33–35.

Calhoun, D. A., Lunoe, M., Du, Y., & Christensen, R. D. (2000). Granulocyte colony-stimulating factor is present in human milk and its receptor is present in human fetal intestine. *Pediatrics, 105*, e7.

Campana, W. M., & Baumrucker, C. R. (1995). Hormones and growth factors in bovine milk. In R. G. Jensen (Ed.), *Handbook of milk composition* (pp. 476–494). Oxford, UK: Academic.

Carlson, S. E. (2001). Docosahexaenoic acid and arachidonic acid in infant development. *Seminars in Neonatology, 6*, 437–449.

Carratù, B., Boniglia, C., Scalise, F., Ambruzzi, A. M., & Sanzini, E. (2003). Nitrogeneous components of human milk: Non-protein nitrogen, true protein and free amino acids. *Food Chemistry, 81*, 357–362.

Carroccio, A., Cavataio, F., Montalto, G., D'Amico, D., Alabrese, L., & Iacono, G. (2000). Intolerence to hydrolysed cow's milk proteins in infants: Clinical characteristics and dietary treatment. *Clinical and Experimental Allergy, 30*, 1597–1603.

Casabiell, X., Pineiro, V., Tome, M. A., Peino, R., Dieguez, C., & Casanueva, F. F. (1997). Presence of leptin in colostrum and/or breast milk from lactating mothers: A potential role in the regulation of neonatal food intake. *The Journal of Clinical Endocrinology and Metabolism, 82*, 4270–4273.

Cashman, K. D. (2006). Milk minerals (including trace elements) and bone health. *International Dairy Journal, 16*, 1389–1398.

Català-Clariana, S., Benavente, F., Giménez, E., Barbosa, J., & Sanz-Nebot, V. (2010). Identification of bioactive peptides in hypoallergenic infant milk formula by capillary electrophoresis mass spectrometry. *Analytica Chimica Acta, 683*, 119–125.

Chabance, B., Jollès, P., Izquierdo, C., Mazoyer, E., Francoual, C., Drouet, L., et al. (1995). Characterization of an antithrombotic peptide from κ-casein in newborn plasma after milk ingestion. *British Journal of Nutrition, 73*, 583–590.

Chabance, B., Marteau, P., Rambaud, J. C., Migliore-Samour, D., Boynard, M., Perrontin, P., et al. (1998). Casein peptide release and passage to the blood in humans during digestion of milk or yoghurt. *Biochimie, 80*, 155–165.

Chang, K. J., Su, Y. F., Brent, D. A., & Chang, J. K. (1985). Isolation of a specific μ-opiate receptor peptide, morphiceptin, from an enzymatic digest of milk proteins. *The Journal of Biological Chemistry, 260*, 9706–9712.

Chatterton, D. E. W., Smithers, G., Roupas, P., & Brodkorb, A. (2006). Review: Bioactivity of β-lactoglobulin and α-lactalbumin—Technological implications for processing. *International Dairy Journal, 16*, 1229–1240.

Chaturvedi, P., Warren, C. D., Altaye, M., Morrow, A. L., Ruiz-Palacios, G., Pickering, L. K., et al. (2001). Fucosylated human oligosaccharides vary between individuals and over the course of lactation. *Glycobiology, 11*, 365–372.

Chin-Dusting, J., Shennan, J., Jones, E., Williams, C., Kingwell, B., & Dart, A. (2006). Effect of dietary supplementation with β-casein A1 or A2 on markers of disease development in individuals at high risk of cardiovascular disease. *British Journal of Nutrition, 95*, 136–144.

Chun, R. F. (2012). New perspectives on the vitamin D binding protein. *Cell Biochemistry and Function, 30*, 445–456.

Cieślińska, A., Kamiński, S., Kostyra, E., & Sienkiewicz-Szlapka, E. (2007). Beta-casomorphin 7 in raw and hydrolysed milk derived from cows of alternative beta-casein genotypes. *Milchwissenschaft, 62*, 125–127.

Cinque, B., Di Marzio, L., Centi, C., Di Rocco, C., Riccardi, C., & Grazia Cifone, M. (2003). Sphingolipids and the immune system. *Pharmacological Research, 47*, 421–437.

Clare, D. A., Catignani, G. L., & Swaisgood, H. E. (2003). Biodefense properties of milk: The role of antimicrobial proteins and peptides. *Current Pharmaceutical Design, 29*, 1239–1255.

Clare, D. A., & Swaisgood, H. E. (2000). Bioactive milk peptides: A prospectus. *Journal of Dairy Science, 83*, 1187–1195.

Collomb, M., Schmid, A., Sieber, R., Wechsler, D., & Ryhänen, E.-L. (2006). Conjugated linoleic acids in milk fat: Variation and physiological effects. *International Dairy Journal, 16*, 1347–1361.

Combs, G. F. (2012). *The vitamins: Fundamental aspects in nutrition and health* (4th ed.). Burlington, MA: Elsevier Academic Press.

Conneely, O. (2001). Anti inflammatory activities of lactoferrin. *Journal of the American College of Nutrition, 20*, 389S–395S.

Creamer, L. K., Loveday, S. M., & Sawyer, L. (2011). Milk proteins: β-Lactoglobulin. In H. Roginski, J. W. Fuquay, & P. F. Fox (Eds.), *Encyclopedia of dairy sciences* (2nd ed., pp. 787–794). Oxford, UK: Elsevier.

Cummings, D. E. (2006). Ghrelin and the short- and long-term regulation of appetite and body weight. *Physiology & Behavior, 89,* 71–84.

Cuthbertson, W. F. J. (1999). Evolution of infant nutrition. *British Journal of Nutrition, 81,* 359–371.

Dalgleish, D. G. (2011). On the structural models of bovine casein micelles—Review and possible improvements. *Soft Matter, 7,* 2265–2272.

Darling, J., Laing, A., & Harkness, R. (1974). A survey of the steroids in cow's milk. *Journal of Endocrinology, 62,* 291–297.

Davies, M. (1988). Infant feeding and childhood lymphomas. *Lancet, 2,* 365–368.

Davis, R. E., & Nichol, D. J. (1988). Folic acid. *International Journal of Biochemistry, 20,* 133–139.

De la Fuente, M. A., Juarez, M. (2012). Bioactive components in sheep milk and products. *Proceedings of Dairy Foods Symposium: Bioactive Components in Milk and Dairy Products: Recent International Perspectives and Progresses in Different Dairy Species, Phoenix, AZ* (pp. 459–460).

De Noni, I. (2008). Release of β-casomorphins 5 and 7 during simulated gastro-intestinal digestion of β-casein variants and milk-based infant formulas. *Food Chemistry, 110,* 897–903.

De Noni, I., & Cattaneo, S. (2010). Occurrence of β-casomorphins 5 and 7 in commercial dairy products and in their digests following *in-vitro* simulated gastro-intestinal digestion. *Food Chemistry, 119,* 560–566.

Deloyer, P., Peulen, O., & Dandrifosse, G. (2001). Dietary polyamines and non-neoplastic growth and disease. *European Journal of Gastroenterology & Hepatology, 13,* 1027–1032.

Donovan, S. M., & Odle, J. (1994). Growth factors in milk as mediators of infant development. *Annual Review of Nutrition, 14,* 147–167.

Drake, F. H., Dodds, R. A., James, I. E., Connor, J. R., Debouck, C., Richardson, S., et al. (1996). Cathepsin K, but not cathepsins B, L, or S, is abundantly expressed in human osteoclasts. *The Journal of Biological Chemistry, 271,* 12511–12516.

Duggan, V. E., Holyoak, G. R., MacAllister, C. G., Cooper, S. R., & Confer, A. W. (2008). Amyloid A in equine colostrum and early milk. *Veterinary Immunology and Immunopathology, 121,* 150–155.

Dunbar, A. J., Priebe, I. K., Belford, D. A., & Goddard, C. (1999). Identification of betacellulin as a major peptide growth factor in milk: Purification, characterization and molecular cloning of bovine betacellulin. *Biochemical Journal, 344,* 713–721.

Dvorak, B. (2010). Milk epidermal growth factor and gut protection. *Journal of Pediatrics, 156,* S31–S35.

Dziuba, B., & Dziuba, M. (2014). Milk proteins-derived bioactive peptides in dairy products: Molecular, biological and methodological aspects. *Acta Scientiarum Polonorum. Technologia Alimentaria, 13,* 5–25.

Dziuba, J., & Minkiewicz, P. (1996). Influence of glycosylation on micelle-stabilizing ability and biological properties of C-terminal fragments of cow's κ-casein. *International Dairy Journal, 6,* 1017–1044.

Elfstrand, L., Lindmark-Månsson, H., Paulsson, M., Nyberg, L., & Åkesson, B. (2002). Immunoglobulins, growth factors and growth hormone in bovine colostrum and the effects of processing. *International Dairy Journal, 12,* 879–887.

Ellinger, S., Linscheid, K. P., Jahnecke, S., Goerlich, R., & Endbergs, H. (2002). The effect of mare's milk consumption on functional elements of phagocytosis of human neutrophils granulocytes from healthy volunteers. *Food and Agricultural Immunology, 14,* 191–200.

Elliott, R. B., Harris, D. P., Hill, J. P., Bibby, N. J., & Wasmuth, H. E. (1999). Type 1 (insulin-dependent) diabetes mellitus and cow milk: Casein variant consumption. *Diabetologia, 42,* 292–296.

Ellison, R. T., & Giehl, T. J. (1991). Killing of gram-negative bacteria by lactoferrin and lysozyme. *Journal of Clinical Investigation, 88,* 1080–1091.

El-Zahar, K., Sitohy, M., Choiset, Y., Métro, F., Haertlé, T., & Chobert, J.-M. (2005). Peptic hydrolysis of ovine β-lactoglobulin and α-lactalbumin. Exceptional susceptibility of native ovine β-lactoglobulin to pepsinolysis. *International Dairy Journal, 15,* 17–27.

Erdman, K., Cheung, B. W. Y., & Schröder, H. (2008). The possible role of food-derived bioactive peptides in reducing the risk of cardiovascular disease. *The Journal of Nutritional Biochemistry, 19,* 643–654.

Eriksson, L., Valtonen, M., Laitinen, J., Paananen, M., & Kaikkonen, M. (1998). Diurnal rhythm of melatonin in bovine milk: Pharmacokinetics of exogenous melatonin in lactating cows and goats. *Acta Veterinaria Scandinavica, 39,* 301–310.

European Food Safety Authority. (2009). Review of the potential health impact of β-casomorphins and related peptides. *EFSA Scientific Report, 231,* 1–107.

Fell, J. M., Paintin, M., Arnaud-Battandier, F., Beattie, R. M., Hollis, A., Kitching, P., et al. (2000). Mucosal healing and a fall in mucosal pro-inflammatory cytokine mRNA induced by a specific oral polymeric diet in paediatric Crohn's disease. *Alimentary Pharmacology & Therapeutics, 14,* 281–288.

Ferranti, P., Traisci, M. V., Picariello, G., Nasi, A., Boschi, V., Siervo, M., et al. (2004). Casein proteolysis in human milk: Tracing the pattern of casein breakdown and the formation of potential bioactive peptides. *Journal of Dairy Research, 71,* 74–87.

Fiat, A. M., & Jollès, P. (1989). Caseins of various origins and biologically active casein peptides and oligosaccharides: Structural and physiological aspects. *Molecular and Cellular Biochemistry, 87,* 5–30.

Fiat, A.-M., Migliore-Samour, D., Jollès, P., Drouet, L., Bal Dit Sollier, C., & Caen, J. (1993). Biologically active peptides from milk proteins with emphasis on two examples concerning antithrombotic and immunomodulating activities. *Journal of Dairy Science, 76,* 301–310.

Filipp, D., Alizadeh-Khiavi, K., Richardson, C., Palma, A., Pareded, N., Takeuchi, O., et al. (2001). Soluble CD14 enriched in colostrum and milk induces B cell growth and differentiation. *Proceedings of the National Academy of Sciences of the United States of America, 98,* 603–608.

Finke, B. M., Mank, H. D., & Stahl, B. (2000). Off-line coupling of low-pressure anion-exchange chromatography with MALDI-MS to determine the elution order of human milk oligosaccharides. *Analytical Biochemistry, 284,* 256–265.

Fitzgerald, R. J. (1998). Potential uses of caseinophosphopeptides. *International Dairy Journal, 8,* 451–457.

Fitzgerald, R. J., & Meisel, H. (2000). Opioid peptides encrypted in intact milk protein sequences. *British Journal of Nutrition, 58,* S27–S31.

Food and Agriculture Organization of the United Nations/World Health Organization. (2008). *Enterobacter sakazakii (Cronobacter spp.) in powdered follow-up formulae* (Microbiological risk assessment series, Vol. 15). Rome: FAO/WHO.

Fosset, S., Fromentin, G., Gietzen, D. W., Dubarry, M., Huneau, J. F., Antoine, J. M., et al. (2002). Peptide fragments released from Phe-caseinomacropeptide in vivo in the rat. *Peptides, 23,* 1773–1781.

Fox, P. F., & Flynn, A. (1992). Biological properties of milk proteins. In P. F. Fox (Ed.), *Advanced dairy chemistry, vol 1, proteins* (pp. 255–284). London, UK: Elsevier Applied Science.

Fox, P. F., & Kelly, A. L. (2003). Developments in the chemistry and technology of milk proteins 2. Minor milk proteins. *Food Australia, 55,* 231–234.

Fox, P. F., & Kelly, A. (2006). Indigenous enzymes in milk: Overview and historical aspects-Part 1. *International Dairy Journal, 16,* 500–516.

Frestedt, J. L., Zenk, J. L., Kuskowski, M. A., Ward, L. S., & Bastian, E. D. (2008). A whey-protein supplement increases fat loss and spares lean muscle in obese subjects: A randomized human clinical study. *Nutrition and Metabolism, 5,* 8–14.

Freudenheim, J. L., Marshall, J. R., Graham, S., Laughlin, R., Vena, J. E., Bandera, E., et al. (1994). Exposure to breast milk in infancy and the risk of breast cancer. *Epidemiology, 5*, 324–331.

Gallier, S., Gragson, D., Jimenez-Flores, R., & Everett, D. (2010). Using confocal laser scanning microscopy to probe the milk fat globule membrane and associated proteins. *Journal of Agricultural and Food Chemistry, 58*, 4250–4257.

Gauthier, S. F., Pouliot, Y., & Saint-Sauveur, D. (2006). Immunomodulatory peptides obtained by the enzymatic hydrolysis of whey proteins. *International Dairy Journal, 16*, 1315–1323.

Ghitis, J., Mandelbaum-Shavit, F., & Grossowicz, N. (1969). Binding of folic acid and derivatives in milk. *British Journal of Nutrition, 31*, 243–257.

Gifford, J. L., Hunter, H. N., & Vogel, H. J. (2005). Lactoferricin: A lactoferrin-derived peptide with antimicrobial, antiviral, antitumor and immunological properties. *Cellular and Molecular Life Sciences, 62*, 2588–2598.

Gillman, M. W. (2010). Early infancy—A critical period for development of obesity. *Journal of Developmental Origins of Health and Disease, 1*, 292–299.

Girardet, J.-M., & Linden, G. (1996). PP3 component of bovine milk: A phosphorylated whey glycoprotein. *Journal of Dairy Research, 63*, 333–350.

Gobbetti, M., Minervini, F., & Rizzello, C. G. (2007). Bioactive peptides in dairy products. In Y. H. Hui (Ed.), *Handbook of food products manufacturing* (pp. 489–517). Hoboken, NJ: John Wiley and Sons, Inc.

Gobbetti, M., Stepaniak, L., De Angelis, M., Corsetti, A., & Di Cagno, R. (2002). Latent bioactive peptides in milk proteins: Proteolytic activation and significance in dairy processing. *Critical Reviews in Food Science and Nutrition, 42*, 223–239.

Graf, M. V., Hunter, C. A., & Kastin, A. J. (1984). The presence of delta sleep-inducing peptide-like material in human milk. *The Journal of Clinical Endocrinology and Metabolism, 59*, 127–132.

Graf, M. V., & Kastin, A. J. (1986). Delta Sleep-inducing peptide (CDSIP): An update. *Peptides, 7*, 1165–1187.

Green, M. R., & Pastewka, J. V. (1978). Lactoferrin is a marker for prolactin response in mouse mammary explants. *Endocrinology, 103*, 1510–1513.

Gregory, I. J. F. (1982). Denaturation of the folacin-binding protein in pasteurized milk products. *Journal of Nutrition, 112*, 1329–1338.

Grosvenor, C. E., Picciano, M. F., & Baumrucker, C. R. (1993). Hormones and growth factors in milk. *Endocrine Reviews, 14*, 710–728.

Groves, M. L. (1969). Some minor components of casein and other phosphoproteins in milk. A review. *Journal of Dairy Science, 52*, 1155–1165.

Groves, M. L., & Greenberg, R. (1982). β2-Microglobulin and its relationship to the immune system. *Journal of Dairy Science, 65*, 317–325.

Guaadaoui, A., Benaicha, S., Elmajdoub, N., Bellaoui, M., & Hamal, A. (2014). What is a bioactive compound? A combined definition for a preliminary consensus. *International Journal of Food Sciences and Nutrition Sciences, 3*, 174–179.

Guo, M. (2012). Bioactive components in buffalo milk and products. *Proceedings of: Dairy Foods Symposium: Bioactive Components in Milk and Dairy Products: Recent International Perspectives and Progresses in Different Dairy Species, Phoenix, AZ* (pp. 459–460).

Guo, H. Y., Pang, K., Zhang, X. Y., Zhao, L., Chen, S. W., Dong, M. L., et al. (2007). Composition, physicochemical properties, nitrogen fraction distribution, and amino acid profile of donkey milk. *Journal of Dairy Science, 90*, 1635–1643.

Haddad, J. G. (1995). Plasma vitamin D-binding protein (Gc-globulin): Multiple tasks. *The Journal of Steroid Biochemistry and Molecular Biology, 53*, 579–582.

Hambræus, L. (1984). Human milk composition. *Nutrition Abstracts and Reviews Series A, 54*, 219–236.

Hambræus, L., & Lönnerdal, B. (2003). Nutritional aspects of milk proteins. In P. F. Fox & P. L. H. McSweeney (Eds.), *Advanced dairy chemistry, vol. 1, proteins* (3rd ed., pp. 605–645). New York, NY: Kluwer Academic/Plenum Publishers.

Hamer, H. M., Jonkers, D., Venema, K., Vanhoutvin, S., Troost, F. J., & Brummer, R. J. (2008). Review article: The role of butyrate on colonic function. *Alimentary Pharmacology & Therapeutics, 27*, 104–119.

Haque, E., & Chand, R. (2008). Antihypertensive and antimicrobial bioactive peptides from milk proteins. *European Food Research and Technology, 227*, 7–15.

Haug, B. E., Strøm, M. B., & Svendsen, J. S. M. (2007). The medicinal chemistry of short lactoferricin-based antibacterial peptides. *Current Medicinal Chemistry, 14*, 1–18.

Hayes, M., Ross, R. P., Fitzgerald, G. F., Hill, C., & Stanton, C. (2005). Casein-derived antimicrobial peptides generated by Lactobacillus acidophilus DPC6026. *Applied and Environmental Microbiology, 72*, 2260–2264.

Heap, R. B., Gwyn, M., Laing, J. A., & Waiters, D. E. (1973). Pregnancy diagnosis in cows; changes in milk progesterone concentration during the oestrous cycle and pregnancy measured by a rapid radioimmunoassay. *Journal of Agricultural Science, 81*, 151–157.

Heegaard, C. W., Rasmussen, L. K., & Andreasen, P. A. (1994). The plasminogen activation system in bovine milk: Differential localization of tissue-type plasminogen activator and urokinase in milk fractions is caused by binding to casein and urokinase receptor. *Biochimica et Biophys Acta, 1222*, 45–55.

Hendricks, G. M., & Guo, M. (2014). Bioactive components in human milk. In M. Guo (Ed.), *Human milk biochemistry and infant formula manufacturing* (pp. 33–54). Cambridge, UK: Woodhead Publishing.

Heremans, K., Van Camp, J., & Huyghebaert, A. (1997). High-pressure effects on proteins. In S. Damodaran & A. Paraf (Eds.), *Food proteins and their applications* (pp. 473–502). New York, NY: Marcel Dekker, Inc.

Hernández-Ledesma, B., Amigo, L., Ramos, M., & Recio, I. (2004). Release of angiotensin converting enzyme-inhibitory peptides by simulated gastrointestinal digestion of infant formulas. *International Dairy Journal, 14*, 889–898.

Hernández-Ledesma, B., Dávalos, A., Bartolomé, B., & Amigo, L. (2005). Preparation of antioxidant enzymatic hydrolyzates from α-lactalbumin and β-lactoglobulin. Identification of peptides by HPLC-MS/MS. *Journal of Agricultural and Food Chemistry, 53*, 588–593.

Hernández-Ledesma, B., Quirós, A., Amigo, L., & Recio, I. (2007). Identification of bioactive peptides after digestion of human milk and infant formula with pepsin and pancreatin. *International Dairy Journal, 17*, 42–49.

Hernández-Ledesma, B., Recio, I., & Amigo, L. (2008). β-Lactoglobulin as a source of bioactive peptides. *Amino Acids, 35*, 257–265.

Hernell, O., & Lönnerdal, B. (2002). Iron status of infants fed low iron formula: No effect of added bovine lactoferrin or nucleotides. *The American Journal of Clinical Nutrition, 76*, 858–864.

Hiraiwa, M., O'Brien, J. S., Kishimoto, Y., Galdzicka, M., Fluharty, A. L., Ginns, E. I., et al. (1993). Isolation, characterization, and proteolysis of human prosaposin, the precursor of saposins (sphingolipid activator proteins). *Archives of Biochemistry and Biophysics, 304*, 110–116.

Horrobin, D. F. (2000). Essential fatty acid metabolism and its modification in atopic eczema. *The American Journal of Clinical Nutrition, 71*, 367S–372S.

Høst, A., & Halken, S. (2004). Hypoallergenic formulas- when, to whom and how long: After more than 15 years we know the right indication. *Allergy, 59*, 45–52.

Houseknecht, K. L., McGuire, M. K., Portocarrero, C. P., McGuire, M. A., & Beerman, K. (1997). Leptin is present in human milk and is related to maternal plasma leptin concentration and adiposity. *Biochemical and Biophysical Research Communications, 240*, 742–747.

Iacono, G., Carroccio, A., Cavataio, F., Montalto, G., Soresi, M., & Balsamo, V. (1992). Use of ass' milk in multiple food allergy. *Journal of Pediatric Gastroenterology and Nutrition, 14*, 177–181.

Ilcol, Y. O., Hizli, Z. B., & Eroz, E. (2008). Resistin is present in human breast milk and it correlates with maternal hormonal status and serum level of C-reactive protein. *Clinical Chemistry and Laboratory Medicine, 46*, 118–124.

Ilcol, Y. O., Hizli, Z. B., & Ozkan, T. (2006). Leptin concentration in breast milk human and its relationship to duration of lactation and hormonal status. *International Breastfeeding Journal, 17*, 1–21.

Innis, S. (2007). Fatty acids and early human development. *Early Human Development, 83*, 761–766.

Iwan, M., Jarmolowska, B., Bielikowicz, K., Kostyra, E., Kostyra, H., & Kaczmarski, M. (2008). Transport of μ-opioid receptor agonists and antagonist peptides across Caco-2 monolayer. *Peptides, 29*, 1041–1047.

Jacobsen, R., Lorenzen, J. K., Toubro, S., Krog-Mikkelsen, I., & Astrup, A. (2003). Effect of short-term high dietary calcium intake on 24-h energy expenditure, fat oxidation, and fecal fat excretion. *International Journal of Obesity, 29*, 292–301.

Jarmolowska, B., Kostyra, E., Krawczuk, S., & Kostyra, H. (1999). β-Casomorphin-7 isolated from Brie cheese. *Journal of the Science of Food and Agriculture, 79*, 1788–1792.

Jarmolowska, B., Sidor, K., Iwan, M., Bielikowicz, K., Kaczmarski, M., Kostyra, E., et al. (2007). Changes of β-casomorphin content in human milk during lactation. *Peptides, 28*, 1982–1986.

Jarmolowska, B., Szlapka-Sienkiewicz, E., Kostyra, E., Kostyra, H., Mierzejewska, D., & Darmochwal-Marcinkiewicz, K. (2007). Opioid activity of human formula for newborns. *Journal of the Science of Food and Agriculture, 87*, 2247–2250.

Jensen, R. G. (1999). Lipids in human milk. *Lipids, 34*, 1243–1271.

Jensen, R. B. (2002). The composition of bovine milk lipids: January 1995 to December 2000. *Journal of Dairy Science, 85*, 295–350.

Jinsmaa, T., & Yoshikawa, M. (1999). Enzymatic release of neocasomorphin and βcasomorphin from bovine β-casein. *Peptides, 20*, 957–962.

Jøhnke, M., Petersen, T. E. (2012). *The alpha-lactalbumin/oleic acid complex and its cytotoxic activity* (pp. 119–144). INTECH, Open Science/Open Minds Publication. Accessed October 21, 2014, from http://cdn.intechopen.com/pdfs-wm/38827.pdf

Jollès, P., & Henschen, A. (1982). Comparison between the clotting of blood and milk. *Trends in Biochemical Sciences, 7*, 325–328.

Jollès, P., Lévy-Toledano, S., Fiat, A.-M., Soria, C., Gillensen, D., Thomaidis, A., et al. (1986). Analogy between fibrinogen and casein. Effect of an undecapeptide isolated from κ–casein on platelet function. *European Journal of Biochemistry, 158*, 379–382.

Jollès, P., Loucheux-Lefebvre, M. H., & Henschen, A. (1978). Structural relatedness of κ-casein and fibrinogen γ-chain. *Journal of Molecular Evolution, 11*, 271–277.

Jones, C. A., Holloway, J. A., Popplewell, E. J., Diaper, N. D., Holloway, J. W., Vance, G. H., et al. (2002). Reduced soluble CD14 levels in amniotic fluid and breast milk are associated with the subsequent development of atopy, eczema, or both. *Journal of Allergy and Clinical Immunology, 109*, 858–866.

Jouan, P.-N., Pouliot, Y., Gauthier, S. F., & LaForest, J.-P. (2006). Hormones in bovine milk and milk products: A survey. *International Dairy Journal, 16*, 1408–1414.

Kalač, P., & Krausová, P. (2005). A review of dietary polyamines: Formation, implications for growth and health and occurrence in foods. *Food Chemistry, 90*, 219–230.

Kamau, S. M., Cheison, S. C., Chen, W., Liu, X. M., & Lu, R. R. (2010). Alpha-lactalbumin: Its production technologies and bioactive peptides. *Comprehensive Reviews in Food Science and Food Safety, 9*, 197–212.

Kamiński, S., Cieślińska, A., & Kostyra, E. (2007). Polymorphism of bovine beta-casein and its potential effect on human health. *The Journal of General and Applied Microbiology, 48*, 189–198.

Kanno, C., Kanehara, N., Shirafuji, K., Tanji, R., & Imai, T. (1991). Binding form of vitamin B2 in bovine milk: Its concentration, distribution, and binding linkage. *Journal of Nutritional Science and Vitaminology, 37*, 15–27.

Kawasaki, Y., Isoda, H., Shinmoto, N., Tanimoto, M., Dosako, S., Idota, T., et al. (1993). Inhibition by κ-casein glycomacropeptide and lactoferrin of influenza virus hemagglutination. *Bioscience, Biotechnology, and Biochemistry, 57*, 1214–1215.

Kawasaki, Y., Isoda, H., Tanimoto, M., Dosako, S., Idota, T., & Ahiko, K. (1992). Inhibition by lactoferrin and κ-casein glycomacropeptide of binding of Cholera toxin to its receptor. *Bioscience, Biotechnology, and Biochemistry, 56,* 195–198.

Keenan, T. W., & Mather, I. H. (2006). Intracellular origin of milk fat globules and the nature of the milk fat globule membrane. In P. F. Fox & P. L. H. McSweeney (Eds.), *Advanced dairy chemistry, vol. 2, lipids* (3rd ed., pp. 137–171). New York, NY: Springer.

Kitts, D. D. (1994). Bioactive peptides in food: Identification and potential uses. *Canadian Journal of Physiology and Pharmacology, 74,* 423–434.

Kitts, D. D., & Weiler, K. (2003). Bioactive proteins and peptides from food sources. Applications of bioprocesses used in isolation and recovery. *Current Pharmaceutical Design, 9,* 1309–1323.

Kizavva, K., Naganuma, K., & Murakami, U. (1995). Calmodulin-binding peptides isolated from α-casein peptone. *Journal of Dairy Research, 62,* 587–592.

Kohmura, M., Nio, N., Kubo, K., Minoshima, Y., Munekata, E., & Ariyoshi, Y. (1989). Inhibition of angiotensin I-converting enzyme by synthetic peptides of human β-casein. *Agricultural and Biological Chemistry, 53,* 2107–2114.

Kohmura, M., Nio, N., & Ariyoshi, Y. (1990). Inhibition of angiotensin-converting enzyme by synthetic peptides of human κ-casein. *Agricultural and Biological Chemistry, 54,* 835–836.

Koldovský, O. (1989). Search for role of milk-borne biologically active peptides for the suckling. *Journal of Nutrition, 119,* 1543–1551.

Koldovský, O., & Štrbák, V. (1995). Hormones and growth factors in human milk. In R. G. Jensen (Ed.), *Handbook of milk composition* (pp. 428–436). Oxford, UK: Academic.

Koletzko, S., Sherman, P., Corey, M., Griffiths, A., & Smith, C. (1989). Role of infant feeding practices in development of Crohn's disease in childhood. *Journal of British Medicine, 298,* 1617–1618.

Kondoh, K., Hineno, T., Sano, A., & Kakimoto, Y. (1991). Isolation and characterization of prosaposin from human milk. *Biochemical and Biophysical Research Communications, 181,* 286–292.

Korhonen, H. (2002). Technology options for new nutritional concepts. *International Journal of Dairy Technology, 55,* 79–88.

Korhonen, H. (2009). Milk-derived bioactive peptides: From science to applications. *Journal of Functional Foods, 1,* 177–187.

Korhonen, H., & Pihlanto, A. (2006). Bioactive peptides: Production and functionality. *International Dairy Journal, 16,* 945–960.

Korhonen, H., & Pihlanto-Leppälä, A. (2004). Milk-derived bioactive peptides: Formation and prospects for health promotion. In C. Shortt & J. O'Brien (Eds.), *Handbook of functional dairy products* (pp. 109–124). Boca Raton, FL: CRC Press.

Korpysa-Dzirba, W., Rola, J. G., & Osek, J. (2007). Enterobacter sakazakii-zagrożenie mikrobiologiczne w żywności (Enterobacter sakazakii– A microbiological threat in food). In Polish, summary in English. *Medycyna Weterynaryjna, 63,* 1277–1280.

Kostyra, E., Sienkiewicz-Szlapka, E., Jarmolowska, B., Krawczuk, S., & Kostyra, H. (2004). Opioid peptides derived from milk proteins. *Polish Journal Of Food And Nutrition Sciences, 13,* 25–35.

Kurzel, M. L., Janusz, M., Lisowski, J., Fischleigh, R. V., & Georgiades, J. A. (2001). Towards an understanding of biological role of colostrinin peptides. *Journal of Molecular Neuroscience, 17,* 379–389.

Kurzel, M. L., Polanowski, A., Wilusz, T., Sokolowska, A., Pacewicz, M., Bednarz, R., et al. (2004). The alcohol-induced conformational changes in casein micelles: A new challenge for the purification of colostrinin. *The Protein Journal, 23,* 127–133.

Labéta, M. O., Vidal, K., Nores, J. E., Aarias, M., Vita, N., Morgan, B. P., et al. (2000). Innate recognition of bacteria in human milk is mediated by a milk-derived highly expressed pattern recognition receptor, soluble CD14. *The Journal of Experimental Medicine, 191,* 1807–1812.

Langer, P. (2009). Differences in the composition of colostrum and milk in eutherians reflect differences in immunoglobulin transfer. *Journal of Mammalogy, 90,* 332–339.

Larson, B. L. (1979). Biosynthesis and secretion of milk proteins: A review. *Journal of Dairy Research, 46,* 161–174.

Larson, M. A., Wei, S. H., Weber, A., Mack, D. R., & McDonald, T. L. (2003). Human serum amyloid A3 peptide enhances intestinal MUC3 expression and inhibits EPEC adherence. *Biochemical and Biophysical Research Communications, 300,* 531–540.

Lawrence, R. A. (1999). Storage of human milk and the influence of procedures on immunological components of human milk. *Acta Paediatrica, 88,* 14–18.

Leitch, E. C., & Willcox, M. D. (1999). Elucidation of the antistaphylococcal action of lactoferrin and lysozyme. *Journal of Medical Microbiology, 48,* 867–871.

Levy, J. (1998). Immunonutrition: The pediatric experience. *Nutrition, 14,* 641–647.

Li-Chan, E., Kummer, A., Losso, J. N., Kitts, D. D., & Nakai, S. (1995). Stability of bovine immunoglobulins to thermal treatment and processing. *Food Research International, 28,* 9–16.

Lindstrom, P., Paulsson, M., Nylander, T., Elofsson, U., & Lindmark-Månsson, H. (1994). The effect of heat treatment on bovine immunoglobulins. *Milchwissenschaft, 49,* 67–71.

Locke, R. (2002). Preventing obesity: The breast milk-leptin connection. *Acta Paediatrica, 91,* 891–894.

Lönnerdal, B. (2003). Nutritional and physiologic significance of human milk proteins. *The American Journal of Clinical Nutrition, 77,* 1537S–1543S.

Lönnerdal, B. J. (2013). Bioactive proteins in breast milk. *Paediatrics and Child Health, 49,* 1–7.

Lönnerdal, B., & Suzuki, Y. A. (2013). Lactoferrin. In P. L. H. McSweeney & P. F. Fox (Eds.), *Advanced dairy chemistry, vol. 1A* (Proteins: Basic aspects, pp. 295–315). New York: Springer.

López-Expósito, I., & Recio, I. (2008). Protective effect of milk peptides: Antibacterial and antitumor properties. *Advances in Experimental Medicine and Biology, 606,* 271–294.

Löser, C. (2000). Polyamines in human and animal milk. *British Journal of Nutrition, 84,* S55–S58.

Mack, D. R., McDonald, T. L., Larson, M. A., Wei, S., & Weber, A. (2003). The conserved TFLK motif of mammary-associated serum amyloid A3 is responsible for up-regulation of intestinal MUC3 mucin expression *in vitro*. *Pediatric Research, 53,* 137–142.

Mainer, G., Dominguez, E., Randrup, M., Sanchez, L., & Calvo, M. (1999). Effect of heat treatment on anti-rotavirus activity of bovine immune milk. *Journal of Dairy Research, 66,* 131–137.

Malacarne, M., Martuzzi, F., Summer, A., & Mariani, P. (2002). Protein and fat composition of mare's milk: Some nutritional remarks with reference to human and cow's milk. *International Dairy Journal, 12,* 869–897.

Malik, S., Fu, L., Juras, D. J., Karmali, M., Wong, B. Y., Gozdzik, A., et al. (2013). Common variants of the vitamin D binding protein gene and adverse health outcomes. *Critical Reviews in Clinical Laboratory Sciences, 50,* 1–22.

Malkoski, M., Dashper, S. G., O'Brien-Simpson, N. M., Talbo, G. H., Macris, M., Cross, K. J., et al. (2001). Kappacin, a novel antibacterial peptide from bovine milk. *Antimicrobial Agents and Chemotherapy, 45,* 2309–2315.

Malven, P. (1977). Prolactin and other hormones in milk. *Journal of Animal Science, 45,* 609–616.

Marchbank, T., Playford, R. J. (2014). Colostrum: Its health benefits in milk and dairy products as functional foods. In: Kanekanian A (Ed.), (pp. 55–83). London, UK: Wiley-Blackwell.

Marinoni, E., Corona, G., Ciardo, F., & Letizia, C. (2010). Changes in the relationship between leptin, resistin and adiponectin in early neonatal life. *Frontiers in Bioscience, 2,* 52–58.

Marten, B., Pfeuffer, M., & Schrezenmeir, J. (2006). Medium-chain triglycerides. *International Dairy Journal, 16,* 1374–1382.

Martin, P., Cebo, G., & Miranda, G. (2013). Inter-species comparison of milk proteins: Quantitative variability and molecular diversity. In P. L. H. McSweeney & P. F. Fox (Eds.), *Advanced dairy chemistry, vol. IA, proteins basic aspects* (4th ed., pp. 387–429). New York, NY: Springer.

Martin, L. J., Woo, J. G., Geraghty, S. R., Altaye, M., Davidson, B. S., Banach, W., et al. (2006). Adiponectin is present in human milk and is associated with maternal factors. *The American Journal of Clinical Nutrition, 83,* 1106–1111.

Maruyama, S., Mitachi, H., Tanaka, H., Tomizuka, N., & Suzuki, H. (1987). Studies on the active site and antihypertensive activity of angiotensin I-converting enzyme inhibitors derivd from casein. *Agricultural and Biological Chemistry, 51*, 1581–1586.

Maruyama, S., Nakagomi, K., Tomizuka, N., & Suzuki, H. (1985). Angiotensin I-converting enzyme inhibitor derived from an enzymatic hydrolysate of casein. II. Isolation of bradykinin-potentiating activity on the uterus and the ileum of rats. *Agricultural and Biological Chem., 49*, 1405–1409.

Maruyama, S., & Suzuki, H. (1982). A peptide inhibitor of angiotensin I-converting enzyme in the tryptic hydrolysate of casein. *Agricultural and Biological Chemistry, 46*, 1393–1394.

Masson, P. L., & Heremans, J. F. (1971). Lactoferrin in milk from different species. *Comparative Biochemistry and Physiology, 39B*, 119–129.

Matsuoka, Y., Serizawa, A., Yoshioka, T., Yamamura, J., Morita, Y., Kawakami, H., et al. (2002). Cystatin C in milk basic protein (MBP) and its inhibitory effect on bone resorption in vitro. *Bioscience, Biotechnology, and Biochemistry, 66*, 2531–2536.

Maubois, J. L., Léonil, J., Trouvé, R., & Bouhallab, S. (1991). Les peptides du lait à activité physiologique III. Peptides du lait à effect cardiovasculaire: activités antithrombotique et antihypertensive. *Lait, 71*, 249–255.

Mazoyer, E., Levy-Toledano, S., Rendu, F., Hermant, L., Lu, H., Fiat, A. M., et al. (1990). KRDS a new peptide derived from lactotransferrin inhibits platelet aggregation and release reactions. *European Journal of Biochemistry, 194*, 43–49.

McDonald, T. L., Larson, M. A., Mack, D. R., & Weber, A. (2001). Elevated extra hepatic expression and secretion of mammary-associated serum amyloid A 3 (M-SAA3) into colostrum. *Veterinary Immunology and Immunopathology, 3*, 203–211.

Meisel, H. (1997). Biochemical properties of bioactive peptides derived from milk proteins: Potential nutraceuticals for food and pharmacological applications. *Livestock Production Science, 50*, 125–138.

Meisel, H. (2004). Multifunctionlal peptides encrypted in milk proteins. *Biofactors, 2*, 55–61.

Meisel, H., & FitzGerald, R. J. (2000). Opioid peptides encrypted in intact milk protein sequences. *British Journal of Nutrition, 84*, 27–31.

Meisel, H., & FitzGerald, R. J. (2003). Biofunctional peptides from milk proteins: Mineral binding and cytomodulatory effects. *Current Pharmaceutical Design, 9*, 1289–1295.

Meisel, H., Fritser, H., & Schlimme, E. (1989). Biologically active peptides in milk proteins. *Zeitschrift fur Ernharungswissenschaft, 28*, 267–278.

Meisel, H., & Schlimme, E. (1990). Milk proteins: Precursors of bioactive peptides. *Trends in Food Science & Technology, 1*, 41–43.

Meisel, H., & Schlimme, E. (1994). Inhibitors of angiotensin I-converting enzyme derived from bovine casein (casokinins). In V. Brantl & H. Teschemacher (Eds.), *β-Casomorphins and related peptides: Recent developments* (pp. 27–33). Germany: VCH-Weinheim.

Meisel, H., & Schlimme, E. (1996). Bioactive peptides derived from milk proteins: Ingredients for functional foods? *Kieler Milchw Forsch, 48*, 343–357.

Michaelidou, A. M. (2008). Factors influencing nutritional and health profile of milk and milk products. *Small Ruminant Research, 79*, 42–50.

Michalidou, A. M. (2008). Factors influencing nutritional and health profile of milk and milk products. *Small Ruminant Research, 79*, 42–50.

Michalski, M. C. (2007). On the supposed influence of milk homogenization on the risk of CVD, diabetes and allergy. *British Journal of Nutrition, 97*, 598–610.

Michalski, M., & Januel, C. (2006). Does homogenization affect the human health properties of cow's milk? *Trends in Food Science & Technology, 17*, 423–437.

Miclo, L., Perrin, E., Driou, A., Papadopoulos, V., Boujrad, N., Vanderesse, R., Boudier, J.-F., Desor, D., Linden, G., Gaillard, J.-L. (2001). Characterization of α-casozepine, a tryptic peptide from bovine αs1-casein with benzodiazepine-like activity. *The FASEB Journal.* http://www.fasebj.org/content/early/2001/08/02/fj.00-0685fje.full.pdf

Mills, S., Ross, R. P., Fitzgerald, G., & Stanton, C. (2009). Microbial production of bioactive metabolites. In A. Y. Tamime (Ed.), *Dairy fats and related products* (pp. 257–285). Oxford, UK: Wiley-Blackwell.

Mills, S., Ross, R. P., Hill, C., Fitzgerald, G., & Stanton, C. (2011). Milk intelligence: Mining milk for bioactive substances associated with human health. *International Dairy Journal, 21*, 377–401.

Miralles, O., Sánchez, J., Palou, A., & Picó, C. (2006). A physiological role of breast milk leptin in body weight control in developing infants. *Obesity, 14*, 1371–1377.

Molenaar, A. J., Kuys, Y. M., Davis, S. R., Wilkins, R. J., Mead, P. E., & Tweedie, J. W. (1996). Elevation of lactoferrin gene expression in developing, ductal, resting, and regressing parenchymal epithelium of the ruminant mammary gland. *Journal of Dairy Science, 79*, 1198–1208.

Möller, N. P., Scholz-Ahrens, K. E., Roos, N., & Schrezenmeir, J. (2008). Bioactive peptides and proteins from foods: Indication for health effects. *European Journal of Nutrition, 47*, 171–182.

Montagne, P., Cuillière, M. L., Molé, C., Béné, M. C., & Faure, G. (1998). Microparticle-enhanced nephelometric immunoassay of lysozyme in milk and other human body fluids. *Clinical Chemistry, 44*, 1610–1615.

Morrissey, P. A., & Hill, T. R. (2009). Fat-soluble vitamins and vitamin C in milk and milk products. In P. L. H. McSweeney & P. F. Fox (Eds.), *Advanced dairy chemistry, vol. 3: Lactose, water, salts and minor constituents* (3rd ed., pp. 527–589). New York, NY: Springer.

Mullally, M. M., Meisel, H., & Fitzgerald, R. J. (1996). Synthetic peptides corresponding to α-lactalbumin and β-lactoglobulin sequences with angiotensin-I-converting enzyme inhibitory activity. *Biological Chemistry Hoppe-Seyler, 377*, 259–260.

Mulvihill, D. M. (1992). Production, functional properties and utilization of milk protein products. In P. F. Fox (Ed.), *Advanced dairy chemistry, vol. 1, proteins* (pp. 369–404). London, UK: Elsevier Applied Science.

Nagaoka, S., Kanamaru, Y., & Kuzuya, Y. (1991). Effects of whey protein and casein on the plasma and liver lipids in rats. *Agricultural and Biological Chemistry, 55*, 813–818.

Naik, L., Sharma, R., Rajput, Y. S., & Manju, G. J. (2013). Application of high pressure processing technology for dairy food preservation: Future perspective: A review. *Journal of Animal production Advances, 3*, 232–241.

Nakamura, Y., Yamamoto, N., Sakai, K., Okubo, A., Yamazaki, S., & Takano, T. (1995). Purification and characterization of angiotensin I-converting enzyme inhibitors from sour milk. *Journal of Dairy Science, 78*, 777–783.

Nakamura, Y., Yamamoto, N., Sakai, K., & Takano, T. (1995). Antihypertensive effect of sour milk and peptides isolated from it are inhibitors to angiotensin-converting enzyme. *Journal of Dairy Science, 78*, 1253–1257.

Neeser, J. R., Golliard, M., Woltz, A., Rouvet, M., Dillmann, M. L., & Guggenheim, B. (1994). In vitro modulation of oral bacterial adhesion to saliva-coated hydroxyapatite beads by milk casein derivatives. *Oral Microbiology and Immunology, 9*, 193–201.

Neville, M. C. (2009). Introduction: Alpha-lactalbumin, a multifunctional protein that specifies lactose synthesis in the Golgi. *Journal of Mammary Gland Biology and Neoplasia, 14*, 211–212.

Nevo, J., Mai, A., Tuomi, S., Pellinen, T., Pentikäinen, O. T., Heikkilä, P., et al. (2010). Mammary-derived growth inhibitor (MDGI) interacts with integrin α-subunits and suppresses integrin activity and invasion. *Oncogene, 29*, 6452–6453.

Newburg, D. S., Pickering, L. K., McCluer, R. H., & Cleary, T. G. (1990). Fucosylated oligosaccharides of human milk protect suckling mice from heat stable enterotoxin of Escherichia coli. *The Journal of Infectious Diseases, 162*, 1075–1080.

Newburg, D. S., & Weiderschain, G. Y. (1997). Human milk glycosidases and the modification of oligosaccharides in human milk. *Pediatric Research, 41*, 86.

Ng, P. C., Lee, C. H., Lam, C. W., Wong, E., Chan, I. H., & Fok, T. F. (2004). Plasma ghrelin and resistin concentrations are suppressed in infants of insulin-dependent diabetic mothers. *The Journal of Clinical Endocrinology and Metabolism, 89*, 5563–5568.

Noh, S. K., & Koo, S. L. (2004). Milk sphingomyelin is more effective than egg sphingomyelin in inhibiting intestinal absorption of cholesterol and fat in rats. *Journal of Nutrition, 134*, 2611–2616.

Nohr, D. (2011). Vitamins: General introduction. In J. Fuquay, P. F. Fox, & P. L. H. McSweeney (Eds.), *Encyclopedia of dairy sciences, vol. 4* (2nd ed., pp. 636–638). Oxford, UK: Elsevier.

Nohr, D., & Biesalski, H. K. (2009). Vitamins in milk and dairy products: B-group vitamins. In P. L. H. McSweeney & P. F. Fox (Eds.), *Advanced dairy chemistry* (Lactose, water, salts and minor constituents 3rd ed., pp. 591–630). New York, NY: Springer.

Norberg, S., O'Connor, P. M., Stanton, C., Ross, R. P., Hill, C., Fitzggerald, G. F., et al. (2011). Altering the composition of caseicins A and B as a means of determining the contribution of specific residues to antimicrobial activity. *Applied and Environmental Microbiology, 77*, 2496–2501.

Nwosu, C. C., Aldredge, D. L., Lee, H., Lerno, L. A., Zivkovic, A. M., German, J. B., et al. (2012). Comparison of the human and bovine milk N-glycome via high-performance microfluidic chip liquid chromatography and tandem mass spectrometry. *Journal of Proteome Research, 11*, 2912–2924.

Nygren-Babol, L., & Jägerstad, M. (2012). Folate-binding protein in milk: A review of biochemistry, physiology, and analytical methods. *Critical Reviews in Food Science and Nutrition, 52*, 410–425.

Nygren-Babol, L., Sternesjö, Å., & Björck, L. (2004). Factors influencing levels of folate-binding protein in bovine milk. *International Dairy Journal, 14*, 761–765.

O'Regan, J., Ennis, M. P., & Mulvihill, D. M. (2009). Milk proteins. In G. O. Philips & P. A. Williams (Eds.), *Handbook of hydrocolloids* (2nd ed., pp. 298–358). Cambridge, UK: Woodhead Publishing Ltd.

Oberkotter, L. V., & Farber, M. (1984). Thyroxine-binding globulin in serum and milk specimens from puerperal lactating women. *Obstetrics & Gynecology, 64*, 244–247.

Ohashi, A., Murata, E., Yamamoto, K., Majima, E., Sano, E., Le, Q. T., et al. (2003). New functions of lactoferrin and beta-casein in mammalian milk as cysteine protease inhibitors. *Biochemical and Biophysical Research Communications, 306*, 98–103.

Oonaka, K., Furuhata, K., Hara, M., & Fukuyama, M. (2010). Powder infant formula contaminated with *Enterobacter Sakazakii*. *Japanese Journal of Infectious Diseases, 63*, 103–107.

Oriquat, G. A., Saleem, T. H., Abdullah, S. T., Soliman, G. T., Yousef, R. S., Adel Hameed, A. M., et al. (2011). Soluble CD14, sialic acid and L-fucose in breast milk and their role in increasing the immunity of breast-fed infants. *American Journal of Biochemistry and Biotechnology, 7*, 21–28.

Oshida, K., Shimizu, T., Takase, M., Tamura, Y., Shimizu, T., & Yamashiro, Y. (2003). Effects of dietary sphingomyelin on central nervous system myelination in developing rats. *Pediatric Research, 53*, 589–593.

Oshida, K., Shimuzu, T., Takase, M., Tamura, Y., Shimizu, T., & Yamashiro, Y. (2003). Effect of dietary sphingomyelin on central nervous system myelination in developing rats. *Pediatric Research, 53*, 580–592.

Ozbay, Y., Aydin, S., Dagli, A. F., Akbulut, M., Dagli, N., Kilic, N., et al. (2008). Obestatin is present in saliva: Alterations in obestatin and ghrelin levels of saliva and serum in ischemic heart disease. *BMB Reports, 41*, 55–61.

Paegelow, I., & Werner, H. (1986). Immunomodulation by some oligopeptides. *Methods and Findings in Experimental and Clinical Pharmacology, 8*, 91.

Pan, Y., Rowney, M., Guo, P., & Hobman, P. (2007). Biological properties of lactoferrin: An overview. *Australian Journal of Dairy Technology, 62*, 31–42.

Papadimitriou, E., Polykratis, A., Courty, J., Koolwijk, P., Heroult, M., & Katsoris, P. (2001). HARP induces angiogenesis in vivo and in vitro: Implication of N or C terminal peptides. *Biochemical and Biophysical Research Communications, 282*, 306–313.

Park, Y. W. (2009). Overview of bioactive components in milk and dairy products. In Y. W. Park (Ed.), *Bioactive components in milk and dairy products* (pp. 3–12). Ames, IA: Wiley-Blackwell.

Park, Y. W. (2012). Bioactive components in goat milk and products. *Proceedings of Dairy Foods Symposium: Bioactive Components in Milk and Dairy Products: Recent International Perspectives And Progresses in Different Dairy Species* (pp. 459–460). Phoenix, AZ.

Parodi, P. W. (1997a). Cows' milk fat components as potential anticarcinogenic agents. *Journal of Nutrition, 127,* 1055–1060.

Parodi, P. W. (1997b). Cow's milk folate binding protein: Its role in folate nutrition. *Australian Journal of Dairy Technology, 52,* 109–118.

Parodi, J. (1999). Conjugated linoleic acid and other anticarcinogenic agents of milk fat. *Journal of Dairy Science, 82,* 1339–1349.

Parodi, P. W. (2001). Cow's milk components with anti-cancer potential. *Australian Journal of Dairy Technology, 56,* 65–73.

Patton, S. (1999). Some practical implications of the milk mucins. *Journal of Dairy Science, 82,* 1115–1117.

Patton, S., Carson, G. S., Hiraiwa, M., O'Brien, J. S., & Sano, A. (1997). Prosaposin, a neurotrophic factor: Presence and properties in milk. *Journal of Dairy Science, 80,* 264–722.

Peaker, M., & Wilde, C. J. (1996). Feedback control of milk secretion from milk. *Journal of Mammary Gland Biology and Neoplasia, 1,* 307–315.

Pellegrini, A., Dettling, C., Thomas, U., & Hunziker, P. (2001). Isolation and characterization of four bactericidal domains in the bovine beta-lactoglobulin. *Biochimica et Biophysica Acta, 1526,* 131–140.

Pellegrini, A., Schumacher, S., & Stephan, R. (2003). In vitro activity of various antimicrobial peptides developed from the bactericidal domains of lysozyme and beta-lactoglobulin with respect to Listeria monocytogenes, Escherichia coli O157. Salmonella spp. and Staphylococcus aureus. *Archiv fuer Lebensmittelhygiene, 54,* 34–36.

Pellegrini, A., Thomas, U., Bramaz, N., Hunziker, P., & Von Fellenberg, R. (1999). Isolation and identification of three bactericidal domains in the bovine alpha-lactalbumin molecule. *Biochimica et Biophysica Acta, 1426,* 439–448.

Pérez, M. D., & Calvo, M. (1995). Interaction of β-lactoglobulin with retinol and fatty acids and its role as a possible biological function for this protein: A review. *Journal of Dairy Science, 78,* 978–988.

Pérez, M. D., Puyol, P., Ena, J. M., & Calvo, M. (1993). Comparison of the ability to bind ligands of β-lactoglobulin and serum albumin from ruminant and non-ruminant species. *Journal of Dairy Research, 60,* 55–63.

Pickering, L., Morrow, A., Ruiz-Palacios, G., & Schanler, R. (2004). *Protecting infants through human milk* (Advances in Experimental Medicine and Biology). New York, NY: Kluwer Academic/Plenum Publishers.

Pihlanto, A., & Korhonen, H. (2003). Bioactive peptides and proteins. In S. Taylor (Ed.), *Advances in food and nutrition research* (Vol. 47, pp. 175–276). San Diego, CA: Elsevier Inc.

Pillai, S. B., Turman, M. A., & Besner, G. E. (1998). Heparin-binding EGF-like growth factor is cytoprotective for intestinal epithelial cells exposed to hypoxia. *Journal of Pediatric Surgery, 33,* 973–978.

Pinotti, L., & Rosi, F. (2006). Leptin in bovine colostrum and milk. *Hormone and Metabolic Research, 38,* 89–93.

Playford, R. J., MacDonald, C. E., & Johnson, W. S. (2000). Colostrum and milk-derived peptide growth factors for the treatment of gastrointestinal disorders. *The American Journal of Clinical Nutrition, 72,* 5–14.

Politis, I., & Chronopoulou, R. (2008). Milk peptides and immune response in the neonate. In Z. Bösze (Ed.), *Bioactive compounds of milk* (Advances in Experimental Medicine and Biology, pp. 253–270). New York, NY: Springer.

Pouliot, Y., & Gauthier, S. F. (2006). Milk growth factors as health products: Some technological aspects. *International Dairy Journal, 16,* 1415–1429.

Raikos, V., & Dassios, T. (2014). Health-promoting properties of bioactive peptides derived from milk proteins in infant food: A review. *Dairy Science & Technology, 94*, 91–101.

Rainhard, P. (1986). Bacteriostatic activity of bovine milk lactoferrin against mastitic bacteria. *Veterinary Microbiology, 11*, 387–392.

Rasmussen, L. K., Due, H. A., & Petersen, T. E. (1995). Human α_{s1}-casein: Purification and characterization. *Comparative Biochemistry and Physiology - Part B, 111*, 75–81.

Rassin, D. K., Sturman, J. A., & Gaull, G. E. (1978). Taurine and other free amino acids in milk of man and other mammals. *Early Human Development, 2*, 1–13.

Rathcliffe, W. A., Green, E., Emly, J., Norbury, S., Lindsay, M., Heath, D. A., et al. (1990). Identification and partial characterization of parathyroid hormone-related protein in human and bovine milk. *Journal of Endocrinology, 127*, 167–176.

Recio, I., & Visser, S. (1999). Two ion-exchange methods for the isolation of antibacterial peptides from lactoferrin—In situ enzymatic hydrolysis on an ion-exchange membrane. *Journal of Chromatography, 831*, 191–201.

Reynolds, E. C. (1997). Remineralization of enamel subsurface lesions by casein phosphopeptide-stabilized calcium phosphate solutions. *Journal of Dental Research, 76*, 1587–1595.

Rhoades, J. R., Gibson, G. R., Formentin, K., Beer, M., Greenberg, N., & Rastall, R. A. (2005). Caseinoglycomacropeptide inhibits adhesion of pathogenic *Escherichia coli* strains to human cells in culture. *Journal of Dairy Science, 88*, 3455–3459.

Rombaut, R., & Dewettinck, K. (2006). Properties, analysis and purification of milk polar lipids. *International Dairy Journal, 11*, 1362–1373.

Rosner, W., Beers, P. C., Awan, T., & Khan, M. S. (1976). Identification of corticosteroid-binding globulin in human milk: Measurement with a filter disk assay. *The Journal of Clinical Endocrinology and Metabolism, 42*, 1064–1073.

Saarinen, U. M., & Kajosaari, M. (1995). Breastfeeding as prophylaxis against atopic disease: Prospective follow-up study until 17 years old. *Lancet, 346*, 1065–1069.

Saito, T., Itoh, T., & Adachi, S. (1987). Chemical structure of three neutral trisaccharides isolated in free from bovine colostrum. *Carbohydrate Research, 165*, 43–51.

Salimei, E. (2011). Animals that produce dairy foods: Donkey. In J. W. Fuquay, P. F. Fox, & P. L. H. McSweeney (Eds.), *Encyclopedia of dairy sciences, vol. 1* (2nd ed., pp. 365–373). Oxford: Elsevier.

Salimei, E., Fantuz, F., Coppola, R., Chiofalo, B., Polidori, P., & Varisco, G. (2004). Composition and characteristics of ass' milk. *Animal Research, 53*, 67–78.

Salimei, E., Varisco, G., & Rosi, F. (2002). Major constituents leptin, and non-protein nitrogen compounds in mares' colostum and milk. *Reproduction Nutrition Development, 42*, 65–72.

Salter, D. N., Scott, K. J., Slade, H., & Andrews, P. (1981). The preparation and properties of folate-binding protein from cow's milk. *Biochemical Journal, 193*, 469–476.

Sarwar, G., Botting, H. G., Davis, T. A., Darling, P., & Pencharz, P. B. (1998). Free amino acids in milk of human subjects, other primates and non-primates. *British Journal of Nutrition, 79*, 129–131.

Savino, F., Costamagna, M., Prino, A., Oggero, R., & Silvestro, L. (2002). Leptin levels in breast-fed and formula-fed infants. *Acta Paediatrica, 91*, 897–902.

Savino, F., Fissore, M. F., Liguori, S. A., & Oggero, R. (2009). Can hormones contained in mothers' mil account for the beneficial effect of breast-feeding on obesity in children? *Clinical Endocrinology, 71*, 757–765.

Savino, F., Liguori, S. A., Fissore, M. F., & Oggero, R. (2009). Breask milk hormones and their protective effect on obesity. *International Journal Pediatric Endocrinology, 2009*, 327505.

Savino, F., Sorrenti, M., Benetti, S., Lupica, M. M., Liguori, S. A., & Oggero, R. (2012). Resistin and leptin in breast milk and infants in early life. *Early Human Development, 88*, 779–782.

Sawyer, L. (2003). Lactoglobulin. In P. F. Fox & P. L. H. McSweeney (Eds.), *Advanced dairy chemistry, vol. 1, proteins* (3rd ed., pp. 319–386). New York, NY: Kluwer Academic/Plenum Publishers.

Schaafma, G. (2003). Vitamin: General introduction. In H. Roginski, J. Fuquay, & P. F. Fox (Eds.), *Encyclopedia of dairy sciences* (pp. 2653–2657). Oxford, UK: Elsevier.

Schaafsma, G. (2003). Nutritional significance of lactose and lactose derivatives. In H. Roginski, J. W. Fuquay, & P. F. Fox (Eds.), *Encyclopedia of dairy science* (pp. 1529–1533). London: Academic.

Scholtz-Ahrens, K. E., & Schrezenmeir, J. (2006). Milk minerals and the metabolic syndrome. *International Dairy Journal, 16*, 1399–1407.

Schrezenmeir, J., Korhonen, H., Willaims, M., Gill, H. S., & Shah, N. P. (2000). Forward. *British Journal of Nutrition, 84*, 1–1.

Schuster, S., Hechler, C., Gebauer, C., Kiess, W., & Kratzsch, J. (2011). Leptin in meternal serum and breast milk: Association with infants' body weight gain in longitudinal study over 6 months of months of lactation. *Pediatric Research, 70*, 63–637.

Schwartzbaum, J. A., George, S. L., Pratt, C. B., & Davis, B. (1991). An exploratory study of environmental and medical factors potentially related to childhood cancer. *Medical and Pediatric Oncology, 19*, 115–121.

Shah, N. P. (2000). Effects of milk-derived bioactives: An overview. *British Journal of Nutrition, 84*, S3–S10.

Shamay, A., Mabjeesh, S. J., & Silanikove, N. (2002). Casein-derived phosphopeptides disrupt tight junction integrity, and precipitously dry up milk secretion in goats. *Life Sciences, 70*, 2707–2719.

Shamay, A., Shapiro, F., Leitner, G., & Silanikove, N. (2003). Infusion of casein hydrolyzates into the mammary gland disrupt tight junction integrity and induce involution in cows. *Journal of Dairy Science, 86*, 1250–1258.

Shimazaki, K. I., Oota, K., Nitta, K., & Ke, Y. (1994). Comparative study of the iron-binding strengths of equine, bovine and human lactoferrins. *Journal of Dairy Research, 61*, 563–566.

Silva, S. V., & Malcata, F. X. (2005). Caseins as source of bioactive peptides. *International Dairy Journal, 15*, 1–15.

Simopoulos, A. P. (2002). The importance of the ratio of omega-6/omega-3 essential fatty acids. *Biomedicine & Pharmacotherapy, 56*, 365–379.

Singh, V. P., Sachan, N., & Verma, A. K. (2011). Melatonin milk; a milk of intrinsic health benefit: A review. *International Journal of Dairy Science, 6*, 246–252.

Skelwagen, K., Davis, S. R., Farr, V. C., Politis, I., Guo, R., & Kindstedt, P. S. (1994). Mammary-derived growth inhibitor in bovine milk: Effect of milking frequency and somatotropin administration. *Canadian Journal of Animal Science, 74*, 695–698.

Smith-Kirwin, S. M., O'Connor, D. M., De Johnston, J., Lancey, E. D., Hassink, S. G., & Funange, V. L. (1998). Leptin expression in human mammary epithelial cells and breast milk. *The Journal of Clinical Endocrinology and Metabolism, 83*, 1810–1813.

Solaroli, G., Pagliarini, E., & Peri, C. (1993). Composition and nutritional quality of mare's milk. *Italian Journal of Food Science, 5*, 3–10.

Spitsberg, V. L. (2005). Bovine milk fat globule membrane as a potential neutraceutical. *Journal of Dairy Science, 88*, 2289–2294.

Spitsberg, V. L., Matitashvili, E., & Gorewit, R. C. (1995). Association of fatty acid binding protein and glycoprotein CD36 in the bovine mammary gland. *European Journal of Biochemistry, 230*, 872–878.

Stan, E. Y., & Chernikov, M. P. (1982). Formation of a peptide inhibitor of gastric secretion from rat milk proteins in vivo. *Bulletin of Experimental Biology and Medicine, 94*, 1087–1089.

Stanton, C., Murphy, J., McGrath, E., & Devery, R. (2003). Aninal feeding strategies for conjugated linoleic acid enrichment of milk. In J. L. Sebedio, W. W. Christie, & R. O. Adlof (Eds.), *Advances in conjugated linoleic acid research* (pp. 123–145). Champaign, IL: AOCS Press.

Steppan, C. M., Bailey, S. T., Bhat, S., Brown, E. J., Banerjee, R. R., Wright, C. M., et al. (2001). The hormone resistin links obesity to diabetes. *Nature, 40*, 307–312.

Stocker, C. J., & Cawthorne, M. A. (2008). The influence of leptin on early life programming of obesity. *Trends in Biotechnology, 26*, 545–551.

Strömqvist, M., Falk, P., Bergström, S., Hansson, L., Lönnerdal, B., Normark, S., et al. (1995). Human milk β-casein and inhibition of *Helicobacter pylori* adhesion to human gastric mucosa. *Journal of Pediatric Gastroenterology and Nutrition, 21*, 288–296.

Strub, J. M., Goumon, Y., Lugardon, K., Capon, C., Lopez, M., Moniatte, M., et al. (1996). Antibacterial activity of glycosylated and phosphorylated chromogranin A-derived peptide 173-194 from bovine adrenal medullary chromaffin granules. *The Journal of Biological Chemistry, 271*, 28533–28540.

Strydom, D. J. (1998). The angiogenins. *Cellular and Molecular Life Sciences, 54*, 811–824.

Sturner, R. A., & Chang, J. K. (1988). Opioid peptide content in infant formulas. *Pediatric Research, 23*, 4–10.

Sun, Z., & Cade, J. R. (1999). A peptide found in schizophrenia and autism causes behaviour changes in rats. *Autism, 3*, 85–95.

Sun, Z., Cade, J. R., Fregly, M., & Privette, R. M. (1999). β-Casomorphin induces Fos-like immunoreactivity in discrete brain regions relevant to schizophrenia and autism. *Autism, 3*, 67–83.

Sun, Z., Zhang, Z., Wang, X., Cade, R., Elmir, Z., & Fregly, M. (2003). Relation of β-casomorphin to apnea in sudden infant death syndrome. *Peptides, 24*, 937–943.

Svedberg, J., de Hass, J., Leimenstoll, G., Paul, F., & Teschemacher, H. (1985). Demonstration of beta-casomorphin immunoreactive materials in in vitro digests of bovine milk and in small intestine contents after bovine milk ingestion in adult humans. *Peptides, 6*, 825–839.

Svensson, M., Fast, J., Mossberg, A.-K., Düringer, C., Gustafsson, L., Hallgren, C., et al. (2003). Alpha-lactalbumin unfolding is not sufficient to cause apoptosis, but is required for the conversion to HAMLET (human alpha-lactalbumin made lethal to tumor cells). *Protein Science, 12*, 2794–2804.

Szwajkowska, M., Wolanciuk, A., Barlowska, J., Kroll, J., & Litwinczuk, Z. (2011). Bovine milk proteins as the source of bioactive peptides influencing the consumers' immune system – A review. *Animal Science Papers and Reports, 29*, 269–280.

Tailford, K. A., Berry, C. L., Thomas, A. C., & Campbel, J. H. (2003). A casein variant in cow's milk is atherogenic. *Atherosclerosis, 170*, 13–19.

Tang, S. Q., Jiang, Q. Y., Zhang, Y. L., Zhu, X. T., Shu, G., Gao, P., et al. (2008). Obestatin: Its physiochemical characteristics and physiological functions. *Peptides, 29*, 639–645.

Tao, N., DePeters, E. J., Freeman, S., German, J. B., Grimm, R., & Lebrilla, C. B. (2008). Bovine milk glycome. *Journal of Dairy Science, 91*, 3768–3778.

Taveras, E. M., Rifas-Shiman, S. L., Sherry, B., Oken, E., Haines, J., Kleinman, K., et al. (2011). Crossing growth percentiles in infancy and risk of obesity in childhood. *Archives of Pediatrics and Adolescent Medicine, 165*, 993–998.

Teschemacher, H. (2003). Opioid receptor ligands derived from food proteins. *Current Pharmaceutical Design, 9*, 1331–1344.

Teschemacher, H., Brantl, V., Henschen, A., & Lottspeich, F. (1990). β-Casomorphins, β-casein fragments with opioid activity: Detection and structure. In V. Brantl & H. Teschemacher (Eds.), *β-Casomorphins and related peptides: Recent developments* (pp. 9–14). Weinheim, Germany: Wiley-VCH Verlag GmbH.

Teschemacher, H., Umbach, M., Hamel, U., Praetorius, K., Ahnert-Hilger, G., Brantl, V., et al. (1986). No evidence for the presence of β-casomorphins in human plasma after ingestion of cows' milk or milk products. *Journal of Dairy Research, 53*, 135–138.

Thomas, T., & Thomas, T. J. (2003). Polyamine metabolism and cancer. *Journal of Cellular and Molecular Medicine, 7*, 113–126.

Thomä-Worringer, C., Sørensen, J., & López-Fandiño, R. (2006). Health effects and technological features of caseinomacropeptide. *International Dairy Journal, 16*, 1324–1333.

Thormar, H., & Hilmarsson, H. (2007). The role of microbicidal lipids in host defense against pathogens and their potential role as therapeutic agents. *Chemistry and Physics of Lipids, 150*, 1–11.

Thorsdottir, I., Birgisdottir, B. E., Johannsdottir, I. M., & Harris, P. (2000). Different (beta-casein) fractions in Icelandic versus Scandinavian cow's milk may influence diabetogenicity of cow's milk in infancy and explain low incidence of insulin-dependent diabetes mellitus in Iceland. *Pediatrics, 106*, 719–724.

Tomita, M., Wakabayashi, H., Yamauchi, K., Teraguchi, S., & Hayasawa, H. (2001). Bovine lactoferrin and lactoferricin derived from milk: Production and applications. *Biochemistry and Cell Biology, 80*, 109–112.

Toyosaki, T., & Mineshita, T. (1988). Antioxidant effects of protein-bound riboflavin and free riboflavin. *Journal of Food Science, 53*, 1851–1853.

Tucker, H. A., & Schwalm, J. W. (1977). Glucocorticoids in mammary tissue and milk. *Journal of Animal Science, 45*, 627–634.

Uauy, R., Birch, D., Birch, E., Tyson, J., & Hoffman, D. (1990). Effect of dietary omega-3 fatty acids on retinal function of very-low –birth-weight neonates. *Pediatric Research, 28*, 485–492.

Uçar, B., Kirel, B., Bör, O., Kilic, F. S., Dogruel, N., Tekin, N., et al. (2000). Breast milk leptin concentrations in initial and terminal milk samples: Relationships to maternal and infant plasma leptin concentrations, adiposity, serum glucose, insulin, lipid and lipoprotein levels. *Journal of Pediatric Endocrinology & Metabolism, 13*, 149–156.

Uniacke-Lowe, T., & Fox, P. F. (2012). Equid milk: Chemistry, biochemistry and processing. In B. Simpson (Ed.), *Food chemistry and food processing* (2nd ed., pp. 491–530). Ames, IA: Wiley-Blackwell.

Uniacke-Lowe, T., Huppertz, T., & Fox, P. F. (2010). Equine milk proteins: Chemistry structure and nutritional significance. *International Dairy Journal, 20*, 609–629.

Urashima, T., Asakuma, S., Kitaoka, K., & Messer, M. (2011). Indigenous oligosaccharides in milk. In J. W. Fuquay, P. F. Fox, & P. L. H. McSweeney (Eds.), *Encyclopedia of dairy sciences* (2nd ed., pp. 241–273). Oxford: Academic.

Urashima, T., Kitaoka, K., Asakuma, S., & Messer, M. (2009). Milk oligosaccharides. In P. L. H. McSweeney & P. F. Fox (Eds.), *Advanced dairy chemistry, volume 3, lactose, water, salts and minor constituents* (3rd ed., pp. 295–349). New York, NY: Springer.

Urashima, T., Messer, M., & Oftedal, O. T. (2014). Comparative biochemistry and evolution of milk oligosaccharides of monotremes, marsupials, and eutherians. In P. Pontarotti (Ed.), *Evolutionary biology: Genome evolution, speciation, coevolution and origin of life* (pp. 3–33). Switzerland: Springer International.

Uysal, F. K., Onal, E. E., Aral, Y. Z., Adam, B., Dilmen, U., & Ardicolu, Y. (2002). Breast milk leptin: Its relationship to maternal and infant adiposity. *Clinical Nutrition, 21*, 157–160.

Vacca-Smith, A. M., Van Wuyckhuyse, B. C., Tabak, L. K., & Bowen, W. H. (1994). The effect of milk and casein proteins on the adherence of *Streptococcus mutans* to saliva-coated hydroxy-apatite. *Archives of Oral Biology, 39*, 1063–1069.

Valtonen, M., Kangas, A.-P., Voutilainen, M., & Eriksson, L. (2003). Diurnal rhythm of melatonin in young calves and intake of melatonin in milk. *Animal Science, 77*, 149–154.

Van Der Kraan, M. I. A., Groenink, J., Nazmi, K., Veerman, E. C. I., Bolscher, J. G. M., & Nieuw Anerongen, A. V. (2004). Lactoferrampin: A novel antimicrobial peptide in the N1-domain of bovine lactoferrin. *Peptides, 25*, 177–183.

Van Der Kraan, M. I. A., Nazmi, K., Teeken, A., Groenink, J., Van Thof, W., Veerman, E. C., et al. (2005). Lactoferrampin, an antimicrobial peptide of bovine lactoferrin, exerts its candidacidal activity by a cluster of positively charged residues at the C-terminus in combination with a helix-facilitating N-terminal part. *The Journal of Biological Chemistry, 386*, 137–142.

Van Der Meer, R., Kleibeuker, J. H., & Lapre, J. A. (1991). Calcium phosphate, bile acids and colorectal cancer. *European Journal of Cancer Prevention, 1*, 55–62.

Vegarud, G., Langsrud, T., & Svenning, C. (2000). Mineral-binding milk proteins and peptides: Occurrence, biochemical and technological characteristics. *British Journal of Nutrition, 84*, s91–s98.

Venn, B. J., Skeaff, C. M., Brown, R., Mann, J. I., & Green, T. J. (2006). A comparison of the effects of A1 and A2 beta-casein protein variants on blood cholesterol concentrations in New Zealand adults. *Atherosclerosis, 188*, 175–178.

Vidal, K., Labéta, M. O., Schiffrin, E. J., & Donnet-Hughes, A. (2001). Soluble CD14 in human breast milk and its role in innate immune responses. *Acta Odontologica Scandinavica, 59*, 330–334.

Wada, Y., & Lönnerdal, B. (2014). Bioactive peptides derived from human milk proteins—Mechanism of action. *The Journal of Nutritional Biochemistry, 25*, 503–514.

Walstra, P. (2003). *Physical chemistry of foods*. New York, NY: Marcel Dekker.

Wang, B., & Brand-Miller, J. (2003). The role and potential of sialic acid in human nutrition. *European Journal of Clinical Nutrition, 57*, 1351–1369.

Welsch, U., Buchheim, W., Schumacher, U., Schinko, I., & Patton, S. (1988). Structural, histochemical and biochemical observations on horse milk-fat-globule membranes and casein micelles. *Histochemistry, 88*, 357–365.

Whigham, L. D., Cook, M. E., & Atkinson, R. L. (2000). Conjugated linoleic acid: Implications for human health. *Pharmacological Research, 42*, 503–510.

Widdowson, E. M. (1984). Lactation and feeding patterns in different species. In D. L. M. Freed (Ed.), *Health hazards of milk* (pp. 85–90). London, UK: Baillière Tindall.

Wiederschain, G. Y., & Newburg, D. S. (2001). Glycoconjugate stability in human milk: Glycosidase activities and sugar release. *The Journal of Nutritional Biochemistry, 12*, 559–564.

Winter, P., Miny, M., Fuchs, K., & Baumgartner, W. (2006). The potential of measuring serum amyloid A in individual ewe milk and in farm bulk milk for monitoring udder health on sheep dairy farms. *Research in Veterinary Science, 81*, 321–326.

Wolford, S. T., & Argoudelis, C. J. (1979). Measuring estrogen in cow's milk, human milk and dietary products. *Journal of Dairy Science, 62*, 1458–1463.

Wynn, P. C., & Sheehy, P. A. (2013). Minor proteins, including growth factors. In P. L. H. McSweeney & P. F. Fox (Eds.), *Advanced dairy chemistry: Vol. 1A: Proteins: Basic aspects* (4th ed., pp. 317–335). New York, NY: Springer.

Xu, R. J. (1998). Bioactive peptides in milk and their biological and health implications. *Food Reviews International, 14*, 1–17.

Xu, R. J., Sangild, P. T., Zhang, Y. Q., & Zhang, S. H. (2002). Bioactive components in porcine colostrum and milk and their effects on intestinal development in neonatal pigs. In R. Zabielski, P. C. Gregory, B. Westrom, & E. Salek (Eds.), *Biology of growing animals, vol. 1* (pp. 169–192). UK: Elsevier.

Xu, R. J., Wang, F., & Zhang, S. H. (2000). Postnatal adaptation of the gastrointestinal tract in neonatal pigs: A possible role of milk-borne growth factors. *Livestock Production Science, 66*, 95–107.

Yamauchi, R., Wada, E., Yamada, D., Yoshikawa, M., & Wada, K. (2006). Effect of β–lactotensin on acute stress and fear memory. *Peptides, 27*, 3176–3182.

Yolken, R. H., Peterson, J. A., Vonderfecht, S. L., Fouts, E. T., Midthun, K., & Newburg, D. S. (1992). Human milk mucin inhibits rotavirus replication and prevents experimental gastroenteritis. *The Journal of Clinical Investigation, 90*, 1984–1991.

Yvon, M., Beucher, S., Guilloteau, P., Le Huerou-Luron, I., & Corring, T. (1994). Effects of caseinomacropeptide (CMP) on digestion regulation. *Reproduction Nutrition Development, 34*, 527–537.

Yvon, M., & Pelissier, J. P. (1987). Characterization and kinetics of evacuation of peptides resulting from casein hydrolysis in the stomach of the calf. *Journal of Agricultural and Food Chemistry, 35*, 148–156.

Zemel, M. B. (2004). Role of calcium and dairy products in energy partitioning and weight management. *The American Journal of Clinical Nutrition, 79*, 907S–912S.

Zemel, M. B. (2005). The role of dairy foods in weight management. *Journal of the American College of Nutrition, 24*, 537S–546S.

Zhang, T., Iguchi, K., Mochizuki, T., Hoshino, M., Yanaihara, C., & Yanaihara, N. (1990). Gonadotropin-releasing hormone-associated peptide immunoreactivity in bovine colostrum. *Experimental Biology and Medicine, 194*, 270–273.

Zinn, S. (1997). Bioactive components in milk: Introduction. *Livestock Production Science, 50*, 101–103.

Suggested Reading

García-Montoya, I. A., Siqueiros Cendón, T., Arévalo-Gallegos, S., & Rasón-Cruz, Q. (2012). Lactoferrin a multiple bioactive protein: An overview. *Biochimica et Biophysica Acta, 1820,* 226–236.

Guo, M. (2014). *Human milk biochemistry and infant formula manufacturing technology.* Cambridge, UK: Woodhead Publishing.

Korhonen, H. J. (2011). Bioactive milk proteins, peptides and lipids and other functional components derived from milk and bovine colostrum. In M. Saarela (Ed.), *Functional foods* (2nd ed., pp. 471–511). Cambridge, UK: Woodhead Publishing.

Park, Y. W. (Ed.). (2009). *Bioactive components in milk and dairy products.* Ames, IA: Wiley-Blackwell.

Pihlanto, A., & Korhonen, H. (2003). Bioactive peptides and proteins. In S. Taylor (Ed.), *Advances in food and nutritional research* (Vol. 47, pp. 175–276). Boston, MA: Academic.

Shortt, C., & O'Brien, J. (2004). *Handbook of functional dairy products.* Boca Raton, FL: CRC Press.

Chapter 12
Chemistry and Biochemistry of Cheese

12.1 Introduction

Cheese is a very varied group of dairy products, produced worldwide; cheesemaking originated in the Middle East during the Agricultural Revolution, about 8,000 years ago. Cheese production and consumption, which vary widely between countries and regions is increasing in traditional producing countries and is spreading to new areas.

Although most traditional cheeses have a rather high fat content, they are rich sources of protein and in most cases of calcium and phosphorus and have anticariogenic properties; some typical compositional data are presented in Table 12.1. Cheese is the classical example of a convenience food: it can be used as the main course in a meal, as a dessert or snack, as a sandwich filler, food ingredient or condiment.

There are probably about 2,000 named cheese varieties, most of which have very limited production. The principal families are Cheddar, Dutch, Swiss and *pasta filata* (e.g., Mozzarella), which together account for the big majority of total cheese production. All varieties can be classified into three super-families based on the method used to coagulate the milk, i.e., rennet coagulation, which represent ~75 % of total production, isoelectric (acid) coagulation and a combination of heat and acid, which represent a very minor group. The diversity of cheese and cheese-related products is summarised in Fig. 12.1.

Production of cheese curd is essentially a concentration process in which the milk-fat and casein are concentrated about tenfold while the whey proteins, lactose and soluble salts are removed in the whey. The acid-coagulated and acid/heat-coagulated cheeses are normally consumed fresh but the vast majority of rennet-coagulated cheeses are ripened (matured) for a period ranging from 2 weeks to >2 years, during which numerous microbiological, biochemical, chemical and physical changes occur, resulting in characteristic flavour, aroma and texture. The biochemistry of cheese ripening is very complex and has been the subject of extensive recent study.

© Springer International Publishing Switzerland 2015
P.F. Fox et al., *Dairy Chemistry and Biochemistry*,
DOI 10.1007/978-3-319-14892-2_12

Table 12.1 Composition of selected cheeses (per 100 g)

Cheese type	Water (g)	Protein (g)	Fat (g)	Cholesterol (mg)	Energy (kJ)
Brie	48.6	19.3	26.9	100	1,323
Caerphilly	41.8	23.2	31.3	90	1,554
Camembert	50.7	20.9	23.1	75	1,232
Cheddar	36.0	25.5	34.4	100	1,708
Cheshire	40.6	24.0	31.4	90	1,571
Cottage	79.1	13.8	3.9	13	413
Cream cheese	45.5	3.1	47.4	95	1,807
Danish blue	45.3	20.1	29.6	75	1,437
Edam	43.8	26.0	25.4	80	1,382
Emmental	35.7	28.7	29.7	90	1,587
Feta	56.5	15.6	20.2	70	1,037
Fromage frais	77.9	6.8	7.1	25	469
Gouda	40.1	24.0	31.0	100	1,555
Gruyere	35.0	27.2	33.3	100	1,695
Mozzarella	49.8	25.1	21.0	65	1,204
Parmesan	18.4	39.4	32.7	100	1,880
Ricotta	72.1	9.4	11.0	50	599
Roquefort	41.3	19.7	32.9	90	1,552
Stilton	38.6	22.7	35.5	105	1,701

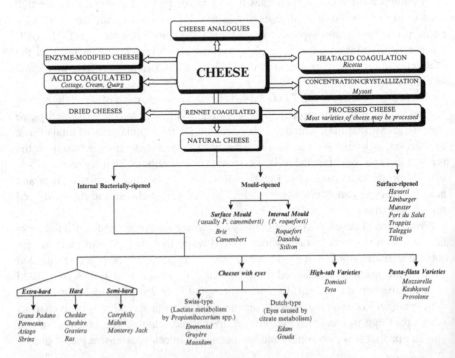

Fig. 12.1 A classification scheme for cheese and related products based principally on method of coagulation of the milk (from McSweeney et al. 2004)

12.2   Rennet-Coagulated Cheeses

The production of rennet-coagulated cheeses can, for convenience, be divided into two phases: (1) conversion of milk to curds and (2) ripening of the curds.

12.2.1   Preparation and Treatment of Cheese Milk

The milk for most cheese varieties is subjected to one or more pre-treatments (Table 12.2). The concentrations of fat and casein and the ratio of these components are two very important parameters affecting cheese quality. While the concentrations of these components in cheese are determined and controlled by the manufacturing protocol, their ratio is regulated by adjusting the composition of the cheese milk. This is usually done by adjusting the fat content by blending whole and skimmed milk in proportions needed to give the desired fat:casein ratio in the finished cheese, e.g., 1.0:0.7 for Cheddar or Gouda. It should be remembered that ~10 % of the fat in milk is lost in the whey while only about 5 % of the casein is lost (unavoidably, see Sect. 12.2.2).

With the recent commercial availability of ultrafiltration, it has become possible to control the actual concentration of casein, and not just its ratio to fat, thus levelling out seasonal variations in milk composition and consequently giving more consistent gel characteristics, cheese quality and better yield.

The pH and the concentration of calcium in milk also vary, with consequential effects on the properties of renneted milk gels. The addition of $CaCl_2$ to cheese milk (0.02 %) is widely practiced and adjustment and standardization of milk pH by using the acidogen, gluconic acid-δ-lactone (GDL), is commercially practised on a limited scale.

Table 12.2 Pre-treatment of cheese milk	*Standardization of fat:protein ratio or concentration*
	Addition of skim milk
	Removal of some fat
	Control of casein level using low concentration factor ultrafiltration
	Addition of CaCl$_2$
	Adjustment of pH (e.g., by gluconic acid-δ-lactone)
	Removal or killing of contaminating bacteria
	Thermization (e.g., 65 °C × 15 s)
	Pasteurization (e.g., 72 °C × 15 s)
	Bactofugation
	Microfiltration

Although raw milk is still widely used for cheese manufacture, e.g., Parmigiano Reggiano (Italy), Emmental (Switzerland), Gruyere de Comté and Beaufort (France) and many less well known varieties, both on a factory and farmhouse scale, most Cheddar and Dutch-type cheeses are produced from pasteurized milk (usually high temperature short time, HTST; $\sim72\,°C \times \sim15$ s). The primary objective of pasteurisation is to kill vegetative pathogens, but it also kills many spoilage organisms and components of the non-starter microflora (see Sect. 12.2.7). However, many desirable indigenous bacteria are also killed by pasteurization and it is generally agreed that cheese made from pasteurized milk ripens more slowly and develops a less intense flavour than raw milk cheese. These differences are caused mainly by the absence of these indigenous bacteria, but the thermal inactivation of certain indigenous enzymes, particularly lipoprotein lipase (see Chap. 10), also contributes. At present, some countries require that all cheese milk should be pasteurized or the cheese aged for at least 60 days (during which time it is hoped that pathogenic bacteria die off). A global requirement for pasteurization of cheese milk has been recommended but would create restrictions for international trade in cheese, especially for the many traditional cheeses from southern Europe made from raw milk and with protected designations of origin. Research is underway to identify the important indigenous microorganisms in raw milk cheese for use as inoculants for pasteurized milk. While recognising that pasteurization is very important in ensuring safe cheese, pH ($< \sim5.2$), water activity (a_w, which is controlled by addition of NaCl), a low level of residual lactose and low oxidation-reduction potential are also critical safety hurdles.

Milk may be thermized ($\sim65\,°C \times 15$ s) on receipt at the factory to reduce bacterial load, especially psychrotrophs, which are heat labile. Since thermization does not kill all pathogens, thermized milk must be fully pasteurized before cheesemaking.

Clostridium tyrobutyricum (an anaerobic sporeformer) causes late gas blowing (through the production of H_2 and CO_2) and off-flavours (butanoic acid) in many hard ripened cheeses; dry-salted varieties such as Cheddar-type cheeses are major exceptions. Contamination of cheese milk with clostridial spores can be avoided or kept to a very low level by good hygienic practises (soil and silage are the principal sources of clostridia) but they are usually prevented from growing through the use of sodium nitrate ($NaNO_3$) or less frequently, lysozyme, and/or removed by bactofugation (centrifugation) or microfiltration.

12.2.2   Conversion of Milk to Cheese Curd

Typically, five steps, or groups of steps, are involved in the conversion of milk to cheese curd: coagulation, acidification, syneresis (expulsion of whey), moulding/shaping and salting (Fig. 12.2). These steps, which partly overlap, enable the cheesemaker to control the composition of cheese, which, in turn, has a major influence on cheese ripening and quality.

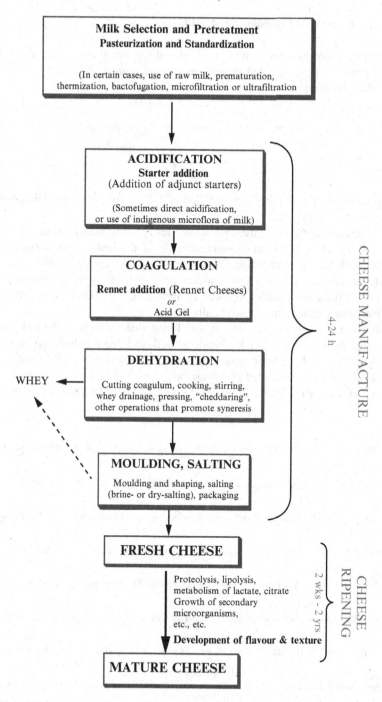

Fig. 12.2 Schematic flow diagram showing the major stages in the manufacture of a typical rennet-coagulated cheese

12.2.2.1   Enzymatic Coagulation of Milk

The enzymatic coagulation of milk involves modification of the casein micelles via limited proteolysis by selected proteinases, called rennets, followed by calcium-induced aggregation of the rennet-altered micelles:

$$\text{Casein} \xrightarrow{\text{Rennet}} \textit{para-}\text{casein} + \text{macropeptides}$$

$$\downarrow \text{Ca}^{2+}, > \sim 20^{\circ}\text{C}$$

$$\text{Gel}$$

If present, the fat globules are occluded in the gel but do not participate in the formation of a gel matrix.

As discussed in Chap. 4, the casein micelles are stabilized by κ-casein, which represents 12–15 % of the total casein and is located mainly on the surface of the micelles such that its hydrophobic N-terminal region reacts hydrophobically with the calcium-sensitive α_{s1}-, α_{s2}- and β-caseins while its hydrophilic C-terminal region protrudes into the surrounding aqueous environment, stabilizing the micelles by a negative surface charge and steric stabilization.

Following its isolation in 1956, it was found that κ-casein is the only casein hydrolysed during the rennet coagulation of milk and that it is hydrolysed specifically at the Phe_{105}-Met_{106} bond, producing *para-*κ-casein (κ-CN f1-105) and macropeptides (f106-169; also called glycomacropeptides since they contain most or all of the sugar groups attached to κ-casein) (Fig. 12.3). The hydrophilic macropeptides

1
Pyro Glu-Glu-Gln-Asn-Gln-Glu-Gln-Pro-Ile-Arg-Cys-Glu-Lys-Asp-Glu-Arg-Phe-Phe-Ser-Asp-

21
Lys-Ile-Ala-Lys-Tyr-Ile-Pro-Ile-Gln-Tyr-Val-Leu-Ser-Arg-Tyr-Pro-Ser-Tyr-Gly-Leu-

41
Asn-Tyr-Tyr-Gln-Gln-Lys-Pro-Val-Ala-Leu-Ile-Asn-Asn-Gln-Phe-Leu-Pro-Tyr-Pro-tyr-

61
Tyr-Ala-Lys-Pro-Ala-Ala-Val-Arg-Ser-Pro-Ala-Gln-Ile-Leu-Gln-Trp-Gln-Val-Leu-Ser-

81
Asn-Thr-Val-Pro-Ala-Lys-Ser-Cys-Gln-Ala-Gln-Pro-Thr-Thr-Met-Ala-Arg-His-Pro-His-

101   105 | 106
Pro-His-Leu-Ser-Phe▼Met-Ala-Ile-Pro-Pro-Lys-Lys-Asn-Gln-Asp-Lys-Thr-Glu-Ile-Pro-

121   Ile (Variant B)
Thr-Ile-Asn-Thr-Ile-Ala-Ser-Gly-Glu-Pro-*Thr*- Ser-*Thr*-Pro-*Thr*-   -Glu-Ala-Val-Glu-
                                                            Thr (Variant A)
141   Ala (Variant B)
Ser-*Thr*-Val-Ala-Thr-Leu-Glu-   -*SerP* - Pro-Glu-Val-Ile-Glu-Ser-Pro-Pro-Glu-Ile-Asn-
                              Asp (Variant A)
161   169
Thr-Val-Gln-Val-Thr-Ser-Thr-Ala-Val.OH

Fig. 12.3 Amino acid sequence of κ-casein, showing the principal chymosin cleavage site (*down arrow*); oligosaccharides are attached at some or all of the threonine residues shown in *italics*

diffuse into the surrounding medium while the *para*-κ-casein remains attached to the micelle core (the macropeptides represent ~30 % of κ-casein, i.e., 4–5 % of total casein; this unavoidable loss must be considered when calculating the yield of cheese). Removal of the macropeptides from the surface of the casein micelles reduces their zeta potential from ~-20 mV to ~-10 mV and removes the steric stabilizing layer. The proteolysis of κ-casein is referred to as the *primary (first) phase* of rennet coagulation.

When ~85 % of the total κ-casein in milk has been hydrolysed, the colloidal stability of the micelles is reduced to such an extent that they coagulate at temperatures > ~20 °C (typically, a coagulation temperature of 30 °C is used in cheesemaking), an event referred to as the *secondary phase* of rennet coagulation. Calcium ions are essential for the coagulation of rennet-altered micelles (although the binding of Ca^{2+} by casein is not affected by renneting).

The Phe_{105}-Met_{106} bond of κ-casein is several orders of magnitude more sensitive to rennets than any other bond in the casein system. The reason(s) for this unique sensitivity has not been fully established but work on synthetic peptides that mimic the sequence of κ-casein around this bond has provided valuable information. The Phe and Met residues themselves are not essential, e.g., both Phe_{105} and Met_{106} can be replaced or modified without drastically changing the sensitivity of the bond—in human, porcine and rodent κ-caseins, Met_{106} is replaced by Ile or Leu and the proteinase from *Cryphonectria parasitica* (see Sect. 12.2.2.2), hydrolyses the bond Ser_{104}-Phe_{105} rather than Phe_{105}-Met_{106}. The smallest κ-casein-like peptide hydrolysed by chymosin is Ser.Phe.Met.Ala.Ile (κ-CN f104-108); extending this peptide from its C and/or N-terminal increases its susceptibility to chymosin (i.e., increases k_{cat}/K_m); the peptide κ-CN f98-111 is as good a substrate for chymosin as is whole κ-casein (Table 12.3). Ser_{104} appears to be essential for cleavage of the Phe_{105}-Met_{106}

Table 12.3 Kinetic parameters for hydrolysis of κ-casein peptides by chymosin at pH 4.7 (compiled from Visser et al. 1976, 1987)

Peptide	Sequence	k_{cat} (S^{-1})	K_M (mM)	k_{cat}/K_M (S^{-1} mM^{-1})
S.F.M.A.I.	104–108	0.33	8.50	0.038
S.F.M.A.I.P.	104–109	1.05	9.20	0.114
S.F.M.A.I.P.P.	104–110	1.57	6.80	0.231
S.F.M.A.I.P.P.K.	104–111	0.75	3.20	0.239
L.S.F.M.A.I.	103–108	18.3	0.85	21.6
L.S.F.M.A.I.P.	103–109	38.1	0.69	55.1
L.S.F.M.A.I.P.P.	103–110	43.3	0.41	105.1
L.S.F.M.A.I.P.P.K.	103–111	33.6	0.43	78.3
L.S.F.M.A.I.P.P.K.K.	103–112	30.2	0.46	65.3
H.L.S.F.M.A.I.	102–108	16.0	0.52	30.8
P.H.L.S.F.M.A.I.	101–108	33.5	0.34	100.2
H.P.H.P.H.L.S.F.M.A.I.P.P.K.	98–111	66.2	0.026	2,509
	98–111[a]	46.2[a]	0.029[a]	1,621[a]
κ-Casein[b]		2–20	0.001–0.005	200–2,000
L.S.F.(NO$_2$)NleA.L.OMe		12.0	0.95	12.7

[a]pH 6.6
[b]pH 4.6

bond by chymosin and the hydrophobic residues, Leu_{103}, Ala_{107} and Ile_{108} are also important. Thus, Phe-Met bond of κ-casein is a chymosin-susceptible bond located in an exposed region of the molecule from residues 98 to 111 that exists as an exposed loop and can fit easily into the active site of the enzyme.

12.2.2.2   Rennets

The traditional rennets used to coagulate milk for most cheese varieties are prepared from the stomachs of young calves, lambs or kids by extraction with NaCl (~15 %) brines. The principal proteinase in such rennets is chymosin; about 10 % of the milk clotting activity of calf rennet is due to pepsin. As the animal ages, the secretion of chymosin declines while that of pepsin increases.

Like pepsin, chymosin is an aspartyl (acid) proteinase, i.e., it has two essential aspartyl residues in its active site which is located in a cleft in the globular molecule (MW ~36 kDa) (Fig. 12.4). Its pH optimum for general proteolysis is ~4, in comparison with ~2 for pepsins from monogastric animals. Its general proteolytic activity is low relative to its milk clotting activity and it has moderately high specificity for bulky hydrophobic residues at the P_1 and P_1^1 positions of the scissile bond. Its physiological function appears to be to coagulate milk in the stomach of the neonate, thereby increasing the efficiency of digestion, by retarding discharge into the intestine, rather than general proteolysis.

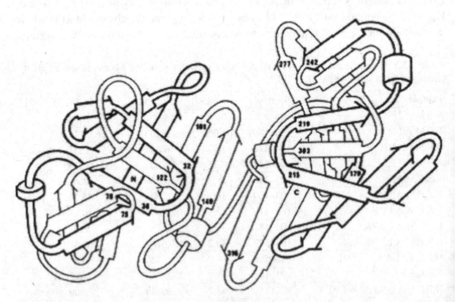

Fig. 12.4 Schematic representation of the tertiary structure of an aspartyl proteinase, showing the cleft which contains the active site; *arrows* indicate β structures and *cylinders* the α-helices (from Foltmann 1987)

Due to increasing world production of cheese and the declining supply of young calf stomachs (referred to as vells), the supply of calf rennet has been inadequate for many years. This has led to a search for suitable substitutes. Many proteinases are capable of coagulating milk but nearly all are too proteolytic relative to their milk clotting activity, leading to a decrease in cheese yield (due to excessive non-specific proteolysis in the cheese vat and loss of peptides in the whey) and defects in the flavour and texture of the ripened cheese, due to excessive or incorrect proteolysis. Only six proteinases have been used commercially as rennet substitutes: porcine, bovine and chicken pepsins and the acid proteinases from *Rhizomucor miehei, R. pusillus* and *C. parasitica*. Chicken pepsin is quite proteolytic and is now used very rarely. Porcine pepsin enjoyed limited success some decades ago, usually in admixtures with calf rennet, but it is very sensitive to denaturation at pH values >6 and may be denatured extensively during cheesemaking, leading to impaired proteolysis during ripening; it is now rarely used as a rennet substitute. Bovine pepsin is quite effective and many commercial calf rennets contain up to 50 % bovine pepsin. *R. miehei* proteinase, the most widely used microbial rennet, gives generally satisfactory results. *C. parasitica* proteinase is, in general, the least suitable of the commercial microbial rennet substitutes and is used only in high-cooked cheeses in which extensive denaturation of the coagulant occurs, e.g., Swiss-type cheeses.

The gene for calf chymosin has been cloned in *Kluyveromyces lactis, Aspergillus niger* and *E. coli*. Fermentation-produced chymosins have given excellent results in cheesemaking trials on various varieties and are now widely used commercially, although they are not permitted in all countries. Significantly, they are accepted for use in vegetarian cheeses and have Kosher and Halal status. Two such coagulants, Maxiren (DSM Food Specialties, Delft, Netherlands) and Chymax (Chr Hansen, Horshølm, Denmark) are available. Recently, the gene for camel chymosin was cloned and the enzyme is now available commercially (Chymax-M; Chr Hansen) and shows promise.

12.2.2.3   Coagulation of Rennet-Altered Micelles

When *ca.* 85 % of the total κ-casein has been hydrolysed, the micelles begin to aggregate progressively into a gel network. Gelation is indicated by a rapid increase in viscosity (η) (Fig. 12.5). The viscosity of milk is reduced marginally during the early stages of renneting as the enzyme removes the glycomacropeptide. In doing so, the effective volume of the casein micelle is reduced by a small reduction in its size and a large reduction in micelle hydration, which is observed as a slight reduction in viscosity. Coagulation commences at a lower degree of hydrolysis of κ-casein if the temperature is increased, the pH reduced or the Ca^{2+} concentration increased.

The actual reactions leading to coagulation are not known. Ca^{2+} are essential but Ca-binding by caseins does not change on renneting. Colloidal calcium phosphate (CCP) is also essential: reducing the CCP concentration by >20 % prevents coagulation. Perhaps, hydrophobic interactions, which become dominant when the surface charge and steric stabilization are reduced on hydrolysis of κ-casein, are

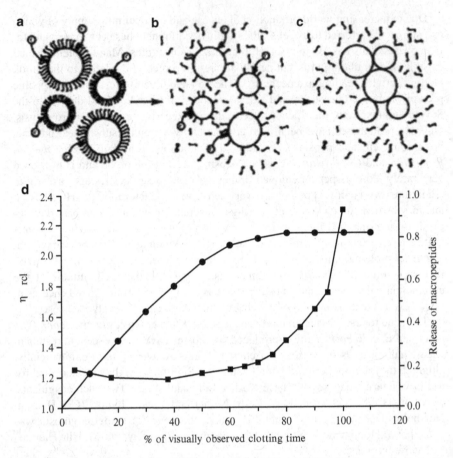

Fig. 12.5 Schematic representation of the rennet coagulation of milk. (**a**) Casein micelles with intact κ-casein layer being attacked by chymosin (C); (**b**) micelles partially denuded of κ-casein; (**c**) extensively denuded micelles in the process of aggregation; (**d**) release of macropeptides (*filled circles*) and changes in relative viscosity (*filled squares*) during the course of rennet coagulation

responsible for coagulation (the coagulum is soluble in urea). The adverse influence of moderately high ionic strength on coagulation suggests that electrostatic interactions are also involved. It is claimed that pH has no effect on the secondary stage of rennet coagulation, which is perhaps surprising since micellar charge is reduced by lowering the pH and should facilitate coagulation. Coagulation is very temperature-sensitive and does not occur < ~18 °C, above which the temperature coefficient, Q_{10}, is ~16. The precise reason why milk does not coagulate in the cold is not known with certainty but may be related to the dissociation of β-casein from the casein micelle at low temperatures (caused by weakened hydrophobic interactions); partially dissociated β-casein may form a protective layer inhibiting aggregation of the renneted micelles.

12.2.2.4 Factors That Affect Rennet Coagulation

The effect of various compositional and environmental factors on the primary and secondary phases of rennet coagulation and on the overall coagulation process are summarized in Fig. 12.6.

Factor	First phase	Second phase	Overall effect, see panel
Temperature	+	++	a
pH	+++	-	b
Ca	-	+++	c
Pre-heating	++	++++	d
Rennet concentration	++++	-	e
Protein concentration	+	++++	f

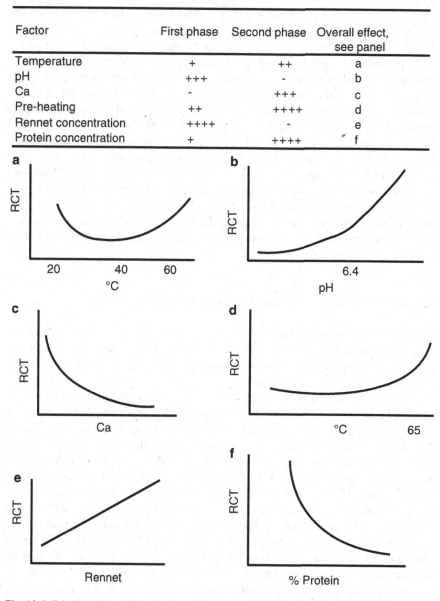

Fig. 12.6 Principal factors affecting the rennet coagulation time (RCT) of milk

No coagulation occurs <20 °C, due mainly to the very high temperature coefficient of the secondary phase. At higher temperatures (>55–60 °C, depending on pH and enzyme) the rennet is denatured. Rennet coagulation is prolonged or prevented by preheating milk at temperatures > ~70 °C (depending on the length of exposure). The effect is due to the interaction of thermally denatured β-lactoglobulin with κ-casein via disulphide (-S-S-) bonds; both the primary and, especially, the secondary phase of coagulation are adversely affected. The effect of pH is mainly on the first (enzymatic) stage of rennet coagulation. As the pH of the milk decreases, the enzyme moves closer to its pH optimum speeding up the reaction. A slight effect on the second stage also occurs as moving the pH closer to the isoelectric point of the caseins reduces the repulsive charge on the micelles and facilitates aggregation. The principal effect of increasing Ca^{2+} concentration is on the second stage of rennet coagulation as calcium is essential for the aggregation of renneted micelles. However, the addition of calcium to milk changes the equilibrium from soluble towards colloidal (casein-bound) calcium phosphate with the production of H^+ and a slight drop in the pH of milk which favours the first stage of rennet coagulation. The major effect of pre-heating milk (e.g., pasteurisation conditions) is on the second stage of rennet coagulation where thermal denaturation of β-lactoglobulin and its interaction with κ-casein at the surface of the micelle via -S-S- bonds leads to adverse effects on the first and second stages of rennet coagulation. Increasing levels of rennet leads to faster coagulation as the first stage is completed more quickly and increasing levels of protein (casein) also speeds up the process due to increased levels of coagulable material.

12.2.2.5 Measurement of Rennet Coagulation Time

A number of principles are used to measure the rennet coagulability of milk or the activity of rennets; most measure actual coagulation, i.e., combined first and second stages, but some specifically monitor the hydrolysis of κ-casein. The most commonly used methods are described below.

The simplest method is to measure the time elapsed between the addition of a measured amount of diluted rennet to a sample of milk in a temperature-controlled water-bath at, e.g., 30 °C. If the coagulating activity of a rennet preparation is to be determined, a "reference" milk, e.g., low-heat milk powder reconstituted in 0.01 % $CaCl_2$, and perhaps adjusted to a certain pH, e.g., 6.5, should be used. A standard method has been published (IDF 1992) and a reference milk may be obtained from Institut National de la Recherche Agronomique, Poligny, France. If the coagulability of a particular milk is to be determined, the pH may or may not be adjusted to a standard value. The coagulation point may be determined by placing the milk sample in a bottle or tube which is rotated in a water bath (Fig. 12.7); the fluid milk forms a film on the inside of the rotating bottle/tube but flocs of protein form in the film on coagulation. Several types of apparatus using this principle have been described.

As shown in Fig. 12.5, the viscosity of milk increases sharply when milk coagulates and may be used to determine the coagulation point. Any type of viscometer

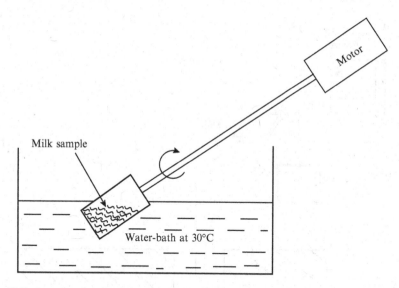

Fig. 12.7 Apparatus for visual determination of the rennet coagulation time of milk based on the International Dairy Federation (Berridge) method

may, theoretically, be used but several dedicated pieces of apparatus have been developed. One such apparatus, although with limited use, is the Formograph (Foss Electric, Denmark), a diagram of which is shown in Fig. 12.8a. Samples of milk to be analysed are placed in small beakers which are placed in cavities in an electrically heated metal block. Rennet is added and the loop-shaped pendulum of the instrument placed in the milk. The metal block is moved back and forth, creating a "drag" on the pendulum in the milk. The arm to which the pendulum is attached contains a mirror from which a flashing light is reflected onto photo-sensitive paper, creating a mark. While the milk is fluid, the viscosity is low and the drag on the pendulum is slight and it scarcely moves from its normal position; hence a single straight line appears on the paper. As the milk coagulates, the viscosity increases and the pendulum is dragged out of position, resulting in bifurcation of the trace. The rate and extent to which the arms of the trace move apart is an indicator of the strength (firmness) of the gel. A typical trace is shown in Fig. 12.8b. A low value of r indicates a short rennet coagulation time while high values of a_{30} and k_{20} indicate a milk with good gel-forming properties. Rheometers are now often used to follow the development of gel structure during coagulation, usually by observing the increase in G' (loss modulus) over time (Fig. 12.9)

One in-vat method for determining the cut time is the hot wire sensor. A diagram of the original assay cell is shown in Fig. 12.10a. A sample of milk is placed in a cylindrical vessel containing a wire of uniform dimensions. A current is passed through the wire, generating heat which is dissipated readily while the milk is liquid. As the milk coagulates, generated heat is no longer readily dissipated and the temperature of the wire increases, causing an increase in its conductivity; a typical

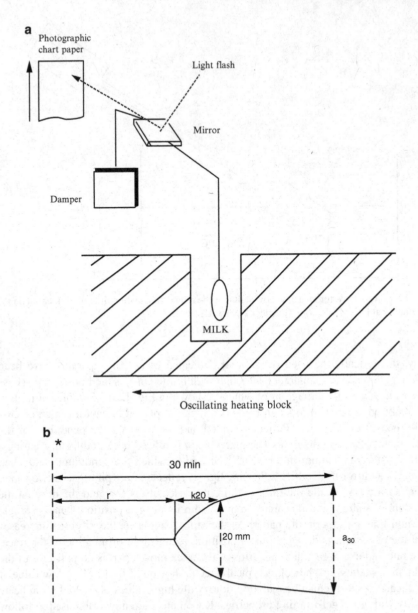

Fig. 12.8 (a) Schematic representation of the Formograph apparatus for determining the rennet coagulation of milk. (b) Typical formogram. *—point of rennet addition, r is rennet coagulation time, k_{20} is the time required from coagulation for the arms of the formogram to bifurcate by 20 mm, a_{30} is the extent of bifurcation 30 min after rennet addition (the approximate time at which the coagulum is cut in cheesemaking)

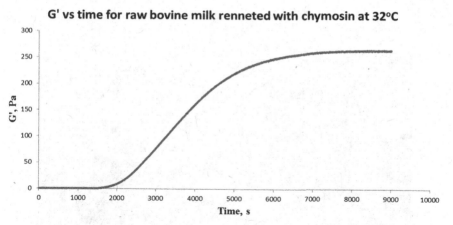

Fig. 12.9 Development of loss modulus (G', Pa) with time during rennet coagulation

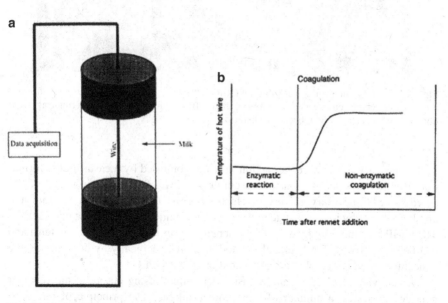

Fig. 12.10 (a) Hot wire sensor for objectively measuring the rennet coagulation of milk. (b) Changes in the temperature of the hot wire during the course of the rennet coagulation of milk

trace is shown in Fig. 12.10b. The wire probe, in a stainless steel shield, is inserted through the wall of the cheese vat. The output from the wire is fed to a computer which can be used to switch on the gel-cutting knife, permitting automation and cutting of the gel at a consistent strength, which is important for maximising cheese yield. Other methods used to determine the cut-time of a cheese vat include near infra-red (NIR) reflectance (Fig. 12.11) which determines micelle aggregation through the reflectance characteristics of the light and the attenuation of ultrasound during aggregation.

Fig. 12.11 Control unit for a near infra-red (NIR) reflectance sensor used to determine the cut-time of a commercial cheese vat. Such sensors shine NIR light into the vat and determine aggregation of casein micelles from reflectance characteristics

The primary phase of rennet action may be monitored by measuring the formation of either product, i.e., *para*-κ-casein or the GMP. *Para*-κ-casein may be measured by SDS-polyacrylamide gel electrophoresis (PAGE), which is slow and cumbersome or by ion-exchange high performance liquid chromatography (HPLC). The GMP is soluble in TCA (2–12 % depending on its carbohydrate content) and can be quantified by the Kjeldahl method or more specifically by determining the concentration of *N*-acetyl neuraminic acid or by RP-HPLC.

The activity of rennets can be easily determined using chromogenic or other peptide substrates, a number of which are available. The principle of one such method is shown in Fig. 12.12; this method can also be used to measure the low levels of rennet that remain in cheese.

12.2.2.6 Gel Strength (Curd Tension)

The gel network continues to develop for a considerable period after visible coagulation (Fig. 12.13). The strength of the gel formed, which is very important from the viewpoints of syneresis (and hence moisture control) and cheese yield, is affected by several factors, the principal of which are summarized in Fig. 12.14. Generally, gel strength is inversely related to RCT (i.e., fast RCT gives a good gel strength) and milks with high casein and Ca^{2+} concentrations have good gel strength. In the pH range 6.7–6.0,

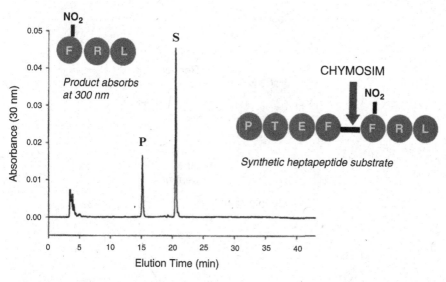

Fig. 12.12 Principle of a method used to measure chymosin activity using a synthetic seven-residue peptide substrate containing a chymosin-susceptible Phe–Phe bond (Hurley et al. 1999). One Phe residue has an –NO₂ group on its benzene ring which absorbs strongly at 300 nm. Substrate (S) and product (P) peptides are separated by HPLC and the area of the product peak is related to chymosin activity

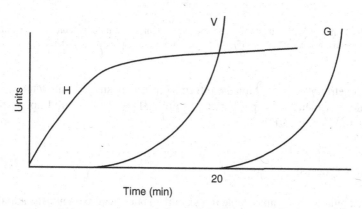

Fig. 12.13 Schematic representation of hydrolysis and gel formation in renneted milk; *H* hydrolysis of κ-casein, *V* changes in the viscosity of renneted milk (second stage of coagulation), *G* changes in the viscoelastic modulus (gel formation)

reducing pH leads to greater strength due to faster chymosin action and a reduction in intermicellar charge leading to better aggregation. High pre-heat treatment (e.g., excessive pasteurisation) is detrimental to gel strength as is homogensation of the milk.

The strength of a renneted milk gel can be measured by several types of viscometers and penetrometers. As discussed in Sect. 12.2.2.5, the Formograph gives a measure of the gel strength but the data cannot be readily converted to rheological

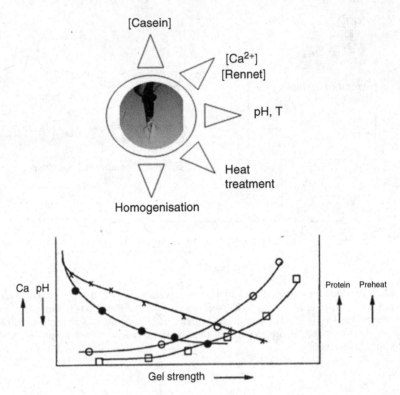

Fig. 12.14 Principal factors that affect the strength of renneted milk gels (curd tension); pH (*filled circle*), calcium concentration (*open circle*), protein concentration (*open square*), preheat treatment (×)

terms. Penetrometers give valuable information but are single point determinations. Dynamic rheometers are particularly useful, allowing the build-up of the gel network to be studied (Fig. 12.9).

12.2.2.7 Syneresis

Renneted milk gels are quite stable if undisturbed but synerese (contract), initially following first order kinetics, when cut or broken. By controlling the extent of syneresis, the cheesemaker can control the moisture content of cheese curd and hence the rate and extent of ripening and the stability of the cheese—the higher the moisture content, the faster the cheese will ripen but the lower its stability. Syneresis is promoted by:

– Cutting the curd finely, e.g., Emmental (fine cut) vs. Camembert (large cut)
– Low pH (Fig. 12.15b)
– Increased concentration of calcium ions
– Increasing the cooking temperature (Camembert, ~30 °C, Gouda, ~36 °C, Cheddar, ~38 °C, Emmental or Parmesan, 52–55 °C) (Fig. 12.15a)

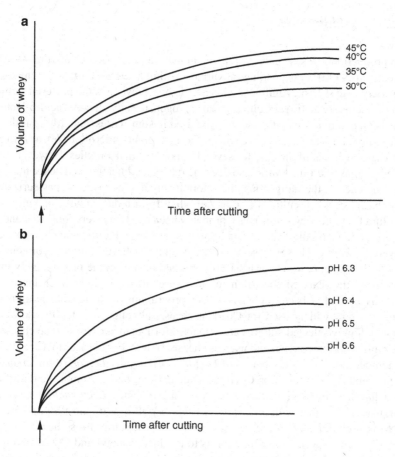

Fig. 12.15 Effect of temperature (**a**) and pH (**b**) on the rate and extent of syneresis in cut/broken renneted milk gels

- Stirring the curd during cooking
- Fat retards syneresis while increasing the protein content up to a point improves it; at high protein concentrations, the gel is too firm and does not synerese (e.g., UF retentate)

Gels prepared from heated milk synerese poorly (assuming that the milk actually coagulates). Such reduced syneresis properties are desirable for fermented milk products, e.g., yoghurt (milk for which is severely heated, e.g., 90 °C × 10 min) but are undesirable for cheese.

Good analytical methods for monitoring syneresis are lacking. Principles that have been exploited include dilution of an added marker, e.g., a dye, which must not adsorb onto or diffuse into the curd particles, measurement of the electrical conductivity or moisture content of the curd or by measuring the volume of whey released (probably the most commonly used method although values obtained are contingent on the method used).

12.2.3   Acidification

Acid production is a key feature in the manufacture of all cheese varieties—the pH
decreases to ~5 (±0.4, depending on variety) within 5–20 h, at a rate depending on
the variety (Fig. 12.16). Acidification is normally achieved via the bacterial fermentation of lactose to lactic acid, although an acidogen, usually gluconic acid-δ-lactone,
alone or in combination with acid, may be used in some cases, e.g., Mozzarella.

Traditionally cheesemakers relied on the indigenous microflora of milk to ferment lactose, as is still the case for several minor artisanal varieties. However, since
the indigenous microflora varies, so does the rate of acidification and hence the quality of the cheese; the indigenous microflora is largely destroyed by pasteurization.
"Slop-back" or whey cultures (the use of whey from today's cheesemaking as an
inoculum for tomorrow's milk) have probably been used for a very long time and are
still used commercially, e.g., for such famous cheese as Parmigiano-Reggiano and
Gruyere de Comté. However, selected "pure" cultures have been used for Cheddar
and Dutch-type cheeses for at least 80 years and have become progressively more
refined over the years. Single-strain cultures were introduced in New Zealand in the
1930s as part of a bacteriophage control programme. Selected phage-unrelated
strains are now widely used for Cheddar cheese; although selected by a different
protocol, highly selected cultures are also used for Dutch and Swiss-type cheeses.

Members of three genera are used as cheese starters. For cheeses that are cooked
to a temperature < ~39 °C, species of *Lactococcus*, usually *Lc. lactis* ssp. *cremoris*,
are principally used, i.e., for Cheddar, Dutch, Blue, surface mould and surface-
smear families. For high-cooked varieties, a thermophilic *Lactobacillus* culture is
used, usually together with *Streptococcus thermophilus* (e.g., most Swiss varieties
and Mozzarella). *Leuconostoc* spp. are included in the starter for some cheese varieties, e.g., Dutch types; their function is to produce diacetyl and CO_2 from citrate
rather than significant acid production.

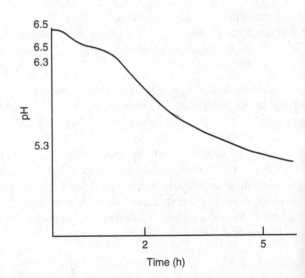

Fig. 12.16 pH profile of
Cheddar during cheese
manufacture

The selection, propagation and use of starters will not be discussed here. The interested reader is referred to Cogan and Hill (1993) and Parente and Cogan (2004).

The primary function of cheese starter cultures is to produce lactic acid at a predictable and dependable rate. The metabolism of lactose is summarized in Fig. 12.17. Most cheese starters are homofermentative, i.e., produce only lactic acid, usually the L-isomer; *Leuconostoc* spp. are heterofermentative. The products of lactic acid bacteria are summarized in Table 12.4.

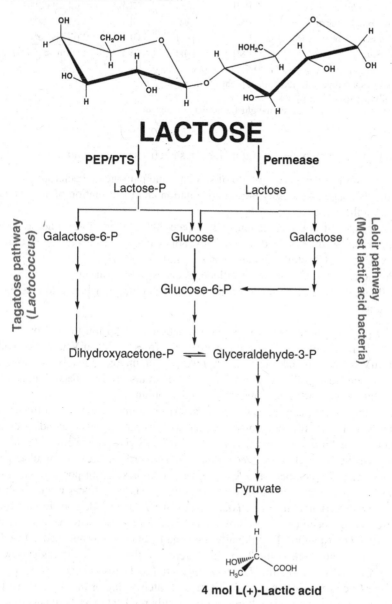

Fig. 12.17 Summary of the metabolic pathways used to ferment lactose to lactic acid by most lactic acid bacteria; many *Lactobacillus* species/strains cannot metabolize galactose

Table 12.4 Salient features of lactose metabolism in starter culture organisms (from Cogan and Hill 1993)

Organism	Transport[a]	Cleavage enzyme[b]	Pathway[c]	Products (mol/mollactose)
Lactococcus spp.	PTS	pβgal	GLY	4 L-Lactate
Leuconostoc spp.	?	βgal	PK	2 D-Lactate + 2 Ethanol + 2CO$_2$
Str. thermophilus	PMF	βgal	GLY	2 L-Lactate[d]
Lb. delbrueckii subsp. *lactis*	PMF?	βgal	GLY	2 D-Lactate[d]
Lb. delbrueckii subsp. *bulgaricus*	PMF?	βgal	GLY	2 D-Lactate[d]
Lb. helveticus	PMF?	βgal	GLY	4 L- (mainly) + D-Lactate

[a]*PTS* phosphotransferase system, *PMF* proton motive force
[b]*pβgal* phospho-β-galactosidase, *βgal* β-galactosidase
[c]*GLY* glycolysis, *PK* phosphoketolase
[d]These species metabolize only the glucose moiety of lactose

Acid production plays several major roles in cheese manufacture:

1. Controls or prevents the growth of spoilage and pathogenic bacteria.
2. Affects coagulant activity during coagulation and the retention of active coagulant in the curd.
3. Solubilizes of colloidal calcium phosphate and thereby affects cheese texture; rapid acid production leads to a low level of calcium in the cheese and a crumbly texture (e.g., Cheshire) and *vice versa* (e.g., Emmental).
4. Promotes syneresis and hence influences cheese composition.
5. Influences the activity of enzymes during ripening, and hence affects cheese quality.

The primary starter performs several functions in addition to acid production, especially reduction of the redox potential (E$_h$, from ~+250 mV in milk to about −300 mV in cheese) and, most importantly, plays a major, probably essential, role in the biochemistry of cheese ripening. Some strains produce bacteriocins which may control the growth of contaminating microorganisms.

The ripening of many varieties is characterized by the action, not of the primary starter, but of other microorganisms, which we will refer to as a secondary culture. Examples are *Propionibacterium freudenreichii* in Swiss-type cheeses, *Penicillium roqueforti* in blue cheeses, *Penicillium camemberti* in surface mould-ripened cheeses, e.g., Camembert and Brie, a very complex Gram-positive microflora including *Brevibacterium linens* and yeasts in surface smear-ripened cheese, citrate-positive (Cit$^+$) strains of *Lactococcus* and *Leuconostoc* spp. in Dutch-type cheeses. The specific function of these microorgansims will be discussed in Sect. 12.2.7 on ripening. Traditionally, a secondary culture was not used in Cheddar-type cheeses but there is much current interest in the use of cultures of selected bacteria, usually mesophilic *Lactobacillus* spp. or lactose-negative *Lactococcus* spp., for Cheddar cheese with the objective of intensifying or modifying flavour or accelerating ripening; such cultures are frequently referred to as "adjunct cultures".

12.2.4 Moulding and Shaping

When the desired pH and moisture content have been achieved, the curds are separated from the whey and placed in moulds of traditional shape and size to drain and form a continuous mass; high moisture curds form a continuous mass under their own weight but low moisture varieties are pressed.

Cheeses are made up in traditional shapes (usually flat cylindrical, but also sausage, pear-shaped or rectangular) and size, ranging from ~250 g (e.g., Camembert) to 60–80 kg (e.g., Emmental; Fig. 12.18). The size of cheese is not just a cosmetic feature; Emmental must be large enough to prevent excessive diffusion of CO_2, which is essential for eye development, while Camembert must be quite small so that the surface does not become over-ripe while the centre is still unripe (this cheese softens from the surface to the centre).

Curds for the *pasta filata* cheeses, e.g., Mozzarella, Provolone and Halloumi, are heated in hot water (70–75 °C), kneaded and stretched when the pH reaches ~5.4; this gives the cheeses their characteristic fibrous structure.

12.2.5 Salting

All cheeses are salted, either by mixing dry salt with the drained curd (confined largely to varieties that originated in England), rubbing dry salt on the surface of the pressed cheese (e.g., Pecorino Romano or Blue cheeses), or by immersion of the pressed cheeses in brine (most varieties). Salt concentration varies from ~0.7 % (~2 % salt-in-moisture) in Emmental to 7–8 % (~15 % salt-in-moisture) in Domiati.

Fig. 12.18 A selection of cheese varieties, showing the diversity of cheese size, shape and appearance

Salt plays a number of important roles in cheese:

1. It is the principal factor affecting the water activity of young cheeses and has a major effect on the growth and survival of bacteria and the activity of enzymes in cheese and hence affects and controls the biochemistry of cheese ripening.
2. Salting promotes syneresis and hence reduces the moisture content of cheese; about 2 kg of water are lost for each kg of salt absorbed.
3. It has a positive effect on flavour.
4. Cheese contributes to dietary sodium, high levels of which have undesirable nutritional consequences, e.g., hypertension and osteoporosis.

12.2.6   Manufacturing Protocols for Some Cheese Varieties

The manufacturing protocol for the various cheese varieties differ in detail but many elements are common to many varieties. The protocols for the principal varieties are summarized in Figs. 12.19a–d.

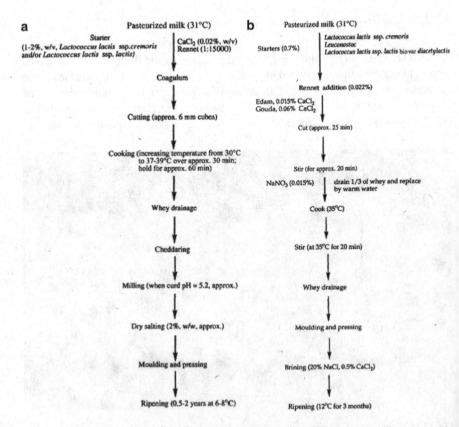

Fig. 12.19 Protocols for the manufacture of (**a**) Cheddar, (**b**) Gouda. (**c**) Emmental and (**d**) Parmigiano-Reggiano

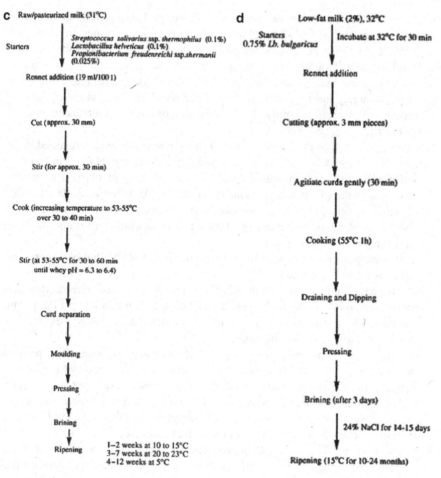

c Raw/pasteurized milk (31°C)

Starters → *Streptococcus salivarius* ssp. *thermophilus* (0.1%)
Lactobacillus helveticus (0.1%)
Propionibacterium freudenreichi ssp.*shermanii* (0.025%)

Rennet addition (19 ml/100 l)

Cut (approx. 30 mm)

Stir (for approx. 30 min)

Cook (increasing temperature to 53-55°C over 30 to 40 min)

Stir (at 53-55°C for 30 to 60 min until whey pH = 6.3 to 6.4)

Curd separation

Moulding

Pressing

Brining

Ripening — 1–2 weeks at 10 to 15°C
3–7 weeks at 20 to 23°C
4–12 weeks at 5°C

d Low-fat milk (2%), 32°C

Starters 0.75% *Lb. bulgaricus* | Incubate at 32°C for 30 min

Rennet addition

Cutting (approx. 3 mm pieces)

Agitiate curds gently (30 min)

Cooking (55°C 1h)

Draining and Dipping

Pressing

Brining (after 3 days)

24% NaCl for 14-15 days

Ripening (15°C for 10-24 months)

Fig. 12.19 (continued)

12.2.7 Cheese Ripening

While rennet-coagulated cheese curd may be consumed immediately after manufacture (and a little is), it is rather flavourless and rubbery. Consequently, rennet-coagulated cheeses are nearly always ripened (matured) for a period ranging from ~3 weeks for Mozzarella to >2 years for Parmesan and extra-mature Cheddar. During this period, a very complex series of biological, biochemical and chemical reactions occur through which the characteristic flavour compounds are produced and the texture altered.

Four, and in some cheeses five or perhaps six, agents are responsible for these changes:

1. *The cheese milk.* As discussed in Chap. 10, milk contains about 60 indigenous enzymes, many of which are associated with the fat globules or casein micelles

and are therefore incorporated into the cheese curd; the soluble enzymes are largely removed in the whey. Many of the indigenous enzymes are quite heat stable and survive HTST pasteurization; at least three of these (plasmin, acid phosphatase and xanthine oxidase) are active in cheese and contribute to cheese ripening; some indigenous lipase may also survive pasteurization. Lipoprotein lipase is largely inactivated by pasteurisation but contributes to the ripening of varieties made from raw milk. The contribution of other indigenous enzymes to cheese ripening is not known.

2. *Coagulant.* Most of the coagulant is lost in the whey but some is retained in the curd. Approximately 6 % of added chymosin is normally retained in Cheddar and similar varieties, including Dutch types; the amount of rennet retained increases as the pH at whey drainage is reduced. As much as 20 % of added chymosin is retained in high-moisture, low-pH cheese, e.g., Camembert. Only about 3 % of microbial rennet substitutes is retained in the curd and the level retained is independent of pH.

   Porcine pepsin is very sensitive to denaturation at pH 6.7 but becomes more stable as the pH is reduced.

   The coagulant is major contributor to proteolysis in most cheese varieties, notable exceptions being high-cooked varieties, e.g., Emmemtal, and Parmesan, and *pasta filata* varieties (e.g., Mozzarella) in which the coagulant is extensively denatured during curd manufacture.

   A good quality rennet extract is free of lipolytic activity but a rennet paste is used in the manufacture of some Italian varieties, e.g., Romano and Provolone. Rennet paste contains a lipase, referred to as pre-gastric esterase (PGE), which makes a major contribution to lipolysis in, and to the characteristic flavour of, these cheeses. Rennet paste is considered unhygienic and therefore semi-purified PGE may be added to rennet extract for such cheeses (see Chap. 10).

3. *Starter bacteria.* The starter culture reaches maximum numbers at the end of the manufacturing phase. Their numbers then decline at a rate depending on the strain, typically by two log cycles within 1 month. At least some of the non-viable cells lyse at a rate dependent on the strain. The only extracellular enzyme in *Lactococcus, Lactobacillus* and *Streptococcus* is a proteinase which is attached to the cell membrane and protrudes through the cell wall; all peptidases, esterases and phosphatases are intracellular and therefore cell lysis is essential before they can contribute to ripening.

4. *Non-starter bacteria.* Cheese made from pasteurized, high quality milk in modern factories using enclosed automated equipment contains very few non-starter bacteria ($<$50 cfu/g) at one day but these multiply to 10^7–10^8 cfu/g within about 2 months (at a rate depending on, especially, ripening temperature and rate of cooling of the cheese block). Since the starter population declines during this period, non-starter bacteria dominate the microflora of cheese during the later stages of ripening.

   Properly made cheese is quite a hostile environment for bacteria due to its low pH, moderate-to-high salt concentration in the moisture phase, anaerobic conditions (except at the surface), lack of a fermentable carbohydrate and perhaps the production of bacteriocins by the starter. Consequently, cheese is a very selective environment and its non-starter microflora is dominated by lactic acid bacteria, principally facultatively heterofermentative (mesophilic) lactobacilli such as *Lb casei* and *Lb paracasei*.

5. *Secondary and adjunct cultures.* As discussed in Sect. 12.2.3, many cheese varieties are characterised by the growth of secondary microorganisms which have strong metabolic activity and dominate the ripening and characteristics of these cheeses.

6. *Other exogenous enzymes.* An exogenous lipase is added to milk for a few varieties, e.g., pre-gastric lipase (in rennet paste) for Romano or Provolone cheese. There has been considerable academic and commercial interest in adding exogenous proteinases (in addition to the coagulant) and/or peptidases to accelerate ripening. The enzymes may be added to the milk or curd in various forms, e.g. free, microencapsulated or in attenuated cells.

The contribution of these agents, individually or in various combinations, has been assessed in model cheese systems from which one or more of the agents was excluded or eliminated, e.g., by using an acidogen rather than starter for acidification or manufacturing cheese in a sterile environment to eliminate NSLAB. Such model systems have given very useful information on the biochemistry of ripening.

During ripening, three primary biochemical events occur: (1) metabolism of residual lactose and of lactate and citrate, (2) lipolysis and metabolism of fatty acids and (3) proteolysis and amino acid catabolism (Fig. 12.20). The products of these

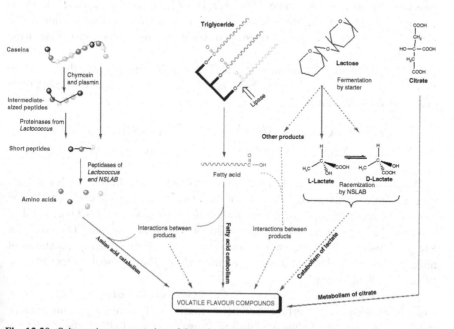

Fig. 12.20 Schematic representation of the principal biochemical events that occur in cheese during ripening (from McSweeney 2004)

primary reactions undergo numerous modifications and interactions. The primary reactions are fairly well characterized but the secondary changes in most varieties are less well known. An overview of the principal biochemical changes follows.

12.2.7.1   Metabolism of Residual Lactose and of Lactate and Citrate

Most (~98 %) of the lactose in cheese milk is removed in the whey as lactose or lactic acid. However, fresh cheese curd contains 1–2 % lactose which is normally metabolised to L-lactic acid by the starter within a short period of time. In most varieties, the L-lactate is racemized to DL-lactate by non-starter lactic acid bacteria (NSLAB) within about 3 months and a small amount is oxidized to acetic acid at a rate dependent on the oxygen content of the cheese and hence on the permeability of the packaging material.

In cheese varieties made using *Streptococcus thermophilus* and *Lactobacillus* spp. as starter, e.g., Swiss types and Parmigiano Reggiano, the metabolism of lactose is more complex than in cheese in which a *Lactococcus* starter is used. In these cheeses, the curd is cooked to 52–55 °C, which is above the growth temperature for both components of the starter; as the curd cools, the *Streptococcus*, which is the more heat-tolerant of the two starters, begins to grow, utilizing the glucose moiety of lactose, with the production of L-lactic acid, but not galactose, which accumulates in the curd. When the curd has cooled sufficiently, the *Lactobacillus* spp. grows, and, if a galactose-positive species/strain is used (which is normal), it metabolises galactose, producing DL-lactate (Fig. 12.21). If a galactose-negative strain of *Lactobacillus* is used, galactose accumulates in the curd and can lead to undesirable secondary fermentations during ripening or contribute to Maillard browning if the cheese is heated.

Swiss-type cheeses are ripened at ~22 °C for a period to encourage the growth of *Propionibacterium* spp. which use lactic acid as an energy source, producing propionic acid, acetic acid and CO_2:

$$3CH_3CHOHCOOH \rightarrow 2CH_3CH_2COOH + CH_3COOH + CO_2 + H_2O$$

    Lactic acid   Propionic acid   Acetic acid

Propionic and acetic acids probably contribute to the flavour of Swiss-type cheeses while the CO_2 is responsible for their large characteristic eyes. Lactic acid may be metabolized by *Clostridium tyrobutyricum* to butyric acid, CO_2 and hydrogen (Fig. 12.22); butyric acid is responsible for off-flavours and the CO_2 and H_2 for late gas blowing. Clostridia are controlled by good hygienic practices, addition of nitrate or lysozyme, bactofugation or microfiltration. The principal sources of clostridia are soil and silage.

In surface mould-ripened cheeses, e.g., Camembert and Brie, *Penicillium camemberti*, growing on the surface, metabolizes lactic acid as an energy source, causing the pH to increase. Lactic acid diffuses from the centre to the surface, where it is catabolized. Ammonia produced by deamination of amino acids contributes to the

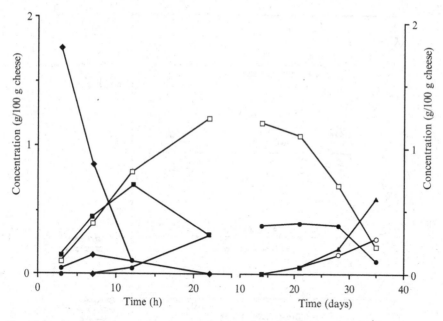

Fig. 12.21 Metabolism of lactose, glucose, galactose, D- and L-lactic acid in Emmental cheese. Cheese transferred to hot room (22–24 °C) at 14 days. *Filled circle*, D-lactate; *open circle*, acetate; *filled square*, galactose; *open square*, L-lactate; *filled diamond*, glucose; *open diamond*, lactose; *filled triangle*, propionate

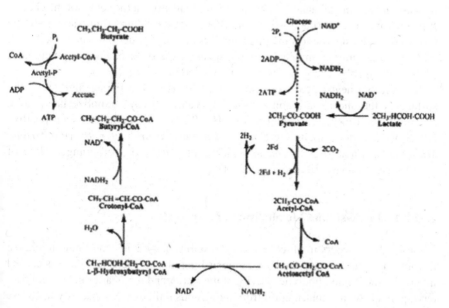

Fig. 12.22 Metabolism of glucose or lactic acid by *Clostridium tyrobutyricum* with the production of butyric acid, CO_2 and hydrogen gas

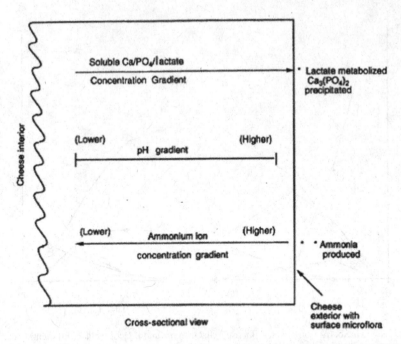

Fig. 12.23 Schematic representation of the gradients of calcium, phosphate, lactic acid, pH and ammonia in ripening of Camembert cheese

increase in pH which reaches ~7.5 at the surface and ~6.5 at the centre of the cheese. Ripening of Camembert and Brie is characterized by softening (liquefaction) of the texture from the surface towards the centre. Softening is due to the increase in pH, proteolysis and migration of calcium phosphate to the surface, where it precipitates due to the high pH. These events are summarized in Fig. 12.23.

In surface smear-ripened cheeses, e.g., Muenster, Limburger and Tilsit, the surface of the cheese is colonized first by yeasts which catabolize lactic acid, causing the pH to increase, and then >pH 5.8 by a very complex Gram-positive bacterial microflora, including *Corynebacterium, Arthrobacter, Brevibacterium, Microbacteriu* and *Staphylococcus* which contribute to the red-orange colour of the surface of these varieties (Fig. 12.24).

12.2.7.2 Lipolysis and Metabolism of Fatty Acids

Some lipolysis occurs in all cheeses; the resulting fatty acids contribute to cheese flavour. In most varieties, lipolysis is rather limited (see Table 12.5) and is caused mainly by the limited lipolytic activity of the starter and non-starter lactic acid bacteria, perhaps with a contribution from indigenous milk lipase, especially in cheese made from raw milk.

Extensive lipolysis occurs in two families of cheese in which fatty acids and/ or their degradation products are major contributors to flavour, i.e., certain Italian

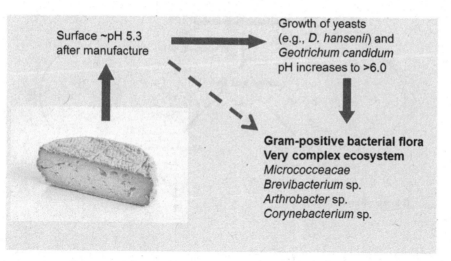

Fig. 12.24 Schematic representation of the development of the surface microflora of a smear-ripened cheese during ripening

Table 12.5 Free fatty acids (mg/kg) in a selection of cheese varieties (Woo and Lindsay 1984; Woo et al. 1984)

Variety	FFA (mg/kg)	Variety	FFA (mg/kg)
Sapsago	211	Gjetost	1,658
Edam	356	Provolone	2,118
Mozzarella	363	Brick	2,150
Colby	550	Limburger	4,187
Camembert	681	Goats'milk	4,558
Port Salut	700	Parmesan	4,993
Monterey Jack	736	Romano	6,743
Cheddar	1,028	Roquefort	32,453
Gruyere	1,481	Blue(US)	32,230

varieties (e.g., Romano and Provolone) and the Blue cheeses. Rennet paste, which contains pregastric esterase (PGE) rather than rennet extract, is used in the manufacture of these Italian cheeses. PGE is highly specific for the fatty acid on the *sn*-3 position of glycerol, which, in the case of milk lipids, is predominantly highly flavoured short-chain fatty acids (butanoic to decanoic). These acids are principally responsible for the characteristic piquant flavour of these Italian cheeses.

Blue cheeses undergo very extensive lipolysis during ripening; up to 25 % of all fatty acids may be released. The principal lipase in Blue cheese is that produced by *Penicillium roqueforti*, with minor contributions from indigenous milk lipase and the lipases of starter and non-starter lactic acid bacteria. The free fatty acids contribute directly to the flavour of blue cheeses but more importantly, they undergo partial β-oxidation to alkan-2-ones (methyl ketones; (R—$C\overset{O}{\diagup}CH_3$)) through the catabolic activity of the mould (Fig. 12.25). A homologous series of alkan-2-ones from C_3 to C_{17} is formed (corresponding to the fatty acids from C_4 to C_{18}), but heptanone and

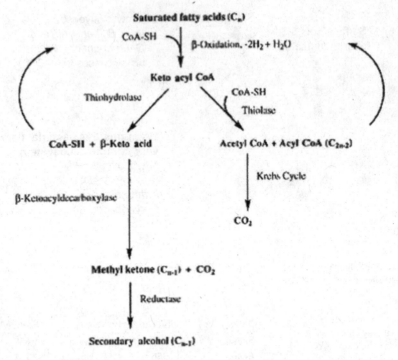

Fig. 12.25 β-Oxidation of fatty acids to methyl ketones by *Penicillium roqueforti* and subsequent reduction to secondary alcohols

Table 12.6 Typical concentrations of alkan-2-ones in blue cheese (from Kinsella and Hwang 1976)

2-Alkanone	μg/10 g dry blue cheese							
	A[a]	B[a]	C[a]	D[b]	E[b]	F[b]	G[c]	H[c]
2-Propanone	65	54	75	210	–	0	60	T
2-Pentanone	360	140	410	1,022	367	51	372	285
2-Heptanone	800	380	380	1,827	755	243	3,845	3,354
2-Nonanone	560	440	1,760	1,816	600	176	3,737	3,505
2-Undecanone	128	120	590	136	135	56	1,304	1,383
2-Tridecanone	–	–	–	100	120	77	309	945
Total	1,940	1,146	4,296	5,111	1,978	603	9,627	9,372

[a]Commercial samples of ripe blue cheese
[b]Samples D, E and F of blue cheese ripened for 2, 3 and 4 months, respectively
[c]Samples G and H of very small batches of experimental blue cheese ripened for 2 and 3 months, respectively

nonanone predominate; typical concentrations are shown in Table 12.6. The characteristic peppery flavour of Blue cheeses is due to alkan-2-ones. Under anaerobic conditions, some of the alkan-2-ones may be reduced to the corresponding alkan-2-ols (secondary alcohols), which cause off-flavours.

12.2.7.3 Proteolysis and Amino Acid Catabolism

Proteolysis is the most complex, and perhaps the most important, of the three primary biochemical events in the ripening of most cheese varieties. In internal, bacterially-ripened cheeses, e.g., Cheddar, Dutch and Swiss varieties, together with solubilisation of casein-bound calcium, it contributes to the textural changes that occur during ripening, i.e., conversion of the tough rubbery texture of fresh curd to the smooth, pliable body of mature cheese. Small peptides and free amino acids contribute directly to cheese flavour and amino acids serve as substrates for a wide range of complex flavour-forming reactions, most commonly initiated by the action of aminotransferases which convert the amino acid to the corresponding α-keto acid (and convert α-ketoglutarate, a co-substrate, to glutamic acids). α-Keto acids are unstable and are degraded enzymatically, and through chemical reactions, to a large number of sapid compounds. Excessive amounts of hydrophobic peptides may be produced under certain circumstances and may lead to bitterness which some consumers find very objectionable; however, at an appropriate concentration and when properly balanced by other compounds, bitter peptides probably contribute positively to cheese flavour.

The level of proteolysis in cheese varies from limited (e.g., Mozzarella) through moderate (e.g., Cheddar and Gouda) to very extensive (e.g., Blue cheeses). The products of proteolysis range from very large polypeptides, only a little smaller than the parent caseins, to amino acids which may, in turn, be catabolized to a very diverse range of sapid compounds, including amines, acids and sulphur compounds.

Depending on the depth of information required, proteolysis in cheese is assessed by a wide range of techniques. Electrophoresis, usually urea-PAGE, is particularly appropriate for monitoring primary proteolysis, i.e., proteolysis of the caseins and the resulting large polypeptides. Quantifying the formation of peptides and amino acids soluble in water, at pH 4.6, in TCA, ethanol or phosphotungstic acid or the measurement of free amino groups by reaction with ninhydrin, 2-phthaldialdehyde, trinitrobenzene or fluorescamine is suitable for monitoring secondary proteolysis. Reversed phase HPLC is especially useful for fingerprinting the small peptide profile in cheese and is now widely used. High performance ion-exchange or size exclusion chromatography are also effective but are less widely used.

Proteolysis has not yet been fully characterized in any cheese variety but considerable progress has been made for Cheddar and as far as is known, generally similar results apply to other low-cook, internal bacterially ripened cheeses (e.g., Dutch types). Proteolysis in Cheddar will be summarized as an example of these types of cheese.

Urea-PAGE shows that α_{s1}-casein is completely hydrolysed in Cheddar within 3–4 months (Fig. 12.26). It is hydrolyzed by chymosin, initially at Phe_{23}-Phe_{24} and later at Leu_{101}-Lys_{102}, and to a lesser extent at Phe_{32}-Gly_{33}, Leu_{98}-Lys_{99} and Leu_{109}-Glu_{110}. Although β-casein in solution is readily hydrolyzed by chymosin, at the ionic strength of the aqueous phase of cheese, β-casein is very resistant to chymosin but is hydrolyzed slowly (~50 % at 6 months) by plasmin at Lys_{28}-Lys_{29}, Lys_{105}-His/Gln_{106} and Lys_{107}-Glu_{108}, producing γ^1, γ^2 and γ^3-caseins, respectively, and the corresponding protease peptones (PP5, PP8 slow and PP8 fast; see Chap. 4). Chymosin and to lesser extent plasmin, are mainly responsible for primary proteolysis, i.e. the formation of water (or pH 4.6)-soluble N, as summarized in Fig. 12.27.

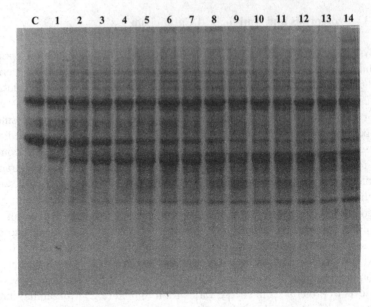

Fig. 12.26 Urea-polyacrylamide gel electrophoretograms of Cheddar cheese after ripening for 0, 1, 2, 3, 4, 6, 8, 10, 12, 14, 16, 18 or 20 weeks (lanes 1–14); *C*, sodium caseinate (supplied by S. Mooney)

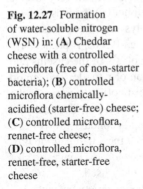

Fig. 12.27 Formation of water-soluble nitrogen (WSN) in: (**A**) Cheddar cheese with a controlled microflora (free of non-starter bacteria); (**B**) controlled microflora chemically-acidified (starter-free) cheese; (**C**) controlled microflora, rennet-free cheese; (**D**) controlled microflora, rennet-free, starter-free cheese

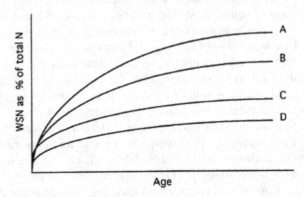

Although *in vitro*, the cell wall-associated proteinase of the *Lactococcus* starters is quite active on β-casein (and that from some strains on α_{s1}-casein also), in cheese, they appear to act mainly on casein-derived peptides, produced by chymosin from α_{s1}-casein or by plasmin from β-casein.

The starter cells begin to die off at the end of curd manufacture (Fig. 12.28); the dead cells may lyse and release their intracellular endopeptidases (PepO, PepF), aminopeptidases (including PepN, PepA, PepC, PepX), tripeptidases and dipeptidases (including proline-specific peptidases) which produce a range of free amino acids (Fig. 12.29). About 150 peptides have been isolated from the water-soluble fraction of Cheddar, and characterized (Fig. 12.30), but it is highly likely that many more peptides remain to be discovered in this variety. These show that both lactococcal proteinase and exopeptidases contribute to proteolysis in cheese. The proteinases and

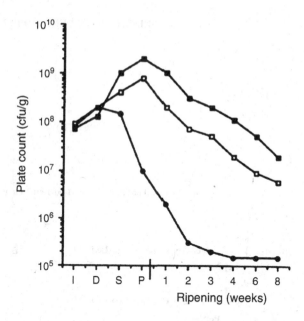

Fig. 12.28 Changes in the population of starter cells in cheese made using different single strain starters. *I*, Inoculation; *D*, whey drainage; *S*, salting; *P*, after pressing

peptidases of the NSLAB (mainly mesophilic lactobacilli) appear to contribute little to proteolysis in Cheddar, except in the production of amino acids.

The principal amino acids in Cheddar are shown in Fig. 12.31.

12.2.8 Cheese Flavour

Although interest in cheese flavour dates from the beginning of this century, very little progress was made until the development of gas liquid chromatography (GC) in the late 1950s and especially the coupling of GC and mass spectrometry (MS). Hundreds of volatile compounds have been identified in cheese by GC-MS. The volatile fraction of cheese may be obtained by taking a sample of headspace but the concentration of many compounds is too low, even for modern GC-MS techniques. The volatiles may be concentrated by solvent extraction or distillation, or more commonly in recent years by solid-phase microextraction, where an adsorbent fibre is exposed to the cheese headspace to trap volatiles which are later released on injection to the GC.

The taste of cheese is concentrated in the water-soluble fraction (peptides, amino acids, organic acids, amines, NaCl) while the aroma is mainly in the volatile fraction. Initially, it was believed that cheese flavour was due to one or a small number of compounds but it was soon realised that all cheeses contained essentially the same sapid compounds. Recognition of this led to the Component Balance Theory, i.e., cheese flavour is due to the concentration and balance of a wide range of compounds. Although considerable information on the flavour compounds in several cheese varieties has been accumulated, it is not possible to describe fully the flavour of any variety, with the possible exception of Blue cheeses, the flavour of which is dominated by alkan-2-ones.

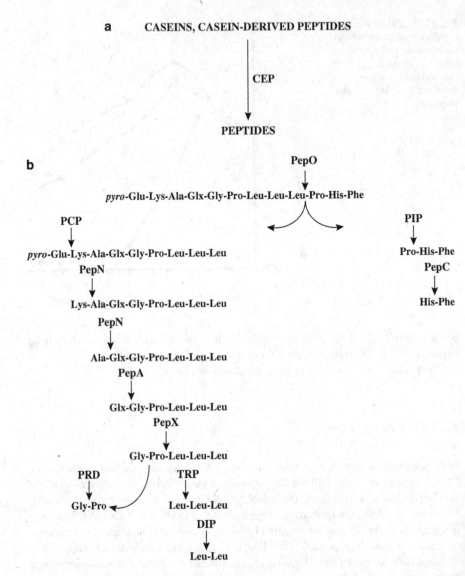

Fig. 12.29 Schematic representation of the hydrolysis of casein (**a**) by lactococcal cell envelope proteinase (CEP), and (**b**) degradation of an hypothetical dodecapeptide by the combined action of lactococcal peptidases: oligopeptidase (PepO), various aminopeptidases (PCP, PepN, PepA, PepX), tripeptidase (TRP), prolidase (PRD) and dipeptidase (DIP)

Many cheeses contain the same or similar compounds but at different concentrations and proportions; the principal classes of components present are aldehydes, ketones, acids, amines, lactones, esters, hydrocarbons and sulphur compounds; the latter, e.g., H_2S, methanethiol (CH_3SH), dimethyl sulphide ($H_3C\text{-}S\text{-}CH_3$) and dimethyl disulphide ($H_3C\text{-}S\text{-}S\text{-}CH_3$) are considered to be particularly important in Cheddar cheese.

12.2.9   Accelerated Ripening of Cheese

Since the ripening of cheese, especially low moisture varieties, is a slow process, it is expensive in terms of controlled atmosphere storage and stocks. Ripening is also unpredictable. Hence, there are economic and technological incentives to accelerate ripening, while retaining or improving characteristic flavour and texture. The principal approaches used to accelerate cheese ripening are:

1. Elevated ripening temperatures, especially for Cheddar which is now usually ripened at 6–8 °C; some other varieties are ripened at a higher temperature, e.g., ~14 °C for Dutch types or 20–22 °C for Swiss types and Parmesan. Elevated temperature ripening (to perhaps 14 °C is the most effective and simplest way to accelerate the ripening of hard cheese, but it is not without risk as increased temperatures can also accelerate the development of off-flavours

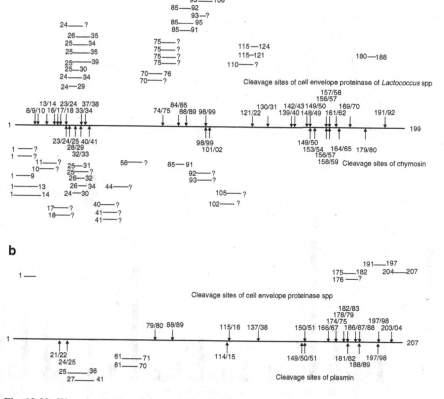

Fig. 12.30 Water-insoluble and water-soluble peptides derived from α_{s1}-casein (**a**), α_{s2}-casein (**b**) or β-casein in (**c**) isolated from Cheddar cheese; *DF* diafiltration. The principal chymosin, plasmin and lactococcal cell-envelope proteinase cleavage sites are indicated by *arrows* (data from T.K. Singh and S. Mooney, unpublished)

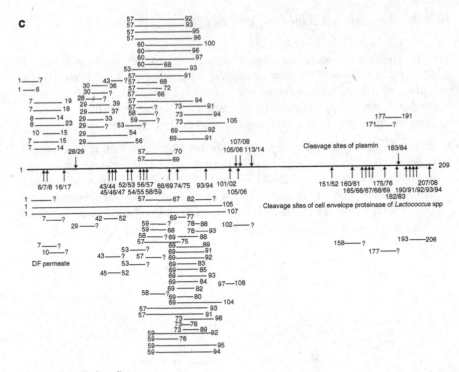

Fig. 12.30 (continued)

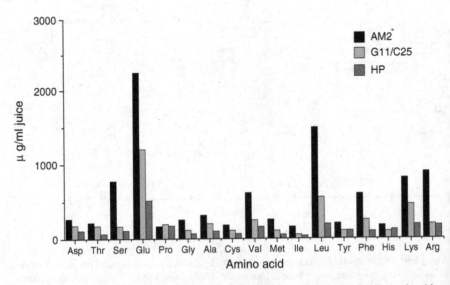

Fig. 12.31 Concentration of individual amino acids in 60-day-old Cheddar cheese, made with a single-strain starter *Lactococcus lactis* ssp. *cremoris* AM₂, G11/C25 or HP (from Wilkinson 1992)

2. Exogenous enzymes, usually proteinases and/or peptidases. For several reasons, this approach has had limited success, except for enzyme-modified cheeses (EMC). These are usually high moisture products which are used as ingredients for processed cheese, cheese spreads, cheese dips or cheese flavourings.
3. Attenuated lactic acid bacteria, e.g., freeze-shocked, heat-shocked or lactose-negative mutants.
4. Adjunct starters.
5. Use of fast lysing starters which die and release their intracellular enzymes rapidly.
6. Genetically modified starters which super-produce certain enzymes; genetically modified starters are not used commercially and have little promise due to consumer resistance and other problems.

Considerable in-depth information on the biochemistry of cheese ripening is now becoming available which will facilitate the genetic engineering of starter cultures with improved cheesemaking properties. Acceleration of cheese ripening has been reviewed by Fox et al. (1996b), Azarnia et al. (2006) and El Soda and Awad (2011).

12.3   Acid-Coagulated Cheeses

On acidification to pH 4.6, the caseins coagulate, which is the principle used to manufacture of a family of cheeses which represent ~20 % of total cheese consumption and are the principal cheeses in some countries. Acidification is traditionally and usually achieved by *in situ* fermentation of lactose by a *Lactococcus* starter but direct acidification by acid or acidogen (gluconic acid-δ-lactone) is also practised. The principal families of acid-coagulated cheeses are illustrated in Fig. 12.32 and a typical manufacturing protocol is shown in Fig. 12.33.

Acid-coagulated cheeses are usually consumed fresh; major varieties include Quarg, Cottage cheese and Cream cheese. These cheeses may be consumed in salads, as food ingredients and serve as the base for a rapidly expanding group of dairy products, i.e., fromage frais-type products.

The casein may also be coagulated at a pH >4.6, e.g., ~5.2, by using a higher temperature, e.g., 80–90 °C. This principle is used to manufacture another family of cheeses, which include Ricotta (and variants thereof), Anari, and some types of Queso Blanco. These cheeses may be made exclusively from whey but usually from a blend of milk and whey and are usually used as a food ingredient, e.g., in Lasagne or Ravioli.

12.4   Processed Cheese Products

Processed cheese is produced by blending shredded natural cheese of the same or different varieties and at different degrees of maturity with emulsifying agents and heating the blend under vacuum with constant agitation until a homogeneous mass is obtained. Other dairy and non-dairy ingredients may be included in the blend. The possibility of producing processed cheese was first assessed in 1895;

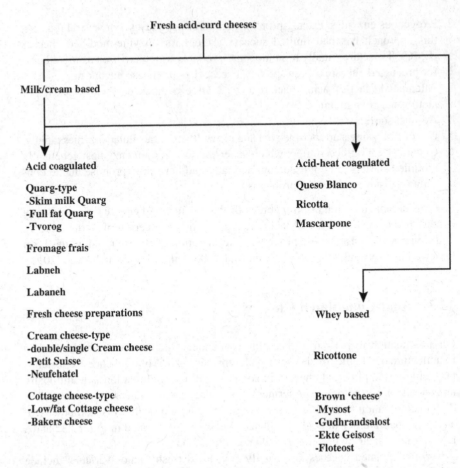

Fig. 12.32 Examples of acid-coagulated or heat-acid coagulated or whey-based cheese varieties (from Fox et al. 1996a)

emulsifying salts were not used and the product was not successful. The first successful product, in which emulsifying salts were used, was introduced in Europe in 1912 and in the USA in 1917 by Kraft. Since then, the market for processed cheese has increased and the range of products expanded.

Although established consumers may regard processed cheeses as inferior products compared to natural cheeses, they have numerous advantages compared to the latter:

1. A certain amount of cheese which would otherwise be difficult or impossible to commercialize may be used, e.g., cheese with deformations, cheese trimmings or cheese after removal of localized mould.
2. A blend of cheese varieties and non-cheese components may be used, making it possible to produce processed cheeses differing in consistency, flavour, shape and size (Table 12.7).
3. They have good storage stability at moderate temperatures, thus reducing the cost of storage and transport.

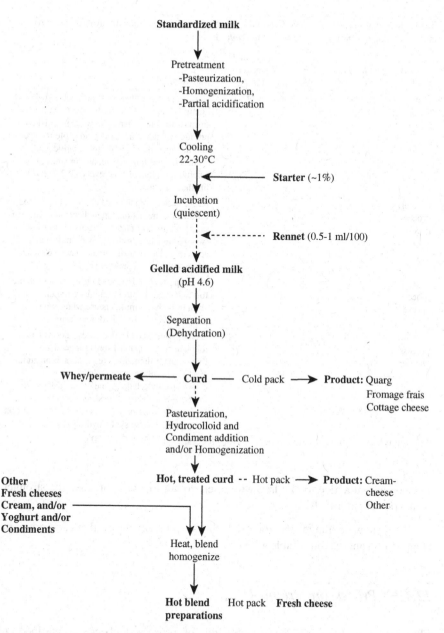

Fig. 12.33 Protocol for the manufacture of fresh acid-coagulated cheese (from Fox et al. 1996)

4. They are more stable than natural cheeses during storage, which results in less wastage, a feature that may be especially important in remote areas and in households with a low level of cheese consumption.
5. They are amenable to imaginative packing in various conveniently sized units.
6. They are suitable for sandwiches and fast food outlets.

Table 12.7 Compositional specifications and permitted ingredients in pasteurized processed cheese products[a] (modified from Fox et al. 1996a)

Product	Moisture (%, w/w)	Fat (%, w/w)	Fat in dry matter (%, w/w)	Ingredients
Pasteurized blended cheese	≤43	–	≥47	Cheese; cream, anhydrous milk fat, dehydrated cream [in quantities such that the fat derived from them is less than 5 % (w/w) in finished product]; water; salt; food-grade colours, spices and flavours; mould inhibitors (sorbic acid, potassium/sodium sorbate, and/or sodium/calcium propionates), at levels ≤0.2 % (w/w) finished product.
Pasteurized process cheese	≤43	–	≥47	As for pasteurized blended cheese, but with the following extra optional ingredients: emulsifying salts [sodium phosphates, sodium citrates; 3 % (w/w) of finished product], food-grade organic acids (e.g., lactic, acetic or citric) at levels such that pH of finished product is ≥5.3.
Pasteurized process cheese	≤44	≥23	–	As for pasteurized blended cheese, but with the following extra optional foodsingredients (milk, skim milk, buttermilk, cheese whey, whey proteins—in wet or dehydrated forms).
Pasteurized process cheese	40-60	≥20	–	As for pasteurized blended cheese, but with the following extra optional spreadsingredients: food-grade hydrocolloids (e.g., carob bean gum, guar gum, xanthan gums, gelatin, carboxymethylcellulose, and/or carageenan) at levels <0.8 % (w/w) of finished products; food-grade sweetening agents (e.g., sugar, dextrose, corn syrup, glucose syrup, hydrolyzed lactose).

[a]Minimum temperatures and times specified for processing are 65.5 °C for 30 s

7. They are attractive to children who often do not like or appreciate the stronger flavour of natural cheeses.

Today, a wide range of processed cheese products is available, varying in composition and flavour (Table 12.7).

12.4.1 Processing Protocol

The typical protocol for the manufacture of processed cheese is outlined in Fig. 12.34.

The important criteria for selecting cheese are type, flavour, maturity, consistency, texture and pH. The selection is determined by the type of processed cheese to be produced and by cost factors.

A great diversity of non-cheese ingredients may be used in the manufacture of processed cheese (Fig. 12.35).

Fig. 12.34 Protocol for the manufacture of processed cheese

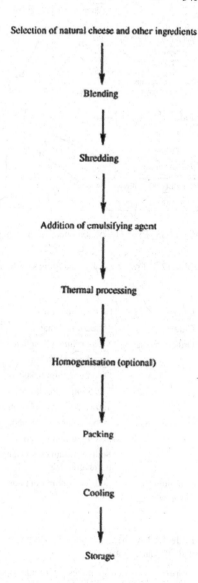

Selection of natural cheese and other ingredients

↓

Blending

↓

Shredding

↓

Addition of emulsifying agent

↓

Thermal processing

↓

Homogenisation (optional)

↓

Packing

↓

Cooling

↓

Storage

Emulsifying salts are critical in the manufacture of processed cheese with desirable properties. The most commonly used salts are orthophosphates, polyphosphates and citrates but several other agents are used (Tables 12.8 and 12.9). Emulsifying salts are not emulsifiers in the strict sense, since they are not surface active. Their essential role in processed cheese is to supplement the emulsifying properties of cheese proteins. This is accomplished by sequestering calcium, solubilizing, dispersing, hydrating and swelling the proteins and adjusting and stabilizing the pH.

The actual blend of ingredients used and the processing parameters depend on the type of processed cheese to be produced; typical parameters are summarized in Table 12.10.

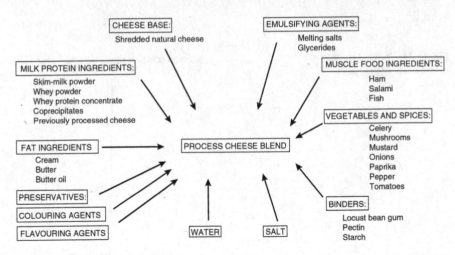

Fig. 12.35 Examples of non-cheese ingredients used in processed cheese (from Caric and Kalab 1987)

Table 12.8 Properties of emulsifying salts for processed cheese products (from Caric and Kalab 1987)

Group	Emulsifying salt	Formula	Solubility (at 20 °C (%))	pH value (1 % solution)
Citrates	Trisodium citrate	$2Na_3C_6H_5O_7.1H_2O$	High	6.23–6.26
Orthophosphates	Monosodium phosphate	$NaH_2PO_4.2H_2O$	40	4.0–4.2
	Disodium phosphate	$Na_2HPO_4.12H_2O$	18	8.9–9.1
Pyrophosphates	Disodium pyrophosphate	$Na_2H_2P_2O_7$	10.7	4.0–4.5
	Trisodium pyrophosphate	$Na_3HP_2O_7.9H_2O$	32.0	6.7–7.5
	Tetrasodium pyrophosphate	$Na_4P_2O_7.10H_2O$	10–12	10.2–10.4
Polyphosphates	Pentasodium tripolyphosphate	$Na_5P_3O_{10}$	14–15	9.3–9.5
	Sodium tetrapolyphosphate	$Na_6P_4O_{13}$	14–15	9.0–9.5
	Sodium hexametaphosphate (Graham's salt)	$Na_{n+2}P_nO_{3n+1}$ (n = 10–25)	Very high	6.0–7.5
Aluminium phosphates	Sodium aluminium phosphate	$NaH_{14}Al_3(PO_4)_8.4H_2O$	–	8.0

Table 12.9 General properties of emulsifying salts in relation to cheese processing (from Fox et al. 1996a, 1996b)

Property	Citrates	Orthophosphates	Pyrophosphates	Polyphosphates	Aluminium
Ion exchange (calcium sequesterization)	Low	Low	Moderate	High–very high	Low
Buffering action in the pH range 5.3–6.0	High	High	Moderate	Low–very low	–
para-Caseinate dispersion	Low	Low	High	Very high	–
Emulsification	Low	Low	Very high	Very high	Very low (n = 3–10) Low
Bacteriostatic	Nil	Low	High	High–very high	–

Table 12.10 Chemical, mechanical and thermal parameters as regulating factors in the cheese processing procedures (from Caric and Kalab 1993)

Process conditions	Processed cheese block	Processed cheese slice	Processed cheese spread
Raw material			
a. Average of cheese	Young to medium ripe, predominantly young	Predominantly young	Combination of young, medium ripe, overipe
b. Water-insoluble N as a % of total N	75–90 %	80–90 %	60–75 %
c. Structure	Predominantly long	Long	Short to long
Emulsifying salt	Structure-building, not creaming, e.g., high molecular weight polyphosphate, citrate	Structure-building, not creaming, e.g., phosphate/citrate mixtures	Creaming, e.g., low and medium molecular weight polyphosphate
Water addition (%)	10–25 (all at once)	5–15 (all at once)	20–45 (in portions)
Temperature (°C)	80–85	78–85	85–98 (150 °C)
Duration of processing, min	4–8	4–6	8–15
pH	5.4–5.7	5.6–5.9	5.6–6.0
Agitation	Slow	Slow	Rapid
Reworked cheese	0–0.2 %	0	5–20 %
Milk powder or whey powder		0	5–12 %
Homogenization	None	None	Advantageous
Filling, min	5–15	As fast as possible	10–30
Cooling	Slowly (10–12 h) at room temperature	Very rapid	10–30 Rapidly (15–30 min) in cool air

One of the major advantages of processed cheese is the flexibility of the finished form, which facilitates usage. The texture may vary from firm and sliceable to soft and spreadable. They may be presented as large blocks (5–10 kg), suitable for industrial catering, smaller blocks, e.g., 0.5 kg, for household use, small unit packs, e.g., 25–50 g, or slices which are particularly suited for industrial catering and fast food outlets.

12.5 Cheese Analogues

Cheese analogues are cheese-like products which probably contain no cheese. The most important of these are Mozzarella (Pizza) cheese analogues which are produced from rennet casein, fat or oil (usually vegetable) and emulsifying salts. The function of emulsifying salts is essentially similar to those in processed cheese, i.e., to solubilize the proteins. The manufacturing protocol is usually similar to that used for processed cheese, bearing in mind that the protein is dried rennet casein rather than a blend of cheeses (Fig. 12.36).

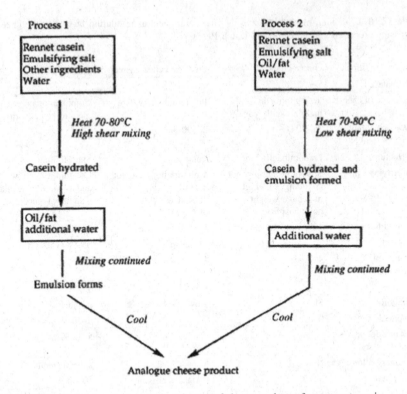

Fig. 12.36 Typical protocols for the manufacture of cheese analogue from rennet casein

The main attributes required of cheese analogues used in pizzas are meltability and stretchability; flavour is provided by other ingredients of the pizza, e.g., tomato paste, sausage, peppers, spices, anchovies, etc. It may be possible to produce analogues of other cheeses by adding biochemically- or chemically-generated cheese flavours. As discussed in Sect. 12.2.8, the flavour and texture of natural cheeses are very complex and cannot be simulated readily.

References

Azarnia, S., Normand, R., & Lee, B. (2006). Biotechnological methods of accelerate Cheddar cheese ripening. *Critical Reviews in Biotechnology, 26*, 121–143.
Caric, M., & Kalab, M. (1987). Processed cheese products. In P. F. Fox (Ed.), *Cheese: Chemistry, physics and microbiology* (Vol. 2, pp. 339–383). London: Elsevier Applied Science.
Caric, M., & Kalab, M. (1993). Processed cheese products. In, *Cheese: Chemistry, physics and microbiology*, Vol. 2, Major Cheese Groups 2nd Edn., P.F. Fox and Chapman and Hall, London, pp. 467–505.
Cogan, T. M., & Hill, C. (1993). Cheese starter cultures. In P. F. Fox (Ed.), *Cheese: Physics, chemistry and microbiology* (2nd ed., Vol. 1, pp. 193–255). London: Chapman & Hall.

El Soda, M., & Awad, S. (2011). Acceleration of cheese ripening. In J. W. Fuquay, P. F. Fox, & P. L. H. McSweeney (Eds.), *Encyclopedia of dairy sciences* (2nd ed., pp. 795–798). Amsterdam: Academic.

Foltmann, B. (1987). General and molecular aspects of rennets. In P. F. Fox (Ed.), *Cheese: Chemistry, physics and microbiology* (Vol. 1, pp. 33–61). London: Elsevier Applied Science.

Fox, P. F., O'Connor, T. P., McSweeney, P. L. H., Guinee, T. P. and O'Brien, N. M. (1996a). Cheese: physical, chemical, biochemical and nutritional aspects. *Advances in Food and Nutrition Research 39* 163–328.

Fox, P. F., Wallace, J. M., Morgan, S., Lynch, S., Niland, E. J., & Tobin, J. (1996b). Acceleration of cheese ripening. *Antonie van Leeuwenhoek, 70*, 271–297.

IDF. (1992). *Bovine Rennets. Determination of total milk-clotting activity, Provisional Standard 157*. Brussels: International Dairy Federation.

Hurley, M. J., O'Driscoll, B. M., Kelly, A. L. and McSweeney, P. L. H. (1999). Novel assay for the determination of residual coagulant activity in cheese. *International Dairy journal 9*, 553–558.

Kinsella, J. E., & Hwang, D. H. (1976). Enzymes of *Penicillium roqueforti* involved in the biosynthesis of cheese flavour. *CRC Critical Reviews in Food Science and Nutrition, 8*, 191–228.

McSweeney, P. L. H., Ottogalli, G., & Fox, P. F. (2004). Diversity of cheese varieties: An overview. In P. F. Fox, P. L. H. McSweeney, T. M. Cogan, & T. P. Guinee (Eds.), *Cheese: Chemistry, physics and microbiology. Volume 2. Major cheese groups* (3rd ed., pp. 1–22). Amsterdam: Elsevier Applied Science.

McSweeney, P. L. H. (2004). Biochemistry of cheese ripening: Introduction and overview. In *Cheese: Chemistry, physics and microbiology. Volume 1. General Aspects*, 3rd edition, P. F. Fox, P. L. H. McSweeney, T. M. Cogen and T. P. Guinee (eds), Elsevier Applied Science, Amsterdam pp. 347–360.

Parente, E., & Cogan, T. M. (2004). Starter cultures: General aspects. In P. F. Fox, P. L. H. McSweeney, T. M. Cogan, & T. P. Guinee (Eds.), *Cheese: Physics, chemistry and microbiology* (3rd ed., Vol. 1, pp. 123–147). Amsterdam: Elsevier.

Visser, S., Slangen, C. J., & van Rooijen, P. J. (1987). Peptide substrates for chymosin (rennin). Interaction sites in kappa-casein-related sequences located outside the (103-108)-hexapeptide region that fits into the enzyme's active-site cleft. *Biochemical Journal, 244*, 553–558.

Visser, S., van Rooijen, P. J., Schattenkerk, C., & Kerling, K. E. (1976). Peptide substrates for chymosin (rennin). Kinetic studies with peptides of different chain length including parts of the sequence 101-112 of bovine κ-casein. *Biochimica et Biophysica Acta, 438*, 265–272.

Wilkinson, M. G. (1992). *Studies on the acceleration of Cheddar cheese ripening*. Cork: National University of Ireland.

Woo, A. H., Kollodge, S., & Lindsay, R. C. (1984). Quantification of major free fatty acids in several cheese varieties. *Journal of Dairy Science, 67*, 874–878.

Woo, A. H., & Lindsay, R. C. (1984). Concentrations of major free fatty acids and flavour development in Italian cheese varieties. *Journal of Dairy Science, 67*, 960–968.

Suggested Reading

Eck, A. (Ed.). (1984). *Le Fromage*. Paris: Diffusion Lavoisier.

Fuquay, J., Fox, P. F., & McSweeney, P. L. H. (Eds.). (2011). *Encyclopedia of dairy sciences*, 4 vols. (2nd ed.). San Diego: Academic.

Fox, P. F. (Ed.). (1993). *Cheese: Chemistry, physics and microbiology* (2nd ed., Vol. 1 and 2). London: Chapman & Hall.

Fox, P. F., Guinee, T. P., Cogan, T. M., & McSweeney, P. L. H. (2000). *Fundamentals of cheese science* (p. 587). Gaithersburg, MD: Aspen Publishers.

Fox, P. F., McSweeney, P. L. H., Cogan, T. M., & Guinee, T. P. (2004a). *Cheese: Chemistry, physics and microbiology. Volume 1. General aspects* (3rd ed., p. 617). Amsterdam: Elsevier Applied Science.

Fox, P. F., McSweeney, P. L. H., Cogan, T. M., & Guinee, T. P. (Eds.). (2004b). *Cheese: Chemistry, physics and microbiology. Volume 2. Major cheese groups* (3rd ed., p. 434). Amsterdam: Elsevier Applied Science.

Frank, J. F., & Marth, E. H. (1988). Fermentations. In N. P. Wong (Ed.), *Fundamentals of dairy chemistry* (3rd ed., pp. 655–738). New York: van Nostrand Reinhold Co.

Kosikowski, F. V. (1982). *Cheese and fermented milk foods* (2nd ed.). Brooktondale, NY: F.V. Kosikowski & Associates.

Law, B. A. (Ed.). (1997). *Advances in the microbiology and biochemistry of cheese and fermented milk*. London: Blackie Academic & Professional.

Berger, W., Klostermeyer, H., Merkenich, K., & Uhlmann, G. (1989). *Die Schmelzkäseherstellung*. Ladenburg: Benckiser-Knapsack GmbH.

Malin, E. L., & Tunick, M. H. (Eds.). (1995). *Chemistry of structure-function relationships in cheese*. New York: Plenum Press.

Robinson, R. K. (Ed.). (1995). *Cheese and fermented milks*. London: Chapman & Hall.

Scott, R. (Ed.). (1986). *Cheesemaking practice* (2nd ed.). London: Elsevier Applied Science Publishers.

Tamime, A. Y., & Robinson, R. K. (1985). *Yoghurt science and technology*. Oxford: Pergamon Press Ltd.

Waldburg, M. (Ed.). (1986). *Handbuch der Käse: Käse der Welt von A-Z; Eine Enzyklopädie*. Kempten, Germany: Volkswirtschaftlicher Verlag GmbH.

Zehren, V. L., & Nusbaum, D. D. (Eds.). (1992). *Process cheese*. Madison, WI: Cheese Reporter Publishing Company Inc.

Chapter 13
Chemistry and Biochemistry of Fermented Milk Products

13.1 Introduction

Milk has always soured spontaneously but at some point in human history, artisans deliberately caused milk to sour or ferment. Fermentation is one of the oldest methods for preserving milk and probably dates back ~10,000 years to the Middle East where the first evidence of organized food cultivation and production is known to have occurred. Traditional fermented milk products have been developed independently worldwide and were, and continue to be, especially important in areas where transportation, pasteurization and refrigeration facilities are inadequate. Nowadays, the primary function of fermenting milk is to extend shelf life, to improve taste, to enhance digestibility and to manufacture a wide range of dairy-based products.

If removed aseptically from a healthy udder, milk is essentially sterile but in practice, milk becomes contaminated by various bacteria, including lactic acid bacteria (LAB) during milking. During storage, these contaminants grow at rates dependent on the temperature. LAB probably dominate the microflora of uncooled milk expressed by hand. Since LAB are well suited for growth in milk, they grow rapidly at ambient temperature, metabolizing lactose to lactic acid and reducing the pH of the milk to the isoelectric point of caseins (~pH 4.6), at which they form a gel under quiescent conditions, thus producing cultured milks. Traditionally, and until relatively recently, fermentation was caused by the indigenous microflora or a "slop-back" culture (some of today's product is used to inoculate fresh milk). The production of fermented milks no longer depends on acid production by the indigenous microflora. Instead, the milk is inoculated with a carefully selected culture of LAB and for some products with LAB plus lactose-fermenting yeasts. The principal function of LAB is to produce acid at an appropriate rate *via* one of the pathways summarized in Chap. 12, Fig. 12.17.

Unlike cheese manufacture, the whey phase is retained within the coagulum of fermented milk products. As a result, fermented milks are high-moisture products (>80 %). Most fermented milks have a low pH (~pH 4.0), too low for most spoilage bacteria and potential pathogens to grow.

© Springer International Publishing Switzerland 2015
P.F. Fox et al., *Dairy Chemistry and Biochemistry*,
DOI 10.1007/978-3-319-14892-2_13

13.1.1 Classification of Fermented Milks

About 400 generic names are applied to traditional and manufactured fermented milk products worldwide although, in reality, the list of products is probably much shorter when divided up by milk type (e.g., cow, goat, sheep, buffalo, camel, yak or horse) or, more commonly, by the dominant microflora. Fermented milks can be divided into three broad categories based on their metabolic products, i.e., lactic fermentations, yeast-lactic fermentations and mould-lactic fermentation. Fermented milks in the lactic fermentation grouping can be sub-divided into mesophilic, thermophilic and therapeutic or probiotic types, depending on the microorganisms involved in the fermentation process. A schematic representing the classification of fermented milks is shown in Fig. 13.1, while Table 13.1 shows the classification of fermented milks based on the dominant microorganisms and their principal metabolites (Robinson and Tamime 1990). Table 13.2 lists many of the fermented milk

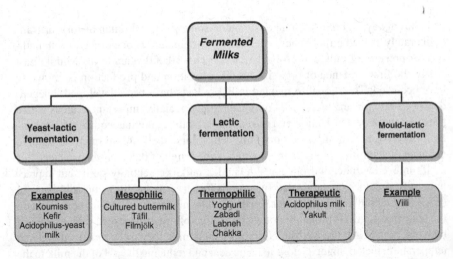

Fig. 13.1 Classification of fermented milks (from Uniacke-Lowe 2011)

Table 13.1 Classification of Fermented Milks based on the dominant microorganisms and their principal metabolites (from Robinson and Tamime 1990)

1. *LAB Fermentations (lactic acid)*
(a) Mesophilic LAB, e.g., cultured buttermilk, filmyölk, tätmjölk, långofil
(b) Thermophilic LAB, e.g., yoghurt, Bulgarian buttermilk, zabadi, dahi
(c) Therapeutic types, e.g., Yakult, Vifit
2. *Yeast-LAB Fermentations (lactic acid-ethanol)*
e.g., Kefir, Koumiss, acidophilus yeast milk
3. *Mould-LAB Fermentations*
e.g., Villi
LAB, lactic acid bacteria

Table 13.2 Origin, characteristics and uses of some important fermented milk products worldwide

Product	Country of origin	Period	Characteristics/uses
Airan	Central Asia, Bulgaria	1253–1255 AD	Cow's milk fermented with *Lb. bulgaricus;* refreshing beverage
Bulgarian milk	Bulgaria	500 AD	Cow's milk fermented with *Lb. bulgaricus* and *S. thermophilus*; very sour fermented milk used as a beverage
Chhash	India	6000–4000 BC	Diluted Dahi product or buttermilk left after churning Dahi; used with or after a meal
Churpi	Nepal	–	Churned fermented milk, buttermilk remaining is heated to form solid curd and partially dried
Cultured cream	Mesopotamia	1300 BC	Naturally soured cream
Dahi	India	6000–4000 BC	Coagulated sour milk; eaten directly or used as intermediate for butter or ghee
Filmjölk	Nordic countries	–	Cow's milk fermented with *L. lactis* and *Leu. mesenteroides;* characteristic taste from diacetyl production; used at breakfast or snack
Kefir	Caucasian	–	Milk fermented with kefir grains; effervescent, acidic and alcoholic
Kishk	Egypt/ Arab countries	–	Dry fermented product from Laban Zeer and parboiled wheat; semi-solid and highly nutritious, eaten as sweet dish with meals
Koumiss (Kumys)	Central Asia (Mongolia/ Russia)	2000 BC	Mare's milk fermented by lactobacilli and yeast; mildly effervescent acidic and alcoholic beverage
Laban Zeer/Khad	Egypt	5000–3000 BC	Sour milk coagulated in earthenware vessels
Langfil/Tattemjölk	Sweden	–	Milk fermented with slime-producing lactococci spp.
Leben	Iraq	ca 3000 BC	Traditional fermented milk; whey partially drained through muslin
Mast	Iran	–	Natural-type yoghurt; firm and cooked
Prostokvasha	Former Soviet Union	–	A fermented milk product (mesophilic lactic acid bacteria)
Shrikhand	India	400 BC	Concentrated soured milk; sweetened and spiced
Skyr	Iceland	870 AD	Ewe's milk partially coagulated using rennet and starter cultures; recently membrane technology used to concentrate product
Taette	Norway	–	Viscous fermented milk known as Cellarmilk

(continued)

Table 13.2 (continued)

Product	Country of origin	Period	Characteristics/uses
Trahana	Greece	–	Traditional Balkan fermented milk from ewe's milk with added wheat flour; dried and semi-solid
Villi	Finland	–	High viscosity fermented milk; LAB-mould
Yakult	Japan	1935 AD	Highly heat-treated milk fermented by *Lb. casei var. Shirota*; used as beverage and health supplement
Ymer	Denmark	–	Protein-fortified milk fermented by leuconostocs and lactococci; whey partially removed
Yoghurt	Turkey	800 AD	Custard-like sour fermented milk
Yoghurt (kisle mliako)	Bulgaria	–	Cow's or ewe's milk fermented using *S. thermophiles* and *Lb. bulgaricus*
Zabadi	Egypt/Sudan	2000 BC	Natural-type yoghurt; firm with 'cooked' flavour

Adapted from Prajapati and Nair 2003

products found worldwide together with their origin, characteristics and uses. Yoghurt, in various forms, is probably the most important type but consumption varies widely (see Table 13.3); other important, widely produced fermented milk products are buttermilk, kefir and koumiss. The characteristics of these four products will be described in this chapter; sour cream is also produced fairly widely and is discussed briefly.

13.1.2   Therapeutic Properties of Fermented Milks

Fermented milk products developed by chance but the increased storage stability and desirable organoleptic properties of such products were soon appreciated. Modern-day interest in the health benefits of fermented milks began with the theory of longevity proposed by the Russian microbiologist Professor Elie Metchnikoff (1845–1916), who proposed that people who consumed fermented milks live longer, as lactic acid bacteria in the fermented product colonized the intestine and inhibited 'putrefaction' caused by harmful bacteria, thereby retarding the aging process. Metchnikoff's theory on longevity led Dr. Minoru Shirota, a Japanese scientist (1899–1982), to isolate a unique strain of LAB, *Lactobacillus casei* subsp. *shirota*, which could survive passage through the acidic environment of the stomach and colonize the intestine and prevent the growth of harmful bacteria. His studies led to a product called Yakult, a fermented milk, which was first marketed in 1935 and is now sold in over 31 countries.

Table 13.3 Consumption of fermented milks (kg/caput/annum)

Country	Fermented milk
European Union	
Austria	21.8
Belgium	10.5
Croatia	16.9
Cyprus	12.4
Czech Republic	16.3
Denmark	48.2
Estonia	8.8
Finland	38.6
France	29.9
Germany	30.5
Greece	6.8
Hungary	13.9
Ireland	11.1
Italy	8.8
Luxemburg	7.0
Netherlands	45.0
Poland	7.8
Portugal	26.6
Slovakia	13.8
Spain	29.1
Sweden	36.4
United Kingdom	10.2
Other European	
Iceland	37.9
Norway	25.5
Switzerland	31.4
Russia	30.0
Ukraine	11.7
Africa and Asia	
China	1.9
India	16.1
Iran	47.3
Israel	28.2
Japan	8.5
Mongolia	50.0
South Africa	3.6
South Korea	9.3
Americas	
Argentina	12.8
Canada	8.2
Chile	4.1
Mexico	5.3
USA	2.1
Oceania	
Australia	7.6
New Zealand	6.7

Data compiled from various sources

In 1953 the term 'probiotics' was introduced to define microorganisms that stimulate the growth of other microorganisms and in 1989, was redefined to include reference to positive health effects, i.e., 'live microbial food supplements which benefit the host by improving its intestinal microbial balance' (Prado et al. 2008). A summary of the benefits attributed to probiotics is presented in Fig. 13.2.

It has been documented that some *Lactobacillus* spp. and in particular *Bifidobacterium* spp. contained in yoghurt can colonize the large intestine, reduce its pH and control the growth of undesirable microorganisms. Some of these bacteria also produce probiotics. Yoghurts containing such cultures, often referred to as bioyoghurt, are enjoying considerable commercial success. Legislation in many countries specifies a minimum number of viable microorganisms in yoghurt.

For some medical conditions fermented milks are preferable to non-fermented milk as they do not act as vectors of infectious diseases due to the low pH, which prevents the growth of many pathogenic organisms. Furthermore, the low pH reduces buffering action in the gastrointestinal tract and is believed to enhance the absorption of calcium.

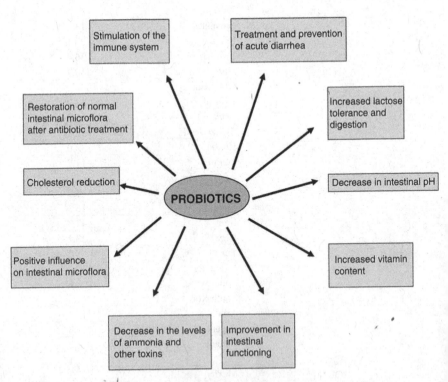

Fig. 13.2 Benefits of probiotics on human health (adapted from Prado et al. 2008)

13.2 Starter Microorganisms

Various bacteria, yeasts and moulds or combinations of these are used in the production of fermented milk products. Table 13.4 summarizes the principal microorganisms, their metabolic products and types of lactose fermentation for some of the most common fermented milk products (from Tamine et al. 2006). Traditional lactic acid bacteria, the principal group of microorganisms used for fermented milk production,

Table 13.4 Some of the principal microorganisms used in the production of fermented milk products

Starter organism	Metabolic product	Lactose fermentation	Examples of fermented milk products
I. Lactic acid bacteria			
Traditional			
Lactococcus lactis biovar. *diacetylactis*	L(+) Lactic acid; diacetyl and CO_2	Homofermentative[a]	Buttermilk, sour cream, ymer, Nordic milks
Leuconostoc mesenteroides subsp. *cremoris*	D(−) Lactic acid; diacetyl, ethanol and CO_2	Heterofermentative[b]	Buttermilk, sour cream, ymer, Nordic milks
Pediococcus acidilactici	DL Lactic acid	Homofermentative	Fermented milk, Kefir
Streptococcus thermophilus	L(+) Lactic acid; diacetyl and acetaldehyde	Homofermentative	Yoghurt, skyr, labneh, sour cream
Lactobacillus delbrueckii spp.	D(−) Lactic acid; diacetyl and acetaldehyde	Homofermentative	Yoghurt, skyr, sour cream
Non-traditional (probiotics)			
Lactobacillus spp. (*acidophilus, gasserie, helviticus, johnsonni*)	DL Lactic acid	Homofermentative	Yoghurt, kefir, buttermilk, sour cream
Lactobacillus spp. (*casei, reuteri, plantarum, rhamnosus*)	DL Lactic acid	Heterofermentative	Yoghurt, kefir
Bifidobacterium spp. (*adolescents, animalis, bifidum, breve, infantis, lactis, longum*)	L(+) Lactic acid, acetic acid	Heterofermentative	Yoghurt, buttermilk, sour cream
Enterococcus spp. (*faecium, faecalis*)	L(+) Lactic acid	Homofermentative	Fermented milk
Acetobacter aceti and *rasens*	Acetic acid, CO_2		Kefir
II. Yeasts			
Candida spp., *Saccharomyces* spp., *Kluyveromyces* spp., *Debaromyces* spp.	Ethanol, CO_2, acetone, amyl-alcohol, propanol		Skyr, kefir
III. Moulds			
Geotrichum candidum	Mould		Villi, kefir

[a]Produce lactic acid from sugars (adapted from Tamine et al. 2006)
[b]Produce lactic acid and alcohol from sugars

include *Lactococcus, Leuconostoc, Pediococcus, Streptococcus* and *Lactobacillus* spp. Several species of microorganism belonging to the *Lactobacillus, Bifidobacterium* and *Enterococcus* genera are used as non-traditional species for several fermented milk products due to the health benefits associated with these products (see Table 13.4). In mixed lactic acid-alcohol fermentations, as in kefir and koumiss (see below), in addition to LAB, yeasts are also used. For reviews on starter cultures used in the production of fermented milk products see Tamine et al. (2006) and Vedamuthu (2013).

13.3 Buttermilk

Originally, buttermilk was a by-product of butter production from ripened (sour) cream acidified by adventitious mesophilic LAB; a similar product is now produced from cream ripened by a culture of mesophilic LAB. However, cultured buttermilk is also produced from skimmed or low-fat milk inoculated with a mesophilic LAB culture; this product is produced mainly in English-speaking countries (USA, Canada, UK, Australia), where most butter is produced from sweet cream. It is primarily a drinking product and is also used in the production of soda bread. Basically similar products, some including an extra-cellular polysaccharide-producing strain of LAB, which increases the viscosity of the product making it ropy, are produced throughout North European countries Such products include Tatmjolk, Surmjolk, Filbunke, Skyr, Langfil, Villi (which contains *Geotricum* spp.), Filmjolk and Ymer (concentrated, 3.5 % fat, 5.6 % protein) (see Tamine 2006).

The characteristic flavour of cultured buttermilk is due mainly to diacetyl which is produced from citrate by *Lactococccus lactis* ssp. *lactis* biovar. *diacetylactis*, which is included in the culture for this product (Fig. 13.3).

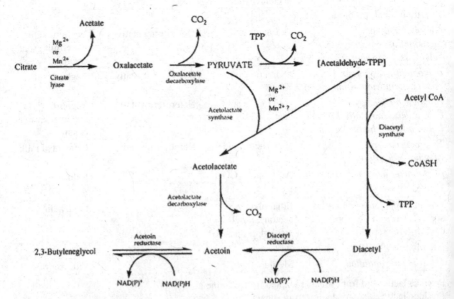

Fig. 13.3 Citrate metabolism by *Lactococcus lactis* ssp. *lactis* biovar. *diacetylactis* or *Leuconostoc* spp. (from Cogan and Hill 1993)

13.4 Yoghurt

Yoghurt is the best known of the fermented milk products and is consumed worldwide. The consistency, flavour and aroma of yoghurt vary between countries from being a highly viscous liquid to a softer gel-like product. Yoghurt may also be produced in frozen form as a dessert or drink. Broadly, yoghurt can be classified as follows:

1. **Set type**, incubated and cooled in its package
2. **Stirred type**, incubated in tanks and cooled before packaging
3. **Drinking type**, similar to the stirred type but the coagulum is broken before packaging
4. **Frozen type**, incubated in tanks and frozen like ice cream
5. **Concentrated yoghurt**, which is incubated in tanks, concentrated and cooled before packaging—also called strained yoghurt, labneh or labaneh.

The yoghurt fermentation is essentially homofermentative, using a mixed culture of *Lb. delbreuckii* and *Str. thermophiles*. The technology of fermented milks will not be discussed in detail and the interested reader is referred to Tamime and Marshall (1997), Marshall and Tamime (1997), Tamime and Robinson (1999) and Tamine (2006). A flow diagram of the manufacturing protocol of yoghurt is presented in Fig. 13.4. Depending on the product, the milk used may be full-fat, partially skimmed or fully skimmed. If it contains fat, the milk is homogenized at 10–20 MPa to prevent creaming during fermentation. For yoghurt, the milk is usually supplemented with skim milk powder to improve gel characteristics. Acid milk gels are quite stable if left undisturbed but if stirred or shaken, they synerese, expressing whey, which is undesirable. The tendency to synerese is reduced by heating the milk at, e.g., 90 °C×10 min or 120 °C×2 min; heating causes denaturation of whey proteins, especially β-lactoglobulin, and their interaction with the casein micelles *via* κ-casein. The whey protein-coated micelles form a finer (smaller whey pockets) gel then that formed from unheated or HTST pasteurized milk, with less tendency to synerese.

In some countries, it is common practice to add sucrose to the milk for yoghurt, production to reduce the acid taste. It is also very common practice to add fruit pulp, fruit essence or other flavouring, e.g., chocolate, to yoghurt, either to the milk (set yoghurt) or to the yoghurt after fermentation (stirred yoghurt).

13.4.1 Concentrated Fermented Milk Products

Throughout the Middle East, concentrated fermented milk products are produced, probably the best known of which is Labneh for which the fermented milk is concentrated by removing part of the serum (whey). This was done traditionally by stirring the yoghurt and transferring it to muslin bags to partially drain. The typical composition of Labneh is: ~25 % total solids, 9–11 % protein, ~10 % fat and ~0.85 % ash (its protein content is similar to that of fresh, acid-curd cheese). This

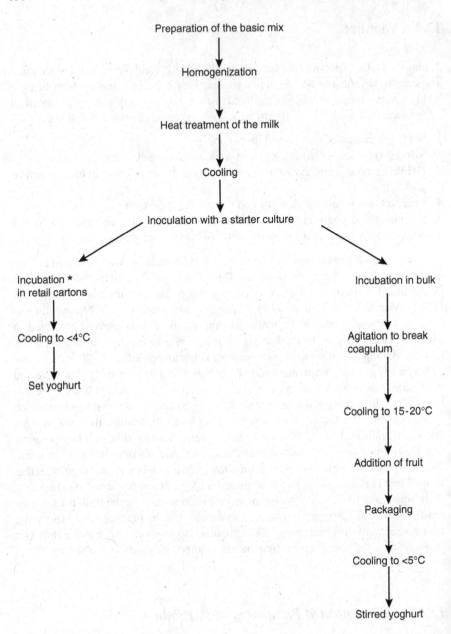

Fig. 13.4 Protocol for the manufacture of yoghurt. *, Sucrose and/or fruit (fruit flavours) may be added at this point. From Robinson and Tamime 1993)

type of concentrated product is known by many names, including Greek-style yoghurt (see Tamine 2006). Labneh-type products are consumed in many forms, directly, as a sandwich spread, as soups or in Turkey, diluted with salted water (Aryan), Concentration can now be achieved by ultrafiltration, before, but preferably after, fermentation (see Tamine 2006).

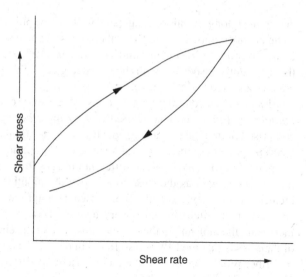

Fig. 13.5 Representation of shear stress as a function of shear rate for yoghurt displaying rheological hysteresis

Shear stress

Shear rate

13.4.2 Novel Yoghurt Products

Since the late twentieth century, a number of yoghurt-based products have been introduced, focussed mainly on children: frozen (ice cream) yoghurt, dried yoghurt (for long-term storage, intended to be rehydrated and to set on rehydration but the quality of the gel is poor), and yogurt-based desserts (mousse).

13.4.3 Rheology of Yoghurt

Fermented milk products exhibit thixotropic rheological properties, i.e., the viscosity (resistance to flow) decreases as the rate of shear increases. When the shear stress is reduced, the strain does not follow the original curve, i.e., it does a hysteresis loop (Fig. 13.5). The rheological properties are major parameters of quality and are controlled by varying the total solids content of the milk, heat treatment and homogenisation of the milk or by the use of hydrocolloids, e.g., gelatin or carrageenan, or including an exocellular polysaccharide-producing strain in the culture.

13.4.4 Exocellular Polysaccharides

Many strains of all species of starter LAB produce exopolysaccharides (EPS) which are responsible for the thickening of yoghurt and give a ropy property to the product; such products include several Scandinavian fermented milk products, e.g., Taette, Skyr and Villi. A simple way to test for EPS-producing cultures is to determine if long strands of coagulated milk can be pulled from milk-grown cultures using an inoculation loop; individual colonies can be tested in a similar manner. The ability to produce EPS is plasmid-encoded and EPS may be produced as capsules that are tightly associated with the producing cell or they may be liberated into the medium as a loose slime.

EPSs are divided into homopolymers, which are produced mainly by *Lc. mesenteroides*, and heteropolymers, produced by the other species. Homopolymers are comprised of only one sugar, e.g., dextran is an α-1,6 linked glucose polymer, while heteropolymers comprise several sugars, most commonly, glucose, galactose and rhamnose in different ratios and different linkages (α or β) depending on the producing strain. As well as their use to improve the mouth-feel and creaminess of fermented milks, they have also been used to improve the texture of reduced-fat cheeses which often have a rubbery texture. They do this by binding water, thus increasing the moisture in the non-fat substance of the cheese. One of the downsides of their use is that the EPS is also found in the whey and clogs the membranes used in further processing of the whey. For reviews on EPS of LAB see de Vuyst et al. (2001, 2011) and Hassan (2008).

13.5   Kefir

Kefir and Koumiss contain ~1 and ~6 % ethanol, respectively, which is produced by lactose-fermenting yeasts, usually *Kluyveromyces marxianus*. The ethanol modifies the flavour of the products and the CO_2 produced in the fermentation affects both their flavour and texture. Kefir, which originated in Northern Caucasus mountains, is most popular in northern and eastern Europe. It is produced mainly from cows' milk but the milk of goats and sheep, or mixtures of the three, are also used.

There are two methods for preparing kefir, (1) using kefir grains and sub-culturing the resultant fermentate or, (2) inoculating milk directly with starter cultures (Rattray and O'Connell 2011). Schematics of both methods are shown in Figs. 13.6 and 13.7.

The traditional culture, "kefir grains", contains a blend of lactic acid bacteria (80–90 %), lactose-fermenting yeast (10–15 %), acetic acid bacteria (*Acetobacter* spp.) and possibly mould (*Geotricum candidum*) which are bound together by exopolysaccharides (Fig. 13.8). Several species of LAB are present, including *Lactococcus* spp. (especially *L. lactis* ssp. *lactis*), *Lactobacillus* spp., *S. thermophilus* and *Leuconostoc* spp. Yeasts include *Kluyveromyces marxianus* var. *lactis*, *Saccharomyces cerevisiae* and *Candida* spp. A symbiotic relationship exists between the yeasts and bacteria in kefir grains; yeasts produce vitamins, amino acids and other growth factors which are essential to maintain the integrity and viability of the microflora, while bacterial end products are used as energy sources by yeasts (Farnworth and Mainville 2003). The grains are up to 2 cm in diameter and contain 10–16 % dry matter, ~3 % protein, 0.3 % fat and ~6 % non-protein nitrogen.

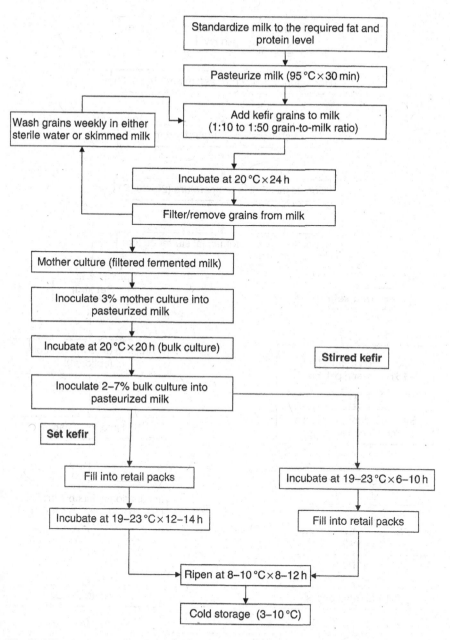

Fig. 13.6 Production of kefir using kefir grains (from Rattray and O'Connell 2011)

The culture for kefir is prepared by inoculating heated (95 °C × 30 min) milk at 20 °C with kefir grains, incubating for ~20 h (to ~0.8 % lactic acid) and ripening at ~10 °C for ~8 h to facilitate the growth of yeast. The grains are then strained off and the "filtrate" used to inoculate fresh milk, at 1–3 %, and incubated to produce kefir or a bulk starter for large operations. The kefir grains are washed and used for the

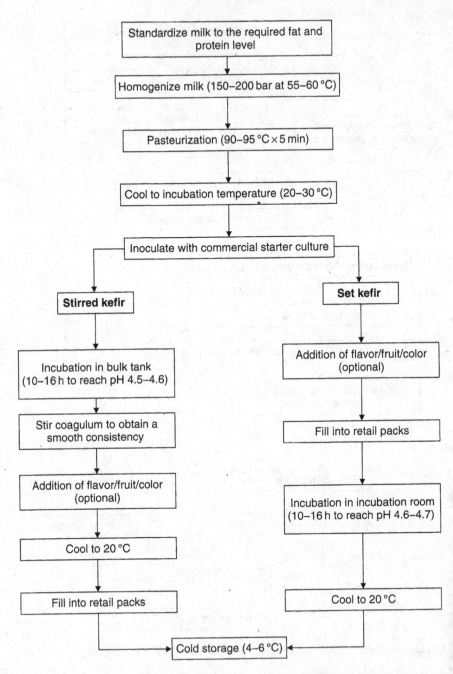

Fig. 13.7 Production of kefir using commercial direct-to-vat cultures (from Rattray and O'Connell 2011)

Fig. 13.8 Image of kefir grains, a yeast/bacterial fermentation culture

next batch of starter, freeze-dried kefir starter cultures, available from culture suppliers, are now used widely.

Kefir is white/yellow in colour with a strong yeasty aroma and an acidic taste and has a thick and slightly elastic texture. The typical composition of kefir in Poland is: not less than 2.7 % protein, 1.5–2 % fat, titratable acidity, not less than 0.6 % lactic acid. It is claimed that the protein in kefir is more digestible than that in milk, much of the lactose is hydrolyzed, making it more suitable for lactose-intolerant people, that it has anti-tumour properties and that sphingolipids in kefir stimulate the immune system (see Tamine 2006).

13.6   Koumiss

Koumiss (Kumys) is a traditional fermented product made from equine milk in Central Asia, Russia, Mongolia, Kazakhstan, etc., and is widely consumed in these regions, primarily for its therapeutic value. Russians, in particular, have long advocated the use of koumiss for a wide variety of illnesses but the variable microbiology of the product has made it difficult to confirm any theoretical basis for the claims (Tamime and Robinson 1999). In Mongolia, koumiss is the national drink (Airag) and a high-alcoholic drink made by distilling koumiss, called Arkhi, is also produced (Kanbe 1992). *Per caput* consumption of koumiss in Mongolia is estimated to be about 50 L *per annum*.

The oldest method for the production of koumiss was by fermentation of lactose by adventitious bacteria and yeasts to lactic acid and ethanol, respectively.

Horses were hand-milked with the foal in close proximity. Traditional koumiss (from fresh raw milk) was usually prepared by seeding milk with a mixture of bacteria and yeasts using part of the previous day's product as an inoculum ('slop-back culture'). The milk was held in a leather sack called, a 'turdusk' (also called a 'saba' or 'burduk') which was made from smoked horsehide taken from the thigh of a horse, i.e., it has a broad bottom and long narrow sleeve, with a capacity of 25–30 L. Fermentation took from 3 to 8 h with a mixed microbial population which consists mainly of *Lb. delbrueckii* subsp. *bulgaricus*, *Lb. casei*, *L. lactis* subsp. *lactis*, *Kluveromyces fragilis* and *Saccharomyces unisporus*. During the agitation and maturation stages of production, more equine milk is added frequently to control the acidity and alcohol level. The whole process was poorly controlled and often resulted in a product with an unpleasant taste, due to the presence of too much yeast or excess acidification. Turdusks, often containing caprine milk from the previous season, were stored in a cool place over winter and the starter culture was reactivated in Spring by gradually filling the turdusk with equine milk over about 5 days. Koumiss is still manufactured in remote areas of Mongolia by traditional methods but with increased demand elsewhere it is now produced under more controlled and regulated conditions.

A standardized protocol for koumiss production is of considerable interest for increasing the market for, and consumption of, equine milk products in countries where it has not normally been consumed. As well as using pasteurised equine milk, pure cultures of lactobacilli, such as *Lb. delbrueckii* subsp. *bulgaricus*, and yeasts are used for koumiss manufacture. *Saccharomyces lactis* is considered best for the production of ethanol and *S. cartilaginosus* is sometimes used for its antibiotic activity against *Mycobacterium tuberculosis*. Other microorganisms such as *Candida* spp., *Torula* spp., *Lb. acidophilus* and *Lb. lactis* may also be used in koumiss production. A schematic of the manufacture of commercial koumiss is shown in Fig. 13.9, which outlines the three stages of production: mother culture preparation, bulk starter preparation and koumiss manufacture. The inoculation level of equine milk with bulk starter at 30 % is probably the highest used in the manufacture of any fermented milk. Agitation is crucial for aeration of the mix which promotes the growth of the yeast. The characteristics of good koumiss are optimal when the lactic and alcoholic fermentations proceed simultaneously so that the products of fermentation occur in definite proportions. As well as lactic acid, ethanol and CO_2, volatile acids and other compounds are formed which are important for aroma and taste and ~10 % of the milk proteins are hydrolysed. Products with varying amounts of lactic acid and ethanol are produced and generally three categories of koumiss are recognised: mild, medium and strong (Table 13.5). Koumiss contains about 90 % water, 2–2.5 % protein (1.2 % casein and 0.9 % whey proteins), 4.5–5.5 % lactose, 1–1.3 % fat and 0.4–0.7 % ash. Viable counts of $\sim4.97 \times 10^7$ cfu ml^{-1} and $\sim1.43 \times 10^7$ cfu ml^{-1} for bacteria and yeast, respectively, have been reported in koumiss.

Lactic acid in koumiss may occur in either the L(+) and D(−) isomer, depending on the type of LAB used (Table 13.6). Both L(+) and D(−) isomers are absorbed from the gastrointestinal tract but differ in the proportions converted to glucose or glycogen in the body. The L(+) isomer is rapidly and completely converted to glycogen whereas

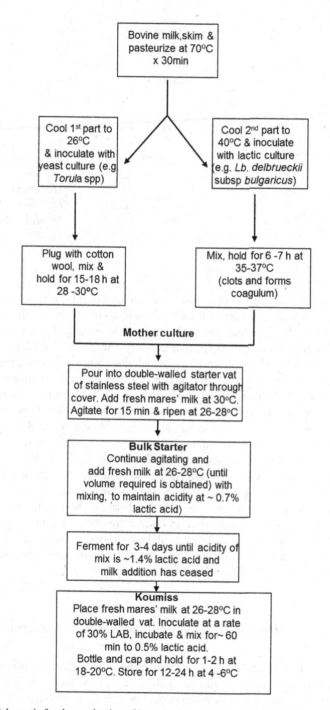

Fig. 13.9 Schematic for the production of koumiss (adapted from Berlin 1962)

Table 13.5 Categories of koumiss

Flavour category	Acidity (%)	Ethanol (%)
Mild	0.6–0.8	0.7–1.0
Medium	0.8–1.0	1.1–1.8
Strong	1.0–1.2	1.8–2.5

Table 13.6 Optical isomers of lactic acid produced by some lactic acid bacteria species used in koumiss production

L(+) Lactic acid (≥95 %)	All *Lactococcus* strains
	Lb. casei
D(−) Lactic acid (100 %)	*Lb. bulgaricus*
	Lb. lactis
	Lb. cremoris
Racemic lactic acid mixture L(+)/D(−)	*Lb. helveticus*
	Lb. acidophilus
	Lb. planatarum
	Lb. brevis

the D(−) isomer is converted more slowly and a significant quantity is excreted in urine. The presence of unmetabolised lactic acid results in metabolic acidosis in infants. Since 1973, fermented milks manufactured commercially use cultures that produce high amounts of the L(+) isomer and very low amounts of the D(−) isomer.

Koumiss is thought to be more effective than raw equine milk in the treatment of various illnesses due to the additional peptides and bactericidal substances from microbial metabolism (Doreau and Martin-Rosset 2002). Nowadays, the main interest in fermented foods such as koumiss is their apparent ability to positively promote functions of human digestion, i.e., to have a probiotic effect (Sahlin 1999).

The low lactose content of koumiss compared to raw equine milk is favourable for those suffering lactose intolerance; ~88 % of Mongolians are lactose intolerant but consume koumiss without ill-effects, probably due to intra-intestinal digestion of lactose by microbial β-galactosidase in koumiss, an enzyme that is not denatured in the acidic environment of the stomach. Furthermore, koumiss is thought to be more effective than raw equine milk in disease treatment due to the presence of additional bioactive peptides and bactericidal substances produced during microbial metabolism while retaining the high levels of lysozyme and lactoferrin of the original milk, which have proven antibacterial activity.

13.6.1 Technological Developments in Koumiss Manufacture

Blends of microorganisms in starter cultures have been developed that enhance flavour development and extend the shelf-life up to 14 days. The presence of a high level of thermo-stable lysozyme in equine milk may interfere with the activity of some starter cultures in the production of fermented products. Equine milk heated to 90 °C for 3 min to inactivate lysozyme has been reported to produce an acceptable fermented milk. In sensory tests, fermented unmodified equine milk has an unacceptable viscosity and scores very low in comparison to fortified products for appearance, consistency and taste. In an attempt to improve the rheological and sensory properties, fortification with sodium caseinate (1.5 g per 100 g), pectin (0.25 g per 100 g) and threonine (0.08 g per 100 g) has been investigated; the resultant products are reported to have good microbiological, rheological and sensory characteristics even after 45 days at 4 °C. Addition of sucrose and sodium caseinate has a positive effect on the rheological properties of the product due to strengthening of the protein network.

13.6.2 Koumiss-Like Products from Non-equine Milk

Koumiss-like products are produced in several areas, e.g., Mongolia, the former USSR, Southern Europe and North Africa from camel milk (shubat), donkey milk (koumiss), goat milk (tarag), ewe's milk (arak or arsa) or buffalo milk (katyk). The physico-chemical and microbiological properties of asinine milk, such as low microbiological load and high lysozyme content, make it a good substrate for the production of fermented products with probiotic *Lactobacillus* strains. Asinine milk has been fermented with the probiotic bacteria, *Lb. rhamnosus* (AT 194, GTI/1, GT 1/3) which is unaffected by the high lysozyme content of the milk and was viable after 15 days at 4 °C and pH 3.7–3.8. *Lb. rhamnosus* inhibits the growth of most harmful bacteria in the intestine and acts as a natural preservative in yoghurt-type products, considerably extending shelf-life. Fermented asinine milk produced using a mixed culture of *Lb. rhamnosus* (AT 194, CLT 2.2) or *Lb. casei* (LC 88) had a high viable bacteria count after storage for 30 days. Some sensory differences have been reported for fermented asinine drinks and those made with the *Lb. casei* strain developed a better and balanced aroma than the boiled vegetable/acidic taste and aroma of the product made with *Lb. rhamnosus* alone.

Due to shortages of equine milk and the cost, when it is available, research has been undertaken to produce koumiss-like products from bovine milk, which must be modified to make it suitable for koumiss production. Koumiss of a reasonable quality has been produced from whole or skimmed bovine milk containing added sucrose using a mixture of *Lb. acidophilus*, *Lb. delbrueckii* ssp. *bulgaricus* and *Kluyveromyces marxianus* var. *marxianus* or *Kluyveromyces marxianus* var. *lactis* as starter culture. Koumiss has also been made from diluted bovine milk with added lactose and, more successfully, from bovine milk mixed with concentrated whey using a starter culture of *Kluyveromyces lactis* (AT CC 56498), *Lb. delbrueckii* subsp. *bulgaricus* and *Lb. acidophilus*. Starter cultures for koumiss manufacture from bovine milk may also include *Saccharomyces lactis* (high antimicrobial activity against *Mycobacterium tuberculosis*) in order to retain the 'anti-tuberculosis image' of equine milk.

13.7 Cultured/Sour Cream

Cultured cream is produced using a culture containing *L. lactis* ssp. *lactis*, *L. Lactis* ssp. *cremoris*, *L. lactis* ssp. *lactis* var. *diacetylactis* and *Leu. mesenteroides* ssp. *cremoris*; the former two are mainly responsible for acid production and the latter two for aroma production (diacetyl). The typical fat content is 10–12 % but may be a high as 30 %.; the pH is about 4.5 but it tastes less acidic than buttermilk or yoghurt, owing to the mellowing effect of the fat. The inoculated cream may be distributed in cartons before fermentation at 22–24 °C until the pH reaches 4.5 in about 20 h and is cooled in the package (set type), or it may stirred during fermentation and then packaged; the former is very viscous. The cream for stirred cultured cream is homogenized at 10–20 MPa. A long-life version of stirred cultured cream can be produced by heat-treating the fermented product at 85–90 °C for a few seconds followed by packaging aseptically.

Cultured cream is used in many dishes, e.g., sauces, soups and dressings; it is popular on baked potato.

References

Berlin, P. J. (1962). *Koumiss* (Bulletin IV, pp. 4–16). Brussels: International Dairy Federation.

Cogan, T. M., & Hill, C. (1993). Cheese started culture. In P. F. Fox (Ed.), *Cheese: Physics, chemistry and microbiology* (2nd ed., Vol. 1, pp. 193–255). London, UK: Chapman and Hall.

De Vuyst, L., de Vin, F., Vaningelgem, F., & Degeest, B. (2001). Recent developments in the biosynthesis and applications of heteropolysaccharides from lactic acid bacteria. *International Dairy Journal, 11*, 687–707.

De Vuyst, L., Weckx, S., Ravyts, F., Herman, L., & Leroy, F. (2011). New insights into the exopolysaccharide production of *Streptococcus thermophilus*. *International Dairy Journal, 21*, 586–591.

Doreau, M., & Martin-Rosset, W. (2002). Dairy animals: Horse. In H. Roginski, J. A. Fuquay, & P. F. Fox (Eds.), *Encyclopedia of dairy sciences* (pp. 630–637). London, UK: Academic Press.

Farnworth, E. R., & Mainville, I. (2003). Kefir: A fermented milk product. In E. R. Farnworth (Ed.), *Handbook of fermented functional foods* (pp. 77–112). Boca Raton, FL: CRC Press.

Hassan, A. N. (2008). Possibilities and challenges of exopolysaccharide-producing lactic cultures in dairy foods. *Journal of Dairy Science, 91*, 1282–1298.

Kanbe, M. (1992). Traditional fermented milk of the world. In Y. Nakazawa & A. Hosono (Eds.), *Functions of fermented milk: Challenges for the health sciences* (pp. 41–60). London: Elsevier Applied Science.

Marshall, V. M. E., & Tamime, A. Y. (1997). Physiology and biochemistry of fermented milks. In B. A. Law (Ed.), *Microbiology and biochemistry of cheese and fermented milk* (2nd ed., pp. 152–192). London, UK: Blackie Academic and Professional.

Prado, F. C., Parada, J. L., Pandey, A., & Soccol, C. R. (2008). Trends in non-dairy probiotic beverages. *Food Research International, 41*, 111–123.

Prajapati, J. B., & Nair, B. M. (2003). The history of fermented foods. In E. R. Farnworth (Ed.), *Handbook of fermented functional foods* (pp. 1–25). London: CRC Press.

Rattray, F. P., & O'Connell, M. J. (2011). Kefir. In J. Fuquay, P. F. Fox, & P. L. H. McSweeney (Eds.), *Encyclopedia of dairy sciences* (2nd ed., Vol. 2, pp. 518–524). Oxford: Academic Press.

Robinson, R. K., & Tamime, A. Y. (1990). Microbiology of fermented milks. In R. K. Robinson (Ed.), *Dairy microbiology* (2nd ed., Vol. 2, pp. 291–343). London, UK: Elsevier Applied Science.

Robinson, R. K., & Tamime, A. Y. (1993). Manufacture of yoghurt and other fermented milks. In R. K. Robinson (Ed.), *Modern dairy technology* (2nd ed., Vol. 2, pp. 1–48). London, UK: Elsevier Applied Science.

Sahlin, P. (1999). *Fermentation as a method of food processing: Production of organic acids, pH-development and microbial growth in fermenting cereals.* Licentiate thesis. Division of Applied Nutrition and Food Chemistry, Lund University.

Tamime, A. Y., & Marshall, V. M. E. (1997). Microbiology and technology of fermented milks. In B. A. Law (Ed.), *Microbiology and biochemistry of cheese and fermented milk* (2nd ed., pp. 57–152). London, UK: Blackie Academic and Professional.

Tamime, A. Y., & Robinson, R. K. (1999). *Yoghurt science and technology* (2nd ed.). Cambridge, UK: Woodhead.

Tamine, A. Y. (2006). *Fermented milks*. Oxford, UK: Blackwell.

Tamine, A. Y., Skriver, A., & Nilsson, L.-E. (2006). Starter cultures. In A. Y. Tamime (Ed.), *Fermented milks* (pp. 11–52). Oxford, UK: Blackwell.

Uniacke-Lowe, T. (2011). Fermented milks, koumiss. In J. W. Fuquay, P. F. Fox, & P. L. H. McSweeney (Eds.), *Encyclopedia of dairy sciences* (2nd ed., Vol. 2, pp. 512–517). Oxford, UK: Academic Press.

Vedamuthu, E. R. (2013). Starter cultures for yoghurt and fermented milks. In R. C. Chandan & A. Kilara (Eds.), *Manufacturing yoghurt and fermented milks* (2nd ed., pp. 115–148). Ames, IA: John Wiley and Sons.

Suggested Reading

Chandan, R. C., & Kilara, A. (2011). *Manufacturing yoghurt and fermented milks* (2nd ed.). West Sussex, UK: John Wiley and Sons.

Farnworth, E. R. (2008). *Handbook of fermented functional foods* (2nd ed.). Boca Raton, FL: CRC Press.

Fuquay, J. W., Fox, P. F., & McSweeney, P. L. H. (2011). Fermented milks. In *Encyclopedia of dairy sciences* (2nd ed., Vol. 2, pp. 470–532). Oxford, UK: Academic Press.

Khurana, H. K., & Kanawjia, S. K. (2007). Recent trends in development of fermented milks. *Current Nutrition and Food Science, 3*, 91–108.

Kurmann, J. A., Rašić, J. L., & Kroger, M. (1992). *Encyclopedia of fermented fresh milk products: An international inventory of fermented milk, cream, buttermilk, whey, and related products.* New York, NY: Van Nostrand Reinhold.

Nakazawa, Y., & Hosono, A. (1992). *Functions of fermented milk: Challenges for the health sciences.* London, UK: Elsevier Applied Science.

Robinson, R. K. (1991). *Therapeutic properties of fermented milks.* London, UK: Elsevier Applied Science.

Surono, I. S., & Hosono, A. (2002). Fermented milks: Types and standards of identity. In H. Roginski, J. A. Fuquay, & P. F. Fox (Eds.), *Encyclopedia of dairy sciences* (pp. 1018–1069). Oxford, UK: Academic Press.

Index

Printed in the United States
By Bookmasters